CENGAGE FOUNDATION SERIES for JEE

Mathematics

BASE Educational Services Pvt. Ltd.

Class X

CENGAGE

Australia • Brazil • India • Mexico • Singapore • United Kingdom • United States

CENGAGE

Mathematics: Class X

ISBN-13: 978-93-5350-004-7

ISBN-10: 93-5350-004-4

Registered Office:
Cengage Learning India Pvt. Ltd.
T-303, 3rd Floor
DAV Complex, Plot 3
Mayur Vihar Phase 1
Delhi-110091

Cengage Learning is a leading provider of customized learning solutions with office locations around the globe, including Australia, Brazil, India, Mexico, Singapore, United Kingdom, and United States. Locate your local office at:
www.cengage.com/global

Cengage Learning products are represented in Canada by Nelson Education, Ltd.

For product information, visit **www.cengage.co.in**

Printed in India at

Table of Contents

Preface .... *iv*

1 Permutations and Combinations 2

2 Binomial Theorem 12

3 Remainder Theorem 20

4 Algebraic Expressions 28

5 Simple and Simultaneous Equations 38

6 Quadratic Equations 52

7 Relations and Functions 72

8 Progressions 86

9 Matrices 102

10 Determinants 120

11 Rational Expressions 130

12 Transformations 142

13 Circles and Cyclic Quadrilaterals 150

14 Tangents to Circles 168

15 Coordinate Geometry 190

16 Area of a Triangle 200

17 Equation of a Straight Line – I 206

18 Equation of a Straight Line – II 216

19 Measurement of Angles 226

20 Trigonometry 234

21 Standard and Allied Angles 242

22 Heights and Distances 254

Hints and Answers to Test Yourself 263

Preface

Students aspiring for a career in Engineering and/or Medical Sciences need to develop a clear understanding of fundamental concepts and problem-solving skills. Cengage Foundation series for Physics, Chemistry, Mathematics, and Biology for Class IX and X is aimed at helping students achieve these objectives.

The books in this series have been written by eminent teachers who have used their rich experience of teaching high school students while developing these books. Each book in this series is organized in a logical sequence to help students' learning for a strong foundation in the subject. The theory and problems are aligned to the curriculum suggested by various high school boards, and these books also deal with some additional topics aimed at preparing students for the entrance examinations.

This series should help in building a strong foundation among students preparing for various engineering and medical entrance examinations and shall bring out the best in them.

Key features

- Each chapter begins with a 'starter' page that offers some important teasers illustrated with photographs.
- In each chapter, the topics are presented in an easy-to-navigate format for enjoyable reading and clear understanding.
- This series takes an illustrative approach that provides explanation for each concept with the help of photographs and figures.
- Interesting information and application boxes highlighted with unique icons have been used throughout the chapter to facilitate deeper understanding of the topics.
- A concept map (mind map) at the end of theory in each chapter assists in recalling the concepts discussed in the chapter.
- End-of-chapter exercises consist of different types of problems to challenge students' understanding of concepts and provide momentum for entrance examination preparation.
- Olympiad and NTSE level exercises are also provided at the end of each chapter.

Acknowledgements

We sincerely thank Mrs Indumathi Raghuraman and Mr H. V. Raghunath for their valuable suggestions and contribution.

BASE Team

1 Permutations and Combinations

Number rules the universe

– Pythagoras

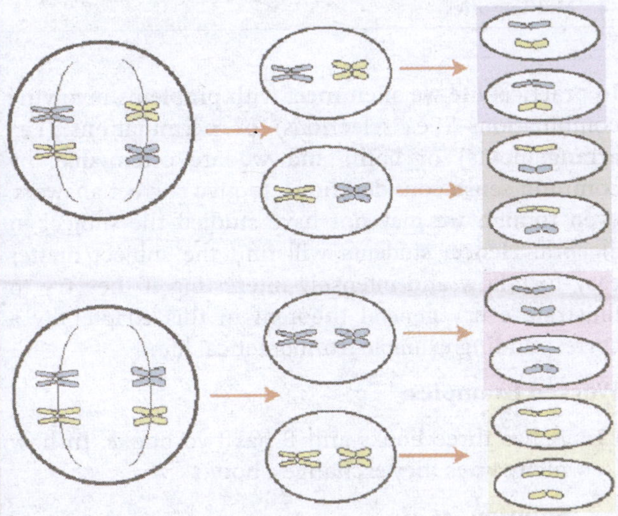

OBJECTIVES

After reading this chapter you should be able to

- learn and understand the concept of permutations and combinations
- learn about the use of factorial notation
- apply permutations and combinations in solving real-life problems

$$^{n}p_{n} = n!$$

$$^{n}p_{r} = \frac{n!}{(n-r)!}$$

$$^{n}C_{r} = \frac{n!}{r!(n-r)!}$$

$$^{n}C_{n} = 1$$

In practical life we often meet with problems involving combinations (i.e., selections) or permutations (i.e., arrangements) or both, and we are often able, by common sense considerations, to give correct answers, even though we may not have studied the subject in algebra. Hence, students will find the subject matter easy to follow and intensely interesting if they try to illustrate every general theorem in this chapter by a corresponding example from practical life.

Worked Examples

1. A has three books and B has two books. In how many ways they exchange a book?

   Solution

   A can give a book to B in three ways (any one of his books may be given), and then B can give one of his books, so that B can give a book in exchange in two ways. Hence, they can exchange a book in 6 ways.

   If A's books are denoted by a, b, c and B's books by d, and e, the exchange may be between the following:

   (a) a and d (b) a and e
   (c) b and d (d) b and e
   (e) c and d (f) c and e

2. There are 10 competitors in a race for three prizes. In how many ways can the prizes be given?

   Solution

   Naturally for the first prize there are 10 ways (any one of the 10 can get the first prize). When it is given away to any one of them, the second prize can be given in nine ways. Hence, the first two prizes can be given 10×9 ways. After the first two prizes have been given in any one of the 90 ways, the third prize can be given in eight ways (to anyone of the remaining eight competitors). Hence, the total number of ways required is $10 \times 9 \times 8 = 720$. It will take too much space to enumerate all these 720 ways by the use of letters $a, b, c, \ldots, j$. These examples illustrate the following principle.

FUNDAMENTAL THEOREM ON COUNTING

If a certain job can be done in 'm' ways, and when it has been done in any one of these ways, a second job can happen in 'n' ways, and then the total number of ways in which these two jobs can be done is $m \times n$.

Suppose the first job is done in any one way. Then corresponding to this one way of doing this first job, we shall have n ways of doing the second job and therefore n ways of doing the two jobs. This applies to each one of the m ways of doing the first job. Hence, the total number of ways of doing two jobs is $n + n + \cdots$ to m terms $= m \times n$.

If there are three different routes from P to Q and two different routes from Q to R, then there are $3 \times 2 = 6$ different routes from P to R passing through Q.

The theorems can obviously be extended. If one thing can be done in m ways and when it is done, a second thing can be done in n ways, and when these two things are done, a third thing can be done in p ways and so on, and then the total number of ways of doing all these things is $m \times n \times p$.

Example

In how many ways can A take one apple, one orange, and one mango from a basket containing four apples, three oranges, and five mangoes?

Solution

The required number is $4 \times 3 \times 5 = 60$.

Worked examples

1. There are six boats plying between two places A and B on either side of a river. In how many ways a man can go from A to B and back to A by a different boat?

   Solution

   There are six ways to go from A to B. With each of these ways, there is a choice of five ways of returning. (Since the man is not to come by the same boat.) Hence, the total number of ways to make the journey is $6 \times 5 = 30$.

2. How many integral number of four digits can be formed with four digits 2, 3, 5, and 7 if no digit is used more than once in each number?

   Solution

   For the units place there are four digits (any one of 2, 3, 5, and 7) and hence four ways. For the tens place there are three digits (as the digit used in units place cannot be used again). Hence, there are $4 \times 3 = 12$ ways for units and tens place. Arguing in the same way, there are two digits for the hundreds place and one digit for the thousands place. Hence, in total we have $4 \times 3 \times 2 \times 1 = 24$ numbers that can be formed using four given digits.

3. How many three-digit numbers can be formed from the digits 1, 2, 3, 4, and 5 assuming that the digits cannot be repeated?

   Solution

   In the units place, we have a choice of any one of the five digits. Once this choice is made, we have any one of the remaining four digits to occupy the tens place. Obviously, therefore, for the hundreds place we have only three choices. We display this as follows:

3	4	5
H	T	U

The three places, in the order, can therefore be filled in $5 \times 4 \times 3 = 60$ ways by the use of fundamental principle of counting. Thus, the number of three-digit numbers from the given five digits is 60.

A very useful notation

Before we introduce the concepts of permutations and combinations, we learn a useful notation which is convenient for representing the product of the first n natural numbers.

The notation is a mark of exclamation (!), suffixed to the highest natural number in the product.

$3! = 3 \times 2 \times 1 = 6$

$5! = 5 \times 4 \times 3 \times 2 \times 1 = 120$

$6! = 6 \times 5 \times 4 \times 3 \times 2 \times 1 = 720$

Each product is called a factorial. We read 3!, as three factorial, 5! as five factorial, and hence n factorial is written as $n! = n(n-1)(n-2) \ldots 3 \times 2 \times 1$.

(a) $n!$ can also be denoted by $\underline{|n}$ read as factorial n.
(b) 0! is defined to be 1.

PERMUTATIONS

A coin is tossed twice and outcomes are recorded. How many possible outcomes are there? Denoting the head of the coin by H and the tail by T, we have HH, HT, TH, and TT as the possible outcomes. The outcome TH is different from the outcomer HT. In other words, the order of occurrence of H and T is to be kept in mind.

Situations, in which the order of occurrence of events is important, give rise to what we call permutation.

Permutation is an arrangement of a number of objects taken, some or all at a time.

In the above coin-tossing problem, outcomes HH, HT, TH, and TT are an example of permutation.

Worked Example

1. Determine the number of permutations of three letters A, B and C.

Solution

Since three letters have to be arranged, the order of occurrence of each letter in a particular arrangement is to be kept in mind. We can fill three boxes.

3	2	1

In the first box, we can write A, B, or C. In other words, we have three choices for writing a letter in the first box.

For the second place, we have a choice of one of the two remaining letters.

And for the third place, we have only one choice of the letter that remains.

By the fundamental principle, the three places can be filled in $3 \times 2 \times 1 = 6$ ways.

Let us write the six permutations. They are ABC, ACB, BCA, BAC, CAB, and CBA. We give two theorems without proof here.

Theorem 1

The number of permutations of n different objects taken all at a time is $P(n, n) = n(n-1)(n-2) \ldots 3 \times 2 \times 1 = n!$

Another notation used for $P(n, n)$ is ^nP_n.

Theorem 2

The number of permutations of n different objects taken r at a time is ^nP_r.

$$^nP_r = n(n-1)(n-2), \ldots, (n-r+1) = \frac{n!}{(n-r)!}$$

Worked Examples

1. Naveen, Prakash, Shaleen, and Rekha equally qualified are to be appointed against four jobs in a factory. Determine the number of ways in which the jobs can be filled:
(a) The first and second jobs
(b) The first three jobs
(c) All the four jobs

Solution

(a) For the first job there are four candidates.
For the second job there are three candidates, and hence for the first two jobs there are $4 \times 3 = 12$ ways.
(b) For three jobs there are $4 \times 3 \times 2 = 24$ ways.

(c) For all the four jobs there are $4 \times 3 \times 2 \times 1 =$ 24 ways.

2. How many of the natural numbers from 1 to 1000 have none of their digits repeated?

Solution

We need to examine one-digit numbers, two-digit numbers, and three-digit numbers from 1 to 1000. See how many of those have distinct digits:

(a) One-digit numbers pose no problem. Clearly, there are nine of them, each distinct digit.

(b) We have to get the number of two-digit numbers which have none of their digits repeated. In the tens place, we can have any one of the digits 1 to 9. (Clearly we cannot have 0 in the tens place.) Thus, there are nine choices for the tens place. For the units place, we can have any one of the remaining eight digits from 1 to 9 or the digit 0. (Since we do not want repetition of digits.) Therefore, there are nine choices again for the units place. The number of two-digit numbers is thus $9 \times 9 = 81$.

In the same way, we can determine the number of three-digit numbers which have none of their digits repeated. The number is $9 \times 9 \times 8 = 648$. Thus, in all, there are $9 + 81 + 648 = 738$ natural numbers, from 1 to 1000, which have none of their digits repeated.

3. Find the value of n such that

$^nP_5 = 42 \times {^nP_3} \qquad n > 4$

Solution

$$\frac{n!}{(n-5)!} = 42\left[\frac{n!}{(n-3)!}\right]$$

$$\frac{(n-3)!}{(n-5)!} = 42$$

$$\frac{(n-3)(n-4)(n-5)!}{(n-5)!} = 42$$

$$n^2 - 7n + 12 = 42$$

$$n^2 - 7n - 30 = 0 \Rightarrow (n-10)(n+3) = 0$$

i.e., $n = 10$ or -3

n cannot be negative and hence $n = 10$.

COMBINATIONS

In permuting n objects, *the order of occurrence* of the objects in a permutation is important. We now consider counting problems where the order of occurrence does not matter. Let us, for instance, determine the number of ways of selecting two persons out of four. Let us denote the four persons as A, B, C, and D. We already know that the number of permutations of four objects taken two at a time $= {^4P_2} = 12$. The permutation are *AB, AC, AD, BC, BD, CD, BA, CA, DA, CB, DB*, and *DC*. But if the order of the occurrence does not matter, then *AB* and *BA* are the same. Similarly, *AC* and *CA* are same and so on. Thus, we have the following selections of two persons out of four: *AB, AC, AD, BC, BD* and *CD*.

We observe that there are six ways of selecting two persons out of four.

A combination is a selection of some or all of a number of different objects. In combinations, the order of the objects is in material.

We give below a formula for determining the number of combinations of n different objects taken r at a time.

Theorem

The number ^nC_r of combinations of n different objects; taken r at a time, is as follows:

$$^nC_r = \frac{^nP_r}{r!} = \frac{n!}{r!(n-r)!}$$

where $r \leq n$.

Proof

We recall that the number of permutations of n different objects, taken r at a time, is as follows:

$$^nP_r = \frac{n!}{(n-r)!}$$

Since the order of arrangement of the objects does not matter, it is clear that to every combination of r objects there corresponds $^rP_r = r!$ arrangements. In other words, $^nP_r = r!\,{^nC_r}$. Thus,

$$^nC_r = \frac{^nP_r}{r!} = \frac{n!}{r!(n-r)!}$$

Hence, the number of combinations of n different objects taken r at a time is as follows:

$$^nC_r = \frac{n!}{r!(n-r)}$$

where $r \leq n$.

Remark 1

$$^nC_n = \frac{n!}{n!(n-n)!} \quad (1)$$

We have already defined 0! = 1.

Eq. (1) simplifies $^nC_n = 1$.

Remark 2

$^nC_r = {}^nC_{n-r}$

Selecting r out of n objects is the same as rejecting $(n-r)$ objects out of n items.

Remark 3

$^nC_0 = 1$

Worked Examples

1. Evaluate

(a) $^{10}C_4$ (b) $^{12}C_7$

(c) $^6C_6$ (d) $^9C_0$

Solution

(a) $$^{10}C_4 = \frac{10!}{4!(10-4)!} = \frac{10!}{4!6!}$$

$$= \frac{10\times9\times8\times7\times6!}{1\times2\times3\times4\times6!} = 210$$

(b) $$^{12}C_7 = \frac{12!}{7!(12-7)!} = \frac{12!}{7!5!}$$

$$= \frac{12\times11\times10\times9\times8\times7!}{1\times2\times3\times4\times5\times7!} = 792$$

(c) $$^6C_6 = \frac{6!}{6!!0!} = 1$$

(d) $$^9C_0 = \frac{9!}{0!9!} = 1$$

2. A school has five good badminton players. A team of four players has to be sent to an interschool tournament. In how many ways can the team be selected?

Solution

We wish to determine the number of ways of selecting four players out of five.

i.e., $^5C_4 = \frac{5!}{4!.1!} = 5$

3. How many lines can be drawn through six points on a circle?

Solution

Since exactly one line can be drawn through two points, the number lies through six points on a circle is as follows:

$$^6C_2 = \frac{6!}{2!4!} = \frac{6\times5\times4!}{1\times2\times4!} = 15$$

4. Determine the number of five card combinations out of a deck of 52 cards.

Solution

The number of five card combinations is as follows:

$$^{52}C_5 = \frac{52!}{5!47!} = \frac{52\times51\times50\times49\times48\times47!}{1\times2\times3\times4\times5\times47!} = 25{,}98{,}960$$

5. Verify that $^8C_4 + {}^8C_3 = {}^9C_4$.

Solution

$$^8C_4 = \frac{8!}{4!4!} = 70$$

$$^8C_3 = \frac{8!}{3!5!} = 56$$

$$^8C_4 + {}^8C_3 = 126$$

$$^9C_4 = \frac{9!}{4!5!} = \frac{9\times8\times7\times6\times5!}{1\times2\times3\times4\times5!} = 126$$

The above problem leads us to an important result.

$$^nC_r + {}^nC_{r-1} = {}^{(n+1)}C_r$$

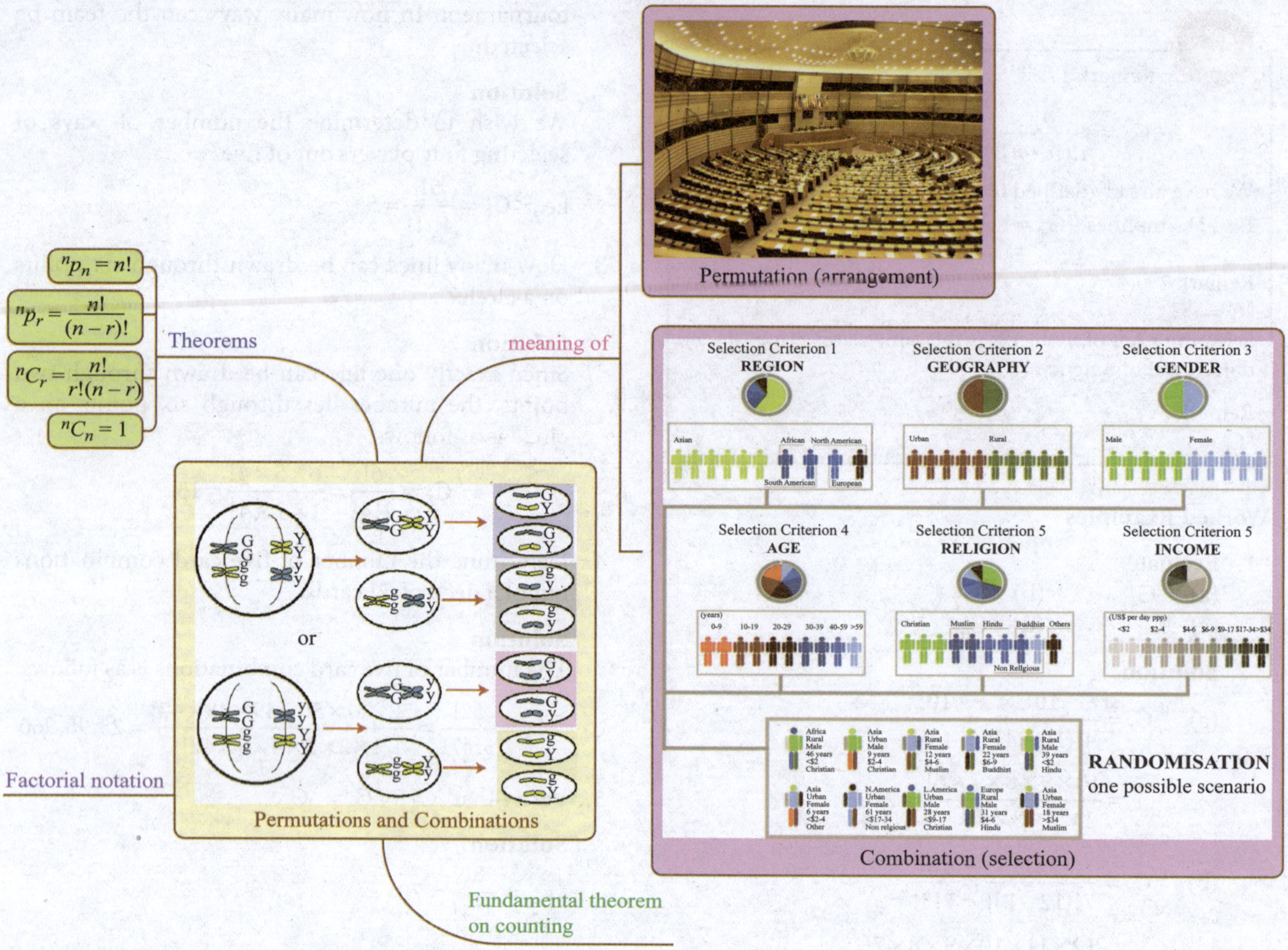

TEST YOURSELF

Level I

1. Evaluate the following:
   (a) $^{10}P_3$
   (b) $^{20}P_4$
   (c) $^{75}P_2$
   (d) $^{9}P_5$
2. Renu wants to arrange three economics, two history, and four language books on a shelf. If the books of the same subjects are different, determine the following:
   (a) The number of possible arrangements
   (b) The number of arrangements if all books of a subject are to be together.
3. Determine the number of five-letter words formed from the letters of the word 'EQUATION'.
4. In how many ways can 10 persons line up at a ticket counter of a cinema hall?
5. Evaluate the following:
   (a) $^{8}C_5$
   (b) $^{15}C_4$

Level 2

6. Find n if
   (a) $^nP_6 = 3 \times {}^nP_5$
   (b) $^nP_4 = 20 \times {}^nP_2$
   (c) $2 \times {}^nP_3 = {}^{n+1}P_3$
7. Twelve students compete in a race. In how many ways can the first three places be taken?
8. In how many ways can seven books be arranged in a shelf? In how many ways can we arrange three particular books always together?

9. Evaluate the following:
(a) ${}^{19}C_{17} + {}^{19}C_{18}$ (b) ${}^{25}C_{22} - {}^{24}C_{21}$
(c) ${}^{31}C_{26} - {}^{30}C_{26}$

10. In a student reunion meeting in a school, 16 students attended. Each shakes hands with each other exactly once. Determine the total number of hand shakes.

11. In how many ways can a team of five students be selected from eight students?

12. There are 10 points on a plane, no three being collinear. By connecting these points in all possible ways, how many line segments can be formed?

13. There are seven candidates vying for two vacancies in a company, all being equally eligible. In how many ways can the vacancies be filled?

14. In how many ways can a cricket team of 11 players be formed from 14 players?

Level 3

15. In how many ways can the letters of the following words be arranged taking all at a time?
(a) LOGARITHM (b) SUNDAY
(c) MATHS

16. How many more numbers can be formed by rearranging the digits of the number 54,679?

17. In how many ways can a committee of three be appointed from 15 members?

18. Find n if ${}^{n}C_5 = 5 \times {}^{n}P_3$ and ${}^{n}C_{n-2} = 3$.

19. Find r if ${}^{15}C_{r+1} = {}^{15}C_{3r-5}$.

Multiple Choice Questions

20. There are five routes for going from station A to station B and four routes for going from station B to station C. Find the number of different ways in which a person can go from A to C via B.
(A) 9 (B) 20
(C) 16 (D) none of these

21. There are 25 students in a class among which 15 are boys and 10 are girls. The class teacher selects either a boy or a girl as monitor of the class. In how many ways the class teacher can make this selection?
(A) 25 (B) 150
(C) 35 (D) none of these

22. Find the value of n if $(n + 2)! = 60 \times (n - 1)!$
(A) $n = 6$ (B) $n = 3$
(C) $n = 9$ (D) $n = 4$

23. If a, b and c are three consecutive positive integers such that $a < b < c$ and $\frac{1}{a!} + \frac{1}{b!} = \frac{v}{c!}$ then the value of $\sqrt{v}$ is
(A) a (B) b
(C) c (D) $a + b + c$

24. If $n!$, $3 \times n!$ and $(n + 1)!$ are in GP then $n!$, $5 \times n!$ and $(n + 1)!$ are in
(A) AP (B) GP
(C) HP (D) none of these

25. If ${}^{56}P_{r+6} : {}^{54}P_{r+3} = 30800 : 1$ then find ${}^{r}P_2$.
(A) 1460 (B) 1840
(C) 1640 (D) none of these

26. If ${}^{n+5}P_{n+1} = \frac{11(n-1)}{2}\,{}^{n+3}P_n$ then find the value of n.
(A) 6, 7 (B) 6, 5
(C) 6, 2 (D) 10, 9

27. If ${}^{m+n}P_2 = 90$ and ${}^{m-n}P_2 = 30$ then find the values of m and n.
(A) $m = 8, n = 2$ (B) $m = 2, n = 8$
(C) $m = 4, n = 4$ (D) none of these

28. Find the value of r if ${}^{8}P_5 + 5 \times {}^{8}P_4 = {}^{9}P_r$.
(A) 3 (B) 4
(C) 5 (D) 6

29. Determine the number of permutations of the letters of the word "SIMPLETON" taken all at a time.
(A) 36280 (B) 362880
(C) 36288 (D) none of these

30. How many different signals can be given using any number of flags from 4 flags of different colours?
(A) 56 (B) 24
(C) 48 (D) 64

31. If ${}^{m+n}P_2 = 56$ and ${}^{m-n}P_3 = 24$ then $\frac{{}^{m}P_3}{{}^{n}P_2}$ equals
(A) 20 (B) 40
(C) 60 (D) 80

32. If ${}^{9}P_5 + 5 \times {}^{9}P_4 = {}^{n}P_r$ then the value of $(n + r)$ is
(A) 13 (B) 14
(C) 15 (D) 16

33. If $^{15}C_{3r} = {}^{15}C_{r+3}$ then find the value of $^rC_2$.

(A) 6 (B) 9
(C) 3 (D) none of these

34. Find the value of $^nC_r + 2 \times {}^nC_{r-1} + {}^nC_{r-2}$.

(A) $^{n+1}C_r$ (B) $^{n+2}C_r$
(C) ^nC_r (D) $^nC_{r+1}$

35. Thirty-six games were played in a football tournament with each team plying once against the other. How many teams were there?

(A) 8 (B) 6
(C) 9 (D) 10

36. If $^nC_3 + {}^nC_4 > {}^{n+1}C_3$ then

(A) $n > 6$ (B) $n < 6$
(C) $n > 7$ (D) $n < 7$

37. If $^nC_r = {}^nC_{r-1}$ and $^nP_r = {}^nP_{r+1}$ then the value of n is

(A) 2 (B) 3
(C) 4 (D) 5

38. If $^nC_{r-1} = 36$, $^nC_r = 84$ and $^nC^{r+1} = 126$ then find r.

(A) 2 (B) 3
(C) 4 (D) 5

OLYMPIAD AND NTSE LEVEL EXERCISES

39. How many numbers divisible by 5 and lying between 3000 and 4000 can be formed from the digits 1, 2, 3, 4, 5 and 6 (repetition is not allowed)?

(A) $\frac{n+r-1}{r}$ (B) $^5P_2$
(C) $^4P_2$ (D) $^6P_3$

40. In how many ways can mn letters be posted in n letter-boxes?

(A) $(mn)^n$ (B) m^{mn}
(C) n^{mn} (D) None of these

41. How many words can be formed by taking 3 consonants and 2 vowels out of 5 consonants and 4 vowels?

(A) $^5C_3 \times {}^4C_2$ (B) $\frac{^5C_3 \times {}^4C_2}{5}$
(C) $^5C_3 \times {}^4C_3$ (D) $(^5C_3 \times {}^4C_2)$ (5)!

42. In how many ways a team of 11 players can be formed out of 25 players, if 6 out of them are always to be included and 5 are always to be excluded?

(A) 2020 (B) 2002
(C) 2008 (D) 8002

43. There are 9 chairs in a room on which 6 persons are to be seated, out of which one is the guest with one specific chair. In how many ways they can sit?

(A) 6720 (B) 60480
(C) 30 (D) 346

44. The number of triangles that can be formed by 5 points in a line and 3 points on a parallel line is

(A) $^8C_3$ (B) $^8C_3 - {}^5C_3$
(C) $^8C_3 - {}^5C_3 - 1$ (D) None of these

45. Consider all possible permutations of the letters of the word INDIANOIL.
Now, match the following lists and then choose the correct option from amongst the given codes.

	List I		List II
(P)	The number of permutations which contain the word INDIA is	(1)	28866
(Q)	The number of permutations which contain the word OIL is	(2)	1374
(R)	The number of permutations which contains neither the word OIL nor the word INDIA is	(3)	1260
(S)	The number of permutations which contains at least one of the words OIL and INDIA is	(4)	120

Codes:

	(P)	(Q)	(R)	(S)
(A)	(4)	(3)	(1)	(2)
(B)	(2)	(1)	(4)	(3)
(C)	(3)	(1)	(4)	(2)
(D)	(2)	(3)	(4)	(2)

46. Read the following statements.

Statement 1: Number of ways in which Indian team (11 players) can bat, if Yuvraj wants to bat before Dhoni and Pathan wants to bat after Dhoni is 11!/3!.

Statement 2: Yuvraj, Dhoni and Pathan can be arranged in batting order in 3! ways.

(A) Both Statement 1 and Statement 2 are true.
(B) Statement 1 is true and Statement 2 is false.
(C) Statement 1 is false and Statement 2 is true.
(D) Both Statement 1 and Statement 2 are false.

47. Match the following columns.

	Column I		Column II
(P)	Number of straight lines joining any two of 10 points of which four points are collinear is	(i)	30
(Q)	Maximum number of points of intersection of 10 straight lines in a plane is	(ii)	60
(R)	Maximum number of points of intersection of 6 circles in a plane is	(iii)	40
(S)	Maximum number of points of intersection of 6 parabolas in a plane is	(iv)	45

Codes:

	P	Q	R	S
(A)	(iii)	(iv)	(ii)	(i)
(B)	(i)	(iv)	(iii)	(ii)
(C)	(iii)	(ii)	(i)	(iv)
(D)	(iii)	(iv)	(i)	(ii)

48. If ${}^{2n}C_r : {}^{n}C_2 = 44:3$ then for which of the following values of r, the value of ${}^{n}C_r$ will be 15?

(A) $r = 3$ (B) $r = 4$

(C) $r = 6$ (D) $r = 5$

2 Binomial Theorem

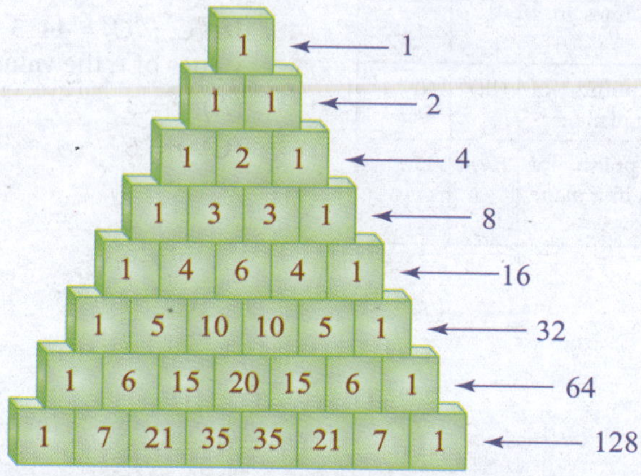

If I have seen farther than other men, it is because I have stood on the shoulders of giants.

– Isaac Newton

1

1 2 1

1 3 3 1

1 4 6 4 1

1 5 10 10 5 1

. . . .

OBJECTIVES

After reading this chapter you should be able to

- understand binomial theorem
- apply the theorem for expansions
- apply the theorem to find particular terms and coefficients

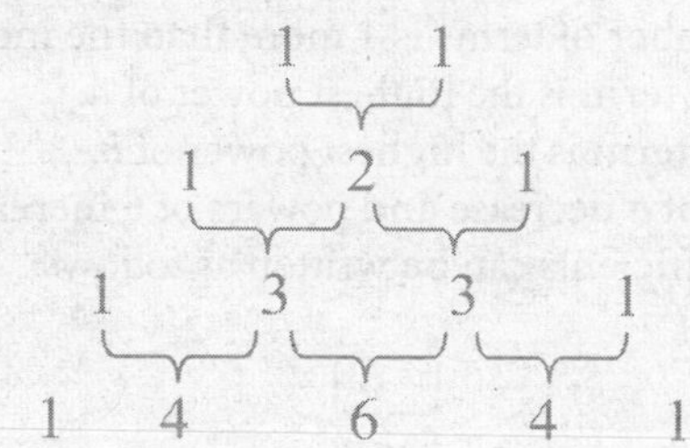

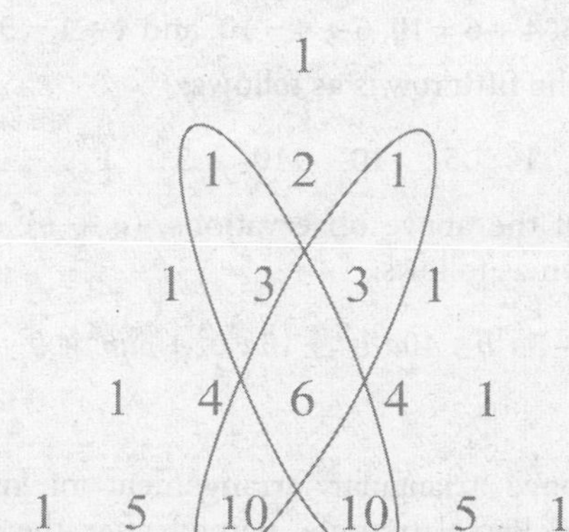

. . . .

INTRODUCTION

As you know the identities $(a+b)^2 = a^2 + 2ab + b^2$ and $(a+b)^3 = a^3 + 3a^2b + 3ab^2 + b^3$, let us find the expansion of $(a+b)^4$:

$$(a+b)^4 = (a+b)^3(a+b) = (a^3 + 3a^2b + 3ab^2 + b^3)(a+b)$$
$$= a^4 + 4a^3b + 6a^2b^2 + 4ab^3 + b^4$$

This method of expansion gets more and more tedious as the power of $(a+b)$ increases.

Let us see how we have to expand bigger powers of $a+b$ using the following pattern:

$$(a+b)^1 = a^1 + b^1$$
$$(a+b)^2 = a^2 + 2ab + b^2$$
$$(a+b)^3 = a^3 + 3a^2b + 3ab^2 + b^3$$
$$(a+b)^4 = a^4 + 4a^3b + 6a^2b^2 + 4ab^3 + b^4$$

The following observations can be made:

(a) The number of terms is 1 more than the index.
(b) The first term is the highest power of a.
(c) The last term is the highest power of b.
(d) Powers of a decrease and powers of b increase.
(e) The coefficients can be written as follows:

1 1
1 2 1
1 3 3 1
1 4 6 4 1

What will be the next row in the above arrangement?

The first and last numbers must be 1. Other numbers are $1+4=5$, $4+6=10$, $6+4=10$, and $4+1=5$.

Therefore, the fifth row is as follows:

1 5 10 10 5 1

Using all the above observations, $(a+b)^5$ can be written down as follows:

$$a^5 + 5a^4b + 10a^3b^2 + 10a^2b^3 + 5ab^4 + b^5$$

Note:

(a) The above triangular arrangement of numbers is called Pascal triangle, named after the famous mathematician Blaise Pascal. (World's first adding machine was invented by Pascal.) Pascal triangle is given as follows:

1
1 2 1
1 3 3 1
1 4 6 4 1
1 5 10 10 5 1
.... and so on

Pascal triangle may also be written as follows:

1
1 2 1
1 3 3 1
1 4 6 4 1
1 5 10 10 5 1
.... and so on

(b) Notice that in each row the coefficients read the same from left to right and from right to left.
(c) These coefficients can be written down using combination notation as follows:

1
$1 \quad {}^2C_1 \quad 1$
$1 \quad {}^3C_1 \quad {}^3C_2 \quad 1$
$1 \quad {}^4C_1 \quad {}^4C_2 \quad {}^4C_3 \quad 1$
.... and so on

Based on all these observations, *binomial theorem* can be written down as follows:

$$(a+b)^n = a^n + {}^nC_1a^{n-1}b + {}^nC_2a^{n-2}b^2 + {}^nC_3a^{n-3}b^3 + \cdots + b^n$$

where n is a natural number.

If we use the summation symbol, the theorem becomes the following:

$$(a+b)^n = \sum_0^n {}^nC_r a^{n-r} b^r$$

(*Proof of the theorem is beyond the scope of this book*)

$$T_{r+1} = {}^nC_r\, a^{n-r} b^r$$

The general term can be used in finding particular terms or coefficients.

When we replace b by $-b$ in the above statement, we get the following:

$$(a-b)^n = a^n - {}^nC_1a^{n-1}b + {}^nC_2a^{n-2}b^2 - {}^nC_3a^{n-3}b^3 + \cdots + (-1)^n b^n$$

Note that the number of terms in binomial expansion is $n+1$, where powers of 'a' decrease and powers of 'b' increase.

Worked Examples

1. Expand $(x+1)^4$.

Solution

We apply binomial theorem with $a = x$, $b = 1$, and $n = 4$ as follows:

$$(x+1)^4 = x^4 + {}^4C_1x^3 \times 1 + {}^4C_2x^2 \times 1^2 + {}^4C_3x \times 1^3 + 1^4$$
$$= x^4 + 4x^3 + 6x^2 + 4x + 1$$

2. Expand $(x - 1/x)^5$.

Solution

Taking n as 5, a as x, and b as $1/x$ in the expansion of $(a - b)^n$, we have the following:

$$\left(x - \frac{1}{x}\right)^5 = x^5 - {}^5C_1x^4 \times \left(\frac{1}{x}\right) + {}^5C_2x^3 \times \left(\frac{1}{x}\right)^2 - {}^5C_3x^2 \times \left(\frac{1}{x}\right)^3 + {}^5C_4x \times \left(\frac{1}{x}\right)^4 - \left(\frac{1}{x}\right)^5$$
$$= x^5 - 5x^3 + 10x - \frac{10}{x} + \left(\frac{5}{x}\right)^3 - \left(\frac{1}{x}\right)^5$$

3. Find the eighth term in the expansion of $(2x+3)^{10}$.

Solution

Here, we need not write all the 11 terms in the expansion. It is possible to find T_8 using the general term formula, $T_{r+1} = {}^nC_ra^{n-r}b^r$.

Taking $r = 7$, $n = 10$, $a = 2x$, and $b = 3$, we have the following:

$$T_8 = {}^{10}C_7(2x)^{10-7}(3)^7 = {}^{10}C_7(2x)^3(3)^7 = 120 \times 8 \times x^3 \times 3^7 - 960 \times 3^7\, x^3$$

4. Find the middle term in the expansion of $(x+2)^{12}$.

Solution

Since $n = 12$, there are 13 terms in the expansion. So, the middle term is T_7.

Taking $n = 12$, $a = x$, $b = 2$, and $r = 6$ in the general term, we have the following:

$$T_7 = {}^{12}C_6x^{12-6}(2)^6 = {}^{12}C_6 \times 64 \times x^6$$

Note: If the number of terms is even, then there will be two middle terms.

5. Find the coefficient of x^4 in the expansion of $(x^4 + 1/x^3)^{15}$.

Solution

As we do not know which term contains x^4, we write down the general term and then calculate the value of r as follows:

$$T_{r+1} = {}^{15}C_r(x^4)^{15-r}\left(\left(\frac{1}{x}\right)^3\right)^r$$
$$= {}^{15}C_rx^{60-4r-3r} = {}^{15}C_rx^{60-7r}$$

Since we need the coefficient of x^4, we equate $60 - 7r$ to 4 and get $r = 8$.

Therefore, the coefficient is ${}^{15}C_8$.

6. In the expansion of $(2x^2 - 1/x)^{12}$, there is a term which does not contain x. Which is that term?

Solution

$$T_{r+1} = {}^nC_ra^{n-r}b^r$$

Here, $n = 12$, $a = 2x^2$, and $b = -1/x$, and we have to find r.

$$T_{r+1} = {}^{12}C_r(2x^2)^{12-r}\left(-\frac{1}{x}\right)^r$$
$$= {}^{12}C_r2^{12-r}\,x^{2(12-r)-r}(-1)^r$$
$$= {}^{12}C_r(-1)^r \times 2^{12-r}x^{24-2r-r}$$
$$= {}^{12}C_r(-1)^r \times 2^{12-r}x^{24-3r}$$

Since we are looking for a term without x, we equate the exponent $24 - 3r$ to 0 and get $r = 8$.

Therefore, the term free from x is the ninth term.

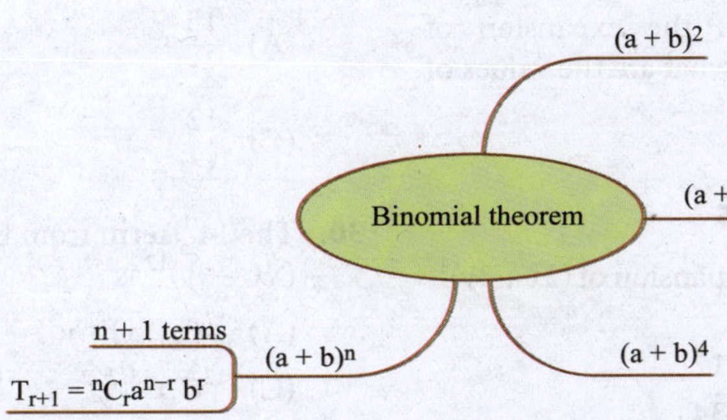

TEST YOURSELF

Level 1

1. Write down the expansion of $(a + b)^6$.
2. Write down the expansion of $(1 + x)^5$.
3. Write down the first three terms in the expansion of $(a + 1/a)^8$.
4. Write down the last two terms in the expansion of $(2x - 1/x)^{10}$.
5. Write down the sum of the first three terms in the expansion of $(1 + 0.02)^{12}$.

Level 2

6. Evaluate $(1.03)^7$ correct to five decimal places using binomial theorem.
7. Evaluate $(0.99)^5$ correct to five decimal places using binomial theorem.
8. Find the seventh term in the expansion of $(x + 2/x)^{10}$.
9. Find the tenth term in the expansion of $(2x + 3/x)^{20}$.
10. Find the middle term in the expansion of $(x + 1/x)^9$.
11. Find the middle terms in the expansion of $(2x^2/3 - 3/2x)^{11}$.

Level 3

12. Find the value of $(2 + \sqrt{3})^5 + (2 - \sqrt{3})^5$.
13. Find the value of $(\sqrt{2} + 1)^6 + (\sqrt{2} - 1)^6$.
14. Find the middle term/terms in the expansion of $(\sqrt{x} - 3/x^2)^{13}$.
15. Find the middle term/terms in the expansion of $(4x^2/3 - 3/2x)^9$.
16. Find the coefficient of x^7 in the expansion of $(x^2 + 1/x)^{11}$.
17. Which term in the expansion of $(2x^2 + 1/3x^3)^{10}$ does not contain x?
18. What is the coefficient of x^{-9} in the expansion of $(x^2/2 + 2/x)^9$?
19. What is the coefficient of x^{-11} in the expansion of $(\sqrt{x} - 2/x)^{17}$?
20. If the first three terms in the expansion of $(1 + ax)^n$ are 1, $6x$, and $16x^2$, what are the values of a and n?

Multiple Choice Questions

21. The number of terms in the expansion of $(2x + 3y)^{17}$ is
    (A) 16 (B) 17
    (C) 18 (D) 34
22. The number of terms in the expansion of $(\sqrt{x} + \sqrt{y})^8 + (\sqrt{x} - \sqrt{y})^8$ is
    (A) 8 (B) 7
    (C) 5 (D) 9
23. The number of terms in the expansion of $\{5x + 2y\}^7 - \{(5x - 2y)^7\}$ is
    (A) 4 (B) 8
    (C) 6 (D) 3
24. The number of terms in the expansion of $\left\{\left(\sqrt{2x}+\sqrt{3y}\right)^{10} - \left(\sqrt{2x}-\sqrt{3y}\right)^{10}\right\}$ is
    (A) 11 (B) 9
    (C) 6 (D) 5
25. The number of terms in the expansion of $\{(x + a)^{16} + (x - a)^{16}\}$ is
    (A) 7 (B) 8
    (C) 9 (D) 17
26. The 10th term in the expansion of $\left(2x^2 + \frac{1}{x^2}\right)^{12}$ is
    (A) $264x^{-16}$ (B) $220x^{-12}$
    (C) $792x^{-14}$ (D) $1760x^{-12}$
27. The 4th term in the expansion of $(x - 2y)^{12}$ is
    (A) $1760\, x^8 y^6$ (B) $-440\, x^7 y^5$
    (C) $-1760\, x^9 y^3$ (D) None of these
28. The 13th term in the expansion of $\left(9x - \frac{1}{3\sqrt{x}}\right)^{18}$, $x \neq 0$ is
    (A) 16854 (B) 18564
    (C) 17954 (D) 18832
29. The 3rd term from the end in the expansion of $\left(x + \frac{1}{x}\right)^6$ is
    (A) $\frac{15}{x^2}$ (B) $\frac{30}{x^3}$
    (C) $\frac{12}{x^2}$ (D) $\frac{24}{x^3}$
30. The 14th term from the end in the expansion of $(\sqrt{x} - \sqrt{y})^{17}$ is
    (A) ${}^{17}C_6(\sqrt{x})^{11} y^3$ (B) $-{}^{17}C_5 x^6(\sqrt{y})^5$
    (C) ${}^{17}C_4 x^{13/2} y^2$ (D) $-{}^{17}C_{13} x^2 y^{13/2}$

31. The middle term in the expansion of $\left(x-\frac{1}{2y}\right)^{10}$ is

(A) $\frac{-63}{8}x^5y^{-5}$ (B) $\frac{-21}{4}x^6y^{-6}$
(C) $\frac{63}{8}x^4y^{-4}$ (D) $\frac{-63}{8}x^4y^{-4}$

32. The middle term in the expansion of $\left(x^2-\frac{2}{x}\right)^{10}$ is

(A) $8064\,x^5$ (B) $-8064\,x^5$
(C) $6720\,x^4$ (D) $-6720\,x^4$

33. The coefficient of x^2 in the expansion of $\left(3x-\frac{1}{x}\right)^6$ is

(A) 405 (B) 1215
(C) 2430 (D) 3645

34. The coefficient of x^6 in the expansion of $\left(3x^2-\frac{1}{3x}\right)^9$ is

(A) 576 (B) 756
(C) 189 (D) 378

35. The term independent of x in the expansion of $\left(\sqrt{x}+\frac{1}{3x^2}\right)^{10}$ is

(A) 135 (B) 132
(C) 15 (D) 5

36. The term independent of x in the expansion of $\left(x-\frac{1}{x}\right)^{12}$ is

(A) 924 (B) 462
(C) 231 (D) 693

37. The coefficient of x^{32} in the expansion of $\left(x^4-\frac{1}{x^3}\right)^{15}$ is

(A) 273 (B) 546
(C) 1365 (D) 1032

38. The term independent of x in the expansion of $\left(\frac{3}{2}x^2-\frac{1}{3x}\right)^9$ is

(A) $\frac{1}{24}$ (B) $\frac{7}{18}$
(C) $\frac{8}{27}$ (D) $\frac{5}{36}$

39. Which term contains x^3 in the expansion of $\left(3x-\frac{1}{2x}\right)^8$?

(A) 2nd (B) 3rd
(C) 5th (D) None of these

40. The coefficient of x^{-4} in the expansion of $\left(\frac{4x}{5}+\frac{5}{2x}\right)^8$ is

(A) 625 (B) 1875
(C) 4375 (D) None

41. The total number of terms in the expansion of $(x+k)^{100}+(x-k)^{100}$ after simplification is

(A) 50 (B) 51
(C) 101 (D) 202

42. If the coefficients of the second, third and fourth terms in the expansion of $(1+x)^n$ are in A.P., then $n =$

(A) 5 (B) 6
(C) 7 (D) 9

43. If p and q are positive integers, then the coefficients of x^p and x^q in the expansion of $(1+x)^{p+q}$ are

(A) equal
(B) equal with opposite signs
(C) reciprocal to each other
(D) none of these

44. Let the coefficient of x^n in the expansion of $(1+x)^{2n}$ be P and the coefficient of x^n in the expansion of $(1+x)^{2n-1}$ be q, then

(A) $P = 2q$ (B) $2P = q$
(C) $2P = 3q$ (D) $3P = 2q$

45. In the expansion of $(1+x)^n$, the binomial coefficient of three consecutive terms are respectively 220, 495 and 792. The value of n is

(A) 10 (B) 11
(C) 12 (D) 13

46. If the coefficients of x^7 and x^8 are equal in the expansion of $\left(2+\frac{x}{3}\right)^n$, then $n =$

(A) 15 (B) 45
(C) 55 (D) 56

OLYMPIAD AND NTSE LEVEL EXERCISES

47. If rth and $(r+1)$th terms in the expansion of $(p+q)^n$ are equal, then $\dfrac{(n+1)q}{r(p+q)}$ is

(A) $\dfrac{1}{2}$ (B) $\dfrac{1}{4}$

(C) 1 (D) 0

48. If the coefficients of T_r, T_{r+1} and T_{r+2} terms of $(1+x)^{14}$ are in A.P., then $r =$

(A) 6 (B) 7

(C) 8 (D) 9

49. The ratio of the coefficient of x^{15} to the term independent of x in the expansion of $\left(x^2+\dfrac{2}{x}\right)^{15}$ is

(A) 1 : 32 (B) 1 : 16

(C) 1 : 12 (D) 1 : 8

50. The approximate value of $(1.0002)^{3000}$ is

(A) 1.6 (B) 1.4

(C) 1.8 (D) 1.2

51. $10^n + 3(4^{n+2}) + 5$ is divisible by $(n \in N)$

(A) 7 (B) 5

(C) 9 (D) 17

52. If n is an odd natural number, then number of zeros at the end of $99^n + 1$ is

(A) $2n$ (B) n

(C) 2 (D) none of these

53. If the three consecutive coefficients in the expansion of $(1+x)^n$ are 28, 56 and 70, then the value of n is

(A) 6 (B) 4

(C) 8 (D) 10

54. The number of integral terms in the expansion of $(5^{1/2} + 7^{1/6})^{642}$ is

(A) 106 (B) 108

(C) 103 (D) 109

55. Match the given columns:

	Column I		Column II
(a)	If ${}^{(n+1)}C_4 + {}^{(n+1)}C_3 + {}^{(n+2)}C_3 > {}^{(n+3)}C_3$, then possible value(s) of n is/are	(p)	4
(b)	The remainder when $(3053)^{456} - (2417)^{333}$ is divided by 9 is less than	(q)	5
(c)	The digit in the units place of the number $183! + 3^{183}$ is greater than	(r)	6
(d)	If the sum of the coefficients of the first, second and third terms in the expansion of $\left(x^2+\dfrac{1}{x}\right)^m$ is 46, then the index of the term that does not contain x is greater than	(s)	7

56. Match the given columns:

	Column I		Column II
(a)	If the coefficients of two consecutive terms in the expansion of $(1+x)^n$ are equal, then n can be	(p)	9
(b)	If $15^n + 23^n$ is divisible by 19, then n can be	(q)	10
(c)	If ${}^{10}C_0\,{}^{20}C_{10} - {}^{10}C_1\,{}^{18}C_{10} + {}^{10}C_2\,{}^{16}C_{10}$ – ... is divisible by 2^n, then n can be	(r)	11
(d)	If the coefficients of T_r, T_{r+1} and T_{r+2} terms of $(1+x)^{14}$ are in A.P., then the sum of possible values of r is more than	(s)	12

GAME TIME

Pascal Triangle

1

1 2 1

1 3 3 1

1 4 6 4 1

1 5 10 10 5 1

....

There are some very interesting number patterns in Pascal triangle. A few are given below. Can you find some more patterns?

1.

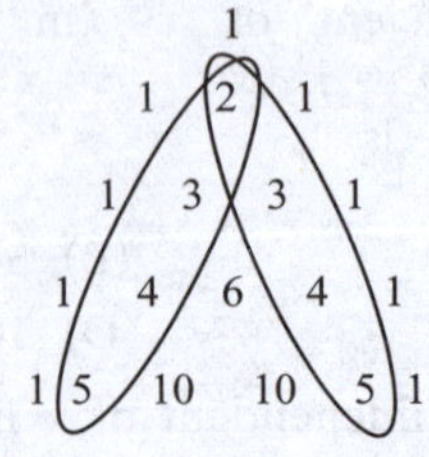

....

Add the numbers in each row; you get 1, 4, 8, 16, 32, etc., which are powers of 2.

Observe the numbers marked above. They are symmetrically placed.

2.

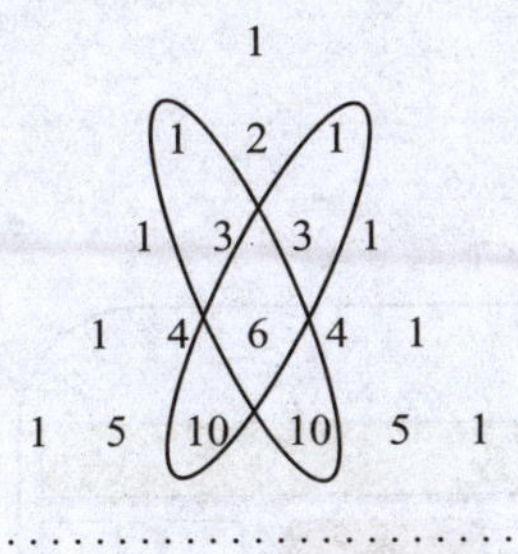

The numbers 1, 3, 6, 10, etc., marked above are triangular numbers. Furthermore, 1, $1 + 3$, $3 + 6$, $6 + 10$, etc., are all square numbers.

There are many more patterns. Try and find them.

3 Remainder Theorem

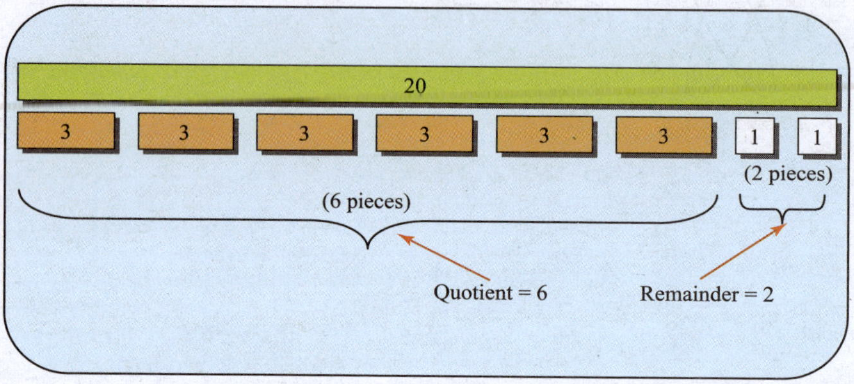

One of the most important concepts in all of mathematics is that of function.

– T.P. Dick and C.M. Patton

−3	1	3	6	−12
	0	−3	0	−18
	1	0	6	−30 R

OBJECTIVES

After reading this chapter you should be able to

- understand remainder theorem and factor theorem
- apply factor theorem to factorise algebraic expressions

−1	1	0	0	0	0	7
	0	−1	+1	−1	+1	−1
	1	−1	+1	−1	+1	6 R

h	a_0	a_1	a_2	…	a_{n-1}	a_n
		hq_0	hq_1	hq_2 …	hq_{n-1}	hq_{n-1}
	q_0	q_1	q_2	q_3 …	q_{n-1}	q_n

Let us divide the expression $f(x) = ax^3 + bx^2 + cx + d$ by $x - m$. We know that $f(x)$ is called the *dividend* and $x - m$ is called the *divisor*. Since the degree of the divisor is 1, the remainder must be independent of x, i.e., a constant.

$$\begin{array}{r|l l}
 & ax^2 + (b + am)x + (am^2 + bm + c) & \\
\hline
x - m & ax^3 + bx^2 + cx + d & \\
 & ax^3 - amx^2 & \\
\hline
 & (b + am)x^2 + cx & \\
 & (b + am)x^2 - (b + am)xm & \\
\hline
 & \quad + (am^2 + mb + c)x + d & \\
 & \quad + (am^2 + mb + c)x - (am^2 + bm + c)m & \\
\hline
 & \qquad (am^2 + bm + c)m + d & \text{is the} \\
 & \qquad = am^3 + bm^2 + cm + d & \text{remainder}
\end{array}$$

The remainder does not contain x. When we compare the function $f(x)$ with the remainder, the remainder appears as though x is replaced by m in the expression.

If $f(x) = ax^3 + bx^2 + cx + d$, then $f(m) = am^3 + bm^2 + cm + d$.

The quotient, $ax^2 + (b + am)x + (am^2 + bm + c)$, is one degree less than $f(x)$. In general, we can state that if $f(x)$ is divided by $x - a$, then the remainder R is $f(a)$.

The relation between the dividend $f(x)$, the quotient $q(x)$, the divisor $x - a$, and the remainder R, is as follows:

$$f(x) = q(x)\,(x - a) + R \qquad (1)$$

Since Eq. (1) is an identity, it is true for any value of x.

If we put $x = a$, we get the following:

$$f(a) = q(a)\,(a - a) + R$$
$$f(a) = 0 + R \Rightarrow R = f(a)$$

(a) If $f(x)$ is divided by $x + a$, then the remainder R is $f(-a)$.
(b) In the most general case, if $f(x) = P_0x^n + P_1x^{n-1} + P_2x^{n-2} + \cdots + P_n$ is divided by $(x - a)$, then the remainder is $f(\ ) = P_0a^n + P_1a^{n-1} + P_2a^{n-2} + \cdots + P_n$.

Let us find the remainder when $f(x)$ is divided by $(x - a)(x - b)$.

As the divisor $(x - a)(x - b)$ is of second degree, the remainder is to be of the form $(\lambda x + \propto)$, where λ and $\propto$ are independent of x. We may express this remainder in the form $l(x - a) + m(x - b,)$ where l and m are independent of x.

Therefore, $f(x) = q(x)\,(x - a)(x - b) + l(x - a) + m(x - b)$.

Now, this is an identity, and hence it must be true for any value of x.

Substituting $x = a$, we obtain the following:

$$f(a) = m(a-b) \Rightarrow m = \frac{f(a)}{a-b}$$

and substituting $x = b$, we obtain the following:

$$f(b) = 1(b-a) \Rightarrow 1 = \frac{f(b)}{b\ a} = -\frac{f(b)}{a-b}$$

Hence, the remainder is $\dfrac{(x-b)f(a)-(x-a)f(b)}{a-b}$.

FACTOR THEOREM

We know that when the function $f(x)$ is divisible by $x - a$, the remainder $f(a)$ is zero. Hence, we deduce an important theorem known as the factor theorem.

If a rational integral function $f(x)$ of x becomes zero when x is replaced by a, then $f(x)$ is divisible by $x - a$ Hence, $x - a$ is a factor.

We may use the factor theorem in finding factors of $f(x)$, where $f(x)$ is a polynomial of degree higher than two.

SYNTHETIC DIVISION

This is a simple method for finding the quotient and the remainder when $f(x)$ is divided by $x - h$.

If $f(x) = a_0x^n + a_1x^{n-1} + \cdots + a_{n-1}x + a_n$ is divided by $x - h$, the quotient will be, $q(x) = q_0x^{n-1} + q_1x^{n-2} + \cdots + q_{n-1}$ and the remainder will be q_n. Then,

$$a_0x^n + a_1x^{n+1} + \cdots + a_{n-1}x + a_n$$
$$= (x - h)\,(q_0x^{n-1} + q_1x^{n-2} + \cdots + q_{n-1}) + q_n$$

Equating the coefficients of like powers of x,

$q_0 = a_0$, $q_1 - hq_0 = a_1$, $q_2 - hq_1 = a_2$, and lastly, $q_n - hq_{n-1} = a_n$.

$\therefore\ q_0 = a_0$

$q_1 = a_1 + hq_0, \cdots, q_{n-1} = a + hq_{n-2}$ and $q_n = a_n + hq_{n-1}$

Writing only the coefficients, we have the following process:

$$\begin{array}{c|ccccccc}
h & a_0 & a_1 & a_2 & & \cdots & a_{n-1} & a_n \\
 & & hq_0 & hq_1 & hq_2 & \cdots & hq_{n-1} & hq_{n-1} \\
\hline
 & q_0 & q_1 & q_2 & q_3 & \cdots & q_{n-1} & q_n
\end{array}$$

Explanation of the Working

$q_0 = a_0$. Below a_1, we write hq_0. We obtain q_1 by adding a_1 to hq_0. Below a_2, we write hq_1. We obtain q_2 by adding a_2 to hq_1, and so on. The last number line in the third line is the remainder, and the previous numbers are the coefficients in the quotient. A few worked examples will make the process of synthetic division clear.

Worked Examples

1. Determine the quotient and the remainder if $f(x) = x^4 + 3x^2 - 2x^2 - 4x - 3$ is divided by $x - 3$.

3	1	3	−2	−4	−3	
	0	+3	+18	+48	132	
	1	+6	+16	+44	+129	R

The quotient is $x^3 + 6x^2 + 16x + 44$.
The remainder is 129.

2. Determine the quotient and the remainder if $f(x) = 3x^2 + 6x - 12 + x^3$ is divided by $x + 3$.

−3	1	3	6	−12	
	1	−3	0	−18	
	1	0	6	−30	R

The quotient is $x^2 + 0x + 6$, i.e., $x^2 + 6$.
The remainder is −30.

If the given equation is incomplete, then care must be taken to supply zero coefficients for the missing terms.

3. Determine the quotient and the remainder if $x^5 + 7$ is divided by $x + 1$.

−1	1	0	0	0	0	7	
	1	−1	+1	−1	+1	−1	
	1	−1	+1	−1	+1	6	R

The quotient is $x^4 - x^3 + x^2 - x + 1$.
The remainder is 6.

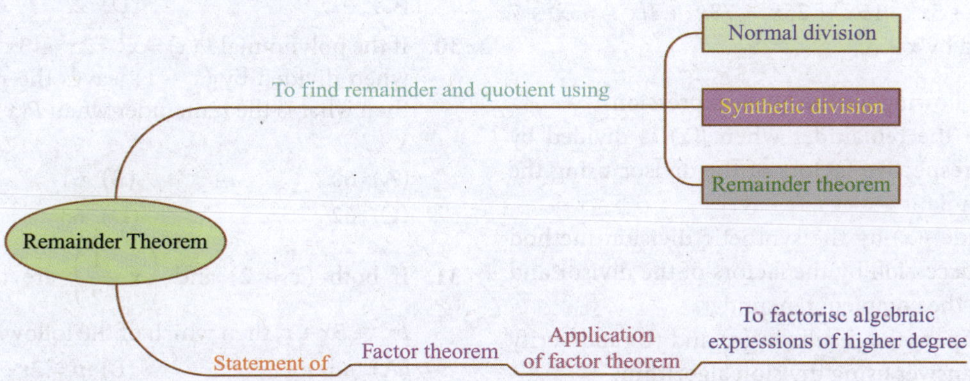

TEST YOURSELF

Level 1

Use synthetic division to find the quotient and remainder and verify by actual division.

1. $x^3 + 6x^2 + 3x + 1 \div x - 2$
2. $2x^3 - 5x + 1 \div x + 1$
3. $x^3 - 11x + 10 \div x + 3$
4. $x^3 - 4x^2 + 3x - 11 \div x - 4$
5. $x^6 - y^6 \div x - y$
6. $x^5 + y^5 \div x + y$
7. $x^5 - 32 \div x - 2$
8. $2x^3 - 13x + 11 \div x - 1$
9. $2x^3 - 4x^2 + 3x - 6 \div x - 2$
10. $1 + x^2 + x^3 + x^4 + x^5 \div x + 1$

Level 2

11. Factorise $2x^2 + 11x + 14$ and deduce the prime factors of 21,114.
12. Determine the value of k for which $x^3 - 6x + k$ may be divisible by $x - 2$, by the synthetic division method.
13. $f(x) \div x^2 - 3x + 2$ leaves the remainder $ax + b$. If $f(1) = 4$ and $f(2) = 2$, determine a and b.

14. If $x - 2$ is a factor of $x^3 - 6x^2 + 11x + k$, determine k.

15. If $x + 1$ divides $5x^3 + 11x^2 + 4x - p$, determine p.

16. If $4x - 1$ divides $4x^3 - 5x^2 - kx + 3$, determine k.

17. Find the remainder when $f(x) = 3x^3 + 6x^2 - 4x - 5$ is divided by $x + 3$.

18. Find the value of l and m if $8x^3 + lx^2 - 27x + m$ is an integral multiple of $2x^2 - x - 6$.

19. Find the remainder and the quotient in the following:
 (a) $3x^2 - 7x^2 + 6x - 2$ is divided by $x - 3$.
 (b) $3x^4 - 20x^2 - 5x - 18$ is divided by $x + 3$.

Level 3

20. Find p and q if $2x^3 + 6px^2 - 2qx - q - 2p + 1$ may contain the factors $2x - 1$ and $x + 3$.

21. Find l and m if $2x^3 - (2l + 1)x^2 + (l + m)x + m$ may be exactly divisible by $2x^2 - x + 3$.

22. Without substituting, find the remainder if $2x^9 - 6x^8 + 5x^7 - 15x^6 + 26x^5 - 78x^4 + 2lx^3 - 63x^2 + 7$ is divided by $x - 3$.

23. Do the following for the given expressions:
 (i) Find the remainder when $f(x)$ is divided by the respective factors of the divisor using the formula.
 (ii) Divide $f(x)$ by the synthetic division method in succession by the factors of the divisor and find the complete remainder.
 (iii) Compare the results in (a) and (b) and verify the answer using division algorithm.
 (a) $x^3 - 8x^2 + 7x + 5 \div (x - 2)(x - 1)$
 (b) $x^3 - 8x^2 + 8x + 5 \div (x + 2)(x + 1)$
 (c) $x^3 + 7x^2 - 8x - 21 \div (x - 2)(x - 3)$
 (d) $2x^3 - 8x^2 + 7x + 20 \div (x + 2)(x + 3)$
 (e) $3x^3 - 4x^2 - 35 \div (x - 2)(x + 3)$

24. $(x + 1)$ and $(x - 1)$ are factors of $x \;\; px^3 + 2x^2 - 3x + q$, then
 (A) $p = -1, q = 1$ (B) $p = 1, q = -2$
 (C) $p = 3, q = -3$ (D) $p = -3, q = 3$

25. When $x^{100} + 1$ is divided by $x^2 - 1$, the remainder is
 (A) $x + 2$ (B) $2x + 1$
 (C) 2 (D) -1

Multiple Choice Questions

26. The remainder when $x^3 - 4x^2 + 3x - 11$ is divided by $x - 4$ is
 (A) 0 (B) 1
 (C) 3 (D) 3

27. Which of the following has $(x - 1)$ as a factor?
 (A) $2x^3 - 13x + 11$
 (B) $2x^3 - 5x + 1$
 (C) $x^2 + x + 1$
 (D) $x^5 + x^4 + x^3 + x^2 + x + 1$

28. What is the value of k for which the given polynomial $x^3 - 6x + k$ is divisible by $x - 2$?
 (A) 1 (B) 2
 (C) 3 (D) 4

29. If the polynomial $ax^3 + 4x^2 + 3x - 4$ and $x^3 - 4x + a$ have the same remainder when divided by $x - 3$, then the value of a is
 (A) 0 (B) 1
 (C) -1 (D) 2

30. If the polynomial $P(x) = x^4 - 2x^3 + 3x^2 - ax + 3a - 7$ when divided by $(x + 1)$ leaves the remainder 19, then what is the remainder when $P(x)$ is divided by $x + 2$?
 (A) 60 (B) 61
 (C) 62 (D) 63

31. If both $(x - 2)$ and $\left(x - \frac{1}{2}\right)$ are the factors of $px^2 + 5x + r$, then which of the following is correct?
 (A) $p = r$ (B) $p = 2r$
 (C) $r = 2p$ (D) None of these

32. When $x^{50} + 1$ is divided by $x^2 - 4$ then remainder will be
 (A) $2x + 2^{50} + 1$ (B) $2^{50} + 1$
 (C) $(2^{50} + 1)x + 5$ (D) none of these

33. If $x^{100} - 1$ is divided by $x^2 - 1$, then remainder will be
 (A) 0 (B) -1
 (C) 1 (D) 2

34. If $(x^2 - 3x + 2)$ is a factor of $(x^4 + ax^3 + 2x^2 + 5x + b)$, then the values of a and b are
 (A) $a = -\frac{26}{7}, b = -\frac{30}{7}$ (B) $a = \frac{26}{7}, b = \frac{30}{7}$

(C) $a=-\frac{26}{7}, b=\frac{30}{7}$ (D) none of these

35. Polynomial $f(x)$ when divided by $x^2 - 3x + 2$ leaves the remainder $3ax + 2b$. If $f(2) = 8$ and $f(1) = 9$, then what are the values of a and b, respectively?

(A) $\frac{1}{3}, 5$ (B) $-\frac{1}{3}, 5$

(C) $5, \frac{1}{3}$ (D) $-5, \frac{1}{3}$

36. If $(x - 2)$ is a factor of $x^4 + ax^3 - bx^2 + 18$, then which of the following is correct?

(A) $2a + b + 17 = 0$ (B) $4a - 2b + 17 = 0$

(C) $2a+b+\frac{17}{2}=0$ (D) $2a-b-\frac{17}{2}=0$

37. $(x + 1)$ is a factor of the polynomial

(A) $x^3 + x^2 - x + 1$

(B) $x^3 + x^2 + x + 1$

(C) $x^4 + x^3 + x^2 + 1$

(D) $4x^4 + 3x^3 + 3x^2 + x + 1$

38. One of the factors of $(25x^2 - 1) + (1 + 5x)^2$ is

(A) $5 + x$ (B) $5 - x$

(C) $5x - 1$ (D) $10x$

39. Find p and q if $2x^3 + 6px^2 - 2qx - q - 2p + 1$ has $(x - 2)$ and $(x - 1)$ as its factors.

(A) $p=-\frac{18}{23}, q=-\frac{1}{23}$ (B) $q=-\frac{2}{23}, p=-\frac{75}{92}$

(C) $p=-\frac{1}{23}, q=-\frac{18}{23}$ (D) $p=-\frac{75}{32}, q=-\frac{1}{23}$

40. When $(x + 1)$ divides $5x^4 - 4x^3 + ax^2 + b$, it leaves remainder 10. Which of the following may be correct?

(A) $a = 10, b = -9$ (B) $a = -10, b = -9$

(C) $a = 10, b = 9$ (D) none of these

41. What is k, if $(4x - 1)$ is an integral multiple of $4x^3 - 5x^2 - kx + 32$?

(A) 27 (B) 127

(C) −127 (D) −27

42. When $P(x) = ax^3 + 8x^2 + 7$ and $Q(x) = 9x^2 + 7x + b$ are divided by $(x - 1)$, both leave the same remainder. Which of the following may be correct?

(A) $a = 4, b = 3$ (B) $a = 3, b = 3$

(C) $a = 3, b = 4$ (D) $a = 4, b = 4$

43. If $ax^3 + bx^2 + x - 6$ has $(x + 2)$ as a factor and leaves a remainder 4 when divided by $(x - 2)$, then what are the values of a and b?

(A) $a = 0, b = 2$ (B) $a = 2, b = 0$

(C) $a = 1, b = 1$ (D) $a = 2, b = 2$

44. R_1 and R_2 are the remainders when the polynomials $x^3 + 2x^2 - 5ax - 7$ and $x^3 + ax^2 - 12x + 6$ are divided by $(x + 1)$ and $(x - 2)$, respectively. If $2R_1 + R_2 = 6$, then what is the value of a?

(A) 1 (B) 0

(C) 2 (D) 4

45. When $P(x) = ax^2 - 3x + 5$ and $Q(x) = 3x^2 + 2a + 5$ are divided by $(x - 1)$ and $(x - 3)$, respectively, both leave the same remainder. What is the value of a?

(A) 30 (B) −28

(C) 28 (D) −30

OLYMPIAD AND NTSE LEVEL EXERCISES

46. What is the remainder when x^{37} is divided by $x^2 - 4x + 3$?

(A) $\left(\frac{3^{37}-1}{3}\right)x+\left(\frac{3-3^{37}}{2}\right)$

(B) $\left(\frac{3^{37}-1}{2}\right)x+\left(\frac{3-3^{37}}{2}\right)$

(C) $\left(\frac{3^{37}-1}{2}\right)x+\left(\frac{3^{37}-3}{2}\right)$

(D) $\left(\frac{3^{37}-1}{2}\right)x-\left(\frac{3-3^{37}}{2}\right)$

47. $5x^2 - 2x + 1$ is a factor of the expression $ax^3 - 19x^2 + bx - 3$. If a and b are integers, then the value of b is

(A) 10 (B) −8

(C) 8 (D) 0

48. $P(x)$ is a polynomial of degree 2. When $P(x)$ is divided by $(x - 1)$ $(x + 2)$ and $(x - 3)$, it leaves 2, 5 and 1 as remainders, respectively. $P(x)$ is

(A) $\frac{1}{10}(x^2-9x-28)$ (B) $\frac{1}{5}(x^2-9x+14)$

(C) $\frac{1}{20}(x^2-9x+28)$ (D) $\frac{1}{10}(x^2-9x+28)$

49. The polynomials $x^2 - x + a$ and $2x^2 + 5x + b$ have a common factor. Which of the following is correct?

(A) $4a^2 + b^2 - 4ab + 35a + 7b = 0$
(B) $4a^2 + b^2 + 4ab + 35a + 7b = 0$
(C) $4a^2 + b^2 - 4ab - 35a + 7b = 0$
(D) $4a^2 + b^2 - 2ab + 35a + 7b = 0$

50. If a third degree polynomial $f(y)$ is divided by $(y + 2)$, the quotient and the remainder obtained are $P(y)$ and 0. Also, if $f(y)$ is divided by $[P(y) + y - 1]$, the quotient and the remainder obtained are $(y - 3)$ and $Q(y)$. Then $Q(y)$ is

(A) $5P(y) - y^2 + 4y$
(B) $5P(y) - y^2 - 4y - 3$
(C) $5P(y) - y^2 + 4y - 3$
(D) $P(y) - y^2 + 4y - 3$

51. When the expression $x^2 - 3x - 3$ is divided by $(x - a)$, the remainder is 1. The value of a is

(A) 1 (B) −1
(C) −2 (D) −3

52. The expression $ax^3 + 3x^2 - 5x + b$ is divisible by the expression $x^2 + 3x + 2$. The values of a and b are

(A) $a = 2; b = -6$ (B) $a = -2; b = -6$
(C) $a = 2/3; b = -22/3$ (D) $a = -6; b = 2$

53. What should be added to the expression $2x^3 - 5x^2 - x + 5$ so that the sum is completely divisible by $2x - 3$?

(A) −1 (B) 1
(C) −2 (D) 2

54. If the expression $12x^2 - 25x + 12$ is divisible by $4x - 3$ then the other factor of the expression is

(A) $3x + 4$ (B) $4x + 3$
(C) $x - 2$ (D) $3x - 4$

55. The expression $px^2 + qx + r$ leaves the remainders 1, 0 and −1 when divided by $(x + 1)$, $(x - 2)$ and $(x - 3)$, respectively. The value of $p + q + r$ is

(A) $\frac{2}{3}$ (B) $\frac{1}{2}$
(C) $\frac{5}{6}$ (D) 0

4 Algebraic Expressions

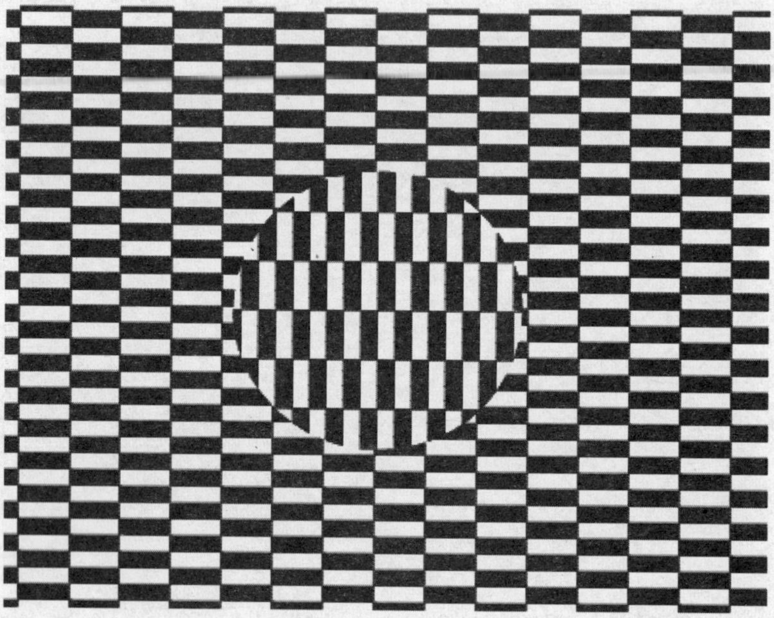

Algebra is the intellectual instrument which has been created for rendering clear the quantitative aspects of the world.

– Alfred North Whitehead

OBJECTIVES

After reading this chapter you should be able to

- define function, integral function, and rational function
- identify homogeneous and symmetric expressions
- understand the properties of symmetric expressions
- factorise cyclic expressions

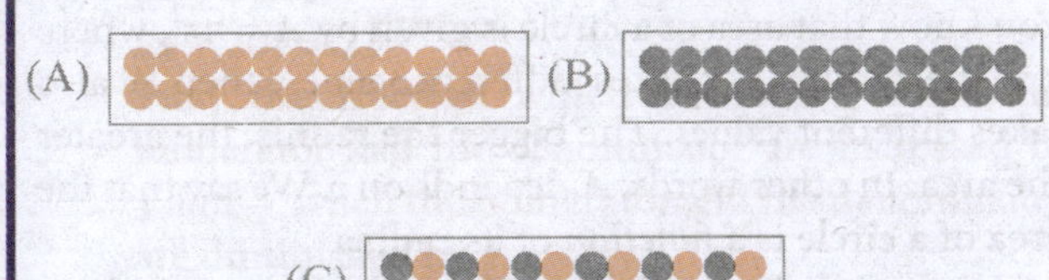

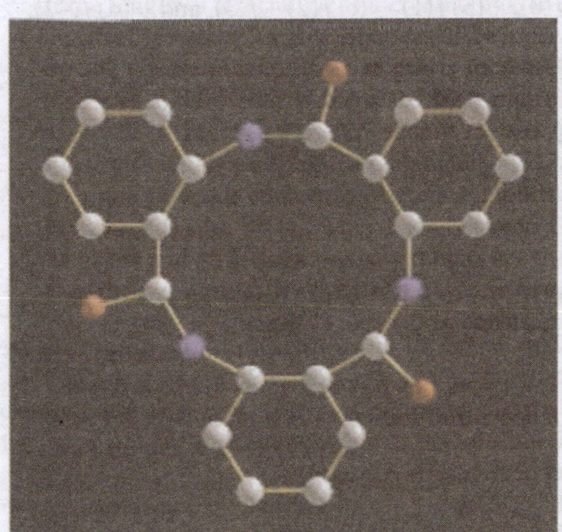

CONSTANTS AND VARIABLES

In algebraic expressions, occasionally, we come across more than one letter. Some of these letters always keep their value unchanged – they represent some numbers. Such letters are called constants, and are usually represented by the first few letters of the alphabet such as *a*, *b*, and *c*. There are others which are given different values, and are called variables. For example, *a*, *b*, and *c* are constants and *x* may have different values in the expression $ax^2 + bx + c$. Thus, when $x = 0$, the expression is equal to *c*, and when $x = 1$, the expression is equal to $a + b + c$.

The variables are usually represented by the last few letters of the alphabet such as *x*, *y*, and *z*. However, this is a matter of convention only, and *a*, *b*, and *c* may also be regarded as variables.

FUNCTION

You know that area of a circle is given by $A = \pi r^2$, where *r* is the radius. If *r* takes different values, then *A* also takes different values. The bigger the radius, the greater the area. In other words, *A* depends on *r*. We say that the area of a circle is a *function* of its radius.

In general, if a variable *y* depends on a variable *x*, then *y* is said to be a function of *x*, written as $y = f(x)$. For example, $y = x^2$, $y = x^3 + 1$ and $y = 7x - 6$ are all function.

Worked Examples

If $f(x)$ stands for $3x^2 + 5x + 7$, determine the value of (a) $f(0)$, (b) $f(1)$, (c) $f(2) + f(3)$, and (d) $f(-2)$.

Solution

We have to replace *x* in $f(x)$ by 0, 1, etc.

(a) $f(0) = 3.0^2 + 5.0 + 7 = 7$

(b) $f(1) = 3.1^2 + 5.1 + 7 = 15$

(c) $f(2) = 3.2^2 + 5.2 + 7 = 12 + 10 + 7 = 29$

$f(3) = 3.3^2 + 5.3 + 7 = 27 + 15 + 7 = 49$

$\therefore f(2) + f(3) = 29 + 49 = 78$

(d) $f(-2) = 3.(-2)^2 + 5(-2) + 7 = 12 - 10 + 7 = 9$

The degree of an expression is the highest degree of its terms. Thus, the degree of $x^7 + 5x^6 + 4x + 2$ is 7 and the degree of $x^5 + 3x^2 + x$ is 5, where *x* is the variable, since the highest degree of the terms is 7 and 5, respectively.

If *x*, *y*, and *z* are all variables, then the expression $xy^3z^2 + 5x^5 + 7x^4 + y^5 + z^3$ is of the sixth degree, and $a^3x^2 + b^3y^2 + c^3z^2$ is of the second degree.

HOMOGENEOUS EXPRESSIONS

An expression is said to be homogeneous when all its terms are of the same degree. For example, $ax + by$ is a homogeneous expression of the first degree in *x* and *y*, $ax^2 + bxy + cy^2$ is a homogeneous expression of the second degree in *x* and *y*, and $ax^3 + by^3 + cz^3 - xyz$ is a homogeneous expression of the third degree.

If a homogeneous expression of the first degree, say $x + y + z$, is multiplied by a homogeneous expression of the second degree, say $ax^2 + by^2 + cz^2$, then the product $ax^3 + by^3 + cz^3 + ax^2y + ax^2z + by^2x + by^2z + cz^2x + cz^2y$ is a homogeneous expression of the third degree ($\because 1 + 2 = 3$). Hence, we have the following law:

The product of two homogeneous expressions of degrees *m* and *n*, respectively, is a homogeneous expression of degrees $m + n$.

SYMMETRIC EXPRESSIONS

An integral function is said to be symmetric with respect to a pair of variables if the expression remains unaltered when the variables are interchanged. Thus, $x^2 + 2xy + y^2$ is symmetric with respect to *x* and *y*, since when *x* and *y* are interchanged, the resulting function, $y^2 + 2yx + x^2$, is the same as the original function. Further, $xy + yz + zx$ is a symmetric expression, since by interchanging *x* and *y*, it becomes $yx + xz + zy$, which is the same as $xy + yz + zx$. Similarly, $xy + yz + zx$ is symmetric with respect to the pair (y, z) and (z, x), and hence $xy + yz + zx$ is symmetric in *x*, *y*, and *z*.

Similarly, $x + y + z$, $x^2 + y^2 + z^2 - xy - yz - zx$, $x^3 + y^3 + z^3 - 3xyz$, $x + y + 1$, and $x^3 + y^3 - x - y$ are symmetric expressions.

Note that $x - y + 1$ is not symmetric in *x* and *y* and $ab + bc - ca$ is not symmetric in *a*, *b*, and *c*.

> A symmetric expression is not necessarily a homogeneous, nor a homogeneous necessarily symmetrical.

For example, $x + y - 1$ is a symmetric function but not a homogenous function. Again, $x^2 + xy$ is a homogeneous function but not a symmetric function.

Properties of Symmetrical Expressions

(a) In a symmetric function, all terms of the same type *must occur* and their coefficients must be *equal.*

Terms which can be obtained from one another by successively interchanging the variables are known as same type terms.

Thus, if x, y, and z are the variables, then y^2z and z^2y are terms of the same type, since the former is changed to the latter by interchanging the pair y, z. Similarly, z^2x, x^2z, y^2x, and x^2y are all terms belonging to the same type as y^2z and z^2y, but x^3, y^3, and z^3 are not terms of the same y^2z type, since the interchanging of the letters x, y, and z will not transform y^2z into x^3, y^3, or z^3.

(b) The sum, difference, product or quotient of two symmetric expressions is a symmetric expression.

If we consider two letters x and y, then, in general, the symmetric homogeneous expressions of the first and second degrees, respectively, are the following:

(i) $k(x + y)$

(ii) $k(x^2 + y^2) + mxy$, where k and m are constants

If we consider three letters x, y, and z, then, in general, the symmetric homogenous expressions of the first and second degrees, respectively, are the following:

(i) $k(x + y + z)$

(ii) $k(x^2 + y^2 + z^2) + m(xy + yz + zx)$, where k and m are any constants.

CYCLIC SYMMETRIC EXPRESSIONS

Factorisation

We shall see how some cyclic expressions can be factorised.

Worked Examples

1. $a^3 + b^3 - c^3 + 3abc$

Solution

We use the identity $a^3 + b^3 + c^3 - 3abc = (a + b + c)(a^2 + b^2 + c^2 - ab - bc - ca)$.

Here, $a = a$, $b = b$, and $c = -c$.

$a^2 + b^2 + (-c)^3 - 3ab(-c)$

$\therefore a^3 + b^3 - c^3 + 3abc = (a + b - c)(a^2 + b^2 + c^2 - ab + bc + ca)$

2. $8a^3 - 27b^3 - c^3 - 18abc$

Solution

Here, a is $2a$, b is $-3b$, and c is $-c$.

$8a^3 - 27b^3 - c^3 - 18abc$

$= (2a)^3 + (-3b)^3 + (-c)^3 - 3(2a)(-3b)(-c)$

$= [(2a) + (-3b) + (-c)][(2a)^2 + (-3b)^2 + (-c)^2 - (2a)(-3b) - (-3b)(-c) - (-c)(2a)]$

$= (2a - 3b - c)(4a^2 + 9b^2 + c^2 + 6ab - 3bc + 2ca)$

3. $8p^3 - 1 - m^3 - 6m$

Solution

$8p^3 - 1 - m^3 - 6m = (2p)^3 + (-1)^3 + (-m)^3 - 3(2p)(-1)(-m)$

Here, a is $2p$, b is -1, and c is $-m$.

$8p^3 - 1 - m^3 - 6pm = (2p - 1 - m)[(2p)^2 + (-1)^2 + (-m)^2 - (2p)(1) - (-1)(-m) - (-m)(2p)]$

$= (2p - 1 - m)(4p^2 + 1 + m^2 - 2p - m + 2pm)$

4. $\Sigma a^2(b - c) = a^2(b - c) + b^2(c - a) + c^2(a - b)$

Solution

$\Sigma a^2(b - c) = a^2b - a^2c + b^2c - b^2a + c^2(a - b)$

$= a^2b - b^2a + b^2c - a^2c + c^2(a - b)$

$= ab(a - b) + c^2(a - b) + c(b^2 - a^2)$

$= (a - b)[c^2 - c(a + b) + ab]$

$= (a - b)[c^2 - ca - cb + ab]$

$= (a - b)[c(c - a) - b(c - a)]$

$= (a - b)(c - a)(c - b)$

$= (a - b)(b - c)(c - a)$

5. $\Sigma x^2y^2(x - y) = x^2y^2(x - y) + y^2z^2(y - z) + z^2x^2(z - x)$

Solution

$\Sigma x^2y^2(x - y) = x^3y^2 - x^2y^3 + y^3z^2 - y^2z^3 + z^2x^2(z - x)$

$= x^3y^2 - y^2z^3 - x^2y^3 + y^3z^2 + z^2x^2(z - x)$

$= y^2(x^3 - z^3) - y^3(x^2 - z^2) + z^2x^2(z - x)$

$= y^2(x - z)(x^2 + xz + z^2) - y^3(x - z)(x + z) - z^2x^2(x - z)$

$= (x - z)[y^2(x^2 + xz + z^2) - y^3(x + z) - z^2x^2]$

$= (x - z)[x^2y^2 + xy^2z + y^2z^2 - y^3x - y^3z - z^2x^2]$

$= (x - z)[(x^2y^2 - y^3x) + (y^2z^2 - z^2x^2) + (xy^2z - y^3z)]$

$= (x - z)[(xy^2(x - y) + z^2(y - x)(y + x) + y^2z(x - y)]$

$= (x - z)(x - y)[xy^2 - z^2(y + x) + y^2z]$

$= (x - z)(x - y)[(xy^2 - z^2x) + y^2z - yz^2]$

$= (x - z)(x - y)[x(y - z)(y + z) + yz(y - z)]$

$= (x - y)(x - z)(y - z)(xy + yz + zx)$

$= (x - y)(y - z)(z - x)(xy + yz + zx)$

6. $\Sigma a^3(b - c)$

Solution

$a^3(b - c) + b^3(c - a) + c^3(a - b)$

$a^3(b - c) + b^3[c - b + b - a] + c^3(a - b)$

$a^3(b - c) + b^3(c - b) + b^3(b - a) + c^3(a - b)$

$(b - c)(a^3 - b^3) + (a - b)(c^3 - b^3)$

$(b - c)(a - b)(a^2 + ab + b^2) + (a - b)(c - b)(c^2 + cb + b^2)$

$(a - b)(b - c)[(a + b)(a - c) + b(a - c)]$

$(a - b)(b - c)(a - c)(a + c + b)$
$= (a - b)(b - c)(c - a)(a + b + c)$

7. Factorise $P + 2abc$ if $P = \Sigma a^2(b + c)$.

Solution

$a^2(b + c) + b^2(c + a) + c^2(a + b) + 2abc$

Arranging the terms in descending power of a, we have the following:

$a^2(b + c) + a(b^2 + c^2 + 2bc) + b^2c + bc^2$
$= a^2(b + c) + a(b + c)^2 + bc(b + c)$
$= (b + c)(a + b)(a + c)$

8. Factorise $P + 3abc$ if $P = \Sigma bc(b + c)$.

Solution

$bc(b + c) + ca(c + a) + ab(a + b) + 3abc$
$= bc(b + c) + abc + ca(c + a) + abc + ab(a + b) + abc$
$= bc(a + b + c) + ca(a + b + c) + ab(a + b + c)$
$= (a + b + c)(ab + bc + ca)$

9. Factorise $(a + b + c)(bc + ca + ab) - abc$.

Solution

$(a + \underline{b + c})(bc + ca + ab) - abc$
$= (a + \underline{b + c})(bc + a(b + c)] - abc$
$= abc + a^2(b + c) + (b + c)bc + (b + c)^2a - abc$
$= (b + c)[a^2 + a(b + c) + bc]$
$= (b + c)(a + b)(a + c)$
$= (a + b)(b + c)(c + a)$

10. Factorise $(a + b + c)^3 - a^3 - b^3 - c^3$.

Solution

$(a + b + c)^3 - a^3 - (b^3 + c^3)$
$(a + b + c - a)[(a + b + c)^2 + (a + b + c)a + a^2] - \{(b + c)[b^2 - bc + c^2]\}$
$= (b + c)[a^2 + b^2 + c^2 + 2ab + 2bc + 2ca + a^2 + ab + ac + a^2 - b^2 + bc - c^2]$
$= (b + c)(3a^2 + 3ab + 3bc + 3ca)$
$= 3(b + c)(a^2 + ab + bc + ca)$
$= 3(b + c)[a(a + b) + c(a + b)]$
$= 3(b + c)(a + b)(a + c)$
$= 3(a + b)(b + c)(c + a)$

11. Factorise $2b^2c^2 + 2c^2a^2 + 2a^2b^2 - a^4 - b^4 - c^4$.

Solution

$2b^2c^2 + 2c^2a^2 + 2a^2b^2 - a^4 - b^4 - c^4$
$= 4b^2c^2 + 2c^2a^2 + 2a^2b^2 - 2b^2c^2 - a^4 - b^4 - c^4$
$= (2bc)^2 - (a^4 + b^4 + c^4 - 2a^2b^2 + 2b^2c^2 - 2a^2c^2)$
$= (2bc)^2 - (a^2 - b^2 - c^2)^2$
$= (2bc + a^2 - b^2 - c^2)(2bc - a^2 + b^2 + c^2)$
$= [a^2 - (b^2 + c^2 - 2bc)][b^2 + c^2 + 2bc - a^2]$
$= [a^2 - (b - c)^2][(b + c)^2 - a^2]$
$= (a + b - c)(a + b + c)(b + c + a)(b + c - a)$
$= (a + b + c)(a - b + c)(a + b - c)(a + b + c)$

12. Show that $a^2(b - c) + b^2(c - a) + c^2(a - b) = -(b - c)(c - a)(a - b)$.

Solution

We have a polynomial expression. We use factor theorem.

Putting $b = c$ in the expression, $a^2(b - c) + b^2(c - a) + c^2(a - b)$, we get the follwoing:

$a^2(0) + c^2(c - a) + c^2(a - c) = a^2 0 + c^2(c - a) - c^2(c - a) = 0$

Therefore, $b - c$ is a factor of the expression.

Similarly, by substituting $c = a$ and $a = b$ successively, we see that $(c - a)$ and $(a - b)$ are also factors. The left-hand side of the expression is homogeneous and symmetric and of the third degree, and it can have only three factors each of the first degree and a constant factor k.

$\therefore a^2(b - c) + b^2(c - a) + c^2(a - b)$
$\equiv k(b - c)(c - a)(a - b)$ (1)

To find k, put convenient values for a, b, and c.

Taking $a = 0$, $b = 1$, and $c = 2$ on both sides of Eq. (1), we have the following:

$0 + 1^2(2 - 0) + 4(0 - 1) = k(1 - 2)(2 - 0)(0 - 1) - 2 = 2k$ or $k = -1$

$\therefore a^2(b - c) + b^2(c - a) + c^2(a - b)$
$= -(b - c)(c - a)(a - b)$

We may also compare the coefficients of any term on both sides and get the value of k. Thus, comparing the coefficients of c^2a, $k = -1$.

13. Factorise $a^3(b - c) + b^3(c - a) + c^3(a - b)$.

Solution

Proceeding as above, we notice that if we put $b = c$ in the given expression, it is easily seen to be zero.

$a^3(0) + c^3(c - a) + c^3(a - c) = 0$

Hence, $b - c$ is a factor.

As the expression is symmetric and homogeneous, $(c - a)$ and $(a - b)$ must also be factors. As the expression is of the fourth degree, symmetric, and homogeneous and the expression $(b - c)(c - a)(a - b)$ formed by factors already obtained is of third degree, there will be one more factor which must be homogeneous, symmetric, and of the first degree, i.e., of the form $k(a + b + c)$, where k is a numerical coefficient.

$\therefore a^3(b - c) + b^3(c - a) + c^3(a - b)$
$\equiv k(a - b)(b - c)(c - a)(a + b + c)$

Since this is an identity, we may compare like terms, say a^3b on both sides, and we get

$k = -1$. Hence,

$a^3(b - c) + b^3(c - a) + c^3(a - b)$
$= -(b - c)(c - a)(a - b)(a + b + c)$

14. Factorise $a^3(b^2 - c^2) + b^3(c^2 - a^2) + c^3(a^2 - b^2)$.

Solution

When we put $b = c$ in the expression, the expression becomes zero.

Therefore, $(b - c)$ is a factor. Similarly, $(c - a)$ and $(a - b)$ are also factors.

As the expression is of the fifth degree symmetric and homogeneous and the expression formed by $(b - c)(c - a)(a - b)$ is of the third degree, there will one more factor of the second degree in a, b, and c which will be symmetric, homogeneous, and of the form $m(a^2 + b^2 + c^2) + n(ab + bc + ca)$. Therefore,

$$\Sigma a^3(b^2 - c^2) \equiv (a - b)(b - c)(c - a)[m(a^2 + b^2 + c^2) + n(ab + bc + ca)]$$

Comparing the terms, say a^3b^2 on both sides, we get the following:

$a^3b^2 = -a^2b \times nab = -na^3b^{2s}$

$\therefore n = -1$

Again comparing the term containing a^4b, we get the following:

$0 = -a^2b\, ma^2 = -ma^4b$

Therefore, $m = 0$ and hence $\Sigma a^3(b^2 - c^2) = -(b - c)(c - a)(a - b)(ab + bc + ca)$.

From these examples it should be clear how to go about factoring polynomials.

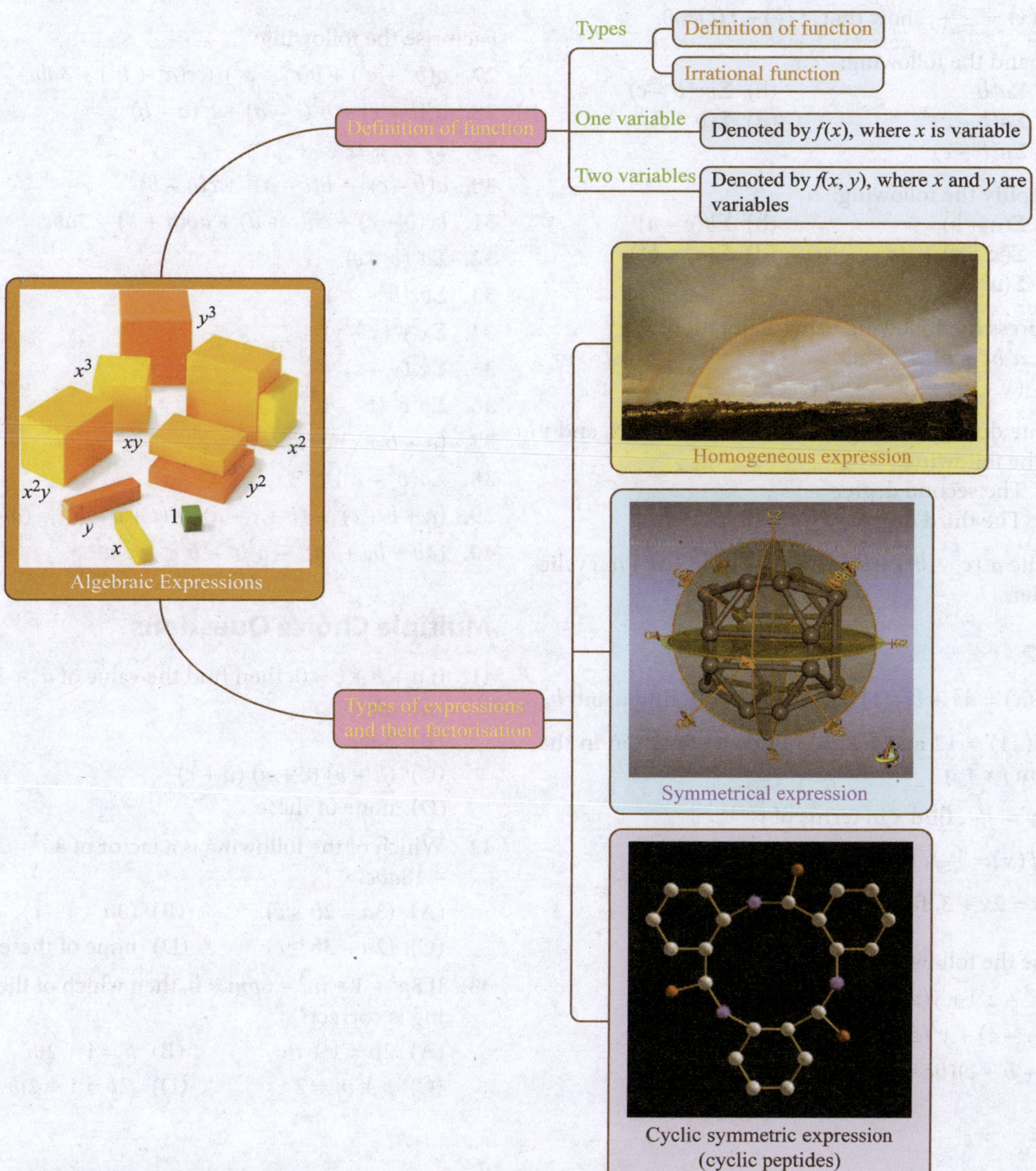

TEST YOURSELF

Level 1

1. If $f(x, y) = 2x^2 - 5xy + 3y^2$, find the following:
   (a) $f(1, -1)$
   (b) $f(0, a)$
2. If $f(x) = 3x^2 + x - 5$, find the following:
   (a) $f(1)$ (b) $f(-1)$
   (c) $f(0)$ (d) $f(a)$
3. If $f(x) = 3x^2 - 4x - 6$, find $f(1) + f(2) + f(3)$.
4. If $f(x) = x^2 + 2x + 1$, find $\frac{f(x+h)-f(x)}{h}$.
5. If $f(x) = \frac{2x}{1-x^2}$, show that $f\left(\frac{1}{t}\right)+f(t)=0$.
6. Expand the following:
   (a) Σa^2b (b) $\Sigma bc(b - c)$
   (c) $\Sigma a(b - c)^2$ (d) $\Sigma(b - c)$
   (e) $\Sigma a(b - c)$
7. Simplify the following:
   (a) $\Sigma(a - b)$ (b) $\Sigma b(c - a)$
   (c) $\Sigma(x + y)$ (d) $\Sigma(a^2 - b^2)$
   (e) $\Sigma(a^n - b^n)$
8. Express the following using Σ notation:
   (a) $a^2bc + b^2ca + c^2ab$
   (b) $(x - y)^2 + (y - z)^2 + (z - x)^2$
9. Write down a homogeneous expression in x, and y of the following:
   (a) The second degree
   (b) The third degree
10. Write $a^2(c^2 - b^2) + b^2(a^2 - c^2) + c^2(b^2 - a^2)$ in cyclic order.

Level 2

11. If $f(x) = ax + b$, $f(1) = 5$ and $f(2) = 7$, find a and b.
12. If $f(-1) = 12$ and $f(2) = 3$ find the function in the form $px + q$.
13. If $y=\frac{2x}{x+1}$, find x in terms of y.
14. If $f(x)=\frac{2x+1}{x-2}$, find $f[f(x)]$.
15. If $y = 2x + 3$, find x in terms of y.

Factorise the following:

16. $x(y^2 - z^2) + y(z^2 - x^2) + z(x^2 - y^2)$
17. $x^2(y - z) + y^2(z - x) + z^2(x - y)$
18. $(a + b + c)(bc + ca + ab) - abc$
19. $xy(x - y) + yz(y - z) + zx(z - x)$
20. $a^2(b + c) + b^2(c + a) + c^2(a + b) + 2abc$
21. $\Sigma a(b - c)^3$
22. If $y=\frac{ax+b}{bx-a}$, find x in terms of y.
23. If $y=\frac{4x+3}{x-3}$, find x in terms of y.

Level 3

24. If $f(a) = 3a^2 + 4a + 1$, show that $\frac{f(a+x)-f(a-x)}{f(x)-f(-x)} = \frac{3a+2}{2}$.
25. If $f(x) = x^2 - x + 1$, find $(x + 1)f(x) - (x - 1)f(x + 1)$.
26. If $m = f(t) = (t - 1)(t - 2)$, find $f(m)$ in terms of t.

Factorise the following:

27. $a(b^2 + c^2) + b(c^2 + a^2) + c(a^2 + b^2) + 3abc$
28. $a^3(b - c) + b^3(c - a) + c^3(a - b)$
29. $2y^2z^2 + 2z^2x^2 + 2x^2y^2 - x^4 - y^4 - z^4$
30. $a(b - c)^3 + b(c - a)^3 + c(a - b)^3$
31. $bc(b + c) + ca(c + a) + ab(a + b) + 3abc$
32. $\Sigma a^4(b - c)$
33. $\Sigma bc(b^3 - c^3)$
34. $\Sigma x^2y^2(x - y)$
35. $\Sigma x^2(y - z)^3$
36. $\Sigma b^2c^2 (b^2 - c^2)$
37. $(a + b + c)^5 - a^5 - b^5 - c^5$
38. $\Sigma a(b^2 - c^2)$
39. $(a + b + c)^3 - (b + c - a)^3 - (c + a - b)^3 - (a + b - c)^3$
40. $(ab + bc + ca)^2 - a^2b^2 - b^2c^2 - c^2a^2$

Multiple Choice Questions

41. If $a + b + c = 0$, then find the value of $a^3 + b^3 + c^3$.
    (A) abc
    (B) $3abc$
    (C) $(a + b) (c + a) (b + c)$
    (D) none of these
42. Which of the following is a factor of $8a^3 - 27b^3 - c^3 - 18abc$?
    (A) $(3a - 2b - c)$ (B) $(3a - b - c)$
    (C) $(2a - 3b - c)$ (D) none of these
43. If $8p^3 - 1 - m^3 - 6pm = 0$, then which of the following is correct?
    (A) $2p = 1 + m$ (B) $p = 1 + 2m$
    (C) $p + m = 2$ (D) $2p = 1 + 2m$

44. The simplest form of the expression $x^2y^2(x-y)+y^2z^2(y-z)+z^2x^2(z-x)$ is

(A) $(x+y)(y+z)(z+x)(xy+yz+zx)$
(B) $(x-y)(y-z)(z-x)(xy-yz-zx)$
(C) $(x-y)(y+z)(z+x)(xy+yz+zx)$
(D) $-(x-y)(y-z)(z-x)(xy+yz+zx)$

45. If a, b and c are all distinct positive, then find the value of $a+b+c$ if $a^3(b-c)+b^3(c-a)+c^3(a-b)=0$.

(A) 0 (B) 1
(C) 2 (D) 3

46. If $f(x, y) = 5x^2 - 10xy + 12y^2$ and $f(a, 3) = f(2, a)$, then which of the following is correct?

(A) $7a^2 + 10a - 88 = 0$ (b) $a^2 + 6a + 8 = 0$
(C) $7a^2 + 10a + 88 = 0$ (d) none of these

47. If $f(x) = 4x^4 + x^2 + 5b$ and $f(1) = f(-1)$, then what is the value of b?

(A) $b \neq 0$
(B) $b \in R$
(C) no value of b is possible
(D) none of these

48. If $f(x) = 2x^2 - 3x + 2$, then the value of $f(0) + f(1) + f(2)$ is

(A) 7 (B) 8
(C) 10 (D) 11

49. If $f(x) = ax^2 + bx + c$ such that $f(-1)$, $f(0)$ and $f(1)$ are in A.P., then which of the following is correct?

(A) $c = b$ (B) $b \neq 0$
(C) $c \neq 0$ (D) $a = 0$

50. Let $f(x) = px^2 + qx + r$, where p, q and r are in A.P. If $f(1) = 3$, then which of the following is correct?

(A) $q = 1$ (B) $p + q = 1$
(C) $p - r = 1$ (D) None of these

51. Which of the following expressions is homogeneous in terms of x and y?

(A) $x^2 + xy + 3$ (B) $x^2 + 2xy + 7y^2$
(C) $x^2 - y^2 + 3x$ (D) $x^2 + y^2 - xy^2$

52. If $f(x)=\dfrac{x+1}{x-1}$, then $f(f(x))=$

(A) x (B) $\dfrac{x-1}{x+1}$
(C) $\dfrac{x^2+1}{x^2-1}$ (D) $\dfrac{1}{x}$

53. If $f(x) = x^2 - x + 1$, then what is the value of $(x + 1) f(x) - (x - 1) f(x + 1)$?

(A) 0 (B) 1
(C) 2 (D) 3

54. If $f(x) = 3x^2 + 4x + 1$, then find $f(x) + f(-x)$.

(A) 2 (B) $6x^2$
(C) $8x$ (D) $2 + 6x^2$

55. If $f(x)=\dfrac{x+1}{x-1}$, then find the value of $f(f(5))$.

(A) 5 (B) 3
(C) 2 (D) 0

56. If $f(x) = x^2 + 4x + 4$, then what is the value of $\dfrac{f(x+a)-f(x)}{(a+2)^2}$?

(A) 1 (B) $(a + 2)$
(C) $(a + 2)^2$ (D) none of these

57. If $a + b + c = 0$ then the value of $(a + b - c)^3 + (a - b + c)^3 + (-a + b + c)^3$ is

(A) $3(a + b - c)(a - b + c)(-a + b + c)$
(B) $(a + b - c)(a - b + c)(-a + b + c)$
(C) 0
(D) none of these

58. $\{a(b-c)\}^3+\{b(c-a)\}^3+\{c(a-b)\}^3$ is equal to

(A) $a^3 + b^3 + c^3$
(B) $3(ab - ac)(bc - ba)(ca - bc)$
(C) 0
(D) none of these

59. Which of the following is equivalent to $2b^2c^2 + 2c^2a^2 + 2a^2b^2 - a^4 - b^4 - c^4$?

(A) $(a + b + c)(a - b + c)(a + b - c)(-a + b + c)$
(B) $(a + b + c)(a - b + c)(a + b - c)$
(C) $(a - b - c)(-a + b + c)(a + b - c)$
(D) none of these

60. If $f(x) = 2x^2 - x + a + b$ such that $f(0) = 1$ and $f(1) = -2$, then the values of a and b are

(A) $a = 0, b = 1$
(B) $a = 4, b = -7$
(C) $a = 2, b = -3$
(D) No possible value

OLYMPIAD AND NTSE LEVEL EXERCISES

61. If $a^2 + b^2 + c^2 - ab - bc - ca = 0$, then

(A) $a + b + c = 0$ (B) $a = b = c$

(C) $a = b = -c$ (D) $a = -b = c$

62. Factorize $a^2 - (b - c)^2$.

(A) $(a + b - c)(a - b + c)$

(B) $(a - b - c)(a - b + c)$

(C) $(a - b - c)(a - b - c)$

(D) $(a + b - c)(a - b + c)$

63. The factorized form of $(x - 1)^2 + 20(x - 1) + 36$ is

(A) $(x + 17)(x + 1)$ (B) $(x - 17)(x + 1)$

(C) $(x - 16)(x + 17)$ (D) $(x + 17)(x - 1)$

64. The factorized form of $x^5 + x^4 - x - 1$ is

(A) $(x + 1)^2 (x + 1)^2 (x - 1)$

(B) $(x^2 + 1)(x^2 + 1)^2 (x - 1)$

(C) $(x^2 + 1)(x + 1)^2 (x - 1)$

(D) $(x^2 + 1)^2 (x - 1)$

65. Factorize $x^6 + 27$.

(A) $(x^2 + 3)(x^2 + 3x + 3)(x^2 - 3x + 3)$

(B) $(x^2 - 3)(x^2 + 3x + 3)(x^2 + 3x + 3)$

(C) $(x^2 + 3)(x^4 - 3x^2 + 9)^2$

(D) $(x^2 + 3)(x^2 - 3x + 3)(x^2 - 3x - 3)$

66. The simplified form of $\frac{a}{x+a} + \frac{b}{x-a} - \frac{c}{x^2-a^2}$ is

(A) $\frac{a(x-a)+b(x+a)-c}{x^2-a^2}$

(B) $\frac{b(x-a)+a(x+a)-c}{(x-a)^2}$

(C) $\frac{b(x-a)+a(x+a)-c}{x^2-a^2}$

(D) $\frac{a(x-a)+b(x+a)+c}{x^2-a^2}$

67. Simplify

$$\frac{1}{(x+1)(x+2)} + \frac{1}{(x+2)(x+3)} + \frac{1}{(x+3)(x+1)}$$

(A) $\frac{1}{(x+1)(x+3)}$ (B) $\frac{2}{(x+1)(x+3)}$

(C) $\frac{3}{(x+2)(x+3)}$ (D) $\frac{3}{(x+1)(x+3)}$

68. Factorize $4(a - b)^2 - 9(a + b)^2$.

(A) $(5a + b)(a + 5b)$ (B) $-(5a + b)(a + 5b)$

(C) $-(5a + b)(a + 5b)$ (D) $-(5a - b)(a - 5b)$

69. The factorized form of $x^2 - x + \frac{1}{4}$ is

(A) $\left(x - \frac{1}{2}\right)\left(x - \frac{1}{2}\right)$ (B) $(2x - 2)\left(x - \frac{1}{2}\right)$

(C) $(2x - 2)(2x - 2)$ (D) $(x - 2)(x - 2)$

70. The factorized form of $a^2 + ac + bc - b^2$ is

(A) $(a - b)(a + b + c)$ (B) $(a + b)(a - b - c)$

(C) $(a + b)(a - b + c)$ (D) $(a - b)(a - b - c)$

GAME TIME

Here are some puzzles for you. Try and solve them in minimum amount of time.

Illustrated Puzzles

1. A frog slips into a well 10 m deep. Each day it climbs 3 m and while resting at night it slips back 2 m. How many days should the frog climb to come out of the well?

Solution

Each day the frog climbs 3 m. Since it slips down by 2 m at night, effective climb is 1 m per day.

In 7 days the frog climbs effectively 7 m and on the next day (eighth day) it climbs 3 m. Once it climbs 10 m, it comes out of the well and there is no question of rest and slipping down.

Thus, the number of days required for the frog to climb and come out of the well is 8 days.

2. A student working in a fish biology laboratory finds that the head of a fish is 10 cm long and the tail is as long as its head plus half its body length. The body is as long as its head plus tail. What is the length of the fish?

Solution

Let the lengths of head, body, and tail be h, b and t, respectively.

It is given that $h = 10$ cm. Also,

$t = h + \frac{1}{2}b$; $t = 10 + \frac{1}{2}b$ (1)

And

$b = h + t$;

$b = 10 + t$ (2)

From Eqs. (1) and (2), we get the following:

$b = 10 + 10 + \frac{1}{2}b$ Þ $b = 40$ cm

Substituting in Eq. (2) we get the following:

$t = 10 + \frac{1}{2}40 = 30$ cm

Length of the fish $= h + b + t$

$= (10 + 40 + 30)$ cm $= 80$ cm.

3. The manager of a game park wanted to check the number of deer and starks. He did this by counting the heads and legs. If the number of legs counted is 130 and the heads 45, how many deer and starks were there?

Solution

Each deer has four legs and stark has two legs. Let the number of deer be x and the number of starks be y.

$x + y = 45$ (1)

$4x + 2y = 130$ (2)

From Eqs. (1) and (2), we get the following:

$4(45 - y) + 2y = 130$

$180 - 4y + 2y = 130;$

$50 = 2y$

$y = 25 \Rightarrow x = 20$

Therefore, the number of deer = 20 and the number of starks = 25.

4. The sum of the ages of x, y and z is 90 years. Ten years ago, their ages were in the ratio 1 : 2 : 3. What is the present age of y?

Solution

$x + y + z = 90$ (1)

$(x - 10) : (y - 10) : (z - 10) : 1 : 2 : 3$

Also,

$(y - 10) = 2(x - 10)$

$(z - 10) = 3(x - 10)$

After adding,

$y + z - 20 = 5x - 50$

$y + z = 5x - 30$ (2)

Substituting from Eq. (2) in Eq. (1), we get the following:

$x + 5x - 30 = 90$

$\therefore x = 20$

$y - 10 = 2(x - 10) = 2(20 - 10) \Rightarrow y = 30$

$x + y + z = 90$

$\therefore 20 + 30 + z = 90 \Rightarrow z = 40$

5. A, B, and C are three boxes. The contents of the boxes are invisible, and no box is empty. Boxes A, B, and C are labelled as containing red balls, blue balls, red and blue balls, respectively. All the boxes are wrongly labelled. By picking a ball from any one of the boxes, identify the contents of each box.

Solution

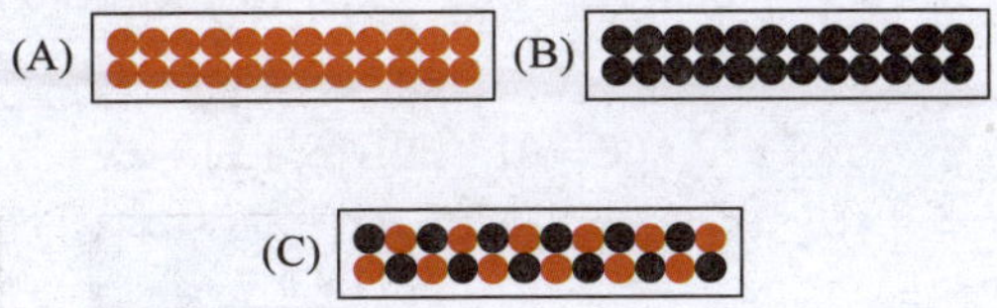

Let us pick a ball from box C.

Assume it to be red. Since each box is wrongly labelled, box C cannot contain both red and blue balls. Hence, box C should have only red balls.

Let us now move on to box B, labelled blue. Since it is wrongly labelled, it should contain either red balls or red and blue balls. Since C contains red balls, box B should contain red and blue balls.

Obviously, box A should have only blue balls.

Challenging Puzzles

1. A ship leaves a port at 1 a.m. at a speed of X km/h. Another ship leaves the same port at 2 a.m. in the same direction. What is the speed in km/h of the second ship if both meet at 5.30 a.m.?
2. In climbing a mountain, a man took half as long again to climb the second third as he did to climb the first third, and a quarter as long again for climbing the last third as for the second third. He altogether took 5 hours 50 minutes. Find the time he spent in minutes on the first third of the journey.
3. 20% of 25% a number is 100. What is the number?
4. A man stated in his will that one-third of his estate would go to his wife and two-third to his child to be born if it were a son. It was further stated that if a daughter were born, two-third of the estate would go to the wife and one-third to the daughter. After the man's death, a boy and a girl were born as twins. How should the estate be divided to fulfil the requirements of the will?
5. A hare pursued by a hound is 50 of her own leaps ahead of him. While the hare takes four leaps, the hound takes three leaps. In one leap the hare goes 1.75 m and the hound 2.75 m. In how many leaps will the hound overtake the hare?

5 Simple and Simultaneous Equations

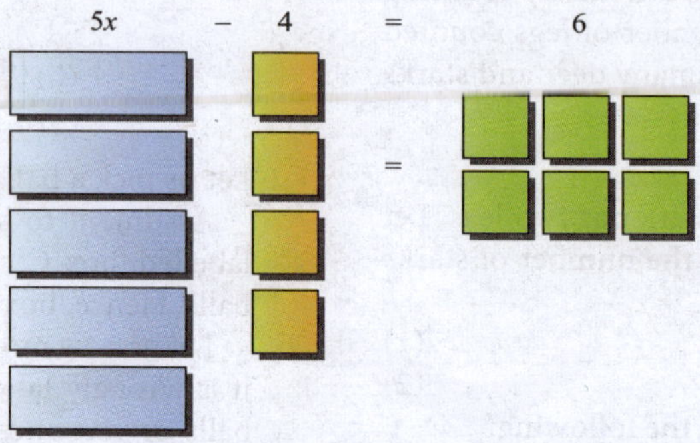

> *A diagram is worth a thousand proofs.*
>
> – Dr. Carl E. Linderholm

OBJECTIVES

After reading this chapter you should be able to

- translate a given statement to algebraic form (equation form)
- solve simple equations
- solve simultaneous equations in two and three variables by different methods:
  - Method of comparison
  - Method of elimination
  - Rule of cross-multiplication
- verify the conditions for two linear equations to have only one, many, and no solution

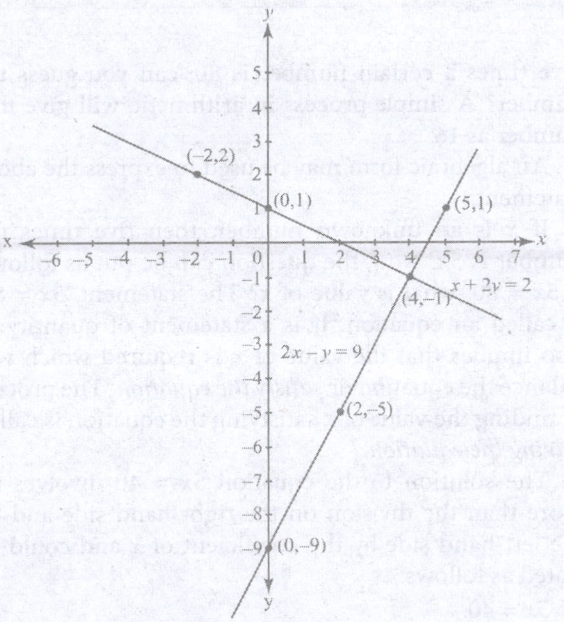

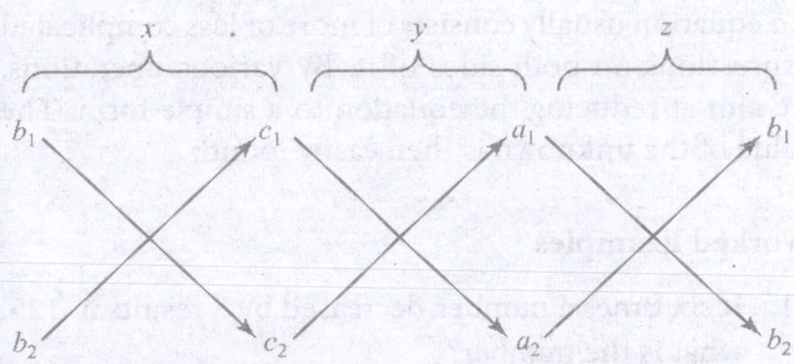

Five times a certain number is 80; can you guess the number? A simple process in arithmetic will give that number as 16.

An algebraic form may be used to express the above statement.

If x is an unknown number, then five times the number is $5x$. Now, the question can be put as follows. If $5x = 80$, what is value of x? The statement '$5x = 80$' is called an equation. It is a statement of quantity; it also implies that the value of x is required which will balance the equation or *satisfy the equation*. The process of finding the value of x satisfying the equation is called *solving the equation*.

The solution to the equation $5x = 40$ involves no more than the division on the right-hand side and on the left-hand side by the coefficient of x and could be stated as follows:

$$5x = 40$$
$$\frac{5x}{5} = \frac{40}{5}$$
$$x = 8$$

The solution to an equation is rarely as simple as this. An equation usually consists of more or less complicated expressions on both sides of it. By various operations, we aim at reducing the equation to a simple form. The value of the unknown is then easily found.

Worked Examples

1. If six times a number decreased by 7 results in 125, what is the number?

Solution

If x is the required number, then six times the number reduced by 7 will be $6x - 7$ which is given as 125.

$\therefore \quad 6x - 7 = 125$

$6x = 125 + 7 = 132$ (adding 7 to both sides)

Dividing both sides by 6, we have the following:

$$\frac{6x}{6} = \frac{132}{6} = 22$$
$$x = 22$$

2. The sum of three consecutive odd numbers is 81. What are the numbers?

Solution

Let the numbers be $2x + 1$, $2x + 3$, and $2x + 5$.

Their sum is $2x + 1 + 2x + 3 + 2x + 5 = 6x + 9$.

This is given to be 81

$\therefore \quad 6x + 9 = 81$

$6x + 9 - 9 = 81 - 9$

$6x = 72$

Dividing both sides by 6, we get $x = 12$.

Therefore, the numbers are $2 \times 12 + 1$, $2 \times 12 + 3$, and $2 \times 12 + 5$; i.e., 25, 27, and 29.

The simple equations considered so far contain only one unknown quantity. Cases may occur in which it will be required to find the value of more than one quantity.

For example, suppose the cost of a table and a chair is ₹ 1550. The cost of the table and the chair cannot be found unless the cost of one of them is given.

If the cost of the chair is ₹ 550, then the cost of the table is ₹ 1000.

If the cost of the chair is ₹ 600, then the cost of the table is ₹ 950.

Like this we can find many pairs of values.

If the cost of the table is taken as x and the cost of the chair is taken as y, then the total cost is $x + y$.

According to the statement, $x + y = 1550$.

Now, it is clear that unless you know the value of one of the items, the value of the other item cannot be found. Many number of pairs of values for x and y can be found if some values of x or y are not given.

Now, consider the equation $x + y = 8$. The values of x and y cannot be found, but when the value of x is known, the value of y can be found.

If $x = 5$, then $x + y = 8$ and $y = 8 - 5 = 3$.

In other words, if two equations are given as above, independent values x and y can be determined.

Now, consider the equation $x + y = 9$ where x and y are two unknown quantities; we can give an unlimited number of pairs of values which will satisfy the equation as follows:

x	1	3	4	5	6	7	8	−1	−2	etc.
y	8	6	5	4	3	2	1	10	11	etc.

Again if we take another equation $x - y = 3$, we can find an unlimited number of pairs of values which will satisfy this equation as follows:

x	2	−4	6	−3	−5	−7	−9	−10	etc.
y	−1	−7	3	−6	−8	−10	−12	−12	etc.

From these patterns, we find that although an unlimited pair of values satisfy the equation $x + y = 9$ and $x - y = 3$, *when considered separately yet there is one value of x (i.e., 6) and one value of y (i.e., 3) for which they are simultaneously true.*

Two or more equations which are all satisfied by the same values of the unknown quantities involved in them are called *simultaneous equations*. They are said to be *simple* or of the first *degree* if each unknown quantity occurs only in the first power and the product of the unknown quantities does not occur.

METHODS OF FINDING SOLUTION

Method of Comparison

Worked Examples

1. Solve the following:

$$2x + 3y = 8 \quad (1)$$
$$x + 2y = 5 \quad (2)$$

Solution

From (1) we have the following:

$$x = \frac{8-3y}{2} \quad (3)$$

From (2) we have the following:

$$x = 5 - 2y$$
$$\therefore \frac{8-3y}{2} = 5-2y$$
$$8 - 3y = 10 - 4y$$
$$4y - 3y = 10 - 8$$
$$y = 2$$

Using this in (3), we get the following:

$$x = \frac{8-3.2}{2} = \frac{8-6}{2} = \frac{2}{2} = 1$$

Thus, the solution is $x = 1$ and $y = 2$.

2. Solve the following:

$$\frac{7+x}{5} - \frac{2x-y}{4} = 3y-5 \quad (1)$$
$$\frac{5y-8}{2} + \frac{4x-3}{6} = 18-5x \quad (2)$$

Solution

Multiplying both sides of (1) by 20, we get the following:

$$4(7 + x) - 5(2x - y) = 20(3y - 5)$$
$$4x - 10x = -5y + 60y - 128$$
$$x = \frac{55y-128}{-6} \quad (3)$$

Multiplying both sides of (2) by 6, we get the following:

$$3(5y - 8) + (4x - 3) = 6(18 - 5x)$$
$$4x + 30x = -15y + 24 + 108$$
$$x = \frac{132-1}{34} \quad (4)$$

Therefore, from (3) and (4), we have the following:

$$\frac{55y-128}{-6} = \frac{132-15y}{34}$$

$$17(55y - 128) + 3(132 - 15y) = 0$$
$$890y = 1780$$
$$\therefore y = 2$$

Using this in (3), we have the following:

$$x = \frac{55 \times 2 - 128}{-6} = 3$$

Thus, the solution is $x = 3$ and $y = 2$.

Method of Substitution

It consists in finding the value of one unknown quantity in terms of the other from one equation and substituting this value in the other equation.

Worked Examples

1. Solve the following:

$$15x - 8y = 29 \quad (1)$$
$$17x + 12y = 75 \quad (2)$$

Solution

From (1) we get the following:

$$y = \frac{15x-29}{8} \quad (3)$$

Using this in (2), we have the following:

$$17x + 12\left(\frac{15x-29}{8}\right) = 75$$
$$34x + 3(15x - 29) = 150$$
$$34x + 45x = 150 + 87$$
$$79x = 237$$
$$x = 3$$

Using the value $x = 3$ in (3), we have the following:

$$y = \frac{15 \times 3 - 29}{8} = 2$$

Thus, the solution is $x = 3$ and $y = 2$.

Method of Equalising Coefficients

It consists in equalising the coefficients of one and the same unknown in both the equations (if they are not already same) and then by subtraction or addition getting an equation, containing only the other unknown quantity.

This method is also known as the method of elimination.

Worked Examples

1. Solve the following:

$$13x + 4y = 51 \quad (1)$$
$$17x - 4y = 39 \quad (2)$$

Solution

As the coefficients of y are equal but opposite in signs, we simply add the equations and get the following:

$$30x = 90$$
$$x = 3$$

Substituting $x = 3$ in (1), we get the following:

$$4y = 51 - 13 \times 3 = 12$$
$$y = 3$$

Thus, the solution is $x = 3$ and $y = 3$.

2. Solve the following:

$$\frac{x+y-3}{2} = \frac{x+2y-4}{3} = \frac{3x+y}{11}$$

Solution

Here, the first equation is written as follows:

$$\frac{x+y-3}{2} = \frac{3x+y}{11}$$
$$11(x + y - 3) = 2(3x + y)$$
$$5x + 9y = 33 \quad (1)$$

The second equation is written as follows:

$$\frac{x+2y-4}{3} = \frac{3x+y}{11}$$
$$11(x + 2y - 4) = 3(3x + y)$$
$$2x + 19y = 44 \quad (2)$$

Multiplying (1) by 2, we get the following:

$$10x + 18y = 66 \quad (3)$$

Multiplying (2) by 5, we get the following:

$$10x + 95y = 220 \quad (4)$$

Subtracting (3) from (4), we have the following:

$$77y = 154$$
$$y = 2$$

Using this value in (2), we get the following:

$$2x = 44 - 19 \times 2 = 6$$
$$x = 3$$

Thus, $x = 3$ and $y = 2$.

3. A bookseller sells books at a cost of ₹ 25 each. After selling a certain number of books at this price, the bookseller sells the remainder at ₹ 20 each and the total receipts are ₹ 1100. If the numbers sold at the two prices were reversed, the bookseller would have received ₹ 1150. How many books were there in all and how many were originally sold at ₹ 25?

Solution

Let x be the number originally sold at ₹ 25 and y be the number sold at ₹ 20.

The amounts received from these were ₹ $25x$ and ₹ $20y$ and their total value was ₹ 1100.

$$\therefore \quad 25x + 20y = 1100 \quad (1)$$

When the numbers are reversed, the bookseller receives ₹ $20x$ and ₹ $25y$ and their total value is ₹ 1150.

$$\therefore \quad 20x + 25y = 1150 \quad (2)$$

Note: In the equations (1) and (2), the coefficients of x and y in one equation are interchanged in the other. You can, in such cases, proceed as follows:

(1) + (2) gives the following:

$$45x + 45y = 2250$$
$$45(x + y) = 2250$$
$$x + y = 50 \quad (3)$$

(1) – (2) gives the following:

$$5x - 5y = -50$$
$$5(x - y) = -50$$
$$x - y = -10 \quad (4)$$

(3) + (4) gives the following:

$$2x = 40$$
$$x = 20$$

(3) – (4) gives the following:

$$2y = 60$$
$$y = 30$$

The total number of books $(x + y)$ was 50. The number originally sold (x) at ₹ 25 is 20.

Note: This method is more commonly used than the first two. In each of the three methods explained above, it is essential to eliminate (i.e., get rid of) one of the two unknown quantities and thus get an equation involving a single unknown quantity and solve it.

Rule of Cross-multiplication

If the equations are given as follows:

$$a_1x + b_1y + c_1z = 0 \quad (1)$$
$$a_2x + b_2y + c_2z = 0 \quad (2)$$

Then, $\dfrac{x}{b_1c_2 - b_2c_1} = \dfrac{y}{c_1a_2 - c_2a_1} = \dfrac{z}{a_1b_2 - a_2b_1}$.

(The students may carefully note that a_1, a_2, b_1, b_2, and c_1, c_2 are different coefficients.)

Proof

Multiplying (1) by c_2 and (2) by c_1, we get the following:

$$c_2a_1x + c_2b_1y + c_1c_2z = 0 \quad (3)$$
$$c_1a_2x + c_1b_2y + c_1c_2z = 0 \quad (4)$$

Subtracting (3) from (4), we have the following:

$$(c_1a_2 - c_2a_1)x + (b_2c_1 - b_1c_2)y = 0$$

or

$$(c_1a_2 - c_2a_1)x = (b_1c_2 - b_2c_1)y$$

That is,

$$\frac{x}{b_1c_2 - b_2c_1} = \frac{y}{c_1a_2 - c_2a_1} \quad (5)$$

Similarly, multiplying (1) by b_2 and (2) by b_1 and subtracting, we get the following:

$$(a_1b_2 - a_2b_1)x + (b_2c_1 - b_1c_2)z = 0$$

$$\frac{x}{b_1c_2 - b_2c_1} = \frac{z}{a_1b_2 - a_2b_1} \quad (6)$$

From (5) and (6), we get the following:

$$\frac{x}{b_1c_2 - b_2c_1} = \frac{y}{c_1a_2 - c_2a_1} = \frac{z}{a_1b_2 - a_2b_1} \quad (7)$$

Note: This important result is known as the rule of Cross-multiplication.

The method of writing down the denominators of *x*, *y*, and *z* in terms of the coefficients *x*, *y*, and *z* is as follows:

(a) Write down the coefficients of *x*, *y*, and *z* in order beginning and ending with those of *y* as shown below:

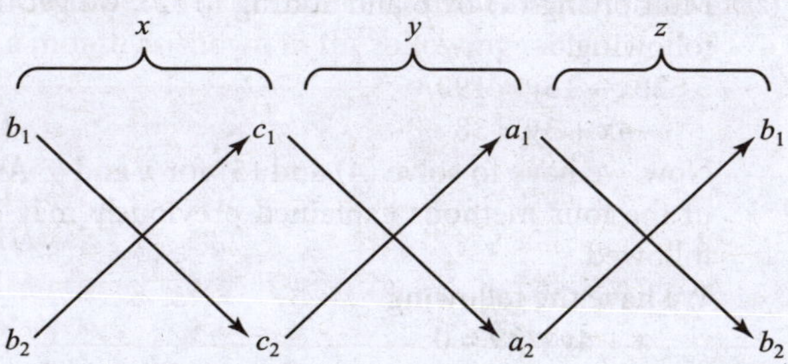

(b) Now, multiply the coefficient across in the way indicated by the arrows and subtract the product formed in ascending form, from the products formed in descending form. Then the expressions $b_1c_2 - b_2c_1$, $c_1a_2 - c_2a_1$, and $a_1b_2 - a_2b_1$ are the denominators of *x*, *y*, and *z*.

Note: In the above equations if we put $z = 1$, we get the following:

$$\frac{x}{b_1c_2 - b_2c_1} = \frac{y}{c_1a_2 - c_2a_1} = \frac{1}{a_1b_2 - a_2b_1}$$

It gives the solution to the equations as follows:

$$a_1x + b_1y + c_1 = 0$$
$$a_2x + b_2y + c_2 = 0$$

Conditions for two linear equations to have only one, many, or no solution at all.

The equations $a_1x + b_1y = c_1$ and $a_2x + b_2y = c_2$ have the following:

(a) Exactly one solution (unique solution) if $a_1/a_2 \neq b_1/b_2$ [i.e., $a_1b_2 - a_2b_1 \neq 0$]

(b) Infinitely many solutions if $a_1/a_2 = b_1/b_2 = c_1/c_2$ [i.e., $a_1b_2 - a_2b_1 = 0$]
Note: In this case, the graphs of the two linear equations are coincident.

(c) No solution if $a_1/a_2 = b_1/b_2 \neq c_1/c_2$ [i.e., $a_1b_2 - a_2b_1 = 0$]
Note: In this case, the graphs of the two linear equations are parallel.

When the equations have unique or infinitely many solutions, they are said to be consistent. If there is no solution, they are inconsistent.

Worked Examples

1. Find the value of *p* if the equations $6x + py = 5$ and $3x + 4y = 2$ have the following:
   (a) Unique solution
   (b) No solution

Solution

(a) The given system of equations will have a unique solution if

$$\frac{a_1}{a_2} \neq \frac{b_1}{b_2}$$

That is,

$$\frac{6}{3} \neq \frac{p}{4}$$
$$3p \neq 24$$
$$p \neq 8$$

(b) The given system of equations will have no solution if

$$\frac{a_1}{a_2} = \frac{b_1}{b_2} \neq \frac{c_1}{c_2}$$

That is,

$$\frac{6}{3} = \frac{p}{4} \neq \frac{5}{2}$$

That is, if $p = 8$.

Worked Examples

1. Solve the following:

$$4x + 5y + 9 = 0$$
$$3x + 4y + 8 = 0$$

Solution

By the rule of cross–multiplication, we get the following:

$$\frac{x}{5\times8-4\times9}=\frac{y}{9\times3-8\times4}=\frac{1}{4\times4-3\times5}$$

$$\frac{x}{4}=\frac{y}{-5}=\frac{1}{1}$$

The solution is $x = 4$ and $y = -5$.

2. Solve the following:

$$8x - 7y = 19$$
$$10x - 9y = 23$$

Solution

$$8x - 7y - 19 = 0$$
$$10x - 9y - 23 = 0$$

By rule of cross-multiplication, we have the following:

$$\frac{x}{(-7)(-23)-(-9)(-19)}=\frac{y}{(-19)10-(-23)8}=\frac{1}{8(-9)-10(-7)}$$

$$\frac{x}{161-171}=\frac{y}{-190+184}=\frac{1}{-72+70}$$

The solution is $x = 5$ and $y = 3$.

Note: Be very careful about the signs of different coefficients.

Simultaneous Equations Involving Three Unknown Quantities

Case I

When two of the equations contain no independent terms.

Worked Examples

1. Solve the following:

$$x + 2y + 3z = 0 \quad (1)$$
$$3x - 6y - 7z = 0 \quad (2)$$
$$2x + 3y - z = 17 \quad (3)$$

Solution

By the rule of cross-multiplication from (1) and (2), we have the following:

$$\frac{x}{-14+18}=\frac{y}{9+7}=\frac{z}{-6-6}$$

$$\frac{x}{4}=\frac{y}{16}=\frac{z}{-12}=k \text{ (say)}$$

$x = 4k$, $y = 16k$, and $z = -12k$

Substituting these values of x, y, and z in (3), we get the following:

$$k(8 + 48 + 12) = 17$$

or

$$68k = 17$$

$$\therefore \quad k=\frac{1}{4}$$

$$\therefore \quad x=4\times\frac{1}{4}=1$$

$$y=16\times\frac{1}{4}=4$$

$$z=-12\times\frac{1}{4}=-3$$

Case II

When the equations are of the following forms:

$$a_1x + b_1y + c_1z = d_1$$
$$a_2x + b_2y + c_2z = d_2$$
$$a_3x + b_3y + c_3z = d_3$$

Worked Examples

1. Solve the following:

$$3x + 2y - z = 20 \quad (1)$$
$$2x + 3y + 6z = 70 \quad (2)$$
$$x - y + 6z = 41 \quad (3)$$

Solution

In such cases, we have to first eliminate one of the unknown quantities. Convenience will be the deciding factors in choosing the particular quantity to be got rid of. Here, we eliminate z first; subtracting (3) from (2), we get the following:

$$x + 4y = 29 \quad (4)$$

Multiplying (1) by 6 and adding to (2), we get the following:

$$20x + 15y = 190$$
$$4x + 3y = 38 \quad (5)$$

Now, we have to solve (4) and (5) for x and y. Any of the four methods explained previously may be followed.

We have the following:

$$x + 4y - 29 = 0$$
$$4x + 3y - 38 = 0$$

By the rule of cross-multiplication, we get the following:

$$\frac{x}{-152+87}=\frac{y}{-116+38}=\frac{1}{3-16}$$

$$\frac{x}{-65}=\frac{y}{-78}=\frac{1}{-13}$$

$$x=\frac{-65}{13}=5 \text{ and } y=\frac{-78}{-13}=6$$

Using these values in (1), we get z as follows:

$3.5 + 2.6 - z = 20$

Therefore, the solution $x = 5$, $y = 6$, and $z = 7$.

Graphical Solution to Simultaneous Linear Equations

Before proceeding to work out particular examples, the students' attention is called to the following principles:

(a) The coordinates of every point on a graph must satisfy the equation of the graph.

(b) The point of intersection of graphs must have coordinates so that they satisfy the equations of both graphs, and hence the solutions to two simultaneous equations are obtained by finding the coordinates of the point of intersection of the graphs of the equations.

To solve simultaneous linear equations:

(i) Draw the graph of both the equations on the same scale of representation.

(ii) Find the coordinates of the point of intersection of the graphs.

(iii) If the point of intersection has the coordinates (h, k), then the solutions are $x = h$ and $y = k$.

Worked Examples

1. Solve graphically the system of the following equations:

$2x - y = 9$ and $x + 2y = 2$.

Solution

$2x - y = 9$

$y = 2x - 9$

x	0	5	2
$y = 2x - 9$	−9	1	−5

$x + 2y = 2$

$2y = 2 - x$

$y = \dfrac{2-x}{2}$

x	0	4	−2
$y = \dfrac{2-x}{2}$	1	−1	2

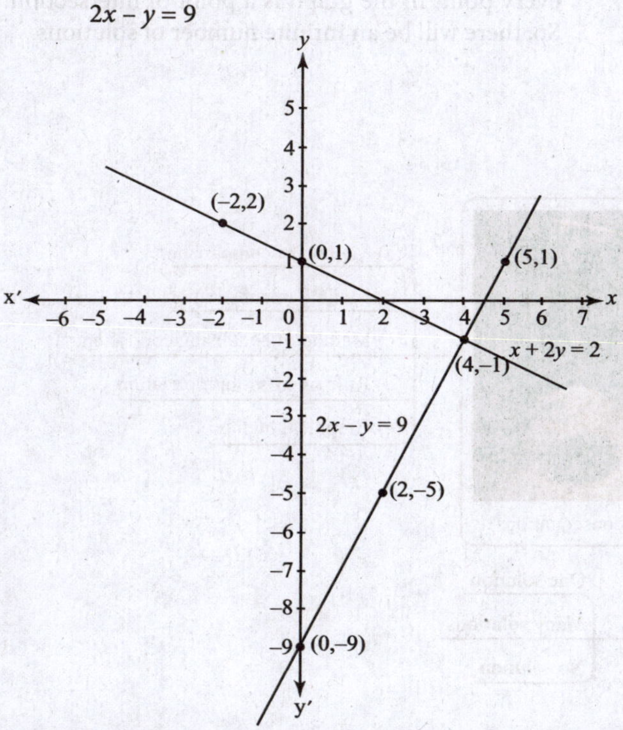

The two graphs intersect at (4, −1).

Therefore, the solution is $x = 4$ and $y = -1$

2. Solve graphically the following: $y = 4x + 3$ and $y = 4x$.

Solution

$y = 4x + 3$

x	0	2	−2
$y = 4x + 3$	3	11	−5

$y = 4x$

x	0	2	−3
$y = 4x$	0	8	−12

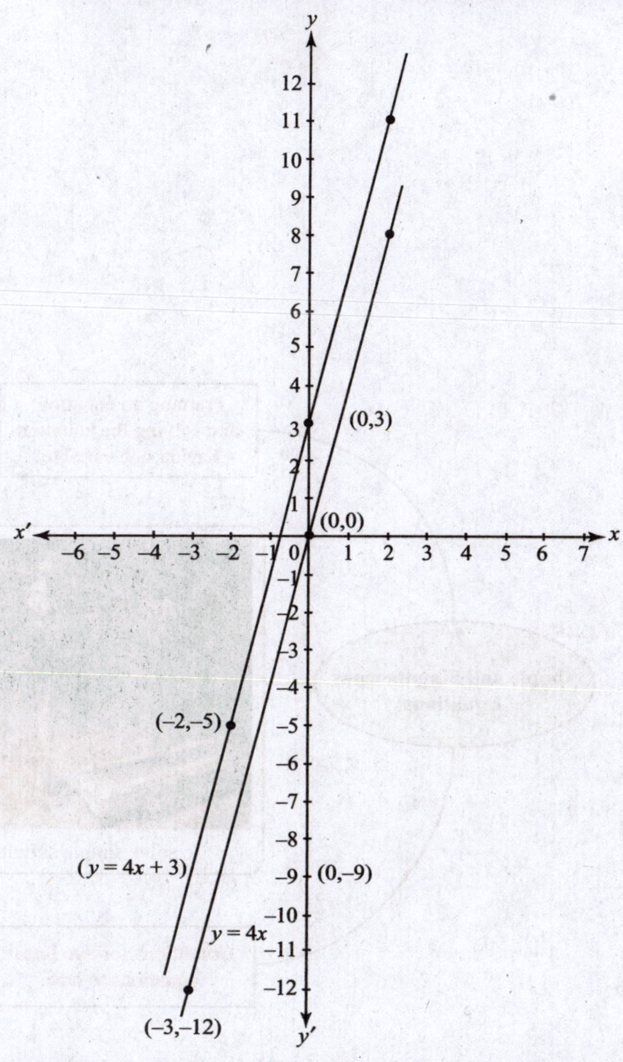

The graphs of the two equations are parallel and hence they do not intersect. So, there will be no solution.

3. Show graphically that the systems of equations $3x - y = 2$ and $9x - 3y = 6$ have an infinite number of solutions.

Solution

$3x - y = 2$

x	1	2	−2
$y = 3x - 2$	1	4	−8

$9x - 3y = 6$

x	0	3	4
$y = \dfrac{9x-6}{3}$	−2	7	10

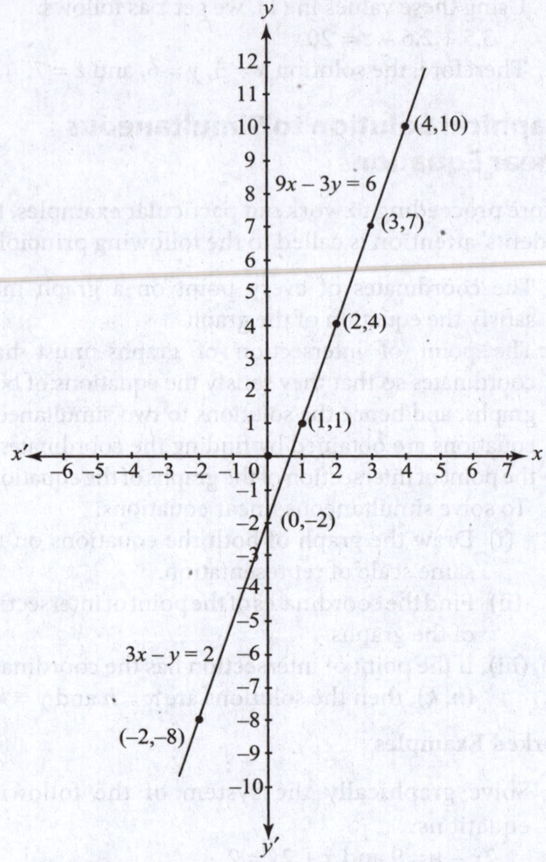

The graphs of both the equations are the same, and every point in the graph is a point of intersection. So, there will be an infinite number of solutions.

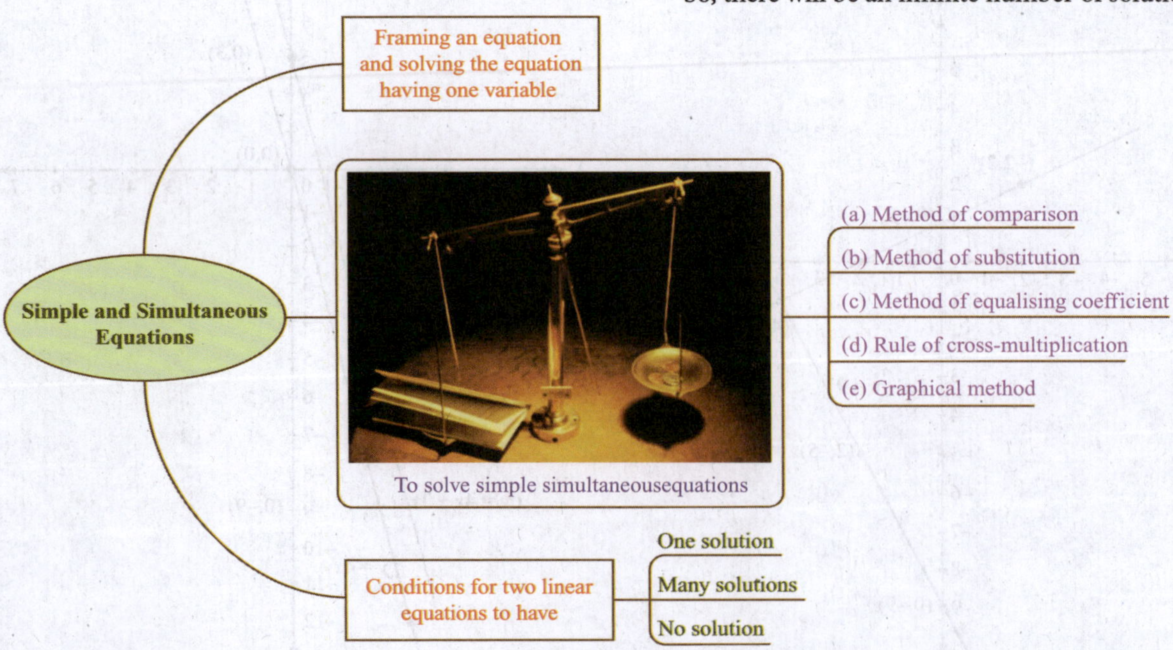

To solve simple simultaneousequations

TEST YOURSELF

Level 1

Write the following statements in the algebraic form:

1. Five times a number is equal to six times another number.
2. Thrice a number is 2 less than twice another number.
3. 10 years later my age will be twice what it was 5 years ago.
4. I choose a number, multiply it by 3, then subtract 7, and I get 8.
5. I think of a number, double it, and then add 6. The result is same as multiplying the original number by 3.
6. The sum of ages of Arun and Babu is 40 and their difference is 10.

Level 2

Solve the following:

7. $x + y = 7$
   $2x + 3y = 18$
8. $x + 4y = 14$
   $7x - 3y = 5$
9. $2x - y = 12$
   $3x + 5y = 31$
10. $4x + 3y = 17$
    $5x - 2y = 4$
11. $7x + 8y = 22$
    $6x + 5y = 17$
12. $2x + y = 18$
    $x - 3y = -33$
13. $x = 2y + 6$
    $y = 2x - 3$
14. $2x + 3y = 32$
    $11y - 9x = 3$

Level 3

15. $\frac{x+y}{2}+\frac{3x-5y}{4}=2$ and $\frac{x}{14}+\frac{y}{18}=1$
16. $\frac{4x+5y}{36}+\frac{x+y-1}{9}$ and $\frac{2x-3y}{3}+3y=10$

Solve the following:

17. $\frac{a}{x}+\frac{b}{y}=m$
    $\frac{b}{x}+\frac{a}{y}=n$
18. $ax + by = c$
    $cx + dy = e$
19. $x + ay = b$
    $ax - by = c$
20. $lx + my = a$
    $mx + ly = a$
21. $\frac{1}{2}x+\frac{2}{3}y+\frac{1}{3}=0$
    $\frac{1}{2}y+\frac{2}{3}x-\frac{5}{12}=0$

Solve the following using the rule of cross-multiplication:

22. $6x - 7y + 25 = 0$
    $5x - 9y + 1 = 0$
23. $2x + 3y - 13 = 0$
    $4x - 9y + 19 = 0$
24. $(a + b)x + (a - b)y = 2a$
    $(a - b)x + (a + b)y = 2b$
25. $3x - 5y = 20$
    $7x + 2y = 17$
26. $6x + 5y = 11$
    $9x + 10y = 21$
27. Find the fraction which becomes $\frac{2}{3}$ when the numerator and the denominator are increased by 1 and $\frac{1}{2}$ when the numerator and the denominator are diminished by 1.
28. A certain number of two digits is four times the sum of its digits. If 9 is added to the number, the digits in the number are reversed. Find the number.
29. Six years ago a man was three times as old as his son. In 6 years, he will be twice as old as his son. Find their present ages.
30. If the length of a rectangle is increased by 8 metres and the breadth by 3 metres, its area will be increased by 200 square metres. If its length is increased by 3 metres and breadth is increased by 8 metres, its area will be increased by 255 square metres. Find the length and breadth of the rectangle.
31. In a pen (a small enclosure) there are rabbits and pheasants (game birds). They have between them 35 heads and 98 feet. How many rabbits are there?
32. Determine graphically whether the system of equations $x - 2y = 2$ and $4x - 2y = 5$ is consistent or inconsistent.
33. Draw the graphs of $2y = 4x - 6$ and $2x = y + 3$ and determine whether this system of linear equations has a unique solution or not.
34. Solve graphically the following: $2(x + y) = 1$ and $2y - 3x = 6$.

Multiple Choice Questions

35. Which is the correct expression for the statement "thrice a number is 10 less than twice the another number"?

(A) $3x = 2y - 10$ (B) $2x = 3y - 10$
(C) $3x - 10 = 2y$ (D) none of these

36. How many solutions are possible for the given set of equations?

$$2x - 3y + 1 = 0$$
$$3x + 2y + 2 = 0$$

(A) one (B) more than one
(C) infinite (D) none

37. Number of solutions of the given set of equation is

$$x + 2y + 3 = 0$$
$$3x + 6y + 9 = 0$$

(A) only one (B) infinite
(C) exactly two (D) none

38. Number of solutions for the given set of equations is

$$2x - 3y + 2 = 0$$
$$6x + 9y + 5 = 0$$

(A) only one (B) infinite
(C) no solution (D) exactly two

39. $x = 1$ and $y = 3$ is a solution of the given set of equation.

$$2x + ky = 2$$
$$x - ny = 2$$

What are the values of n and k?

(A) $n = 3, k = 2$ (B) $n = \frac{-1}{3}, k = \frac{1}{3}$
(C) $n = \frac{1}{3}, k = \frac{-1}{3}$ (D) none of these

40. If $x \in A$ and $y \in A$ such that A ={1, 2, 3, 4}, then find the solution set for $x + y = 6$.

(A) {(1, 5), (2, 4), (3, 3), (4, 2)}
(B) {(1, 5), (2, 4), (3, 3)}
(C) {(2, 4), (3, 3), (4, 2)}
(D) {(3, 3), (4, 2)}

41. What is the solution of the given set of equations?

$$0.2u + 0.3v = 1.3$$
$$0.4u + 0.5v = 2.3$$

(A) $u = 2, v = 3$ (B) $u = -2, v = 3$
(C) $u = 2, v = 2$ (D) $u = 3, v = 3$

42. Find r and s for the following set of equations.

$$\sqrt{2}r + \sqrt{3}s = 0$$
$$\sqrt{3}r - \sqrt{8}s = 0$$

(A) $r = 0, s = 1$ (B) $r = 0, s = 0$
(C) $r = 1, s = 1$ (D) $r = 2, s = 0$

43. Find the values of x and y for the following set of equations.

$$\frac{x}{2} + \frac{2y}{3} + 1 = 0$$
$$x - \frac{y}{3} - 3 = 0$$

(A) $x = 2, y = -2$ (B) $x = 3, y = -3$
(C) $x = -2, y = -3$ (D) $x = 2, y = -3$

44. For what values of p and q, the given set of equations will have infinite solutions?

$$2x + 3y = 7$$
$$(p - q)x + (p + q)y = 3p + q - 2$$

(A) $p = 5, q = 1$ (B) $p = -5, q = -1$
(C) $p = 5, q = -1$ (D) $p = -1, q = 5$

45. By solving the following set of equations find u and v (if $u \neq 0$ and $v \neq 0$).

$$6u + 3v = 6uv$$
$$2u + 4v = 5uv$$

(A) $u = 1, v = 2$ (B) $u = 2, v = 1$
(C) $u = -1, v = 2$ (D) $u = -1, v = -2$

46. Solve for x and y.

$$\frac{2}{x+y} + \frac{3}{x-y} = 2$$
$$\frac{5}{x+y} + \frac{10}{x-y} = \frac{35}{6}$$

(A) $x = 5, y = -1$ (B) $x = \frac{5}{2}, y = \frac{-1}{2}$
(C) $x = \frac{1}{2}, y = \frac{-1}{2}$ (D) none of these

47. Find the values of u and v for the following equations.

$$\frac{1}{3u+v} + \frac{1}{3u-v} = \frac{3}{4}$$
$$\frac{1}{2(3u+v)} - \frac{1}{2(3u-v)} = \frac{-1}{8}$$

(A) $u = 1, v = -1$ (B) $u = -1, v = -1$
(C) $u = 1, v = 1$ (D) $u = -1, v = 1$

48. Solve for x and y.

$ax + by = c$

$bx + ay = 1 + c$

(A) $x = \frac{bc - ac + b}{b^2 - a^2}, y = \frac{bc - ac - a}{b^2 - a^2}$

(B) $x = \frac{b}{b^2 - a^2}, y = \frac{a}{b^2 - a^2}$

(C) $x = a, y = b$

(D) $x = b, y = \frac{bc - ac + b}{b^2 - a^2}$

49. Solve for x and y.

$ax + by = a - b$

$bx - ay = a + b$

(A) $x = 1, y = 1$ (B) $x = -1, y = 1$

(C) $x = 1, y = -1$ (D) $x = -1, y = -1$

50. Solve for x and y.

$x - y = 0.9$

$\frac{11}{2(x + y)} = 1$

(A) $x = \frac{-16}{5}, y = \frac{23}{10}$ (B) $x = \frac{16}{5}, y = \frac{23}{10}$

(C) $x = \frac{23}{10}, y = \frac{16}{5}$ (D) none of these

51. Solve for x and y.

$x + y = a + b$

$ax - by = a^2 - b^2$

(A) $x = b, y = -b$ (B) $x = b, y = a$

(C) $x = -a, y = -b$ (D) $x = a, y = b$

52. Find x and y.

$\frac{x}{a} + \frac{y}{b} = a + b$

$\frac{x}{a^2} + \frac{y}{b^2} = 2$

(A) $x = a, y = b$ (B) $x = a^2, y = b$

(C) $x = a, y = b^2$ (D) $x = a^2, y = b^2$

53. Find the value of p for which the given set of equations will have a unique solution.

$px + 2y = 5$

$3x + y = 1$

(A) $p = 6$

(B) all values

(C) any real values except 6

(D) none of these

54. Find the value of t for which no solution of given set of equations is possible.

$3x + y = 1$

$(2t - 1)x + (t - 1)y = 2t + 1$

(A) $t = 2$ (B) $t = -2$

(C) $t = 1/2$ (D) $t = -1/2$

55. Find the value of k for which given set of equations will have infinitely many solutions.

$kx + 3y = k - 3$

$12x + kx = k$

(A) $k = 6$ (B) $k = -6$

(C) $k \neq 6$ (D) not possible

56. Find the value of k for which the following two lines are coincident.

$2x + 3y = 4$

$(k + 2)x + 6y = 3k + 2$

(A) $k \neq 2$ (B) $k = 2$

(C) $k = -2$ (D) $k \neq -2$

57. Find the values of α and β for which the given set of equations will have infinitely solutions.

$2x + 3y = 7$

$2\alpha x + (\alpha + \beta)y = 28$

(A) $\alpha = -4, \beta = -8$ (B) $\alpha = -4, \beta = 8$

(C) $\alpha = 4, \beta = -8$ (D) $\alpha = 4, \beta = 8$

58. Find the value of m for which the given set of equations has no solution.

$3x + y = 1$

$(2m - 1)x + (m - 1)y = 2m + 1$

(A) $m = 1$ (B) $m = -1$

(C) $m = 2$ (D) $m = -2$

59. Find the values of x and y (if $p \neq 0$ and $q \neq 0$).

$p(x + y) + q(x - y) = p^2 - pq + q^2$

$p(x + y) - q(x - y) = p^2 + pq + q^2$

(A) $x = \frac{q^2}{2p}, y = p - \frac{q^2}{2p}$ (B) $x = \frac{q^2}{2p}, y = \frac{p^2}{2q}$

(C) $x = \frac{p}{q}, y = \frac{p}{q^2}$ (D) $x = \frac{q^2}{2p}, y = p + \frac{q^2}{2p}$

60. There is a certain two-digit number sum of whose digits is 11. If 45 is added to the number, the digits in the number are reversed. The original nu[mber is]

(A) 83 (B) 38

(C) 56 (D) 65

61. *ABCD* (taken in order) is a cyclic quadrilateral in which $\angle A = (x + y + 10)°$, $\angle B = (y + 20)°$, $\angle C = (x + y - 30)°$, and $\angle D = (x + y)°$. Find the values of x and y.

(A) $x = 40°, y = 60°$ (B) $x = 60°, y = 40°$
(C) $x = 20°, y = 80°$ (D) none of these

62. Ratio of present ages of a father and his son is 3:1. After 14 years, the ratio of their ages will be 2:1. What are the present ages of the father and his son, respectively?

(A) 42, 14 (B) 30, 10
(C) 45, 15 (D) 60, 20

63. If the length of a rectangle is increased by 3 metres and the breadth is increased by 2 metres, its area will increase by 67 sq. metres. If its length is decreased by 5 metres and breadth is increased by 3 meters, its area will decrease by 9 sq. metres. Find the area of the rectangle (in sq. metres).

(A) 150 (B) 153
(C) 162 (D) 154

64. In a triangle *PQR*, $\angle R$ is greater than $\angle Q$ by 9°. If $\angle P = \alpha°$, $\angle Q = (3\alpha - 2)°$, and $\angle R = \beta°$, then find all the angles of the triangle.

(A) 30°, 60°, 90° (B) 45°, 45°, 90°
(C) 25°, 73°, 82° (D) none of these

OLYMPIAD AND NTSE LEVEL EXERCISES

65. Which of the following is correct?

Column I	Column II
(i) $3u - 5v = 20$, $6u - 10v = 40$	(p) only one solution is possible
(ii) $x - 3y = 3$, $3x - 9y = 2$	(q) infinitely many solution
(iii) $3x - 5y = 25$, $7x + 2y = 15$	(r) No solution

(A) (i) – q, (ii) – r, (iii) – p
(B) (i) – q, (ii) – p, (iii) – r
(C) (i) – q, (ii) – q, (iii) – p
(D) (i) – r, (ii) – r, (iii) – q

66. If x men can do a piece of work in y days, in how many days will x men do the same work?

(A) $\frac{xz}{y}$ (B) $\frac{xy}{z}$
(C) $\frac{yz}{x}$ (D) xyz

67. Find the values of x and y if

$$\frac{5}{y} - \frac{2}{x} = \frac{7}{6} \text{ and } \frac{36}{x} - \frac{24}{y} = 1$$

(A) $x = 4, y = 3$ (B) $x = -4, y = +3$
(C) $x = -4, y = -3$ (D) $x = 4, y = -3$

68. A fraction becomes 2 when 9 is added to its numerator and 1 when 2 is subtracted from its denominator. Then the fraction is

(A) 6/8 (B) 7/9
(C) 5/7 (D) 4/7

69. What should be the value of p if $3x + 2y = 8$ and $6x + 4y = p$ have infinitely many solutions?

(A) 3 (B) 16
(C) 5 (D) 6

70. What should be value of m in the pair of equations to have unique solution?

$4x + my + 9 = 0$
$3x + 4y + 8 = 0$

(A) $m \neq 16$ (B) $m \neq 15$
(C) $m \neq \frac{16}{3}$ (D) $m \neq \frac{15}{4}$

71. If the sum of two numbers is 640 and their difference is 280, then the numbers are

(A) 140, 500 (B) 180, 460
(C) 130, 510 (D) 150, 490

72. A certain number of two digits is four times the sum of the digits. If 9 is added to the number, the digits in the number are reversed. Find the number.

(A) 13 (B) 15
(C) 12 (D) 14

73. Six years ago a man was three times as old as his son. In 6 years' time, he will be twice as old as his son. Find their present ages.

(A) 30, 15 (B) 40, 20
(C) 42, 18 (D) 41, 19

74. The total salary of 15 men and 8 women is ₹3050. The difference of salaries of 5 women and 3 men is ₹50. Find the sum of the salaries of 3 men and 3 women

(A) ₹ 900 (B) ₹ 850
(C) ₹ 950 (D) ₹ 1000

6 Quadratic Equations

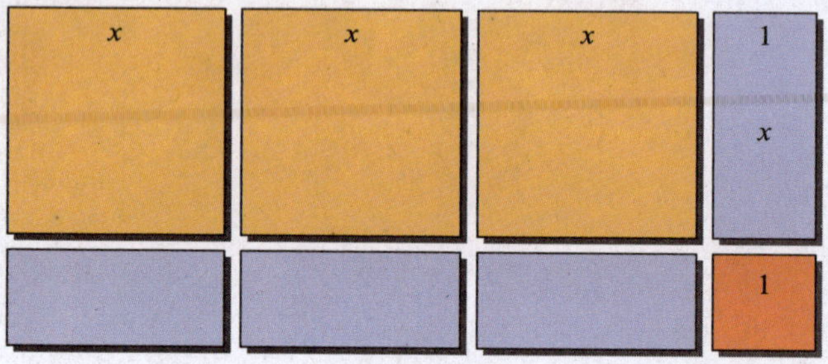

$= 3x^2 + 4x + 1$

> Algebra is generous; she often gives more than is asked of her.
>
> – D'Alembert

OBJECTIVES

After reading this chapter you should be able to

- learn about zeros of polynomials
- define quadratic equations
- learn about two types of quadratic equations
- solve quadratic equations by one of the four methods:
- Factorisation
    - ➢ Completing the square
    - ➢ Using formulae
    - ➢ Graphing
- solve equations reducible to quadratics
- solve equations of the following types:
    - ➢ $(x + a)(x + b)(x + c)(x + d) = R$
    - ➢ Reciprocal equations
    - ➢ Equations involving radicals
- understand theory of quadratic equations
- understand the nature of the roots of quadratic equations
- solve problems leading to quadratic equations given in the verbal form

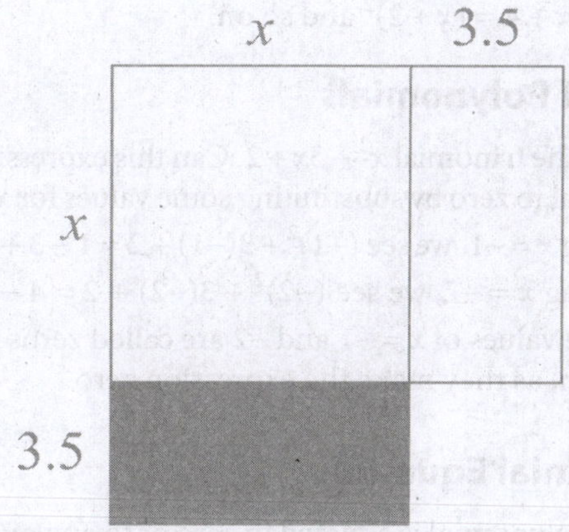

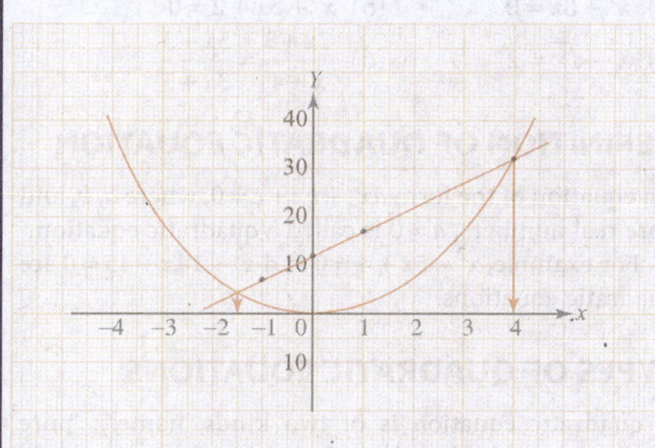

INTRODUCTION

In the first volume you have learnt about product of algebraic expressions and factorisation of algebraic expressions, whose samples are given below.

Product

(a) $x \times x = x^2$
(b) $x \times (x+a) = x^2 + ax$
(c) $(x+a)(x+b) = x^2 + (a+b)x + ab$
(d) $(x+a)^2 = x^2 + 2ax + a^2$ and so on

Factorisation

(a) $x^2 = x \times x$
(b) $x^2 + 8x = x(x+8)$
(c) $x^2 + 3x + 2 = (x+2)(x+1)$
(d) $x^2 + 4x + 4 = (x+2)^2$ and so on.

Zeros of Polynomials

Consider the trinomial $x^2 + 3x + 2$. Can this expression be made equal to zero by substituting some values for x? Yes.

Putting $x = -1$, we see $(-1)^2 + 3(-1) + 2 = 1 - 3 + 2 = 0$, and putting $x = -2$, we see $(-2)^2 + 3(-2) + 2 = 4 - 6 + 2 = 0$. These values of $x = -1$ and -2 are called zeros of the expression, as they make the expression zero.

Polynomial Equation

When a polynomial is equated to zero or to any number, or in fact to any polynomial, it is called an equation.

In general, we can write a polynomial equation as $f(x) = 0$.

Examples

(a) $x^2 + 8x = 0$ (b) $x^2 + 3x + 2 = 0$

(c) $x - \sqrt{x} = 2$ (d) $\dfrac{2x+3}{x+4} = \dfrac{5x+3}{3x+5}$

DEFINITION OF QUADRATIC EQUATION

An equation of the form $ax^2 + bx + c = 0$, where a, b, and c are real numbers, $a \neq 0$ is called a quadratic equation.

For example, $x^2 - 5x + 4 = 0$ and $x^2 - 14x + 45 = 0$ are quadratic equations.

TYPES OF QUADRATIC EQUATIONS

A quadratic equation is of two kinds, namely, 'pure quadratic' and 'adfected quadratic' (complete quadratic equation).

(a) A quadratic equation which contains only the second power of the unknown quantity (and not the first power) is called a *pure quadratic equation*.

Examples

(i) $3x^2 = 147$ (ii) $8x^2 - 47 = 100 + 5x^2$

(b) A quadratic equation which contains the second as well as the first power of the unknown quantity is called an 'adfected quadratic equation'.

Examples

(i) $x^2 - 14x = 45$ (ii) $2x^2 + 7x + 6 = 0$

GENERAL FORM OR STANDARD FORM

$ax^2 + bx + c = 0$ is known as the general form of a quadratic equation, where a, b, and c are real numbers and $a \neq 0$.

(a) When $b = 0$, it becomes a pure quadratic equation.
(b) When $c = 0$ and $b \neq 0$, it is an adfected quadratic equation.
(c) When $c \neq 0$ and $b \neq 0$, it becomes an adfected quadratic equation.

ROOTS

The values of the unknown quantity for which the equation is satisfied are called its *roots* and the process of finding them is called *solving the equation.*

An Important Result to be Kept in Mind

If the product of two factors is zero, then at least one of them must be zero. For example, if $ab = 0$, either $a = 0$ or $b = 0$ or both a and $b = 0$.

Proof

Consider the following possibilities:

(a) If $a = 0$, then there is nothing to prove.

(b) If $a \neq 0$, then there exists a number $\frac{1}{a}$.

$$ab = 0 \Rightarrow \frac{1}{a}(ab) = \frac{1}{a}(0)$$

(Multiplying both sides by $\frac{1}{a}$)

$$\therefore \left(\frac{1}{a} \times a\right) \times b = 0$$

$$\therefore 1 \times b = 0$$

$$\therefore b = 0$$

(c) If $a=0$ and $b=0$, then there is nothing to prove. Hence, either $a=0$ or $b=0$ or both a and b are equal to zero.

Solutions

A quadratic equation can be solved by any one of the following four ways:

(a) Factorisation
(b) Completing the square
(c) Using the formula
(d) Graphing

SOLUTION TO PURE QUADRATIC EQUATIONS

A pure quadratic equation can always be reduced to the simple form $x^2 = k$, and then the values of x can be determined by taking the square root on both sides. The roots evidently will be $+\sqrt{k}$ or $-\sqrt{k}$ written as $\pm\sqrt{k}$.

Worked Examples

1. $8x^2 - 47 = 100 + 5x^2$

Solution

By transposing we get the following:

$$8x^2 - 5x^2 = 100 + 47$$
$$3x^2 = 147$$
$$x^2 = \frac{147}{3} = 49$$
$$x = \pm 7$$

By taking the square root on both sides, we get $\pm x = \pm 7$. But $+x = +7$ and $-x = -7$ give the same value. Similarly, $+x = -7$ and $-x = +7$ give the same value. Hence, we write only two values, i.e. $x = +7$ and $x = -7$, and not four.

2. $(x+5)(x-5) = 39$

Solution

$$x^2 - 25 = 39$$
$$x^2 = 64$$
$$x = \pm 8$$

3. $\dfrac{x+2}{x-2} + \dfrac{x-2}{x+2} = \dfrac{5}{2}$

Solution

Multiplying both sides by $2(x^2 - 4)$, we get the following:

$$2(x+2)^2 + 2(x-2)^2 = 5(x^2 - 4)$$
$$2(x^2 + 4x + 4) + 2(x^2 - 4x + 4) = 5(x^2 - 4)$$

On simplification, we get the following:

$$4x^2 + 16 = 5x^2 - 20,\ x^2 = 36,\ x = \pm 6$$

4. $(x+3)(x+4) + (x-2)(x-5) = 30$

Solution

$$x^2 + 7x + 12 + x^2 - 7x + 10 = 30$$
$$2x^2 + 22 = 30$$
$$x^2 = 4$$
$$x = \pm 2$$

5. $\sqrt{8x^2 - 12x + 29} = 3x - 2$

Solution

Squaring both sides, we have the following:

$$8x^2 - 12x + 29 = (3x-2)^2$$
$$8x^2 - 12x + 29 = 9x^2 - 12x + 4$$
$$x^2 = 25$$
$$x = \pm 5$$

SOLUTION TO ADFECTED QUADRATIC EQUATIONS

Method of Factorisation

Worked Examples

1. Solve the following equation: $x^2 - 5x + 6 = 0$.

Solution

On factorising, we get the following:

$$(x-3)(x-2) = 0$$

Now, $x - 3 = 0$ gives $x = 3$, or $x - 2 = 0$ gives $x = 2$. Therefore, the roots are 3 and 2.

2. Solve the following equation: $ab(x^2 + 1) = (a^2 + b^2)x$.

Solution

$$ab(x^2 + 1) = (a^2 + b^2)x$$

On transposing, we get the following:

$$abx^2 - a^2x - b^2x + ab = 0$$
$$ax(bx - a) - b(bx - a) = 0$$
$$(bx - a)(ax - b) = 0$$
$$\therefore\ bx - a = 0 \quad \text{or} \quad ax - b = 0$$
$$\therefore\ x = \frac{a}{b} \quad \text{or} \quad x = \frac{b}{a}$$

Hence, the roots are $\dfrac{a}{b}$ and $\dfrac{b}{a}$.

Method of Completing the Square

The general form of a quadratic equation is as follows:

$$ax^2+bx+c=0$$

$$ax^2+bx=-c$$

Dividing by a (since $a\neq 0$), we have the following:

$$x^2+\frac{b}{a}\times x=-\frac{c}{a}$$

$$x^2+2\times x\times\left(\frac{b}{2a}\right)+\frac{b^2}{4a^2}=\frac{b^2}{4a^2}-\frac{c}{a}$$

Adding $\frac{b^2}{4a^2}$ to both sides, we get the following:

$$\left(x+\frac{b}{2a}\right)^2=\frac{b^2-4ac}{4a^2}$$

Taking square root, we have the following:

$$x+\frac{b}{2a}=\pm\frac{\sqrt{b^2-4ac}}{2a}$$

$$\therefore x=\frac{-b\pm\sqrt{b^2-4ac}}{2a}$$

Thus, the roots are as follows:

$$\frac{-b+\sqrt{b^2-4ac}}{2a} \quad\text{and}\quad \frac{-b-\sqrt{b^2-4ac}}{2a}$$

Worked Examples

1. Solve the following equation: $2x^2-7x+6=0$.

Solution

Here, $a=2$, $b=-7$, and $c=6$.

$$\therefore x=\frac{7\pm\sqrt{(-7)^2-4(2)(6)}}{2.2}$$

$$=\frac{7\pm\sqrt{49-48}}{4}=\frac{7\pm1}{4}$$

$$=\frac{7+1}{4}=\frac{8}{4}=2$$

$$=\frac{7-1}{4}=\frac{3}{2}$$

2. Solve for $x\times x^2-5x+6=0$ by completing the square.

Solution

$$x^2-2\times x\times\frac{5}{2}+\left(\frac{5}{2}\right)^2=-6+\left(\frac{5}{2}\right)^2$$

$$\left(x-\frac{5}{2}\right)^2=-6+\frac{25}{4}=\frac{-24+25}{4}=\frac{1}{4}$$

$$x-\frac{5}{2}=\pm\frac{1}{2}$$

$$x=\frac{5}{2}\pm\frac{1}{2}$$

$$x=\frac{5}{2}+\frac{1}{2}=3 \quad\text{or}\quad x=\frac{5}{2}-\frac{1}{2}=2$$

3. Solve the following by completing the square: $(x+4)(x+5)=3(x+1)(x+2)+2x$

Solution

$$(x+4)(x+5)=3(x+1)(x+2)+2x$$

$$x^2+9x+20=3x^2+9x+6+2x$$

$$2x^2+2x-14=0$$

$$x^2+x-7=0$$

$$x^2+2\times x\times\frac{1}{2}+\left(\frac{1}{2}\right)^2=7+\left(\frac{1}{2}\right)^2$$

$$\left(x+\frac{1}{2}\right)^2=7+\frac{1}{4}=\frac{29}{4}$$

$$x+\frac{1}{2}=\pm\frac{\sqrt{29}}{2}$$

$$x=-\frac{1}{2}+\frac{\sqrt{29}}{2} \quad\text{or}\quad -\frac{1}{2}-\frac{\sqrt{29}}{2}$$

Solving the Quadratic Equation Using the Formula

While solving the quadratic equation, $ax^2+bx+c=0$, by completing square, we get the result as follows:

$$x=\frac{-b\pm\sqrt{b^2-4ac}}{2a}$$

i.e., $x=\dfrac{-b+\sqrt{b^2-4ac}}{2a}$

or $x=\dfrac{-b-\sqrt{b^2-4ac}}{2a}$

a, b, and c of the above equation may be positive or negative quantities, and hence we may regard the above as formulae, from which we get the roots of any equation by substituting the value of a, b, and c from the given equation.

The formula may be written as follows:

$$x=\frac{-(\text{coefficient of } x)\pm\sqrt{(\text{coefficient of } x)^2-4(\text{coefficient of } x^2)(\text{absolute term})}}{2(\text{coefficient of } x^2)}$$

Worked Examples

Solve using the formula.

1. $3x^2=2x+1$

Solution

$3x^2 = 2x + 1$

$3x^2 - 2x - 1 = 0$

Here, $a = 3$, $b = -2$, and $c = -1$.

Therefore,

$$x = \frac{-b \pm \sqrt{b^2 - 4ac}}{2a}$$

$$x = \frac{2 \pm \sqrt{4+12}}{6}$$

$$x = \frac{2 \pm 4}{6}$$

$$x = \frac{2+4}{6} = \frac{6}{6} = 1$$

or $x = \frac{2-4}{6} = \frac{-2}{6} = -\frac{1}{3}$

2. $x^2 - x - 1 = 0$

Solution

$x^2 - x - 1 = 0$

Here, $a = 1$, $b = -1$, and $c = -1$.

Hence,

$$x = \frac{-b \pm \sqrt{b^2 - 4ac}}{2a}$$

$$x = \frac{1 \pm \sqrt{1+4}}{2} = \frac{1 \pm \sqrt{5}}{2}$$

i.e., $x = \frac{1+\sqrt{5}}{2}$ or $\frac{1-\sqrt{5}}{2}$

EQUATIONS REDUCIBLE TO A QUADRATIC EQUATION

Sometimes equations of a higher than the second degree can be solved by reducing them to quadratic equations. We shall consider some important types of such equations.

Worked Examples

Solve the following reducing them to a quadratic equation.

1. $4x^4 - 25x^2 + 36 = 0$

Solution

$4x^4 - 25x^2 + 36 = 0$

Putting $x^2 = m$, we get the following:

$4m^2 - 25m + 36 = 0$

$4m^2 - 16m - 9m + 36 = 0$

$4m(m-4) - 9(m-4) = 0$

$(m-4)(4m-9) = 0$

$m = 4$ or $m = \frac{9}{4}$

$x^2 = 4 \Rightarrow x = \pm 2$ or $x^2 = \frac{9}{4} \Rightarrow x = \pm\frac{3}{2}$

2. $8x^{3/2} - 8x^{-3/2} = 63$

Solution

$8x^{3/2} - 8x^{-3/2} = 63$

Putting $x^{3/2} = m$, we have the following:

$8m - 8m^{-1} - 63 = 0$

$8m - \frac{8}{m} - 63 = 0$

Multiplying throughout by m, we have the following:

$8m^2 - 63m - 8 = 0$

$(8m+1)(m-8) = 0$

$m = 8$ or $m = -\frac{1}{8}$

$x^{3/2} = -\frac{1}{8}$

$x^3 = \left(-\frac{1}{8}\right)^2 = \frac{1}{64}$

$\therefore\ x = \frac{1}{4}$ or $x^{3/2} = 8$

$x^3 = 8^2 = 64$

$x = 4$

$\therefore x = 4$ or $x = \frac{1}{4}$

3. $4.2^{2x+1} - 9.2^x + 1 = 0$

Solution

$4.2^{2x+1} - 9.2^x + 1 = 0$

$\Rightarrow 4.2^{2x} \times 2^1 - 9.2^x + 1 = 0$

$\Rightarrow 8.2^{2x} - 9.2^x + 1 = 0$

Putting $2^x = m$, we get the following:

$8m^2 - 9m + 1 = 0$

$(8m-1)(m-1) = 0$

$m = \frac{1}{8}$ or $m = 1$

i.e., $2^x = \frac{1}{8} = 2^{-3}$

$x = -3$ or $2^x = 1 = 2^0$

$x=0$

$\therefore\ x=0 \quad \text{or} \quad x=-3$

4. $\sqrt{\dfrac{x}{1-x}}+\sqrt{\dfrac{1-x}{x}}=2\dfrac{1}{6}$

Solution

$$\sqrt{\frac{x}{1-x}}+\sqrt{\frac{1-x}{x}}=2\frac{1}{6}$$

Putting $\sqrt{\dfrac{x}{1-x}}=m$, we get the following:

$$m+\frac{1}{m}=\frac{13}{6}$$

Multiplying by 6m throughout, we get the following:

$$6m^2-13m+6=0$$

$$(2m-3)(3m-2)=0$$

$$m=\frac{3}{2} \quad \text{or} \quad m=\frac{2}{3}$$

i.e., $\sqrt{\dfrac{x}{1-x}}=\dfrac{3}{2}$ or $\sqrt{\dfrac{x}{1-x}}=\dfrac{2}{3}$

$$\frac{x}{1-x}=\frac{9}{4}$$

$$9-9x=4x$$

$$13x=9$$

$$x=\frac{9}{13} \quad \text{or} \quad \sqrt{\frac{x}{1-x}}=\frac{2}{3}\Rightarrow\frac{x}{1-x}=\frac{4}{9}$$

$$9x=4-4x$$

$$13x=4$$

$$x=\frac{4}{13}$$

$$\therefore\ x=\frac{4}{13} \quad \text{or} \quad \frac{9}{13}$$

EQUATION OF THE TYPE $(x+a)(x+b)(x+c)(x+d)=r$

Factors should be combined in such a manner that it is possible to make convenient substitution at a later stage.

Let us take a numerical example.

Solve the following for x:

$$(x-1)(x+3)\,(x-2)(x-6)+36=0$$

$$\underline{(x-1)(x-2)}\ \underline{(x+3)(x-6)}+36=0$$

Note the combination:

$$(x^2-3x+2)(x^2-3x-18)+36=0$$

Now, putting $x^2-3x+2=y$, we get the following:

$$y(y-20)+36=0$$

$$y^2-20y+36=0$$

$$(y-18)\,(y-2)=0$$

$$\therefore\ y=18 \quad \text{or} \quad y=2$$

$$x^2-3x+2=18 \quad \text{or} \quad x^2-3x+2=2$$

$$x^2-3x-16=0 \quad \text{or} \quad x^2-3x=0$$

$$\frac{+3\pm\sqrt{9+64}}{2} \qquad x(x-3)=0$$

$$\underline{x=0 \text{ or } x=3}$$

Hence, the values of x are 0, 3, $\frac{3+\sqrt{73}}{2}$, and $\frac{3-\sqrt{73}}{2}$.

RECIPROCAL EQUATIONS

An equation which remains unaltered when x is replaced by its reciprocal $1/x$ is called a *reciprocal equation*. Thus, if m is a root of the reciprocal equation then $1/m$ is also its root; i.e., the roots of a reciprocal equations occur in reciprocal pairs.

Again as we know that the number of roots of an equation is equal to the degree of that equation. If a quadratic equation is reciprocal, then one root must be the reciprocal of the other. It will be observed that in a reciprocal equation, the coefficients of the terms equidistant from the beginning and the end are equal and of the same sign.

EQUATIONS INVOLVING RADICALS

An equation which contains a variable or variables under radical sign is called a *radical equation*. In the process of solving such equations, sometimes some roots are introduced in the equation, which do not satisfy the original equation. Such roots are called *extraneous roots*. The occurrence of such roots is not because we have made an error but of squaring both sides of the equation. Thus, it is absolutely necessary to check the roots obtained in solving an equation, where the process of squaring both sides of an equation is performed so as to eliminate the extraneous roots.

Worked Examples

1. Solve the following: $x^2-12x+\sqrt{x^2-12x+81}-9$

Solution

Write the equation as follows:

$$x^2-12x+81+\sqrt{x^2-12x+81}-81-9=0$$

Putting $\sqrt{x^2-12x+81}=m$, we have the following:

$$y^2+y-90=0$$

$$(y+10)(y-9)=0$$

$$y=-10 \quad \text{or} \quad y=9$$

When $y=9$,

$$\sqrt{x^2-12x+81}=9$$

$$x^2-12x+81=81$$

$$x(x-12)=0$$

$$x=0 \quad \text{or} \quad x=12$$

When $y=-10$,

$$\sqrt{x^2-12x+81}=-10$$

$$x^2-12x+81=100$$

$$x^2-12x-19=0$$

$$x=\frac{12\pm\sqrt{144+76}}{2}$$

$$=6\pm\sqrt{55}$$

Therefore, the roots are 0, 12, $6+\sqrt{55}$, and $6-\sqrt{55}$.

$6+\sqrt{55}$ and $6-\sqrt{55}$ will not satisfy the original equation and hence they are extraneous roots and therefore the only roots are 0 and 12.

2. Solve the following: $3x^4-20x^3-94x^2-20x+3=0$

Solution

Observing the coefficients, we see that the given equation is reciprocal. Dividing by x^2 and grouping the terms, we get the following:

$$3\left(x^2+\frac{1}{x^2}\right)-20\left(x+\frac{1}{x}\right)-94=0$$

Putting $x+\frac{1}{x}=m$, hence $x^2+\frac{1}{x^2}=m^2-2$, we get the following:

$$3\left(m^2-2\right)-20m-94=0$$

$$3m^2-20m-100=0$$

$$(3m+10)(m-10)=0$$

$$\therefore m=-\frac{10}{3} \quad \text{or} \quad m=10$$

When $m=-\frac{10}{3}$, we have the following:

$$x+\frac{1}{x}=-\frac{10}{3}$$

i.e., $3x^2+10x+3=0$

$$(3x+1)(x+3)=0$$

$$x=-\frac{1}{3} \quad \text{or} \quad -3$$

When $m=10$, we have the following:

$$x+\frac{1}{x}=10$$

$$x^2-10x+1=0$$

$$x=\frac{10\pm\sqrt{100-4}}{2}=\frac{10\pm2\sqrt{24}}{2}=5\pm\sqrt{24}$$

Therefore, the four roots of the given reciprocal equation are -3, $-\frac{1}{3}$, $5+\sqrt{24}$, and $5-\sqrt{24}$.

USE OF IDENTITY FOR RADICAL EQUATION

Solve $\sqrt{x^2-3x+16}-\sqrt{x^2-3x+9}=1$.

Squaring and solving this equation will be tedious. Hence, we apply a different method to solve this radical equation.

To solve this type of equation, we use an identity depending on the problem. Here, for this problem, we use the identity.

$$\left(x^2-3x+16\right)-\left(x^2-3x+9\right)=7$$

$$=\left(\sqrt{x^2-3x+16}\right)^2-\left(\sqrt{x^2-3x+9}\right)^2=7 \qquad \text{(A)}$$

$$=\sqrt{x^2-3x+16}-\sqrt{x^2-3x+9}=1 \qquad \text{(B)}$$

Dividing Eq. (A) by Eq. (B), we get the following:

$$\sqrt{x^2-3x+6}+\sqrt{x^2-3x+9}=7 \qquad \text{(C)}$$

Adding Eqs. (B) and (C), we get the following:

$$2\sqrt{x^2-3x+16}=8$$

$$\sqrt{x^2-3x+16}=4$$

On squaring, we have the following:

$$x^2-3x+16=16$$

$$x^2-3x=0$$

$$x(x-3)=0$$

Therefore, $x=0$ or 3.

THEORY OF QUADRATIC EQUATION

A quadratic equation cannot have more than two roots. The standard quadratic equation is $ax^2+bx+c=0$. If possible, let the equation $ax^2+bx+c=0$ have *three different roots*, α, β, and γ. Since they satisfy the equation, we have

$$a\alpha^2 + b\alpha + c = 0 \quad (1)$$

$$a\beta^2 + b\beta + c = 0 \quad (2)$$

$$a\gamma^2 + b\gamma + c = 0 \quad (3)$$

Subtracting Eq. (2) from Eq. (1), we have the following:

$$a(\alpha^2 - \beta^2) + b(\alpha - \beta) = 0$$

or $(\alpha - \beta)[a(\alpha + \beta + b)] = 0$

Therefore, either $\alpha - \beta = 0$ or $a(\alpha + \beta) + b = 0$.

But $\alpha - \beta$ cannot be zero as α and β are different.

$$a(\alpha + \beta) + b = 0 \quad (4)$$

Similarly, using Eqs. (2) and (3), we get the following:

$$a(\beta + \gamma) + b = 0 \quad (5)$$

Subtracting Eq. (5) from Eq. (4), we have the following:

$$a(\alpha - \gamma) = 0 \quad (6)$$

Therefore, either $a = 0$ or $\alpha - \gamma = 0$. But a is not equal to zero; also $\alpha - \gamma$ is not equal to zero as α and γ are different. Therefore, our assumption is wrong. Hence, a quadratic equation cannot have more than two roots.

NATURE OF THE ROOTS OF A QUADRATIC EQUATION

We know that the quadratic equation, $ax^2 + bx + c = 0$, has roots $\frac{-b\pm\sqrt{b^2-4ac}}{2}$. The nature of the roots depends on the expression $b^2 - 4ac$ under the radical sign, called *discriminant*. The following facts are quite obvious:

(a) If $b^2 - 4ac = 0$, the roots are real, rational, and equal.
(b) If $b^2 - 4ac$ is positive and a perfect square, the roots are real and unequal.
(c) If $b^2 - 4ac$ is positive and not a perfect square, the roots are real and irrational.
(d) If $b^2 - 4ac$ is negative, the roots are (not real) imaginary and unequal.

> Irrational and imaginary roots of a quadratic equation will always occur in conjugate pairs.

SUM AND PRODUCT OF THE ROOTS OF A QUADRATIC EQUATION

If α and β are the roots of the equation $ax^2 + bx + c = 0$,

$$\alpha = \frac{-b+\sqrt{b^2-4ac}}{2a} \quad \text{and} \quad \beta = \frac{-b-\sqrt{b^2-4a}}{2a}$$

On adding, we get the following:

$$\alpha + \beta = \frac{-2b}{2a} = -\frac{b}{a} = -\frac{\text{coefficient of } x}{\text{coefficient of } x^2}$$

On multiplying, we have the following:

$$\alpha \times \beta = \frac{(-b)^2 - (b^2 - 4ac)}{4a^2} = \frac{c}{a} = \frac{\text{absolute term}}{\text{coefficient of } x^2}$$

To Form an Equation when Roots are Given

Let the required equation be $ax^2 + bx + c = 0$, whose roots are α and β.

The equation can be written as follows:

$$x^2 + \frac{b}{a}x + \frac{c}{a} = 0 \quad (1)$$

Since $\alpha + \beta = -\frac{b}{a}$ and $\alpha\beta = \frac{c}{a}$,

Eq. (1) becomes $x^2 - (\alpha + \beta)x + \alpha\beta = 0$, which is the required equation.

> The equation can be remembered as x^2 – (sum of the roots)x + product of the roots = 0.

Worked Examples

1. Find the nature of the roots of $x^2 - 5x - 2 = 0$.

Solution

The discriminant $b^2 - 4ac = (-5)^2 - 4(-2) = 33$.

Therefore, roots are real, unequal, and irrational.

2. For what value of m will the equation $x^2 + mx - (m^2 + 3m - 32) = 0$ have equal roots?

Solution

The discriminant $b^2 - 4ac$

$$= m^2 + 4(m^2 + 3m - 32)$$

$$= 5m^2 + 12m - 128$$

Since the roots are equal, the discriminant should be zero.

$$\therefore 5m^2 + 12m - 128 = 0$$

$$\therefore m = \frac{-12 \pm \sqrt{144 + 2560}}{2 \times 5}$$

$$= \frac{-12 \pm \sqrt{2704}}{2 \times 5} = \frac{-12 \pm 52}{10}$$

i.e., $m = 4$ or $-6\frac{2}{5}$

3. Form the equation whose roots are −2 and 5.

Solution

The sum of the roots is −2 + 5 = 3, and the product of the roots is (−2)(5) = −10.

The required equation is as follows:

x^2 − (sum of the roots) x + product of the roots = 0

$x^2 - 3x - 10 = 0$

4. If α and β are the roots of $lx^2 + mx + n = 0$, find the value of $\alpha^2 + \beta^2$ and $\alpha^3 + \beta^3$.

Solution

(a) $\alpha^2 + \beta^2$

Since α and β are roots of $ln^2 + mx + n = 0$, we have the following:

$$\alpha+\beta=-\frac{m}{l} \quad \text{and} \quad \alpha\times\beta=\frac{n}{l}$$

$$\alpha^2+\beta^2=(\alpha+\beta)^2-2\alpha\beta$$

$$=\left(\frac{-m}{l}\right)^2-2\times\frac{n}{l}$$

$$=\frac{m^2}{l^2}-\frac{2n}{l}=\frac{m^2-2nl}{l^2}$$

(b) $\alpha^3+\beta^3=(\alpha+\beta)^3-3\alpha\beta(\alpha+\beta)$

$$=\left(-\frac{m}{l}\right)^3-3\frac{n}{l}\left(-\frac{m}{l}\right)$$

$$=-\frac{m^3}{l^3}+\frac{3mn}{l^2}$$

$$=\frac{-m^3+3mnl}{l^3}$$

5. If the roots of $x^2 - px + q = 0$ are two consecutive integers, prove that $p^2 = 4q + 1$.

Solution

Given equation is $x^2 - px + q = 0$.

Let the roots be m and $(m + l)$.

The sum of the roots is as follows:

$$m+m+l=p \quad (1)$$

The product of the roots is as follows:

$$m(m+1)=q \quad (2)$$

$$2m+1=p$$

$$m=\frac{p-1}{2}$$

Using this in Eq. (2), we have the following:

$$\left(\frac{p-1}{2}\right)\left(\frac{p-1}{2}+1\right)=q$$

$$(p-1)(p-1+2)=4q$$

$$(p-1)(p+1)=4q$$

$$p^2-1=4q$$

$$p^2=4q+1$$

6. If α and β are the roots of the equation $4x^2 - 6x + 1 = 0$, find the following:

(a) $\alpha\beta^3+\alpha^3\beta$ (b) $\dfrac{\alpha^2}{\beta}+\dfrac{\beta^2}{\alpha}$

(c) $\alpha^4+\beta^4$

Solution

$$\alpha+\beta=-\frac{b}{a}=\frac{+6}{4}=\frac{3}{2}$$

$$\alpha\beta=\frac{c}{a}=\frac{1}{4}$$

(a) $\alpha\beta^3+\alpha^3\beta=\alpha\beta(\beta^2+\alpha^2)$

$$=\alpha\beta[(\alpha+\beta)^2-2\alpha\beta]$$

$$=\frac{1}{4}\left[\frac{9}{4}-2\times\frac{1}{4}\right]=\frac{1}{4}\times\frac{7}{4}=\frac{7}{16}$$

(b) $\alpha^3+\beta^3=(\alpha+\beta)^3-3\alpha\beta(\alpha+\beta)$

$$=\frac{27}{8}-3\times\frac{1}{4}\left(\frac{3}{2}\right)$$

$$=\frac{27}{8}-\frac{9}{8}=\frac{18}{8}=\frac{9}{4}$$

$$\frac{\alpha^3+\beta^3}{\alpha\beta}=\frac{9/4}{1/4}=\frac{9}{4}\times4=9$$

(c) $(\alpha^2)^2+(\beta^2)^2$

$$=(\alpha^2+\beta^2)^2-2\alpha^2\beta^2$$

$$=[(\alpha+\beta)^2-2\alpha\beta]^2-2\alpha^2\beta^2$$

$$=\left[\frac{9}{4}-2\times\frac{1}{4}\right]^2-2\times\frac{1}{16}$$

$$=\left(\frac{7}{4}\right)^2-\frac{1}{8}=\frac{49}{16}-\frac{1}{8}=\frac{49-2}{16}=\frac{47}{16}$$

7. Find the quadratic equation whose roots are α and β having given the following: $5\alpha+\beta=9$ and $\alpha-4\beta=11$.

Solution

$$5\alpha+\beta=9 \quad (1)$$

$\alpha - 4\beta = 11$ (2)

Multiplying Eq. (1) by 1, we get the following:

$5\alpha + \beta = 9$ (3)

Multiplying Eq. (2) by 5, we get the following:

$5\alpha - 20\beta = 55$ (4)

Subtracting Eq. (3) from (4), we get the following:

$21\beta = -46$

$\Rightarrow \beta = \frac{-46}{21}$

$\alpha + 4\alpha + 4\left(\frac{46}{21}\right) = 11$

$21\alpha + 184 = 231$

$21\alpha = 47$

$\alpha = \frac{47}{21}$

$\alpha + \beta = \frac{47}{21} - \frac{46}{21} = \frac{1}{21}$

$\alpha\beta = \left(\frac{47}{21}\right) \times \left(\frac{-46}{21}\right) = \frac{-2162}{441}$

The equation is $9261x^2 - 441x - 2162 = 0$.

PROBLEMS LEADING TO QUADRATIC EQUATIONS

1. Find two even positive consecutive integers whose product is 288.

Note:

(a) Any general even number is $2n$.

(b) Any general odd number is $2n + 1$ or $2n - 1$.

Solution

Let the two consecutive positive even integers be $2n$ and $2n + 2$.

Their product is $2n(2n + 2)$.

As per the problem,

$2n(2n+2) = 288$

$n(2n+2) = 144$

$n(n+1) = 72$

$n^2 + n - 72 = 0$

$(n+9)(n-8) = 0$

Therefore, $n = 8$ or -9.

Neglecting -9, the first even number is $2n = 16$; next even number is $2n + 2 = 16 + 2 = 18$.

2. Find the length and breadth of a rectangular field, whose perimeter is 40 m and area is 96 m^2.

Solution

Let x be the breadth and y be the length.

The area is $x \times y$ and the perimeter $2(x + y)$ as per the problem is as follows:

$2(x + y) = 40$

$x + y = 20$

$x = (20 - y)$

The area is as follows: $(20 - y) \times y = 96$

$20y - y^2 = 96$

$y^2 - 20y + 96 = 0$

i.e., $(y - 8)(y - 12) = 0$

$y = 8$ or 12

Its length is 12, and breadth is $x = 20 - 12 = 8$.

If length is 8, then breadth $x = 20 - 8 = 12$, which is not applicable as length is smaller than breadth.

Therefore, the length is 12 m and breadth is 8 m.

3. The digit in the tens place of a two-digit number is equal to the square of the digit in the units place. If 54 is subtracted from that number, its digits are interchanged. Find the number.

Solution

Let the digit in the units place with two-digit number be x.

As given in the problem, the tens place will be x^2.

Therefore, the number is $10x^2 + x$.

If the digits are interchanged, the new number will be $10x + x^2$.

As per the given condition,

$(10x^2 + x) - 54 = 10x + x^2$

$9x^2 - 9x - 54 = 0$

$x^2 - x - 6 = 0$

On simplifying, we get the following:

$(x - 3)(x + 2) = 0$

$\therefore x = 3$ or $x = -2$

Rejecting the negative number, we have $x = 3$.

Therefore, the two-digit number is as follows:

$10x^2 + x$.

$= 10 \times 3^2 + 3$

$= 90 + 3 = 93$

4. The product of four consecutive natural numbers is 360. Find the numbers.

Solution

Let the four consecutive natural numbers be $x, x+1, x+2$, and $x+3$.

The product $x(x+1)(x+2)(x+3) = 360$

$$\left(x^2+3x\right)\left(x^2+3x+2\right)=360$$

Putting $x^2 + 3x = m$, we get the following:

$$m(m+2)-360=0$$

$$m^2+2m-360=0$$

$$m^2+20m-18m-360=0$$

$$(m+20)(m-18)=0$$

Therefore, $m=-20$ or $m=18$.

$$x^2+3x-18=0 \quad (1)$$

$$x^2+3x+20=0 \quad (2)$$

Using Eq. (1), we have the following:

$(x+6)(x-3)=0$

Hence, $x=3$ or $x=-6$.

Roots of Eq. (2) are not real as $b^2-4ac = 9-80 = -79$ is negative and numbers are natural, and therefore $x \neq -6$. Therefore, the required numbers are 3, 4, 5, and 6.

Solving the Quadratic Equation Using Graph

Before proceeding to work out particular examples the students' attention is called to the following principles. They are emphasised owing to their fundamental importance in the theory of graph in general and in the graphical solution to equations in particular.

(a) The coordinates of every point on a graph must satisfy the equation of the graph.

The point of intersection of graphs must have coordinates so that they satisfy the equations of both graphs.

(b) Hence, the solutions to two simultaneous equations are obtained by finding the coordinates of the points of intersection of the graph of the equations.

(c) To find the point where the graph crosses the x-axis, put $y=0$ in the equation, and to find the point where the graph crosses y-axis, put $x=0$ in the equation.

Let us trace the graph of the expression $2x^2-5x-12$.

Let y represent the value of the expression. We then have to trace the graph of the following equation:

$y=2x^2-5x-12$

We tabulate the corresponding values of x and y as follows:

x	0	1	2	3	4	.....	−1	−2	.....
$y=2x^2$	0	2	8	18	32	.....	2	8	.....
$-5x-12$	−12	−17	−22	−27	−32	.....	−7	−2	.....
$y=2x^2$ $5x-12$	−12	−15	−14	−9	0	.....	−5	6	.....
	(0, −12)	(1, −15)	(2, −14)	(3, −9)	(4, 0)		(−1, −5)	(−2, 7)	

We plot the points (0, −12), (1, −15), (2, −14) ...... (−2, 6) .... taking 5 mm as the unit for abscissa and ordinate.

A free-hand curve is drawn passing through all the points. The nature of the curve has to be adapted to the general trend of these points. It will be observed that the graph is similar in shape to the graph $y=x^2$.

Solve the equation graphically.

$$2x^2-5x-12=0 \quad (1)$$

First Method

Draw the graph of the following equation:

$$y=2x^2-5x-12 \quad (2)$$

x	6	−4	−2	1
y	30	40	5	−15

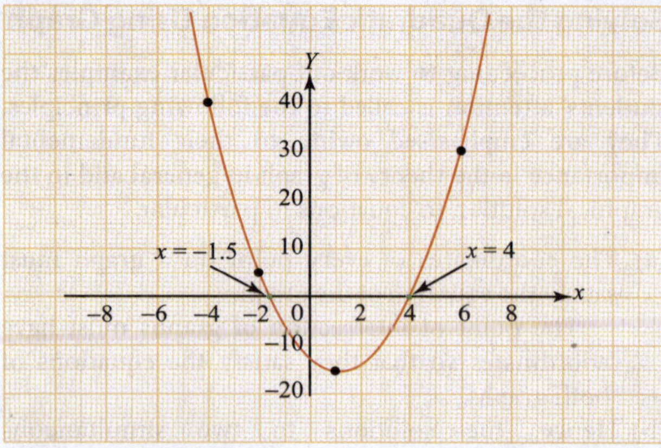

This graph is same as the one we traced. To solve Eq. (1), we have to find values of x, which make $2x^2 - 5x - 12 = 0$, i.e., values of x which make $y = 0$. Now, when $y = 0$, the graph of Eq. (2) crosses the x-axis. In the figure, we see that this graph crosses x-axis at the points P and Q. The values of x at these points are OP and OQ.

Hence, the roots of Eq. (1) are 4 and –1.5.

Second Method

The equation $2x^2 - 5x - 12 = 0$ is the same as the equation $2x^2 = 5x + 12$.

Draw the graph of the following:

$y = 2x^2$ (1)

$y = 5x + 12$ (2)

For Eq. (1), when $y = 2x^2$

x	5	6	–4	–6
y	50	72	32	+72

For Eq. (2), when $y = 5x + 12$

x	0	1	+4	–1
y	12	17	32	7

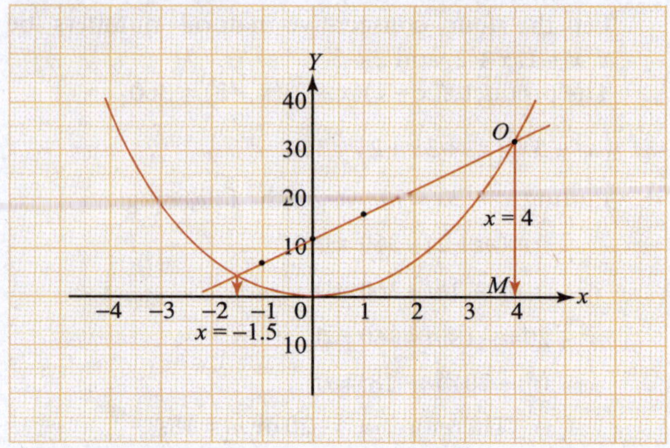

The graphs intersect at the points P and Q. The values of x at the point P give the same value of y, namely MP for both the graphs, i.e., $x = OM$. We have $2x^2 = 5x + 12$. Therefore $x = OM = 4$ is a solution to the equation.

Similarly, $x = ON = -1.5$ is another solution to the equation. In other words, the abscissae of the points of intersection of the graph $y = 2x^2$ and $y = 5x + 12$ is the roots of the equation $2x^2 = 5x + 12$. We conclude in general by saying that we can find the roots of $ax^2 + bx + c = 0$.

(a) By finding the abscissae of the points where the graph of the equation $y = ax^2 + bx + c$ crosses the x-axis

OR

(b) By finding the abscissae of the points of intersection of the graph of the equation

$y = ax^2$ (1)

$y = -bx - c$ (2)

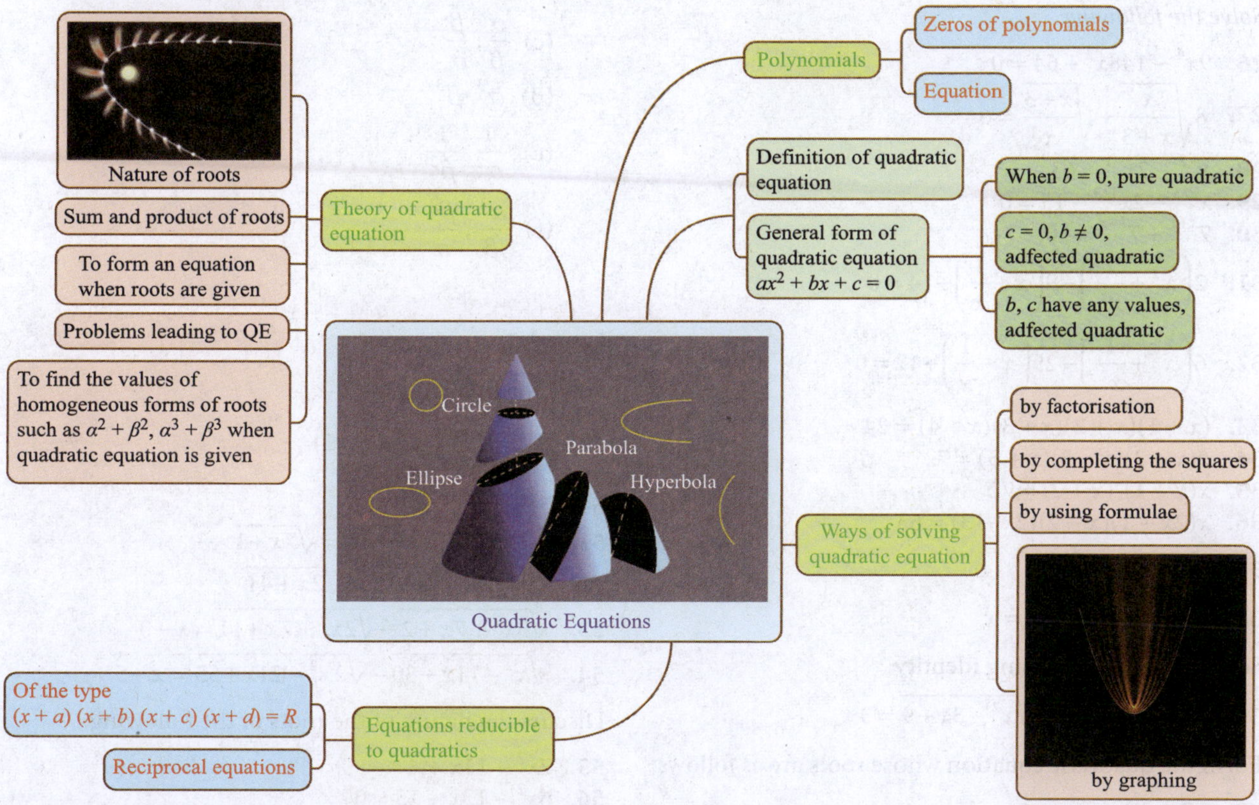

TEST YOURSELF

Level 1

Solve the following equations:

1. $5x^2 - 8 = 37$
2. $2x^2 - 2 = 18 - 3x^2$
3. $(x+3)(x-3) = 7$
4. $(5+x)(6-x) + (6+x)(5-x) = 0$
5. $(x+3)(x+4) + (x-2)(x-5) = 0$
6. $\sqrt{8x^2 - 12x + 29} = 3x - 2$

Solve the following equations by factorisation:

7. $9x^2 + 15x - 14 = 0$
8. $3x^2 + 16x + 5 = 0$
9. $x + \frac{1}{x} = 5\frac{1}{5}$
10. $ab(x^2+1) = (a^2+b^2)x$
11. $x^2 + 2bx - 1 = -b^2$

Level 2

12. $\frac{x-a}{x-b} + \frac{x-b}{x-a} = \frac{a}{b} + \frac{b}{a}$
13. $\frac{16}{x^{3/2}} + \frac{x^{1/2}}{2} = \frac{6}{x^{1/2}}$

Solve the following equations by completing the square:

14. $x^2 - 2x = 15$
15. $x^2 + 10x = 24$
16. $(x-1)(x-2) = 20$
17. $\frac{8-x}{2} - \frac{2x-11}{x-3} = \frac{x-2}{6}$
18. $\frac{(x-1)}{(x-2)} - \frac{x-3}{x-4} = \frac{x-5}{x^2-6x+5}$
19. $ax(x-a) = (a-x)$
20. $x^2 - 2ax = 1 - a^2$

Solve the following using formula:

21. $x^2 + 3x - 4 = 0$
22. $x^2 = 4(x-1)$
23. $2x^2 + \frac{3}{4}x = \frac{5}{16}$
24. $12x^2 + 9x = 9^2$
25. $(a-b)^2 x^2 - (a-b)^2 x - ab = 0$

Solve the following:

26. $9x^4 - 148x^2 + 64 = 0$

27. $8\sqrt{\frac{x}{x+3}} + \sqrt{\frac{x+3}{x}} = 6$

28. $x^{-2} - 2x^{-1} = 63$

29. $x^{2/3} - 2x^{1/3} - 15 = 0$

30. $7^{x+1} + 7^{1-x} = 50$

31. $2\left(x^2+\frac{1}{x^2}\right)-9\left(x+\frac{1}{x}\right)+14=0$

32. $6\left(x^2+\frac{1}{x^2}\right)-25\left(x-\frac{1}{x}\right)+12=0$

33. $(x+1)(x+2)(x+3)(x+4) = 24$

34. $x(x+2)(x+3)(x+5) = 72$

35. $x(x+1)^2(x+2) = 72$

36. $x(2x+1)(x-2)(2x-3) = 63$

37. $\sqrt{2x+1}-\sqrt{x}=1$

38. $1-\sqrt{7x^2-9x+3}=x$

Solve the following using identity:

39. $\sqrt{x^2-3x+36}-\sqrt{x^2-3x+9}=3$

Form the quadratic equation whose roots are as follows:

40. $-2, -3$

41. $a, -a$

42. $\frac{9}{2}, \frac{9}{2}$

43. $1\pm\sqrt{2}$

44. One of the roots is $2+\sqrt{3}$.

Level 3

45. If α and β are the roots of the equation $4x^2 - 6x + 1 = 0$, find the following:
(a) $\alpha^3 + \beta^3$
(b) $\alpha^4\beta^5 + \alpha^5\beta^4$

46. Find the quadratic equation whose roots are α and β having given the following:
(a) $\alpha+\beta=9$, $\alpha\beta=-5$
(b) $\alpha+\beta=5$, $\frac{1}{\alpha}+\frac{1}{\beta}=8$

47. If α and β are roots of $x^2 - 5x + 6 = 0$, form the equation whose roots are as follows:
(a) α^2, β^2
(b) $\frac{1}{\alpha}, \frac{1}{\beta}$
(c) $\frac{\alpha}{\beta}, \frac{\beta}{\alpha}$
(d) α^3, β^3
(e) $\frac{1}{\alpha^2}, \frac{1}{\beta^2}$
(f) $\frac{\alpha^2}{\beta}, \frac{\beta^2}{\alpha}$

Level 4

Solve the following:

48. $2^{2x+3} - 57 = 65(2^x - 1)$

49. $3^{x+2} + 3^{-x} + 10 = 0$

50. $5^{x+1} + 5^{2-x} = 5^3 + 1$

51. $\sqrt{2x+7}+\sqrt{3x-18}=\sqrt{7x+1}$

52. $\sqrt{x+5}+\sqrt{x+12}=\sqrt{2x+41}$

53. $\sqrt{3x^2+7x+2}-\sqrt{2x^2+7x+11}=x-3$

54. $\sqrt{x^2-11x+30}-\sqrt{2x^2-21x+55}=x-5$

Discuss the nature of the roots of the following:

55. $7x^2 - 11x + 4$

56. $6x^2 - 13x - 15 = 0$

57. $x^2 - 7x + 8 = 0$

58. $x^2 + 9x + 27 = 0$

59. $4x^2 - 12x + 15 = 0$

60. For what value of k will $18x^2 - kx + 2 = 0$ have equal roots?

61. Show that $3x^2 - 7x + 8 = 0$ cannot be satisfied by any real value of x.

62. If p and q are the roots of $x^2 - ax + b = 0$, find the value of $\frac{1}{p^3}+\frac{1}{q^3}$.

63. Find the quadratic equation one of whose roots is $p+\sqrt{q}$.

64. If one of the roots of $ax^2 + bx + c = 0$ is twice the other, show that $2b^2 = 9ac$.

65. The sum of the squares of two consecutive natural numbers is 145. Find the numbers.

66. The length of a rectangle is greater than twice its breadth by 2 cm. The length of its diagonal is 13 cm. Find the length and breadth of the rectangle.

67. In a rectangular mango grove, the number of trees lengthwise is 5 more than the number of trees breadthwise. If the total number of trees is 1400, find the number of trees lengthwise and breadthwise.

68. The sum of the reciprocals of two consecutive odd natural numbers is $\frac{12}{35}$. Find the numbers.

69. A car covers a distance of 300 km with the same speed; it will cover the same distance in 1 hour less if its speed is increased by 10 km/h. Find the speed of the car.

70. A cistern can be filled by two pipes in $33\frac{1}{3}$ minutes; if the larger pipe takes 15 minutes less than the smaller to fill the cistern, find in what time it will be filled by each pipe separately.

71. One-fourth of a herd of camels was seen in the forest. Twice the square root of the herd had gone to mountains and the remaining 15 camels were seen on the bank of a river. Find the total number of camels.

72. A motorboat, whose speed in still water is 15 km/h, goes 30 km downstream and returns to the starting point in a total time of 4 hours 30 minutes. Find the speed of the stream.

73. A passenger train takes 3 hours less for a journey of 360 km. If its speed is increased by 10 km/h from its usual speed, what is the usual speed?

74. In a flight of 600 km, an aircraft was slowed down due to bad weather. Its average speed for the trip was reduced by 200 km/h and the time of flight increased by 30 minutes. Find the duration of flight.

Find the roots of the following, drawing the graphs in two ways:

75. $x^2 + 4x \quad 12 - 0$

76. $2 - 2x + x^2 = 0$

77. $x^2 - 4x + 3 = 0$

78. $x^2 + x - 6 = 0$

79. $2x^2 - 5x - 3 = 0$

Find the minimum value of the following and find the value of x for which the expression takes the minimum value:

80. $4x^2 + 8x + 25$

81. $9x^2 - 8x - 8$

82. $16x^2 - 24x - 26$

Multiple Choice Questions

83. Sum of roots of the equation $4^x - 3(2^{x+3}) + 128 = 0$, is

(A) 0 (B) 7
(C) 5 (D) 8

84. If α and β are the roots of the polynomial $f(x) = x^2 - 5x + k$, such that $\alpha - \beta = 1$, then the value of k is

(A) 8 (B) 4
(C) 13/2 (D) 6

85. If $P(x) = ax^2 + bx + c$ and $Q(x) = -ax^2 + dx + c$, where $ac \neq 0$, then $P(x) \times Q(x) = 0$ has at least

(A) four real roots
(B) two real roots
(C) four imaginary roots
(D) only one real root

86. If roots of the equation $(q - r)x^2 + (r - p)x + (p - q) = 0$ are equal, then p, q and r are in

(A) AP (B) GP
(C) HP (D) AGP

87. If $3p^2 = 5p + 2$ and $3q^2 = 5q + 2$ where $p \neq q$, then pq is equal to

(A) $\frac{2}{3}$ (B) $\frac{-2}{3}$
(C) $\frac{3}{2}$ (D) $\frac{-3}{2}$

88. If α and β are the roots of the quadratic equation $x^2 + bx - c = 0$, then the equation whose roots are b and c, is

(A) $x^2 + \alpha x - \beta = 0$
(B) $x^2 - [(\alpha + \beta) + \alpha\beta]\, x - \alpha\beta\,(\alpha - \beta) = 0$
(C) $x^2 + [(\alpha + \beta) + \alpha\beta]\, x + \alpha\beta\,(\alpha + \beta) = 0$
(D) $x^2 + [(\alpha + \beta) + \alpha\beta]\, x - \alpha\beta\,(\alpha + \beta) = 0$

89. Let $p, q \in \{1, 2, 3, 4\}$. The number of equations of the form $px^2 + qx + 1 = 0$ having real roots, is

(A) 15 (B) 9
(C) 8 (D) 7

90. Find the number of rational roots of the equation $(x + 2)(x + 3)(x + 8)(x + 12) = 4x^2$.

(A) 4 (B) 2
(C) 0 (D) 1

91. Solutions of the equation $(6 - x)^4 + (8 - x)^4 = 16$ are

(A) 8, 6 (B) −8, −6
(C) 8, 9 (D) −8, 6

92. Solutions of the equation $\sqrt{2x^2 + 5x - 2} - \sqrt{2x^2 + 5x - 9} = 1$ are

(A) −2, 9/2 (B) 2, −9/2
(C) −2, −9/2 (D) 2, 9/2

93. Solutions of the equation $64 \times 9^x - 84 \times 12^x + 27 \times 16^x = 0$ are

(A) $-1, -2$ (B) $-1, 2$
(C) $2, 3$ (D) $1, 2$

94. Let α, β be the roots of the equation $x^2 - x + p = 0$ and γ, δ be the roots of the equation $x^2 - 4x + 9 = 0$. If α, β, γ and δ are in GP then the integral values of p and q, respectively, are

(A) $-2, -32$ (B) $-2, 3$
(C) $-6, 3$ (D) $-6, -32$

95. If α, β are the roots of the equation $x^2 - 3x + 5 = 0$ and γ, δ are the roots of the equation $x^2 + 5x - 3 = 0$, then the equation whose roots are $\alpha\gamma + \beta\delta$ and $\alpha\delta + \beta\gamma$, is

(A) $x^2 - 15x - 158 = 0$
(B) $x^2 + 15x - 158 = 0$
(C) $x^2 - 15x + 158 = 0$
(D) $x^2 + 15x + 158 = 0$

96. If $\cos^4\theta + p$, $\sin^4\theta + p$ are the roots at the equation $x^2 + a(2x + 1) = 0$ and $\cos^2\theta + q$, $\sin^2\theta + q$ are the roots of the equation $x^2 + 4x + 2 = 0$ then a is equal to

(A) -2 (B) $-1/2$
(C) 1 (D) 2

97. If α and β are roots of the equation $x^2 - p(x+1) - c = 0$ hen $\dfrac{\alpha^2 + 2\alpha + 1}{\alpha^2 + 2\alpha + c} + \dfrac{\beta^2 + 2\beta + 1}{\beta^2 + 2\beta + c} =$

(A) 1 (B) -1
(C) 2 (D) 3

98. The quadratic equation, product of whose roots x_1 and x_2 is equal to 4 and are related as $\dfrac{x_1}{x_1 - 1} + \dfrac{x_2}{x_2 - 1} = 2$, is

(A) $x^2 + 2x + 4 = 0$ (B) $x^2 - 4x + 4 = 0$
(C) $x^2 - 2x + 4 = 0$ (D) $x^2 + 4x + 4 = 0$

99. If one root of the equation $x^2 - dx + 12 = 0$ is even prime, while $x^2 + dx + \mu = 0$ has equal roots then μ is

(a) 8 (B) 20
(C) 32 (D) 16

100. The graph of a quadratic polynomial $y = ax^2 + bx + c$; $a, b, c \in R$ is as shown.

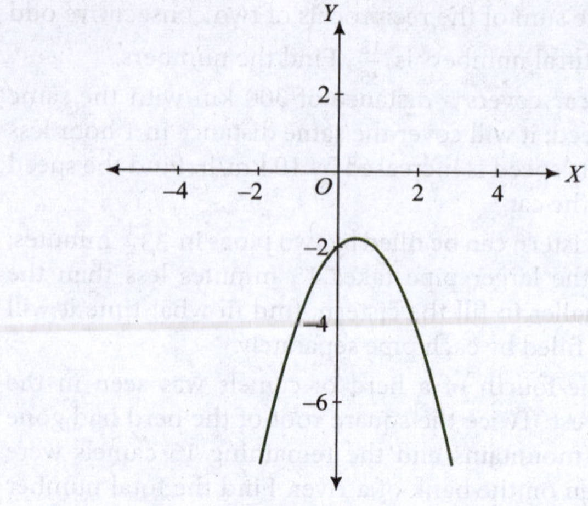

Which one of the following is NOT correct?

(A) $b^2 - 4ac < 0$

(B) $\dfrac{c}{a} < 0$

(C) c is negative

(D) abscissa corresponding to the vertex is $\left(\dfrac{-b}{2a}\right)$

101. If $x + \dfrac{1}{x} = 5$ then $x^3 - 5x^2 + x + \dfrac{1}{x^3} - \dfrac{5}{x^2} + \dfrac{1}{x} =$

(A) -5 (B) 0
(C) 5 (D) 10

102. If $(a^2 - 1)x^2 + (a - 1)x + (a^2 - 4a + 3) = 0$ is an identity in x, then the value of a is/are

(A) -1 (B) 3
(C) 1 (D) $-1, 1, 3$

103. If α and β are the roots of $ax^2 + bx + c = 0$, then which of the following are the roots of the equation $ax^2 - bx(x - 1) + c(x - 1)^2 = 0$?

(a) $\dfrac{\alpha}{\alpha + 1}, \dfrac{\beta}{\beta + 1}$ (B) $\dfrac{\alpha + 1}{\alpha}, \dfrac{\beta + 1}{\beta}$

(C) $\dfrac{\alpha}{\alpha - 1}, \dfrac{\beta}{\beta - 1}$ (D) $\dfrac{\alpha - 1}{\alpha}, \dfrac{\beta - 1}{\beta}$

104. The equation for x: $\sqrt[3]{2x - 1} + \sqrt[3]{x - 1} = 1$ is

(a) 0 (B) 0, 1
(C) 1 (D) None of these

105. The equation $\sqrt{x+1}-\sqrt{x-1}=\sqrt{4x-1}$ has
(A) no solution
(B) one solution
(C) two solution
(D) more than two solutions

OLYMPIAD AND NTSE LEVEL EXERCISES

106. If $a, b, c \in Q$ then roots of the equation $(b+c-2a)x^2 + (c+a-2b)x + (a+b-2c) = 0$ are
(A) irrational (B) non-real
(C) rational (D) equal

107. Solutions of the equation $(12x-1)(6x-1)(4x-1)(3x-1) = 5$ are
(A) $-\frac{1}{12}, \frac{1}{2}$ (B) $-\frac{1}{12}, -\frac{1}{2}$
(C) $\frac{1}{12}, \frac{1}{2}$ (D) $\frac{1}{12}, -\frac{1}{2}$

108. If α and β are the roots of the equation $3x^2+2x+1=0$ then the value of $\left(\frac{1-\alpha}{1+\alpha}\right)^3 + \left(\frac{1-\beta}{1+\beta}\right)^3$ is
(A) 10 (B) –9
(C) 9 (D) –10

109. The square of difference of the roots of the equation $x^2 + px + 45 = 0$ is 144. The roots are
(a) –3, –15 (B) 2, 5
(C) –12, 12 (D) 18, –18

110. Read the following statements.

Statement 1: If $\sqrt{\sqrt{\sqrt{x}}} = \sqrt[4]{\sqrt[4]{\sqrt[4]{3x^4+4}}}$ then the value of x^4 is 4.

Statement 2: If a and b are positive numbers and each of the equations $x^2 + ax + 2b = 0$ and $x^2 + 2bx + a = 0$ has real roots then the smallest possible value of $(a + b)$ is 10.
(A) Both Statement 1 and Statement 2 are true.
(B) Statement 1 is true and Statement 2 is false.
(C) Statement 1 is false and Statement 2 is true.
(D) Both Statement 1 and Statement 2 are false.

111. If $x, y \in R$ satisfy the equation $x^2 + y^2 - 4x - 2y + 5 = 0$ then the value of the expression
$$\frac{\left(\sqrt{x}-\sqrt{y}\right)^2 + 4\sqrt{xy}}{x+\sqrt{xy}}$$ is
(A) $\sqrt{2}+1$ (B) $\frac{\sqrt{2}+1}{2}$
(C) $\frac{\sqrt{2}-1}{2}$ (D) $\frac{\sqrt{2}+1}{\sqrt{2}}$

112. If one root of the equation $x^2 + px + 12 = 0$ is 4, while the equation $x^2 + px + q = 0$ has equal roots then the value of q is
(A) 4 (B) 49/4
(C) 4/49 (D) None of these

113. If α and β are the roots of $x^2 - 2x + 4 = 0$ then $\alpha^5 + \beta^5$ is equal to
(A) 32 (B) 64
(C) 16 (D) None of these

114. Match the following columns:

	Column I		Column II
(a)	If a, b, c and d are four non-zero real numbers such that $(d+a-b)^2 + (d+b-c)^2 = 0$ and the roots of the equation $a(b-c)x^2 + b(c-a)x + c(a-b) = 0$ are real and equal then	(p)	$a+b+c=0$
(b)	If a, b and c are non-zero positive real numbers such that log a, log b and log c are in A.P. then	(q)	a, b, c are in AP
(c)	If the equation $ax^2 + bx + c = 0$ and $x^3 - 3x^2 + 3x - 1 = 0$ have a common real root then	(r)	a, b, c are in GP
(d)	If a, b and c are positive real numbers such that the expression $bx^2 + \left(\sqrt{(a+c)^2 + 4b^2}\right)x + (a+c)$ is non-negative $\forall x \in R$ then	(s)	a, b, c are in HP

115. Column I contains rational algebraic expression and column II contains possible integers which lie in their ranges. Match the columns.

	Column I		Column II
(a)	$y = \frac{x^2-2x+4}{x^2+2x+4}, x \in R$	(p)	1
(b)	$y = \frac{x^2-3x-2}{2x-3}, x \in R$	(q)	4
(c)	$y = \frac{2x^2-2x+4}{x^2-4x+3}, x \in R$	(r)	–3
(d)	$x^2 - (a-3)x + 2 < 0, \forall x \in (-2, 3)$	(s)	–10

GAME TIME

Solving a quadratic equation using cardboard.

Solve the following: $x^2 + 7x + 12 = 0$.

Materials required: A piece of cardboard of size about 10 cm × 5 cm and scissors.

Step 1: Write the equation as $x(x + 7) = -12$.

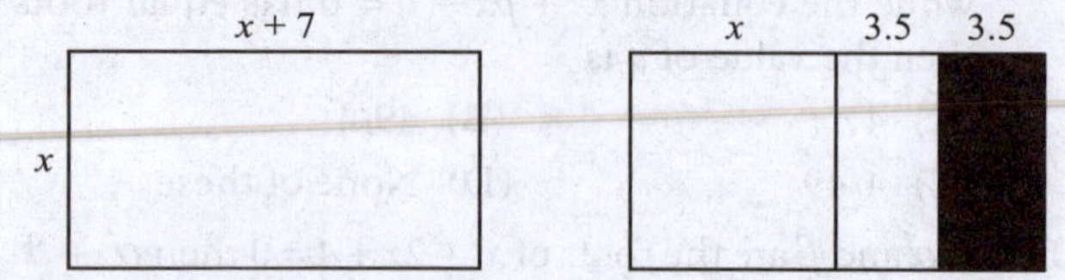

Mark the length of the card as $x + 7$ and the width as x so that the area is $x(x + 7)$.

Step 2: Measure x along the length and draw a line as shown above, dividing the rectangle into a square ($x \times x$) and a rectangle.

Step 3: Divide this rectangle into two equal rectangles of length x and width 3.5, and cut the card into three pieces.

Step 4: Now place one of the rectangles as shown below. You will get an incomplete square.

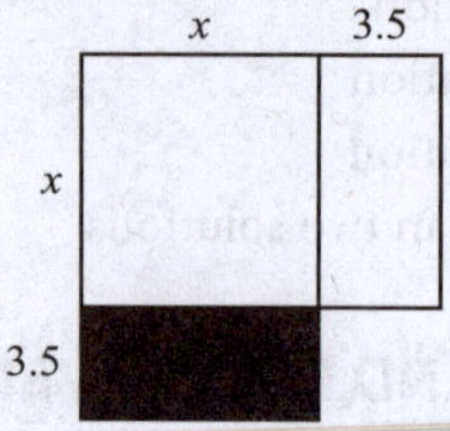

Step 5: Complete the square by adding a small square of side 3.5.

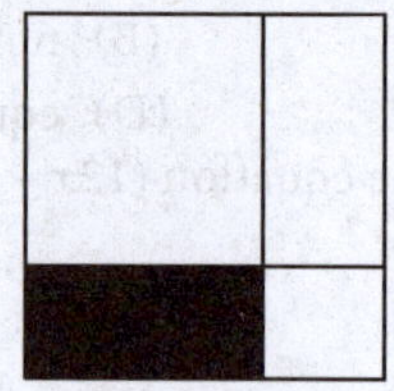

Area of the whole square $= (x + 3.5)^2 = -12 + (3.5)^2 = 0.25$.

Therefore, $x + 3.5 = \pm 0.5$ and hence $x = -3$ or -4.

7 Relations and Functions

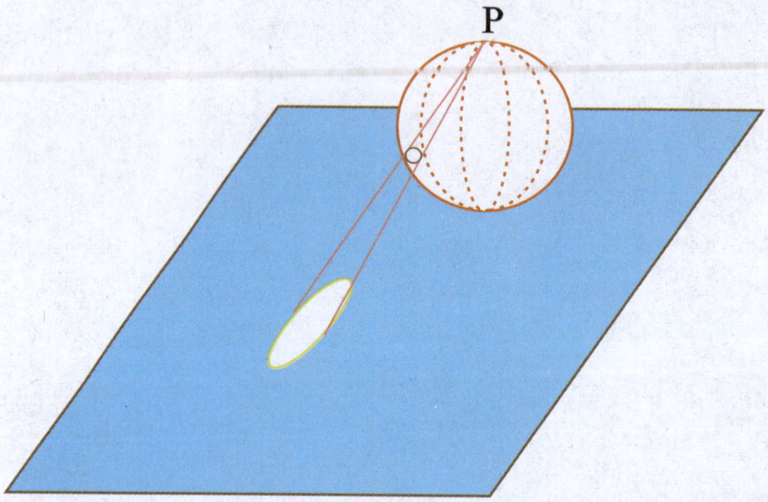

What is the relation between the circle on the sphere and the ellipse on the plane?

OBJECTIVES

After reading this chapter you should be able to

- find the Cartesian product of two sets
- represent the Cartesian product graphically
- find the cardinal number of the Cartesian product
- identify relation as a subset of the Cartesian product
- differentiate domain, range, and co-domain of a relation
- write the relation in roster form, arrow diagram, and lattice diagram
- identify the function as a subset of relation
- know and differentiate the types of functions
- draw Venn diagrams for different types of functions

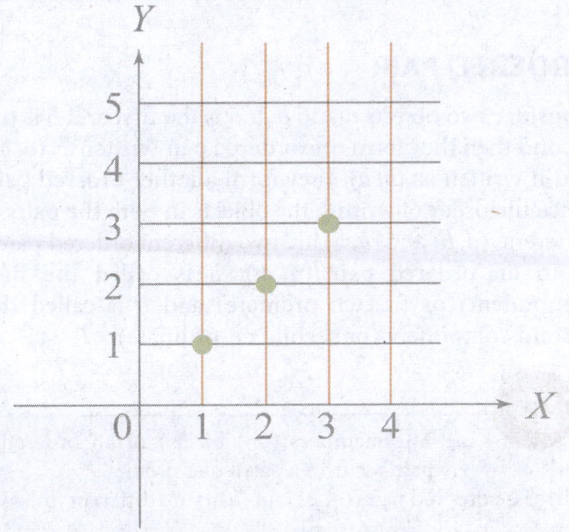

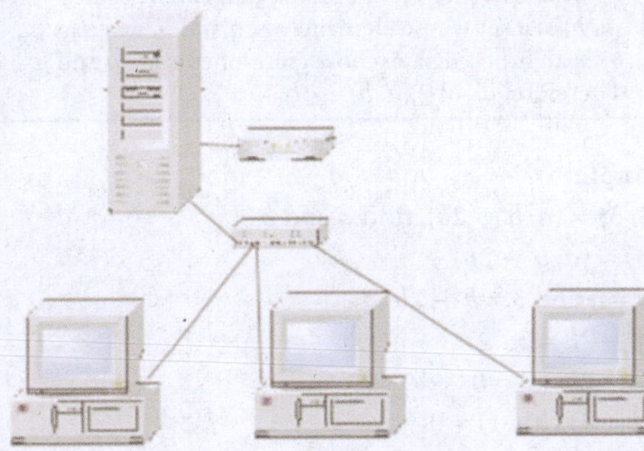

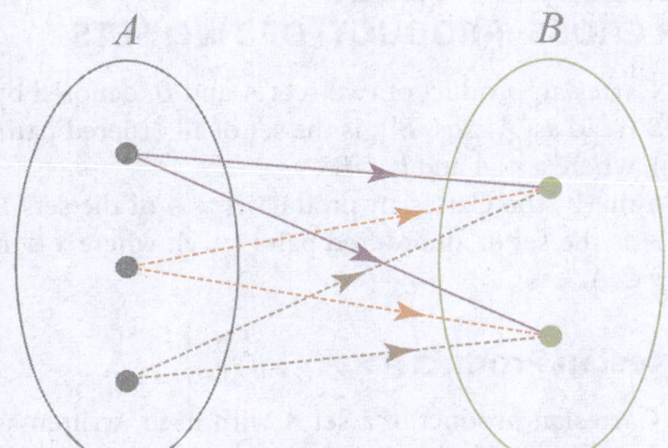

ORDERED PAIR

Consider two objects a and b. If a is the first and b is the second then they form one ordered pair written as (a, b), and if written as (b, a), they form another ordered pair. Since the order of writing the objects in both the pairs is different, (a, b) and (b, a) are two different ordered pairs.

In an ordered pair (a, b), a is called the first component (or first co-ordinate) and b is called the second component (or second co-ordinate).

(a) The members (or objects) of an ordered pair occur in a particular order.
(b) The ordered pairs (a, b) and (b, a) are different unless $a = b$; i.e., $(5, 6) \neq (6, 5)$.
(c) Ordered pairs (a, b) and (c, d) are equal, if and only if $a = c$ and $b = d$.
For example, $(x, y) = (3, 5) \Rightarrow x = 3$ and $y = 5$.
(d) The components of an ordered pair can be same.
(e) $\{a, b\}$ is a set whose elements are a and b, whereas (a, b) is an ordered pair whose components are a and b. Therefore, $\{a, b\} \neq (a, b)$.

Example

If $(8, 3) = (a, b^2 - 2b)$, find a and b.

$(8, 3) = (a, b^2 - 2b)$

$\Rightarrow 8 = a$ and $3 = b^2 - 2b$

$\Rightarrow b^2 - 2b = 3$

$\Rightarrow b^2 - 2b - 3 = 0$

$\Rightarrow (b - 3)(b + 1) = 0$

$\Rightarrow b = 3$ or -1

Therefore, $a = 8$, $b = 3, -1$.

CARTESIAN PRODUCT (OR CROSS–PRODUCT) OF TWO SETS

The Cartesian product of two sets A and B, denoted by $A \times B$ (read as 'A cross B'), is the set of all ordered pairs (a, b), where $a \in A$ and $b \in B$.

Similarly, the Cartesian product $B \times A$ of the sets B and A is the set of all ordered pairs (x, y), where $x \in B$ and $y \in A$.

Cartesian Product $A \times A$

The Cartesian product of a set A with itself, written as $A \times A$, is the set of all ordered pairs (p, q), where $p \in A$ and $q \in A$.

Example

Find the Cartesian products $A \times B$, $B \times A$, and $A \times A$ of the sets A and B, where $A = \{1, 2\}$, $B = \{x | x^2 = 4\}$.

$A = \{1, 2\}$. Writing B in the tabular form, we have the following:

$B = \{-2, 2\}$

By definition,

$A \times B = \{(1, -2), (1, 2), (2, -2), (2, 2)\}$

$B \times A = \{(-2, 1), (-2, 2), (2, 1), (2, 2)\}$

$A \times A = \{(1, 1), (1, 2), (2, 1), (2, 2)\}$

Representation of the Cartesian Product

(a) In the set-builder form
$A \times B = \{(a, b) | a \in A, b \in B\}$
Similarly, $B \times A = \{(x, y) | x \in B, y \in A\}$

(b) Method of writing the set $A \times B$ in the roster form:
(i) Write the sets A and B in the roster form.
(ii) Take an element of A as the first member and all the elements of B as the second member one by one, and form the ordered pairs.
(iii) Repeat the above step for all the elements of A.
(iv) Write a set with all the ordered pairs obtained using the above two steps.

(c) Arrow diagram of $A \times B$

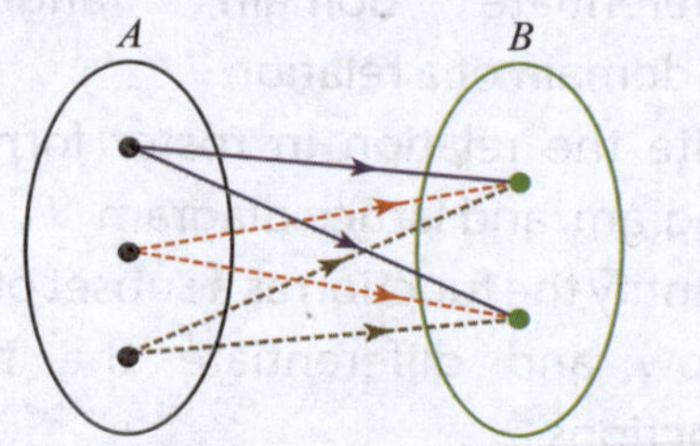

In the figure, arrows are drawn from the elements of A to the elements of B to indicate how their pairing is being done.

Let $A = \{1, 2, 3\}$, $B = \{-1, 1\}$

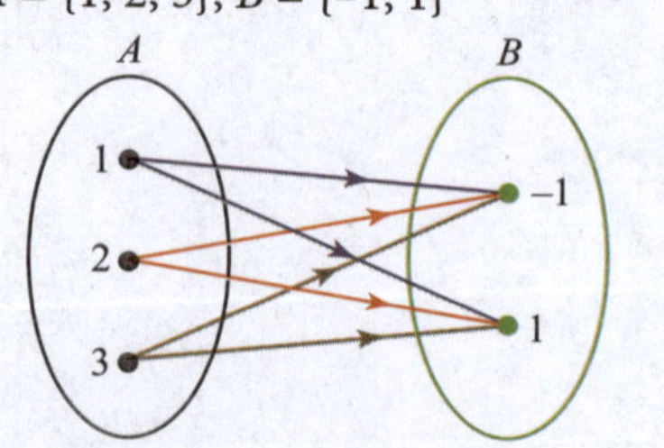

Graphical Representation of $A \times B$

Let $A = \{1, 2\}$ and $B = \{1, 3\}$

$A \times B = \{(1, 1), (1, 3), (2, 1), (2, 3)\}$

$B \times A = \{(1, 1), (1, 2), (3, 1), (3, 2)\}$

Take two mutually perpendicular lines OX and OY. Using a scale represent the numbers 1 and 2 (members of A) on OX and the numbers 1 and 3 (members of B) on OY. Draw lines parallel to OY through the points 1 and 2 (members of A). Again draw lines parallel to OX through the points 1 and 3 (members of B). Write the ordered pairs of numbers at the points of intersection of the lines drawn. Similarly, we can represent graphically $B \times A$ as shown in the figure below.

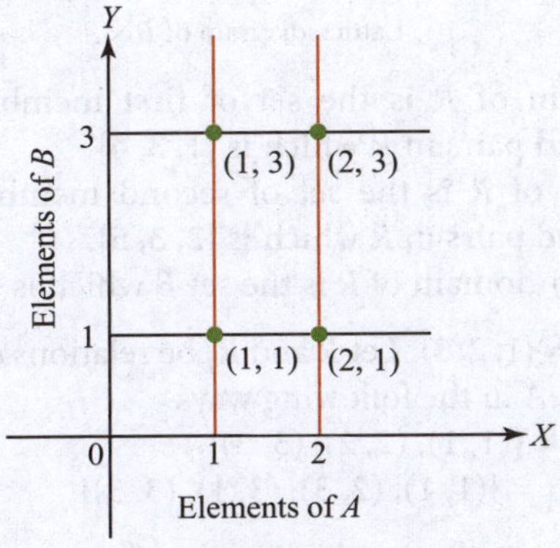

Lattice diagram of $A \times B$

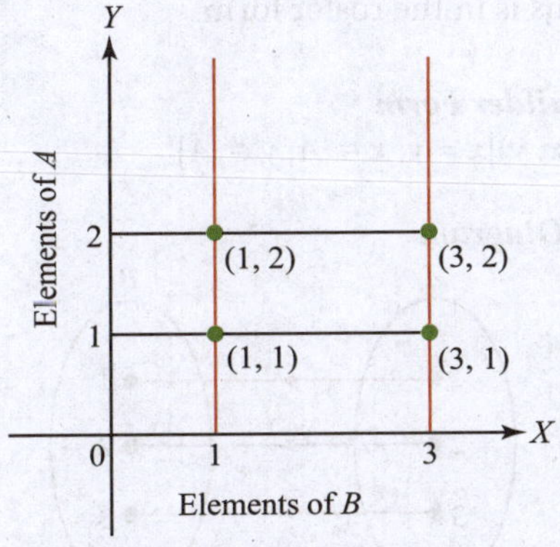

Lattice diagram of $B \times A$

Cardinal Number of the Cartesian Product

The cardinal number of the Cartesian product $A \times B$, denoted by $n(A \times B)$, is the number of elements of $A \times B$ and $n(A \times B) = n(A) \times n(B)$, where $n(A)$ is the cardinal number of A and $n(B)$ is the cardinal number of B.

$A \times B = \varnothing \Leftrightarrow A = \varnothing$ or $B = \varnothing$.

RELATIONS

Relation as Association Between Two Sets

In our day-to-day life, we use sentences such as 'Ram is the brother of Shyam', '6 is less than 9', 'Bangalore is the capital of Karnataka' and 'wood is lighter than iron'. The phrases *brother of, less than, capital of, lighter than* describe the association between the members of two sets or members of the same set.

Let A and B denote the sets of all male and female members in the royal family of and Dasarath's kingdom.

Clearly A = {Dasratha, Rama, Lakshmana, Bharatha, and Shatrugna}.

B = {Kaushalya, Kaikeyi, Sumitra, Sita, Urmila, Mandovi, and Shrutakirthi}

If we write R for the relation 'is the husband of', then we have Dasaratha R Kaushalya, Dasaratha R Kaikeyi, Dasaratha R Sumitra, Rama R Sita, Lakshma R Urmila, Bharatha R Mandovi, and Shatrugna R Shrutakirthi.

Now, if we omit the letter R between the pairs of names and write them as ordered pairs, then the above can also be written as a set R of ordered pairs, where R = {(Dasaratha, Kausalya), (Dasaratha, Kaikeyi), (Dasaratha, Sumitra), (Rama, Sita), (Lakshmana, Urmila), (Bharatha, Mandovi), and (Shatrugna, Shrutakirthi)}.

Clearly $R \subseteq A \times B$

If $(a, b) \in R$, where $a \in A$ and $b \in B$, we say a is 'R-related' to b and write as $a\,R\,b$.

Example

Let $P = \{2, 4\}$ and $Q = \{1, 3, 5\}$. Let R be the relation 'is less than'. Find R.

Now $2(\in P) < 3(\in Q) \Rightarrow 2\,R\,3$

$2(\in P) < 5(\in Q) \Rightarrow 2\,R\,5$

$4(\in P) < 5(\in Q) \Rightarrow 4\,R\,5$

Thus, $R = \{(2, 3), (2, 5), (4, 5)\}$

Relation as a Subset of the Cartesian Product

If A and B are two non-empty sets, then a subset of $A \times B$ is called a relation from the set A to the set B.

For example, if $A = \{1, 3\}$ and $B = \{2, 3, 4\}$, $A \times B = \{(1, 2), (1, 3), (1, 4), (3, 2), (3, 3), (3, 4)\}$

According to the above definition, every subset of $A \times B$ is a relation from A to B.

Consider the following:

(a) $R_e = \{(1, 2), (1, 3), (1, 4), (3, 4)\}$

(R_1 may be understood as the relation 'is less than')

(b) $R_2 = \{(1, 4), (3, 2)\}$
(R_2 may be understood as the relation '$x + y = 5$')

(c) $R_3 = \{(1, 2), (1, 4), (3, 2), (3, 3)\}$
R_3 cannot be described as an association, but it is also a relation.

Thus, it is very important to note that it may or may not be possible to express a relation given as a subset of $A \times B$ in terms of an association between the sets A and B.

Relation on a Set

If A is a non-empty set and $R \subseteq A \times A$, then A is called a relation on the set A or a relation in A.

Domain, Range, and Co-domain of a Relation

If R is a relation from a set A to the set B, i.e., $R \subseteq A \times B$,

(a) The set of first components of all ordered pairs in R is called its domain clearly, the domain of $R \subseteq A$.
(b) The set of second components of all ordered pairs in R is called its range, clearly, the range of $R \subseteq B$.
(c) The set B is called the co-domain of relation R.

Representation of a Relation

Example

(a) Let $A = \{1, 2, 3, 4\}$ and $B = \{2, 3, 5\}$. If $R = \{(x, y) | x + y = 6\}$ is a relation from the set A to set B, write the relation in other forms. Also, write the domain, range, and the co-domain of the relation R. Now,
$A \times B = \{(1, 2), (1, 3), (1, 5), (2, 2), (2, 3), (2, 5), (3, 2), (3, 3), (3, 5), (4, 2), (4, 3), (4, 5)\}$
$R = \{(x, y) | x \in$ A and $y \in B$ and $x + y = 6\}$
This is in the set-builder form.

Roster Form
$R = \{(1, 5), (3, 3), (4, 2)\}$

Venn Diagram

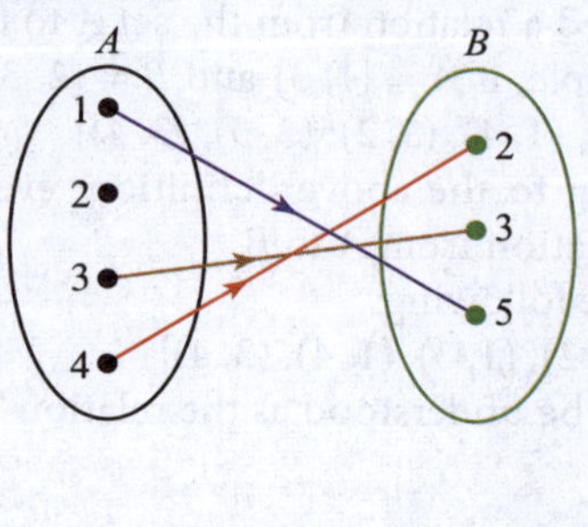

Lattice Diagram

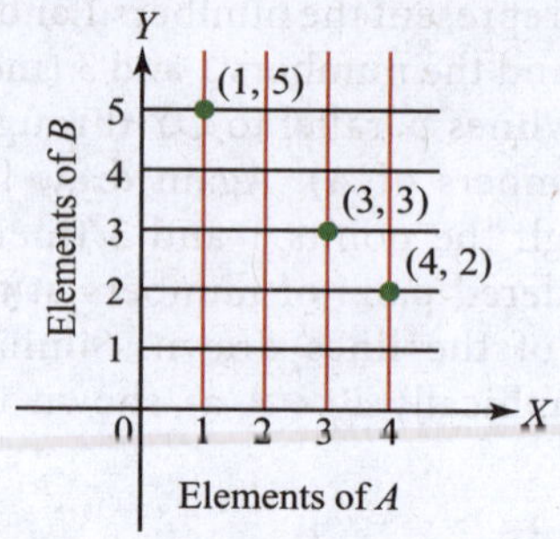

Lattice diagram of $B \times A$

Domain of R is the set of first members of the ordered pairs in R which is $\{1, 3, 4\}$.
Range of R is the set of second members of the ordered pairs in R which is $\{2, 3, 5\}$.
The co-domain of R is the set B which is $\{2, 3, 5\}$.

(b) Let $A = \{1, 2, 3\}$. Let R and R_1 be relations defined on the set A in the following ways
(i) $R = \{(1, 1), (2, 2), (3, 3)\}$
(ii) $R_1 = \{(1, 1), (2, 3), (3, 1), (3, 3)\}$

Now, write the relation in other form.
(i) $R = \{(1, 1), (2, 2), (3, 3)\}$
This is in the roster form.

Set-Builder Form
$R = \{(x, y) | x = y, x \in A, y \in A\}$

Venn Diagram

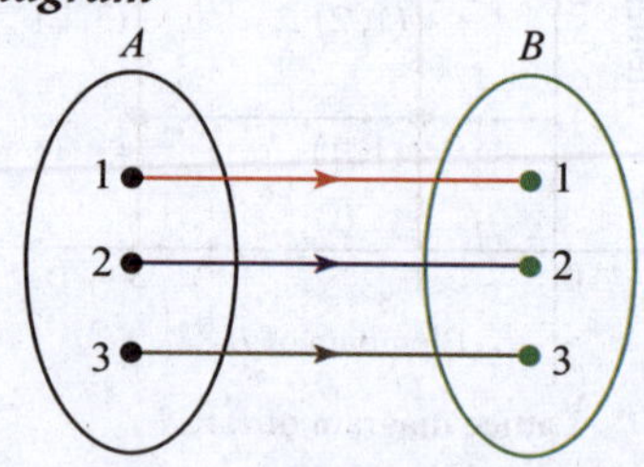

Lattice Diagram

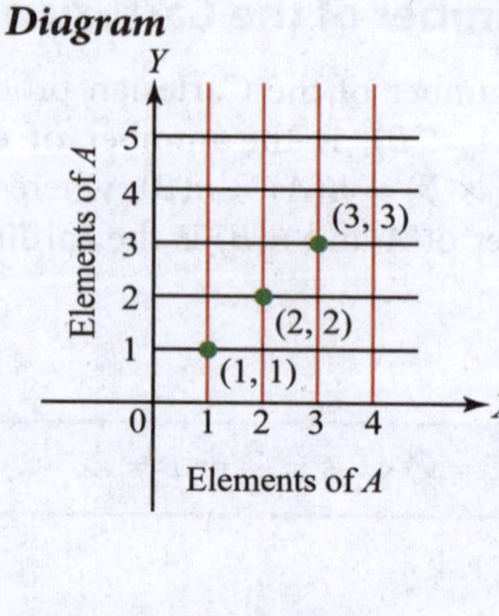

(ii) $R_1 = \{(1, 1), (2, 3), (3, 1), (3, 3)$

This is in the roster form.

As there is no common association between the first member and the second member in the ordered pairs, R_1 cannot be represented in the set-builder form. But it can be represented by diagrams as shown below.

Venn Diagram

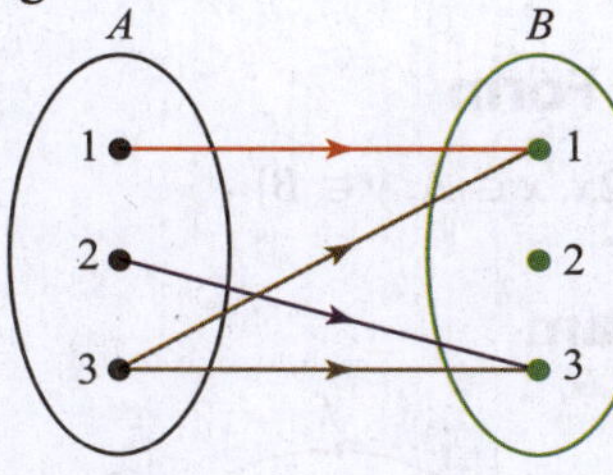

Lattice Diagram

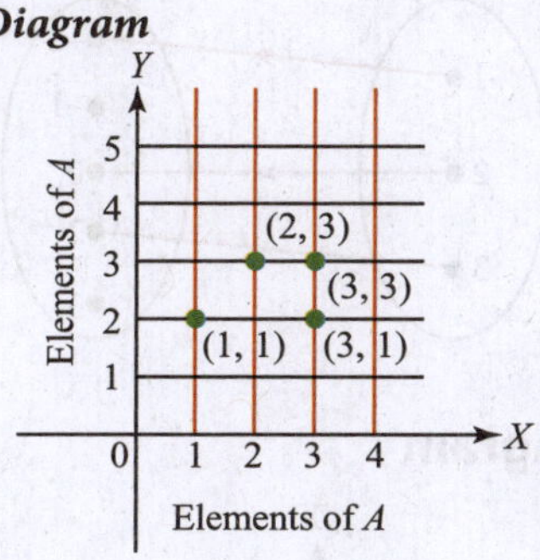

Graph

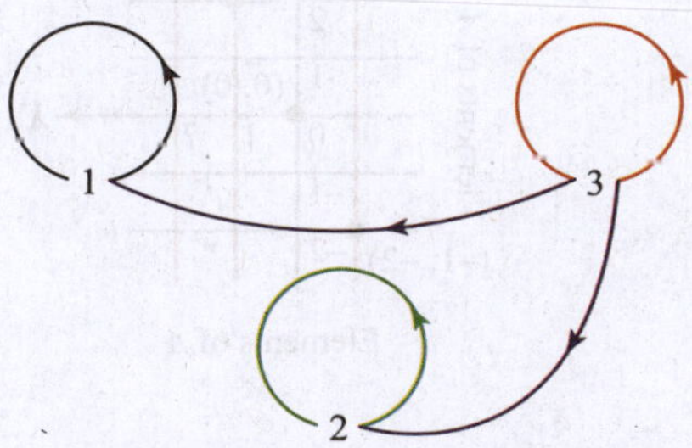

TYPES OF RELATION ON A SET

Void Relation

Let A be a set. Then $\emptyset \subseteq A \times A$ and so it is a relation on A. This relation is called the void or empty relation on A.

Universal Relation

Let A be a set. Then $A \times A \subseteq A \times A$ and so it is a relation on A. This relation is called the universal relation on A.

The void and universal relations on a set A are respectively the smallest and largest relations on A.

A relation R on a set A may have some special properties. We now classify a relation on the basis of such properties.

Identity Relation

Let A be a set.

Then the relation $I_A = \{(a, a) | a \in A\}$ on A is called the identity relation on A.

In other words, a relation I_A on A is called an identity relation if every element of A is related to itself *only*.

Example

The relation $I_A = \{(1, 1), (2, 2), (3, 3)\}$ is the identity relation on $A = \{1, 2, 3\}$. But $R_1 = \{(1, 1), (2, 2)\}$ and $R_2 = \{(1, 1), (2, 2), (3, 3), (1, 3)\}$ are not identity relations on A, for $(3, 3) \notin R_1$ and in R_2 '1' is related to elements '1' and '3'.

Reflexive Relation

A relation R on a set A is said to be reflexive if every element of A is *related to itself*, if R is reflexive and only (a, a) if $\in R$ for all A.

Example

Let L be the set of all straight lines in a plane. Then the relation R 'is parallel to' on L is a reflexive relation, since every line is parallel to itself.

> The identity relation on a non-empty set A is always reflexive on A. However, a reflexive relation on A is not necessarily the identity relation on A.
> For example, $\{(a, a), (a, b), (b, b), (c, c), (b, a)\}$ is a reflexive relation on $A = \{a, b, c\}$ but it is not the identity relation on A.

Again the universal relation on a non-empty set A is reflexive, but it is not the identity relation on A.

Symmetric Relation

A relation R on a set A is said to be a symmetric relation if $1(a, b) \in R \Rightarrow (b, a) \in R$ for all $a, b \in A$,

i.e., $a\,R\,b \Rightarrow b\,R\,a$ for all $a, b \in A$.

Example

Let L be the set of all straight lines in a plane. Then the relation R 'is parallel to' on L is a symmetric relation because whenever a line l_1 is parallel to l_2, l_2 is parallel to l_1.

Further, the identity and universal relations on a non-empty set are symmetric.

Transitive Relation

Let A be any set. A relation R on A is said to be a transitive relation if $(a, b) \in R$ and $(b, c) \in R$ imply $(a, c) \in R$ for all $a, b, c \in A$.

Hence, $a\,R\,b$ and $b\,R\,c \Rightarrow a\,R\,c$ for all $a, b, c \in A$.

Example

Let L be the set of all straight lines in a plane. Then the relation R 'is parallel to' on L is a transitive relation because for any $l_1, l_2, l_3 \in L$.

$(l_1, l_2) \in R \Rightarrow l_1 \parallel l_2$

$(l_2, l_3) \in R \Rightarrow l_2 \parallel l_3$

$l_1 \parallel l_2$ and $l_2 \parallel l_3 \Rightarrow l_1 \parallel l_3 \Rightarrow (l_1, l_3) \in R$

Further, the identity and universal relations on a non-empty set are transitive.

Equivalence Relation

A relation R on a set A is said to be an equivalence relation on A if it is

(a) reflexive, i.e., $(a, a) \in R$ for all $a \in A$

(b) symmetric, i.e., $(a, b) \in R \Rightarrow (b, a) \in R$ for all $a, b \in A$

(c) transitive, i.e., $(a, b) \in R$ and $(b, c) \in R \Rightarrow (a, c) \in R$ for all $a, b, c \in A$

Example

Let L be the set of all straight lines in a plane. Then the relation R defined by 'is parallel to' on L is an equivalence relation (which is clear from the above discussion).

FUNCTIONS

Definition of a Function as a Special Relation

A relation R from a set A to a set B is called a function 'f' from A to B, written as $f: A \to B$, if R has the following properties:

(a) Every element of A appears as the first member in an ordered pair belonging to R, i.e., $(x, y) \in R$ for all $x \in A$.

(b) Two ordered pairs belonging to R having different second members do not have the same first member, i.e., if $(x, y) \in R$ then $(x, z) \notin R$, where $y \neq z$.

Thus, a relation R from a set A to a set B is called a function from A to B, if for each element x in A there exists an *unique* element y in B such that $(x, y) \in R$ is a rule which associates *every element* of A to a unique element in B.

REPRESENTATION OF A FUNCTION

Let $A = \{-1, 0, 1\}$ and $B = \{-2, -1, 0, 2, 3\}$, and let f be defined by 'x is half of y' where $x \in A$ and $y \in B$.

Now, $f: A \to B$ can be represented in different forms.

Roster Form

$F = \{(-1, -2), (0, 0), (1, 2)\}$

Set-Builder Form

$F = \{(x, y) | y = 2x, x \in A, y \in B\}$

Venn Diagram

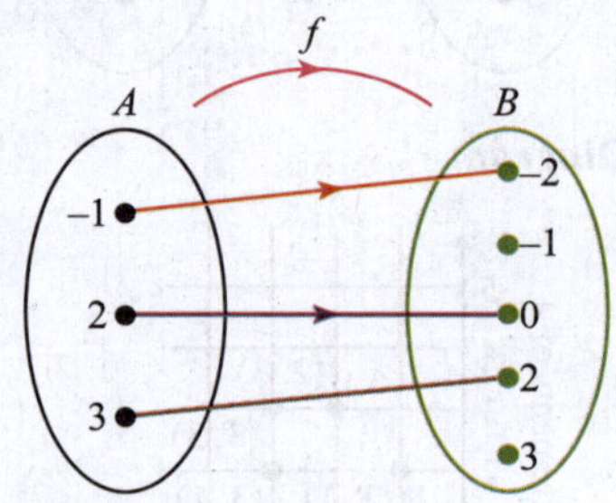

Lattice Diagram

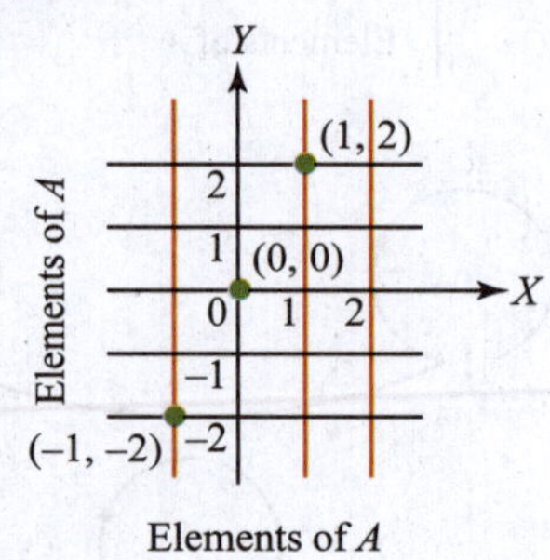

By Formula

An expression for y in terms of x is also used to define a function from a subset of real numbers to another subset of real numbers.

Here, $y = 2x$ represents the function since for each value of x there exists a unique value of y.

Value-Table Form

From the equation $y = 2x$, we can form a table.

x	-1	0	1
$y = 2x$	-2	0	2

By Graphs

A graph can also describe a function from a subset of real numbers to another subset of real numbers.

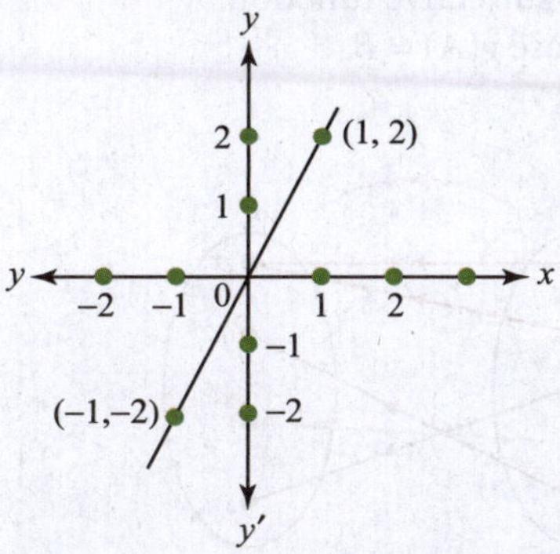

The line PQ is the graph of the function $y = 2x$.

> Since for each value of x, there corresponds a unique value of y, the vertical line through a point on OX should not cut the graph at more than one point.

For example, consider the following graphs:

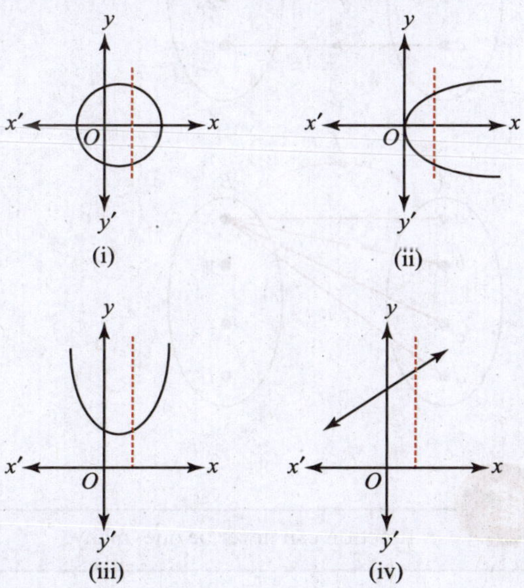

Graphs (i) and (ii) do not represent functions. Graphs (iii) and (iv) represent functions.

Image and Pre-image

If $f: A \to B$ is a function for which $x \in A$ corresponds to $y \in B$, then y is called the image of x, and x is called the pre-image of y.

If $(x, y) \in f$ then y is called the image of x, and x is called the pre-image of y.

Domain, Range, and Co-domain of a Function

Domain

If $f: A \to B$ is a function from the set A to the set B, then the set A is called the domain of f.

Range

The set of images of x is called the range of f.
The range of f is denoted by $f(A)$ and $f(A) \subseteq B$.

Co-domain

The set B is called the co-domain of the function.

Example

Let $A = \{-1, 2, -4\}$ and $B = \{4, -1, 5, 1, 2\}$.
Let $f = \{(-1, -1), (2, 1), (-4, 2)\}$.

Clearly, $f \subseteq A \times B$ and all the elements of A appear as first members in the ordered pairs in f and they appear only once. So f is a function from A to B.

Now, domain of f = the set of first members of the ordered pairs in $f = \{-1, 2, -4\}$.

Range of f = the set of second members of the ordered pairs in $f = \{-1, 1, 2\}$.

Co-domain of f = set $B = \{4, -1, 5, 1, 2\}$.

Value of a Function at a Point

Let $y = f(x)$ be a function from $A = \{1, 2, 3\}$ to $B = N$ defined by $y = 2x + 1$.

Then the image of 1 is $2(1) + 1 = 3$.

This is also expressed as 'the value of the function at x equal to 1 is 3' and written as $f(1) = 3$.

The value of the function at $x = 2$ is $f(2) = 2(2) + 1 = 5$ and $f(3) = 2.3 + 1 = 7$.

TYPES OF FUNCTIONS

One–One Function

A function $f: A \to B$ such that no element of B has more than one pre-image in A is called a one–one function or injective function.

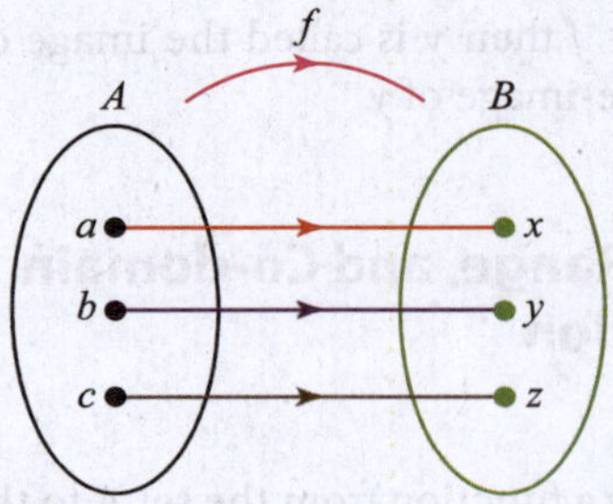

Many–One Function

A function $f : A \to B$ such that at least one element of B has more than one pre-image in A, is called a many–one function.

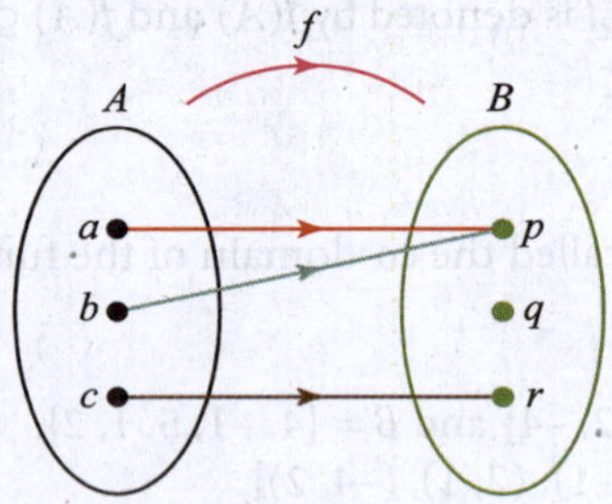

Into Function

If $f : A \to B$ is a function such that there is at least one element in B which has no pre-image in A, then f is called an into function.

In this case $f(A) \subset B$.

(i)

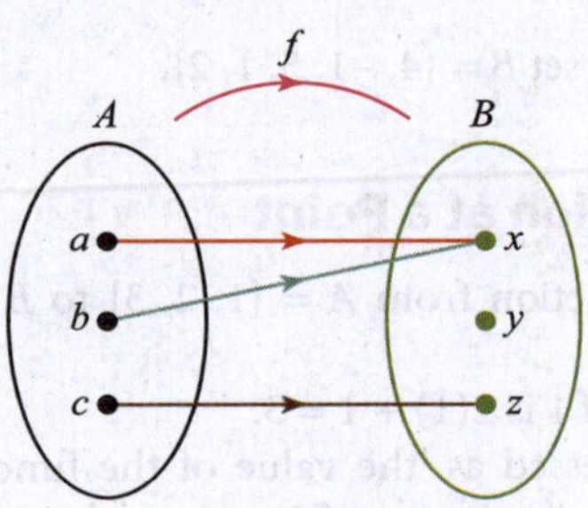

(ii)

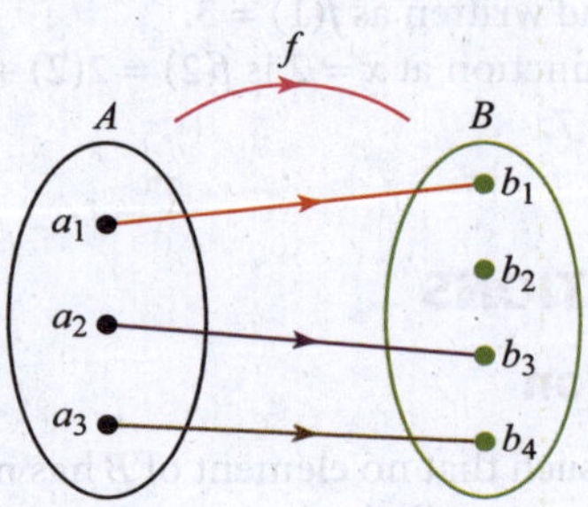

Onto Function

If $g: A \to B$ is a function such that every element of B has at least one pre-image in A, then g is called an onto function or surjective function.

In this case $g(A) = B$.

(i)

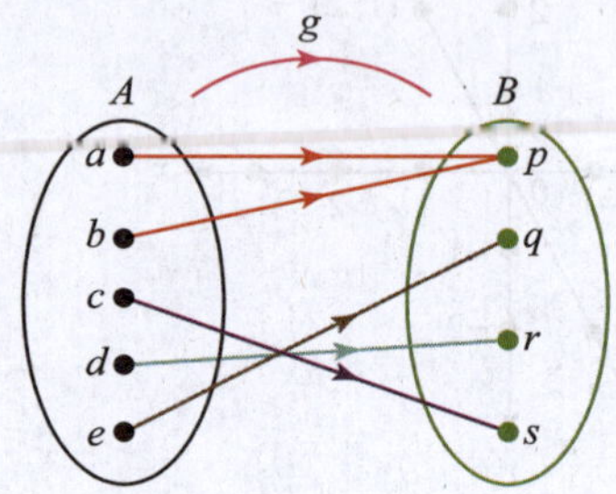

(ii)

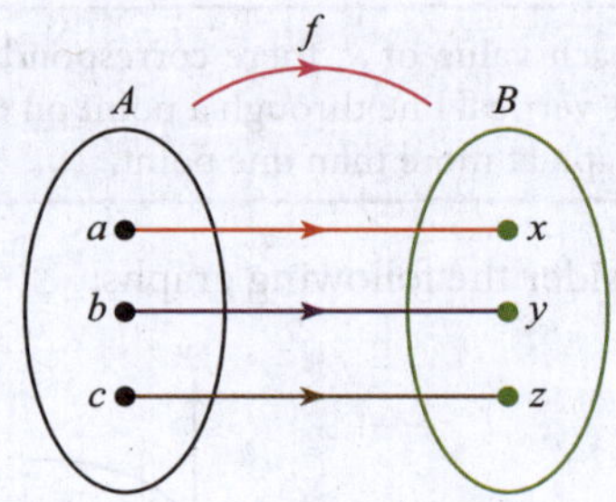

(iii)

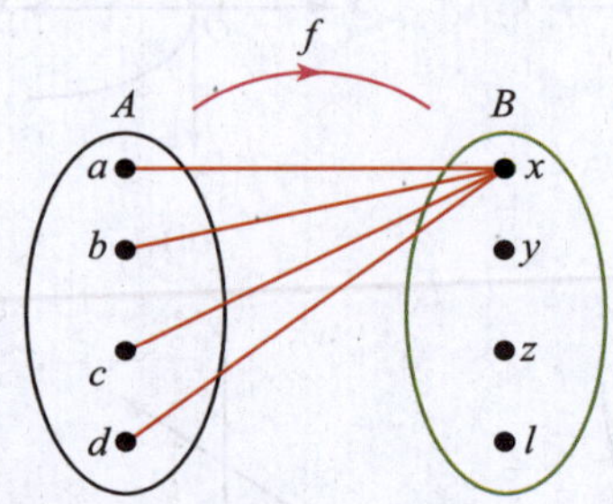

A function can never be one–many.

Bijective Function

A function $f : A \to B$ is said to be bijective if f is both one–one and onto. In the above example, f in (ii) is bijective.

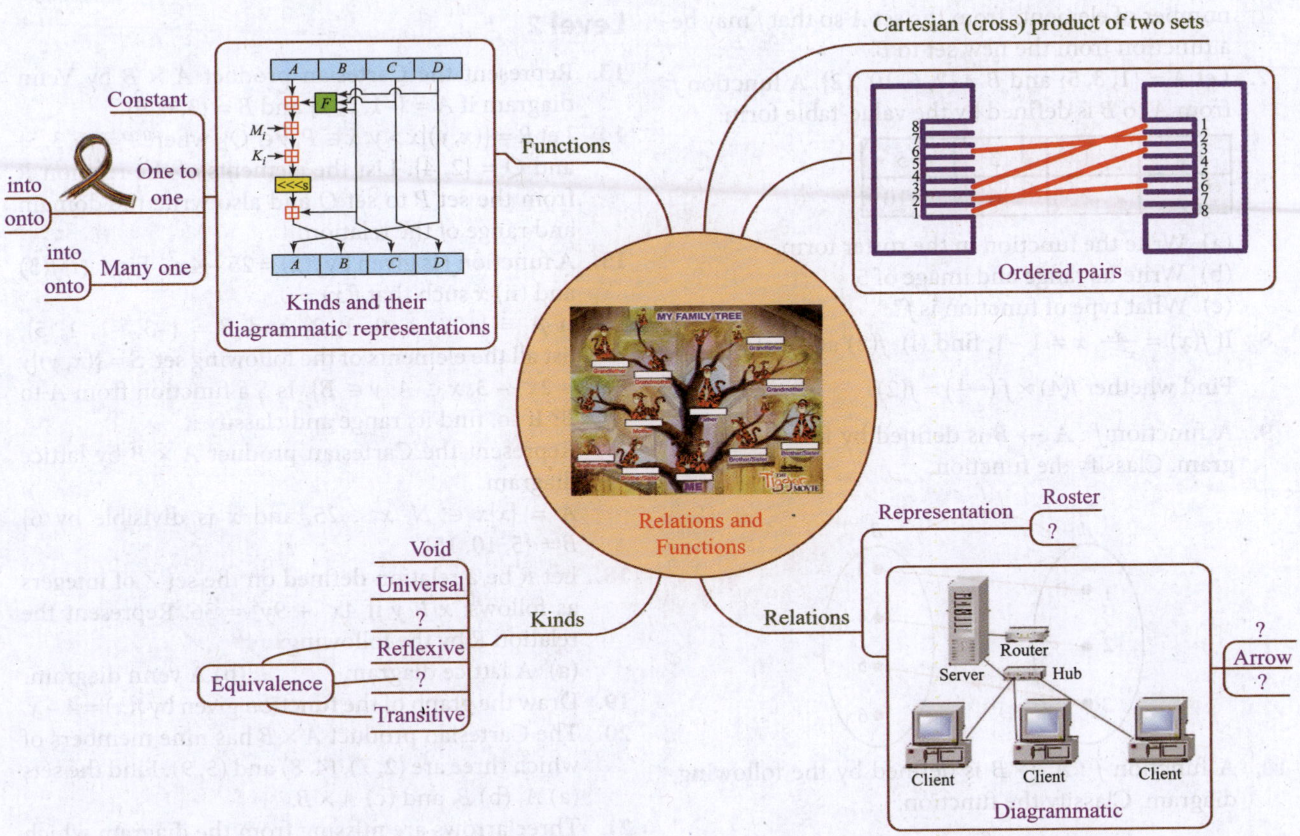

TEST YOURSELF

Level 1

1. Write the set in the roster form: $\{(x, y)|x \in N, y \in N, x + 2y = 7\}$.
2. If $A=\{-2,-1,0\}$, $B=\{-1,1,2\}$, and $C=\{x|x^2-3x+2=0\}$, verify $A \times (B \cup C) = (A \times B) \cup (A \times C)$.
3. Write the Cartesian product $X \times Y$ for the following diagram.

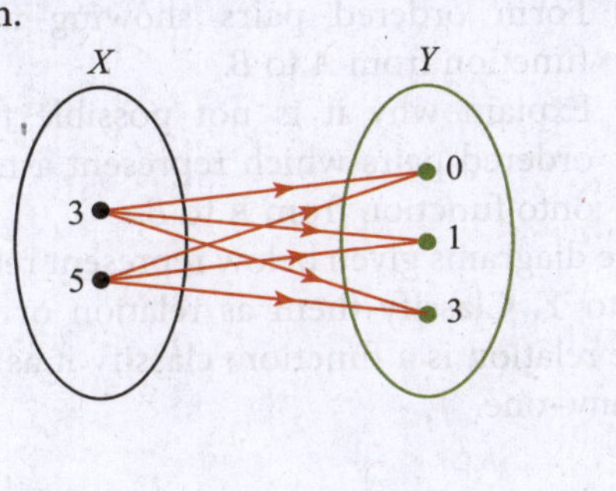

4. Write the Cartesian product $X \times Y$ for the following diagram.

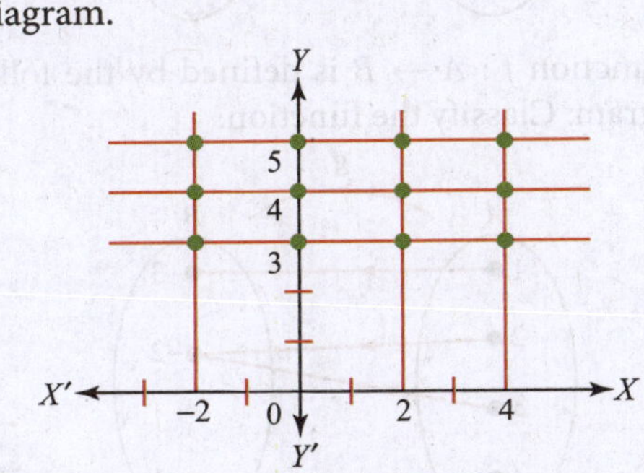

5. Let $R = \{(x, y)|2x + 3y < 10, x \in N, y \in N\}$ be a relation on N. Write the relation in the roster form.
6. Let $A = \{-2, -1, 0, 1\}$, $B = \{1, 2, 3, 4\}$, and f be a subset of $A \times B$ given by $f = \{(x, y)|x + y = 1\}$. Is f a function from A to B? If not, remove minimum

number of elements from the set A so that f may be a function from the new set to B.

7. Let $A = \{1, 3, 5\}$ and $B = \{2, 6, 10, 12\}$. A function f from A to B is defined by the value-table form:

x	1	3	5
y	2	6	10

(a) Write the function in the roster form.
(b) Write its range and image of 5.
(c) What type of function is f?

8. If $f(x) = \frac{x}{x^2-1}$, $x \neq 1$ –1, find (i) $f(2)$ and (ii) $f\left(-\frac{1}{2}\right)$. Find whether $f(4) \times f\left(-\frac{1}{2}\right) = f(2)$.

9. A function $f : A \to B$ is defined by following diagram. Classify the function.

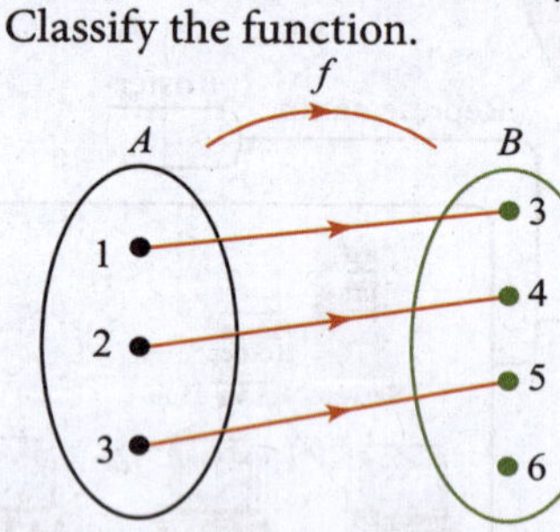

10. A function $f : A \to B$ is defined by the following diagram. Classify the function.

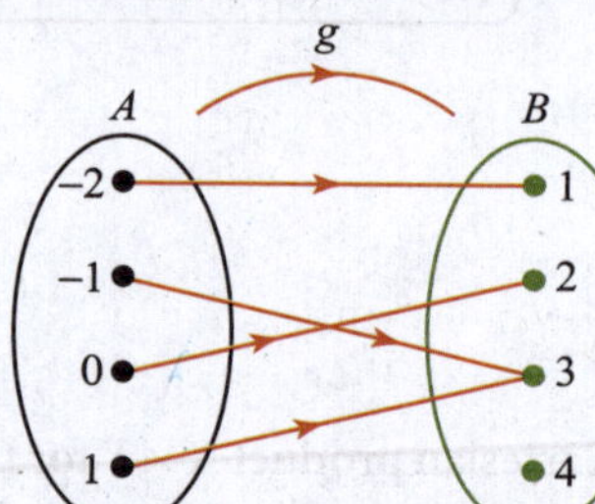

11. A function $f : A \to B$ is defined by the following diagram. Classify the function.

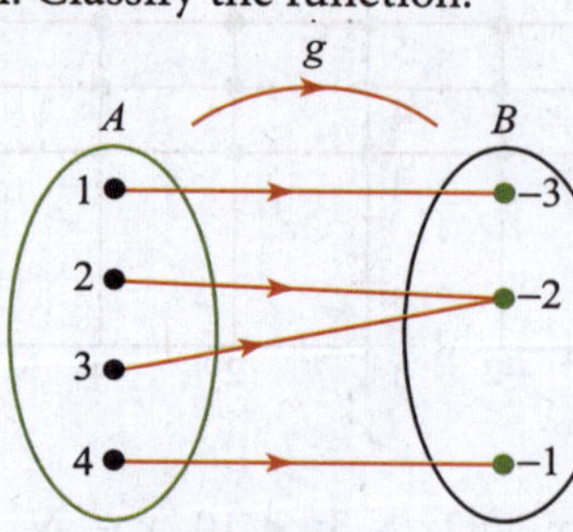

12. The Cartesian product $A \times B$ of two sets has six elements of which five are given below: $(2, a)$, $(3, b)$, $(2, b)$, $(3, a)$, and $(3, c)$. Find the sixth ordered pair, and the sets A and B.

Level 2

13. Represent the Cartesian product $A \times B$ by Venn diagram if $A = \{-1, 0, 1\}$ and $B = \{3, 6, 9\}$.

14. Let $R = \{(x, y) | x > y, x \in P, y \in Q\}$, where $P = \{1, 3, 5\}$ and $Q = \{2, 4\}$. List the elements of the relation R from the set P to set Q and also write the domain and range of the relation.

15. A function f is given by $f(x) = 25 - 4x^2$. Find (i) $f(3)$ and (ii) x such that $f(x) = 0$.

16. If $A = \{-2, -1, 0, 1, 2\}$ and $B = \{-3, -1, 1, 5\}$, list all the elements of the following set: $S = \{(x, y) | y = 2x^2 - 3, x \in A, y \in B\}$. Is S a function from A to B? If so, find its range and classify it.

17. Represent the Cartesian product $A \times B$ by lattice diagram.
$A = \{x | x \in N, x < 25$, and x is divisible by $6\}$
$B = \{5, 10, 15\}$

18. Let R be a relation defined on the set Z of integers as follows: $x \, R \, y$ if $4x^2 + 9y^2 = 36$. Represent the relation R by the following:
(a) A lattice diagram (b) A venn diagram.

19. Draw the graph of the function given by $f(x) = 4 - x$.

20. The Cartesian product $A \times B$ has nine members of which three are $(2, 7)$, $(4, 8)$ and $(5, 9)$. Find the sets (a) A, (b) B, and (c) $A \times B$.

21. Three arrows are missing from the diagram which partly show the relation 'is greater than' on the set $\{a, b, c, d\}$ of integers. Complete the Venn diagram. State which is the smallest of the four integers.

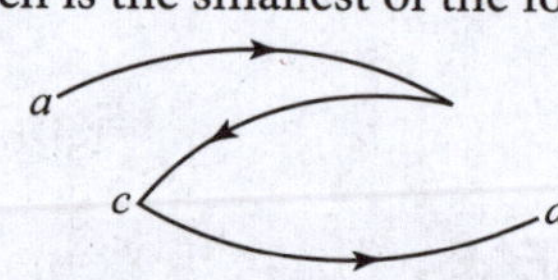

22. Let '<' be the relation of 'less than' in the set R of real numbers. Is the relation (a) reflexive, (b) symmetric, or (c) transitive? Justify your answer.

23. Given: $A = \{a, b, c, d\}$ and $B = \{1, 2, 3, 4\}$.
(a) Form ordered pairs showing a one–one function from A to B.
(b) Form ordered pairs showing a many–one function from A to B.
(c) Explain why it is not possible, to construct ordered pairs which represent a many-to-one onto function from A to B.

24. The diagrams given below represent relations from X to Y. Classify them as relation or function. If the relation is a function, classify it as one–one or many–one.

(a)
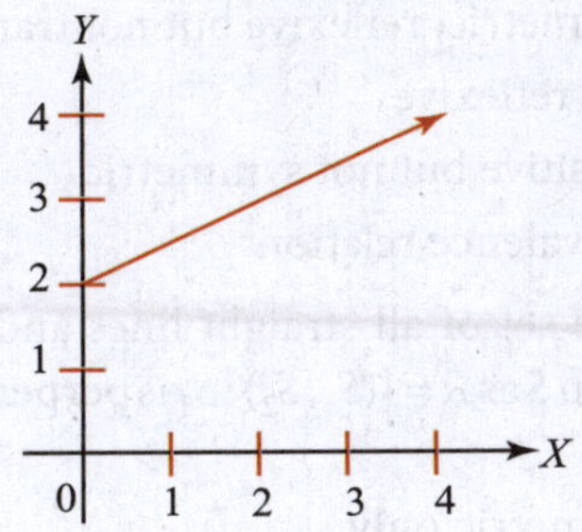

(b)
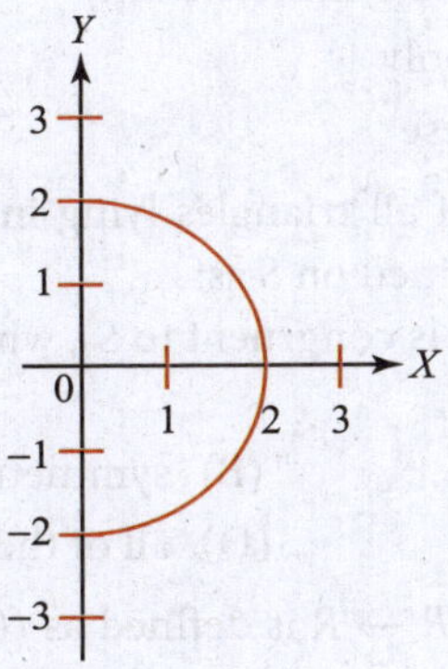

(c)
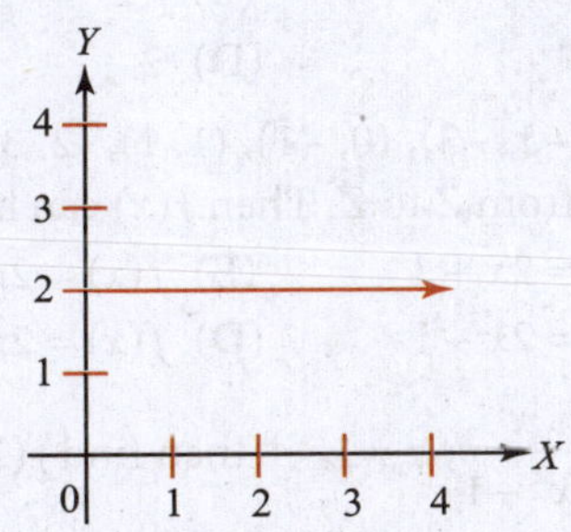

Level 3

25. On the set L of all lines on a plane, a relation, R is given by '$x\,R\,y$' if and only if '$x \perp y$'. Is R (a) reflexive, (b) symmetric, or (c) transitive?

26. $A = \{1, 2, 3, 4, 5, 6\}$. Which of the following relations on A are (a) reflexive, (b) symmetric, or (c) transitive?

(A) $\{(x, y) : x - y > 0\}$ (B) $\{(x, y) : x \le y\}$
(C) $\{(x, y) : x - y < 2\}$ (D) $\{(x, y) : x > y\}$

27. Which of the following are equivalence relations on $A = \{4, 6, 8\}$?

(A) $\{(4, 4), (6, 6), (8, 8), (4, 6), (6, 4)\}$
(B) $\{(4, 4), (6, 6), (8, 8), (4, 6)\}$
(C) $\{(4, 4), (6, 6), (8, 8), (4, 8), (8, 6)\}$
(D) $\{(4, 4), (6, 6), (8, 8), (4, 8), (8, 4)\}$

28. A set of cones have a fixed height of 12 cm and varying radii, the maximum radius being 36 cm. Define the relation between the radius and the volume of cones. What are the domain and the range of the relation? Is it a function?

29. Which of the following are functions? If so, what type of function is it?

(A) P = set of people in Bangalore, Q = set of all surnames, and $f_1 = \{(x, y) : y$ is the surname of $x, x \in P, y \in Q\}$.
(B) G = set of measurements of length in cm, H = set of measurements of area in cm^2, and $f_2 = \{(x, y) : y$ is the area of a square of side x, $x \in G, y \in H\}$.
(C) P = set of parents, C = set of children, and $f_3 = \{(x, y) : x$ is a parent of $y, x \in P, y \in C\}$.
(D) B = set of buses, P = set of passengers, and $f_4 = \{(x, y) : y$ is a passenger on $x, x \in B, y \in P\}$.

30. If $f : R \to R$ is given by $f(x) = x^3 - 1$, find x where $f(x) = 215$.

Multiple Choice Questions

31. If $(a^2 + ab + b^2, b) = (8, 2)$ then (a, b) is given by

(A) $(-1 + \sqrt{5}, -1 - \sqrt{5})$ (B) $(-1 \pm \sqrt{5}, 2)$
(C) $(2, 3)$ (D) none of these

32. If $A \times B = \phi$ then which of the following is necessarily true?

(A) $A = \phi$ or $B = \phi$ (B) $A = \phi$ and $B = \phi$
(C) Both (A) and (B) (D) None of these

33. If $A = \{1, 2\}$ and $B = \{3, 4\}$ then $A \times B =$

(A) $A \times B = \{(1, 1), (2, 2), (3, 3), (4, 4)\}$
(B) $A \times B = \{(1, 1), (1, 4), (3, 1), (3, 4)\}$
(C) $A \times B = \{(3, 1), (3, 2), (4, 1), (4, 2)\}$
(D) $A \times B = \{(1, 3), (1, 4), (2, 3), (2, 4)\}$

34. A and B are two sets given in such a way that $(A \times B)$ contains 6 elements. If three elements of $(A \times B)$ are $(1, 3)$, $(2, 5)$ and $(3, 3)$ then its remaining elements are

(A) $(1, 1)$, $(1, 5)$ and $(2, 3)$
(B) $(1, 5)$, $(2, 3)$ and $(3, 5)$
(C) $(1, 5)$ $(1, 2)$ and $(3, 5)$
(D) $(2, 1)$, $(2, 3)$ and $(3, 5)$

35. If $A \times A$ has 9 elements two of which are $(-1, 0)$ and $(0, 1)$ then the set A is

(A) $\{-1, 0\}$ (B) $\{-1, 1\}$
(C) $\{-1, 0, 1\}$ (D) $\{0, 1\}$

36. If $A \times A$ has 9 elements two of which are $(-1, 0)$ and $(0, 1)$ then what are the other elements?

(A) $(-1,-1), (-1, 1), (0, -1), (0, 0), (1, -1), (1, 2), (2, 1)$

(B) $(-1,-2), (-1, 1), (0, -1), (0, 0), (0, 2), (0, 3), (0, 4)$

(C) $(-1,-1), (0, -1), (1, 1), (0, 0), (1, 9), (9, 1), (9, 0)$

(D) $(-1,-1), (-1, 1), (0, -1), (0, 0), (1, -1), (1, 0), (1, 1)$

37. 'R' is a relation from A to B and the mapping is:

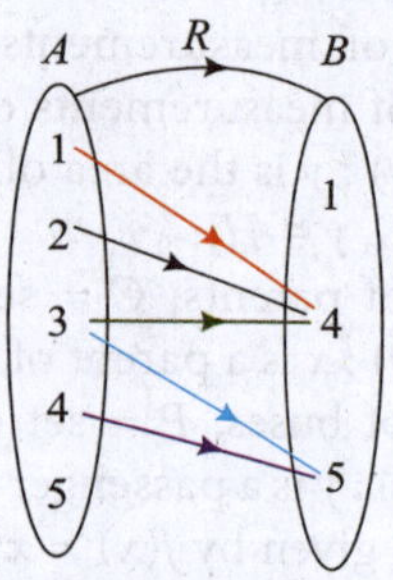

Then which of the following is correct?

(A) Domain of $R = \{1, 2, 3, 4, 5\}$

(B) Range of $R = \{1, 4, 5\}$

(C) Domain of $R = \{1, 2, 3, 4\}$

(D) Range of $R = \{1, 4\}$

38. Let $A = \{1, 3\}$ and $B = \{1, 2, 3, 4\}$. Then how many relations from A to B are possible?

(A) 2^8 (B) 2^7

(C) 2^6 (D) 2^5

39. Let N be the set of all natural numbers and let $R = \{(a, b): a, b \in N \text{ and } 2a + b = 10\}$. The domain and range of R is

(A) domain $(R) = \{1, 2, 3, 4\}$ and range $= \{2, 4, 6, 8\}$

(B) domain $(R) = \{1, 2, 3\}$ and range = all natural numbers

(C) domain $(R) = \{1, 2, 3, 4\}$ and range $= \{2, 4, 6, 8\}$

(D) none of these

40. Let $R = \{(x, x^3): x \text{ is prime}, x < 8\}$. Which of the following is correct representation of R in roster form?

(A) $R = \{(3, 3^3), (5, 5^3), (7, 7^3)\}$

(B) $R = \{(2, 2^3), (3, 3^3), (5, 5^3), (7, 7^3)\}$

(C) $R = \{(5, 5^3), (7, 7^3)\}$

(D) $R = \{(1, 1^3), (2, 2^3), (3, 3^3), (5, 5^3), (7, 7^3)\}$

41. Let R be a relation on Z, defined as $R = \{(a, b) : a, b \in Z \text{ and } a - b \in Z\}$. Then which of the following is correct about R?

(A) Symmetric, reflexive but not transitive

(B) only reflexive

(C) transitive but not symmetric

(D) equivalence relation

42. If S is the set of all straight lines and relation R is defined on S as $R = \{(S_1, S_2) : S_1 \text{ is perpendicular to } S_2\}$ then R is

(A) symmetric only

(B) reflexive and symmetric only

(C) transitive only

(D) none of these

43. If S is the set of all triangles lying in a plane and relation R is defined on S as:

$R = \{(S_1, S_2) : S_1$ is congruent to S_2, where S_1 and S_2 belong to $S\}$

(A) transitive (B) symmetric

(C) reflexive (D) all of these

44. If a function $f : R \to R$ is defined as $f(x) = 2x + 3a$ and $f(1) = 2f(2)$ then what is the value of a?

(A) -1 (B) -2

(C) 1 (D) 2

45. Let $f = \{(-1, -3), (0, -1), (1, 1), (2, 3)\}$ be a linear function from Z to Z. Then $f(x)$ can be defined as

(A) $f(x) = 2x + 1$ (B) $f(x) = 2x + 2$

(C) $f(x) = 2x - 1$ (D) $f(x) = 2x - 2$

46. If $f(x) = \dfrac{x}{x^2 - 1}, x \neq 1, -1$ then find $f(2) \times f\left(\dfrac{-1}{2}\right)$.

(A) $\dfrac{2}{3}$ (B) $\dfrac{3}{2}$

(C) $\dfrac{-3}{2}$ (D) $\dfrac{4}{9}$

47. A function $f : A \to B$ is defined as:

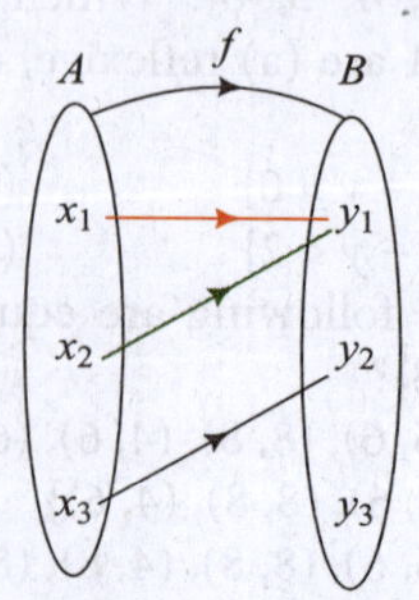

Which of the following is true?

(A) one-one function
(B) one-one and onto function
(C) many-one and into function
(D) many-one and onto function

48. Which of the following are equivalence relations on A = {4, 6, 8}?
(A) {(4, 4), (6, 6), (8, 8), (4, 6), (6, 8)}
(B) {(4, 4), (6, 6), (8, 8), (4, 6), (4, 8)}
(C) {(4, 4), (6, 6), (8, 8), (4, 8), (8, 6)}
(D) {(4, 4), (6, 6), (8, 8), (4, 8), (8, 4)}

49. If $f(x) = x^3$ then what is the value of $\frac{f(5)-f(1)}{(5-1)}$?
(A) 30 (B) 31
(C) 32 (D) 33

OLYMPIAD AND NTSE LEVEL EXERCISES

50. If $f(x)=\frac{x}{x-1}$ then $\frac{f(a)}{f(a+1)}=$
(A) $f(-a)$ (B) $f\left(\frac{1}{a}\right)$
(C) $f(a^2)$ (D) $f\left(\frac{-a}{a-1}\right)$

51. If $y=f(x)=\frac{ax+b}{cx-a}$ then x is equal to
(A) $1/f(x)$ (B) $1/f(y)$
(C) $yf(x)$ (D) $xf(y)$

52. If the domain of function $f(x) = x^2 - 6x + 7$ is $(-\infty, \infty)$, then the range of function is
(A) $(-\infty, \infty)$ (B) $[-2, \infty)$
(C) $(-2, 3)$ (D) $(-\infty, -2)$

53. The domain of the function $f(x) = \log_{3+x}(x^2 - 1)$ is
(A) $(-3, -1) \cup (1, \infty)$
(B) $[-3, -1) \cup [1, \infty)$
(C) $(-3, -2) \cup (-2, -1) \cup (1, \infty)$
(D) $[-3, -2) \cup (-2, -1) \cup [1, \infty)$

54. The domain of the function
$f(x)=\sqrt{x-x^2}+\sqrt{4+x}+\sqrt{4-x}$ is
(A) $[-4, \infty)$ (B) $[-4, 4]$
(C) $[0, 4]$ (D) $[0, 1]$

55. The largest possible set of real numbers which can be the domain of $f(x)=\sqrt{1-\frac{1}{x}}$ is
(A) $(0, 1) \cup (0, \infty)$ (B) $(-1, 0) \cup (1, \infty)$
(C) $(-\infty, -1) \cup (0, \infty)$ (D) $(-\infty, 0) \cup (1, \infty)$

56. If $f(x) = \log_a x$ and $F(x) = a^x$, then $F[f(x)]$ is
(A) $f[F(x)]$ (B) $f[F(2x)]$
(C) $F|f(2x)|$ (D) $F[(x)]$

57. Let two functions $f(x)$ and $g(i)$ are defined on $R \to R$ such that

$$f(x)=\begin{cases} x^2, & x \text{ is irrational} \\ 2-x^2, & x \text{ is rational}\end{cases} \quad \text{and}$$

$$g(x)=\begin{cases} 2-x^2, & x \text{ is irrational} \\ x^2, & x \text{ is rational}\end{cases}$$

Then the function $f+g : R \to R$ is
(A) injective as well as surjective.
(B) injective but not surjective.
(C) surjective but not injective.
(D) neither surjective nor injective

58. Range of the function $f(x)=\frac{x^2}{x^2+1}$ is
(A) $(-1, 0)$ (B) $(-1, 1)$
(C) $[0, 1)$ (D) $(1, 1)$

59. The domain of the function
$f(x) = \log(\sqrt{x-4}+\sqrt{6-x})$ is
(A) $[4, \infty)$ (B) $(-\infty, 6]$
(C) $[4, 6]$ (D) None of these

8 Progressions

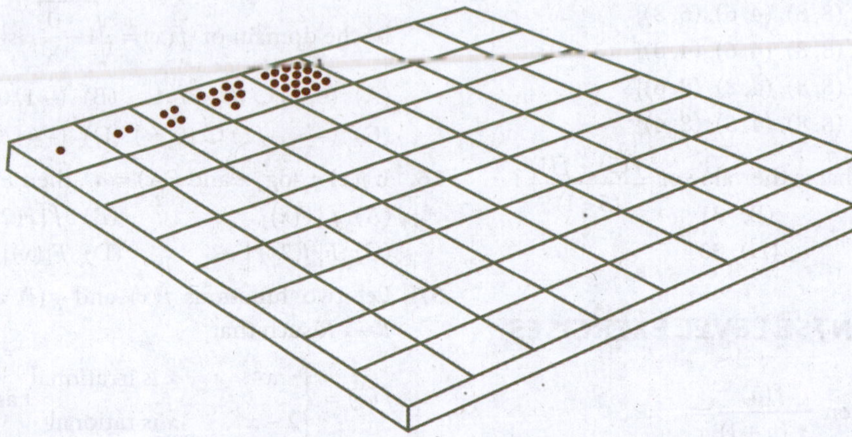

A single grain of wheat placed on the first square of the chess board is doubled on the second square; this number is doubled on the third, and so on, presumably for all 64 squares. Note that each square contains one more grain than all the preceeding squares combined. Does enough wheat exist in the world to fill all 64 squares in this manner?

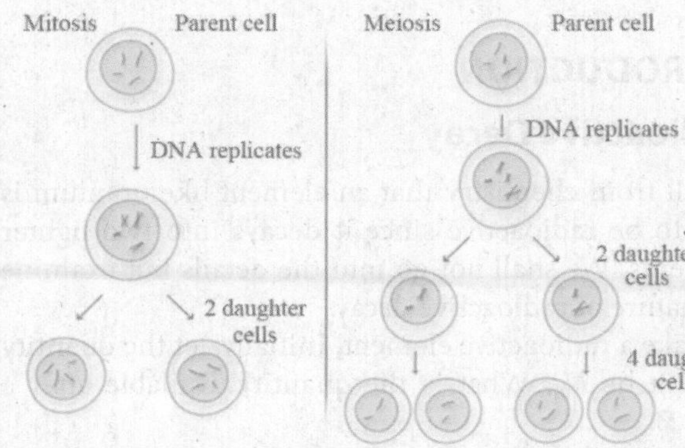

OBJECTIVES

After reading this chapter you should be able to

- understand the meaning of sequence of numbers, series of numbers
- define arithmetic progression
- find the sum to *n* terms of an arithmetic progression
- find arithmetic mean
- define geometric progression
- find *n*th term of a geometric progression
- find the sum to *n* terms of a geometric progression
- define harmonic progression
- find harmonic mean
- show $A > G > H$

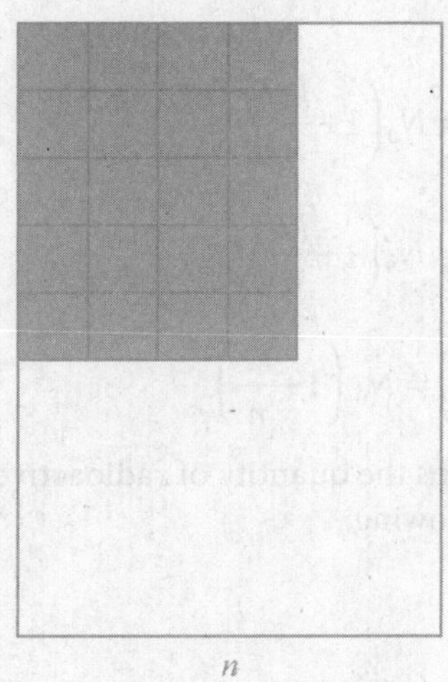

INTRODUCTION

Radioactive Decay

Recall from chemistry that an element like uranium is said to be radioactive since it decays into two lighter elements. We shall not go into the details but examine the nature of radioactive decay.

Take a radioactive element. Initially, let the quantity present be N_0. What is the quantity available after a time t?

Let N_n be the quantity remaining after time t. Now suppose that the total time t be broken up into n steps.

At the first step we have N_1, but $N_1 - N_0$ is proportional to the initial quantity, namely N_0, as well as the time interval $\frac{t}{n}$.

Therefore, we have the following:

$$N_1 - N_0 \propto N_0 \frac{t}{n} \quad \text{or} \quad N_1 - N_0 = kN_0 \frac{t}{n}$$

where k is a constant.

Similarly,

$$N_2 - N_1 = kN_1 \frac{t}{n}$$

$$N_3 - N_2 = kN_2 \frac{t}{n}$$

$$N_n - N_{n-1} = kN_{n-1} \frac{t}{n}$$

These equations may be rewritten as follows:

$$N_1 = N_0\left(1+\frac{kt}{n}\right)$$

$$N_2 = N_1\left(1+\frac{kt}{n}\right) = N_0\left(1+\frac{kt}{n}\right)^2$$

$$N_3 = N_2\left(1+\frac{kt}{n}\right) = N_0\left(1+\frac{kt}{n}\right)^3$$

$$N_n = N_{n-1}\left(1+\frac{kt}{n}\right) = N_0\left(1+\frac{kt}{n}\right)^n$$

That is, after n steps the quantity of radioactive element left over is the following:

$$N_n = N_0\left(1+\frac{kt}{n}\right)^n$$

You learn later in the chapter that this is a 'geometric progression'.

SEQUENCES AND SERIES

Karl Friedrich Gauss (1777–1855)

Gauss was a German mathematician. He demonstrated a method for constructing a heptadecagon using only a straightedge and compass. He also showed that only regular polygons with a certain number of sides could be constructed in that manner. He proved the fundamental theorem of algebra, which states that every polynomial has a root of the form $a + ib$. He proved the fundamental theorem of arithmetic, which states that every natural number can be represented as the product of primes in only one way.

At 24 years of age, Gauss published one of the most brilliant achievements in mathematics, *Disquisitiones Arithmeticae*, in which he systematized the study of number theory (properties of the integers). He proved that every number is the sum of at most three triangular numbers and developed the algebra of congruencies. Gauss developed the method of least squares fitting which enabled him to calculate the orbit of the asteroid Ceres. He published his monumental treatise on celestial mechanics, *Theoria Motus*, in 1806. He developed the magnetometer and, with Wilhelm Weber, measured the intensity of magnetic forces. With Weber, he also built the first successful telegraph.

Sequence

A set of numbers arranged in some definite order formed according to some definite rule is called a *sequence*. The elements of the set are called *terms* of the sequence.

Consider the following sets of numbers:

(a)	1	11	5	18	25	…
(b)	1	5	11	18	25	…
(c)	1	2	3	4	5	…
(d)	23	18	13	8	3	…
(e)	1	5	25	125	625	…

We observe that terms in (a) are not arranged in any particular order and as such it is not a sequence. In (b) terms are arranged in ascending order, but they do not follow any rule. Therefore, this is again not a sequence.

In the case of (c), (d), and (e), we see that the terms are arranged in some order and obey a definite rule. In (c) each term is obtained by adding 1 to the preceding term, while in (d) each succeeding term is obtained by subtracting 5 from each preceding one. In (e) we multiply a term by 5 to obtain the next term.

Series

The sum of the terms of the sequence is defined as a series. Some examples of series are as follows:

(a) $1 + 2 + 3 + 4 + 5 + \dots$
(b) $23 + 18 + 13 + 8 + 3 + \dots$
(c) $1 + 5 + 25 + 125 + \dots$

A series which can be written to its last term is called a *finite series*. A series which cannot be written to the last term is called an *infinite series*.

Arithmetic Progression

Consider the following sequences:

(a)	1	3	5	7	9
(b)	1	7	13	19	25
(c)	$\frac{1}{2}$	$\frac{3}{2}$	$\frac{5}{2}$	$\frac{7}{2}$	$\frac{9}{2}$
(d)	25	20	15	10	5

In each of the above sequences, the difference between any term and the term that precedes it remains constant. In (a), difference is 2, in (b) it is 6, in (c) it is 1, and in (d) it is –5. These four sequences are called *arithmetic progressions. Thus, a sequence in which the difference between any two consecutive terms remains constant is called an arithmetic sequence or an arithmetic progression.* It is briefly written as AP. The difference is called the common difference of the AP The first term of an AP is generally denoted by *a* and the common difference by *d*, the general form of an AP, being $a, (a + d), (a + 2d) \dots$. An AP is completely determined if one of the terms and the common difference are known.

General Term of an AP

If *a* is the first term and *d* is the common difference of an AP, then the first, second, and third terms are as follows:

$a = a + (1 - 1)d$
$a + d = a + (2 - 1)d$
$a + 2d = a + (3 - 1)d$

We observe that in each of the above terms the coefficient of the *a* is unity and the coefficient of *d* is one less than the serial number of the term in the series.

Thus, *n*th term is as follows:

$t_n = a + (n - 1)d = l$

where *l* is the last term.

> The above formula involves four quantities *a*, *n*, *d*, and t_n. If any three of them are known, the fourth can be determined.

Worked Examples

1. Find the *n*th term and hence the 50th term of the AP 2, 5, 8, 11, …

Solution

Here, 2 is *a* and 3 is *d*.

$$\begin{aligned} t_n &= a + (n - 1)d \\ &= 2 + (n - 1)3 \\ &= 2 + 3n - 3 \\ &= 3n - 1 \end{aligned}$$

The 50th term $3 \times 50 - 1 = 150 - 1 = 149$.

2. Which term of the sequence 37, 32, 27, 22 … is –103?

Let the *n*th term $t_n = -103$.

Here, $a = 37, \quad d = -5, \quad t_n = -103$.

$$\begin{aligned} -103 &= 37 + (n - 1)(-5) \\ -103 &= 37 - 5n + 5 \\ 5n &= 145 \\ n &= 29 \end{aligned}$$

Hence, 29th term of the given sequence is –103.

3. If the 5th and 17th terms of an AP are 7 and 25, respectively, find the 13th term.

Solution

Let *a* be the first term and *d* the common difference of the given AP.

$$t_5 = a + (5 - 1)d = 7 \quad (1)$$
$$t_{17} = a + (17 - 1)d = 25 \quad (2)$$

Subtracting Eq. (1) from Eq. (2), we get the following:

$$12d = 18 \Leftrightarrow d = \frac{18}{12} = \frac{3}{2}$$

Again from Eq. (1) we have the following:

$$a = 7 - 4 \times \frac{3}{2} = 1$$

$$\begin{aligned} \therefore t_{13} &= a + (13 - 1)d \\ &= 1 + 12 \times \frac{3}{2} = 19 \end{aligned}$$

4. Find the sequence whose nth term is $5n - 12$.

Solution

Putting $n = 1, 2, 3, 4 \ldots$ in succession, we get the following:

$t_1 = 5 \times 1 - 12 = -7$

$t_2 = 5 \times 2 - 12 = -2$

$t_3 = 5 \times 3 - 12 = 3$

$t_4 = 5 \times 4 - 12 = 8$

Hence, the sequence is $-7, -2, 3, 8 \ldots$.

5. If pth term of an AP is q and qth term is p, find rth term.

Solution

$t_p = a + (p - 1)d = q$ (1)

$t_q = a + (q - 1)d = p$ (2)

Subtracting Eq. (2) from Eq. (1), we get the following:

$(p - q)d = q - p$

$$d = \frac{-(p-a)}{(p-a)} = -1$$

Again from Eq. (1), we have the following:

$a = q - (p - 1)(-1)$

$= q + p - 1$

$\therefore t_r = a + (r - 1)d$

$= (q + p - 1) + (r - 1)(-1)$

$= p + q - r$

Sum to *n* Terms of an AP

Consider a problem of finding the sum of natural numbers from 1 to 100:

$1 + 2 = 3$

$1 + 2 + 3 = 6$

$1 + 2 + 3 + 4 = 10$

If we try to complete in this way, would be very long and tedious task to find the sum up to 100. To avoid this, we devise a procedure which gives the desired result more conveniently.

If the sum of the series is denoted by S, then we have the following:

$S = 1 + 2 + 3 \ldots 98 + 99 + 100$

Writing in descending order, we get the following

$S = 100 + 99 + 98 \ldots 3 + 2 + 1$

On adding, we have the following:

$2S = 101 + 101 + 101 \ldots 101 + 101 + 101$

$2S = 101 \times 100$

Thus, 101×100 is equal to twice the sum of natural numbers from 1 to 100.

$\therefore 1 + 2 + 3 + \cdots + 98 + 99 + 100$

$$= \frac{101 \times 100}{2} = 5050$$

The above procedure can be generalised for all series of this type.

Let a be the first term, d common difference, l the last term, and n be the number of terms, of an AP. Then the AP is $a, a + d \ldots a + 2d \ldots l - 2d, l - d, l$.

If S_n denotes the sum to n terms of the AP,

$S_n = a + (a + d) + (a + 2d) \cdots + a + (n - 1)d$ (1)

$S_n = a + (n - 1)d + a + (n - 2)d + a + (n - 3)d \ldots a$ (2)

Adding Eqs (1) and (2), we get the following:

$2S_n = 2a + (n - 1)d + 2a + (n - 1)d$

$+ \cdots + 2a + (n - 1)d \ldots n$ terms

$2S_n = n[2a + (n - 1)d]$

$$S_n = \frac{n}{2}[2a + (n-1)d] \quad \text{(A)}$$

$$S_n = \frac{n}{2}[a + a + (n-1)d]$$

Replacing $a + (n - 1)d$ by l we get the following:

$$S_n = \frac{n}{2}(a + l) \quad \text{(B)}$$

> The above formula involves four quantities a, n, d, and S_n. Thus, if any three of these are given, the fourth can be found.

Worked Examples

1. The sum the sequence $25, 21, 17 \cdots$ to 16 terms.

Solution

Here, $a = 25$, $d = -4$, and $n = 16$.

$$S_{16} = \frac{16}{2}[2 \times 25 + (16-1)(-4)]$$

$= 8(50 - 60) = -80$

Hence, the required sum is -80.

2. How many terms of the series $-9\ -3\ +3\ +9\ +15 \ldots$ must be taken to make the sum 63?

Solution

Here $a = -9$, $d = 6$, and $S_n = 63$.

Let the number of terms be n.

$$S_n = \frac{n}{2}[2a + (n-1)d]$$

$$63 = \frac{n}{2}[2 \times (-9) + (n - 1)6]$$

$63 = n(3n - 12) \quad \Rightarrow \quad 3n^2 - 12n - 63 = 0$

$n^2 - 4n - 21 = 0$

$(n - 7)(n + 3) = 0$

$\therefore n = 7$ or -3

Since n is the number of terms, it must be positive. Hence, the required number of terms is 7.

> If we take seven terms of the series, we have -9 -3 $+3$ $+9$ $+15$ $+21$ $+27$ whose sum is 63. The sum of last three terms taken in the reverse order, i.e., $27 + 21 + 15$, is 63. This gives us an interpretation of the negative value of n.

3. Find the sequence whose sum to n terms is $an^2 + bn + c$.

Solution

$S_n = an^2 + bn + c$

$S_{n-1} = a(n - 1)^2 + b(n - 1) + c$

$t_n = S_n - S_{n-1}$

$= (an^2 + bn + c) - (a(n - 1)^2 + b(n - 1) + c)$

$= 2an - a + b$

Putting $n = 1, 2, 3, 4$, we have the following:

$t_1 = 2a \times 1 - a + b = a + b$

$t_2 = 2a - 2 - a + b = 3a + b$

$t_3 = 2a \times 3 - a + b = 5a + b$

$t_4 = 3a \times 4 - a + b = 7a + b$

Hence, the required sequence is $a + b$, $3a + b$, $5a + b$, $7a + b$ … which is an AP with $a + b$ as the first term and $2a$ as common difference.

Sum of First *n* Natural Numbers

If S_n denotes the sum of first n natural numbers, then $S_n = 1 + 2 + 3 + \cdots + n$.

This is an AP having n terms whose first term is 1 and the common difference is 1.

$$\therefore S_n = \frac{n}{2}[2+(n-1)\times 1] = \frac{n(n+1)}{2}$$

Sum of the Squares of First *n* Natural Numbers

If S_{n^2} denotes the sum of the squares of n natural numbers, then $S_{n^2} = 1^2 + 2^2 + 3^2 + \cdots + n^2$.

Consider the identity

$x^3 - (x - 1)^3 = 3x^2 - 3x + 1$

Putting $x = 1, 2 \ldots (n - 1)$ and n successively, we get the following:

$1^3 - 0^3 = 3.1^2 - 3.1 + 1$

$2^3 - 1^3 = 3.2^2 - 3.2 + 1$

— — — — — — — — — —

— — — — — — — — — —

— — — — — — — — — —

$(n - 1)^3 - (n - 2)^3 = 3(n - 1)^2 - 3(n - 1) + 1$

$n^3 - (n - 1)^3 = 3n^2 - 3n + 1$

Adding both sides column-wise, we have the following:

$n^3 = 3(1^2 + 2^2 + \cdots n^2) - 3(1 + 2 + \cdots + n)$
$+ (1 + 1 + 1 \ldots n \text{ terms})$

$$\therefore 3S_{n^2} = n^3 + \frac{3}{2}[n(n+1)] - n$$

$$= \frac{1}{2}[2n^3 + 3n^2 + 3n - 2n] = \frac{n}{2}(2n^2 + 3n + 1)$$

$$3S_{n^2} = \frac{n}{2}(n+1)(2n+1)$$

$$\therefore S_{n^2} = \frac{n(n+1)(2n+1)}{6}$$

Sum of the Cubes of First *n* Natural Numbers

If S_{n^3} denotes the sum of the cubes of first n natural numbers, then $S_{n^3} = 1^3 + 2^3 + 3^3 + \cdots + n^3$.

Consider the identity

$x^4 - (x - 1)^4 = 4x^3 - 6x^2 + 4x - 1$

Putting $x = 1, 2, 3, \cdots (n - 1)$ and n successively, we get the following:

$1^4 - 0^4 = 4.1^3 - 6.1^2 + 4.1 - 1$

$2^4 - 1^4 = 4.2^3 - 6.2^2 + 4.2 - 1$

— — — — — — — — — —

— — — — — — — — — —

— — — — — — — — — —

$(n - 1)^4 - (n - 2)^4$
$= 4(n - 1)^3 - 6(n - 1)^2 + 4(n - 1) - 1$

$n^4 - (n - 1)^4 = 4n^3 - 6n^2 + 4n - 1$

Adding both sides column-wise, we have the following:

$n^4 = 4(1^3 + 2^3 + \cdots n^3) - 6(1^2 + 2^2 + \cdots n^2)$
$+ 4(1 + 2 - n) - n$

$$n^4 = 4S_{n^3} - 6\frac{n(n+1)(2n+1)}{6} + \frac{4n(n+1)}{2} - n$$

$$4S_{n^3} = n^4 + n(n + 1)(2n + 1) - 2(n)(n + 1) + n$$

$$4S_{n^3} = n(n^3 + 1) + n(n + 1)(2n + 1) - 2n(n + 1)$$

$$= n(n + 1)[n^2 - n + 1 + (2n + 1) - 2]$$

$$= n(n + 1)(n^2 + n)$$

$$= [n(n + 1)]^2$$

$$S_{n^3} = \left(\frac{n(n+1)}{2}\right)^2 = (S_n)^2$$

It may be observed that the sum of the cubes of first n natural numbers is equal to the square of the sum of first n natural numbers.

Sigma Notation

For the sake of brevity, the sum of the series is denoted by writing the Greek letter 'Σ' (Sigma) before the general term, for example,

$1 + 2 + 3 + \cdots + n$ is denoted by Σn

$1^2 + 2^2 + 3^2 + \cdots + n^2$ is denoted by Σn^2

$1^3 + 2^3 + 3^3 + \cdots + n^3$ is denoted by Σn^3

Thus, using the above notation, we have the following:

$$\Sigma n = \frac{n(n+1)}{2}$$

$$\Sigma n^2 = \frac{n(n+1)(2n+1)}{6}$$

$$\Sigma n^3 = \left\{\frac{n(n+1)}{2}\right\}^2 = [\Sigma n]^2$$

Arithmetic Mean

When three quantities are in AP, the middle one is defined as arithmetic mean (AM) between the other two.

If a, A, and b are in AP, then A is termed as the AM between a and b.

If A is the arithmetic mean between a and b, then by definition a, A, and b are in AP.

$\therefore A - a = b - A$

$2A + a + b$

$$A = \frac{a+b}{2}$$

Thus, arithmetic mean between two quantities a and b is equal to half of their sum.

To insert 'n' number of arithmetic means between two given quantities a and b, $d = \frac{b-a}{n+1}$.

GEOMETRIC PROGRESSION

Geometric Progression

Consider the following sequences:

(a) 1, 3, 9, 27, 81 … …. ….

(b) 2, −8, 32, −128, 512 ……. ….

(c) 0.3, 0.03, 0.003, 0.0003 ……. ….

(d) a, ar, ar^2, ar^3 … …. ….

In each of the above sequences, the ratio of any term to the term that precedes it remains constant throughout. For example, in (a) the ratio is 3, in (b) −4, in (c) 0.1, and in (d) it is r. Such sequences are called *geometric progressions*. Thus, a sequence in which the ratio of any term to the term that precedes it remains constant is called geometric progression and is briefly written as GP. The ratio of any term to the one that precedes it is called common ratio. The first term of GP is generally denoted by a and the common ratio by r. The general form of a GP is a, ar, ar^2, ar^3 ….

General Term of a GP

If a is the first term and r the common ratio, then the GP is a, ar, ar^2, ar^3 ….

Here, the first term $= a = a\,r^{1-1}$

The second term $= ar = ar^{2-1}$

The third term $= ar^2 = ar^{3-1}$

We observe that in each of the above terms the index is one less than the serial number of the term in the series. Thus, the nth term will be $t_n = ar^{n-1}$.

The above formula has four quantities a, n, r and t_n. Thus, the fourth can be determined if any three of them are known.

Worked Examples

1. Find the nth term and hence the 7th term of the GP 3, −6, 12, −24 …

Solution

Here, $a = 3$ and $r = -2$.

$t_n = a \times r^{n-1}$

$\therefore t_n = 3(-2)^{n-1}$

$t_7 = 3(-2)^{7-1} = 3(-2)^6 = 192$

2. What term of the sequence 16, 12, 9 … is $\frac{243}{64}$?

Solution

Here, $a = 16$, $r = \frac{3}{4}$ and $t_n = \frac{243}{64}$.

$$\therefore 16\left(\frac{3}{4}\right)^{n-1} = \frac{243}{64}$$

$$\left(\frac{3}{4}\right)^{n-1} = \left(\frac{3}{4}\right)^5$$

$n - 1 = 5$

$n = 6$

Hence, the sixth term of the given sequence is $\frac{243}{64}$.

3. Find the sequence whose fifth term is 48 and ninth term is 768.

Solution

If a is the first term and r is the common ratio, then we have the following:

$$ar^4 = 48 \quad (1)$$
$$ar^8 = 768 \quad (2)$$

Dividing Eq. (2) by Eq. (1), we get the following:

$$r^4 = \frac{768}{48} = 16 = 2^4$$
$$r = 2$$

Again $a = \frac{48}{2^4} = 3$

Hence, the sequence is 3, 6, 12, …

4. Find the sequence whose nth term is $3 \times 2^{3n-2}$.

Solution

$t_n = 3 \times 2^{3n-2}$

Putting n = 1 2 3 …, we get the following:

$t_1 = 3 \times 2^1 = 6$

$t_2 = 3 \times 2^4 = 48$

$t_3 = 3 \times 2^7 = 384$

The series is 6, 48, 384 ….

Sum to *n* Terms of a GP

If a is the first term, r is the common ratio, and S_n is the sum to n terms, then we have the following:

$$S_n = a + ar + ar^2 + \cdots + ar^{n-1} \quad (1)$$
$$r\,S_n = ar + ar^2 + \cdots + ar^{n-1} + ar^n \quad (2)$$

Subtracting Eq. (2) from Eq. (1), we get the following:

$$S^n(1 - r) = a - ar^n = a(1 - r^n)$$
$$\therefore S_n = a \times \frac{1-r^n}{1-r} \quad (A)$$
$$S_n = \frac{a(r^n - 1)}{(r-1)} \quad (B)$$

Either of the two Eqs. (A) or (B) can be used to find the sum to n terms of a GP. It is, however, convenient to use the Eq. (A) when $r < 1$ and Eq. (B) when $r > 1$.

> (a) If $r = 1$, both (A) and (B) fail as we get $S_n = \frac{0}{0}$, which is meaningless.
> (b) In this case, we write $S_n = a + a + \cdots a$ n terms $= na$.

Worked Examples

1. Find the sum of the series to six terms $\frac{2}{9} - \frac{1}{3} + \frac{1}{2} + \cdots$

Solution

$$a = \frac{2}{9};\; r = -\frac{1/3}{2/9} = -\frac{3}{2};\; n = 6$$

$$S_n = \frac{\frac{2}{9}\left[\left(-\frac{3}{2}\right)^6 - 1\right]}{-\frac{3}{2} - 1}$$

On simplification, we get $-\frac{133}{144}$.

2. How many terms of the series 2 + 6 + 18 + ⋯ must be taken to make the sum 2186?

Solution

Here, $a = 2$, $r = 3$, and $S_n = 2186$, where n is the number of terms.

$$\frac{2(3^n - 1)}{(3-1)} = 2186$$
$$3^n - 1 = 2186$$
$$3^n = 2186 + 1 = 2187 = 3^7$$
$$\therefore n = 7$$

Thus, the number of terms is 7.

Geometric Mean

When three quantities are in GP, middle one is defined as geometric mean (GM) between the two. If a, G, and b are in GP, then G is termed as the GM between a and b.

Then by definition a, G, and b are in GP.

$$\frac{G}{a} = \frac{b}{G}$$
$$G^2 - ab$$
$$G = \sqrt{ab}$$

Thus, geometric mean between two quantities a and b is equal to the square root of their product.

> To insert n geometric means between two given quantities a and b, $r = \left(\frac{b}{a}\right)^{1/(n+1)}$.

HARMONIC PROGRESSION

Consider the following sequences:

(a) $\frac{1}{3}, \frac{1}{5}, \frac{1}{7}, \frac{1}{9}, \ldots$

(b) $1, \frac{1}{7}, \frac{1}{13}, \frac{1}{19}, \ldots$

(c) $\frac{1}{2}, \frac{1}{3}, \frac{1}{4}, \frac{1}{5}, \ldots$

(d) $\frac{1}{a}, \frac{1}{a+d}, \frac{1}{a+2d}, \frac{1}{a+3d}, \ldots$

In each of the above sequences, we observe that the sequences formed by taking the reciprocals of each term are an AP. Such sequences are called *harmonic progressions*. Thus, sequences of numbers whose reciprocals form an arithmetical progression are called harmonic progressions.

Most of the results in HP are obtained with the help of this reciprocal property. There is no general formula for the sum of any number of terms in HP.

General Term of an HP

Let the given HP be $\frac{1}{a}, \frac{1}{a+d}, \frac{1}{a+2d}, \frac{1}{a+3d}, \ldots$.
Therefore, the corresponding AP is $a, a+d, a+2d, \ldots$.

By definition, the *n*th term of an HP is the reciprocal of the *n*th term of the corresponding AP.

i.e., T_n in HP $= \dfrac{1}{T_n \text{ of the corresponding AP}}$

But the *n*th term of the corresponding AP is $a+(n-1)d$.

$$\therefore T_n \text{ of HP} = \frac{1}{a+(n-1)d}$$

Worked Examples

1. Find the *n*th term of the following sequence:

$\frac{4}{7}, \frac{12}{25}, \frac{12}{29}, \ldots$

Solution

The reciprocals of every term of the given sequence are as follows:

$\frac{7}{4}, \frac{25}{12}, \frac{29}{12}, \ldots$

This is in AP with the common difference $\frac{1}{3}$. Thus, the given sequence is in HP.

Therefore, the *n*th term of the above AP is as follows:

$$= \frac{7}{4} + (n-1)\frac{1}{3}$$

$$= \frac{17+4n}{12}$$

Hence, the *n*th term of the given HP $= \frac{12}{17+4n}$.

2. The 6th and 12th terms of a sequence in HP are 12 and 6, respectively. What is the 18th term?

Solution

If the given HP is

$\frac{1}{a}, \frac{1}{a+d}, \frac{1}{a+2d}, \frac{1}{a+3d}, \ldots$, then we have the following:

$$T_n = \frac{1}{a+(n-1)d}$$

$$T_6 = \frac{1}{a+5d} = 12$$

$$T_{12} = \frac{1}{a+11d} = 6$$

$$a+5d = \frac{1}{12} \quad (1)$$

$$a+11d = \frac{1}{6} \quad (2)$$

Subtracting Eq. (1) from Eq. (2), we get the following:

$$6d = \frac{1}{12}$$

$$d = \frac{1}{72}$$

Substituting it in Eq. (1), we get the following:

$$a = \frac{1}{72}$$

$$\therefore T_{18} = \frac{1}{a+17d} = \frac{1}{\frac{1}{72}+\frac{17}{72}} = 4$$

3. The *m*th term of an HP is *n* and the *n*th term is *m*. Find the $(m+n)$th term.

Solution

If the given HP is $\frac{1}{a}, \frac{1}{a+d}, \frac{1}{a+2d}, \frac{1}{a+3d}, \ldots$, then we have the following:

$$T_m = \frac{1}{a+(m-1)d} = n$$

$$T_n = \frac{1}{a+(n-1)d} = m$$

$$a+(m-1)d = \frac{1}{n} \quad (1)$$

$$a+(n-1)d = \frac{1}{m} \quad (2)$$

Subtracting Eq. (2) from Eq. (1), we get the following:

$$(m-n)d = \frac{1}{n} - \frac{1}{m}$$

$$(m-n)d = \frac{m-n}{mn}$$

$$\Rightarrow d = \frac{1}{mn}$$

Substituting this value of *d* in Eq. (1), we get the following:

$$a+(m-1)\frac{1}{mn}=\frac{1}{n}$$

or $$a=\frac{1}{n}-\frac{1}{n}+\frac{1}{mn}$$

$$a=\frac{1}{mn}$$

Again, $$T_{m+n}=\frac{1}{a+(m+n-1)d}$$

$$=\frac{1}{\frac{1}{mn}+(m+n-1)\frac{1}{mn}}$$

$$=\frac{1}{\frac{1}{mn}+\frac{m+n}{mn}-\frac{1}{mn}}$$

i.e., $$T_{m+n}=\frac{mn}{m+n}$$

4. If a, b, and c are in HP, then show that $\frac{a-b}{b-c}=\frac{a}{c}$.

Solution

Since a, b, and c are in HP, we have $\frac{1}{a}$, $\frac{1}{b}$, and $\frac{1}{c}$ in AP.

$$\frac{1}{b}-\frac{1}{a}=\frac{1}{c}-\frac{1}{b}$$

$$\frac{a-b}{ab}=\frac{b-c}{bc}$$

$$\frac{a-b}{b-c}=\frac{ab}{bc}=\frac{a}{c}$$

> This result leads us to another definition of HP; i.e., if any three consecutive terms a, b, and c of a sequence satisfy the relation $\frac{a-b}{b-c}=\frac{a}{c}$, then a, b, and c are in HP.

HARMONIC MEAN

When three quantities are in HP, the middle one is defined to be the harmonic mean (HM) between the other two. If a, H, and b are in HP, then H is termed as the HM between a and b.

Let H be the harmonic mean between a and b. Then by definition a, H, and b are in HP; i.e., $\frac{1}{a}$, $\frac{1}{H}$, and $\frac{1}{b}$ are in AP.

$$\frac{1}{H}-\frac{1}{a}=\frac{1}{b}=\frac{a+b}{ab}$$

$$\Rightarrow H=\frac{2ab}{a+b}$$

Thus, the harmonic mean between two quantities a and b is equal to twice the product of two quantities divided by their sum.

Important Result Between Arithmetic Mean (A), Geometric Mean (G), and Harmonic Mean (H) Between a and b

(a) A, G, and H are in GP.
(b) $A > G > H$.

Proof

If A, G, and H are the arithmetic, geometric, and harmonic, respectively, between two real positive quantities a and b, then we have the following:

$$A=\frac{a+b}{2}$$

$$G=\sqrt{ab}$$

$$H=\frac{2ab}{a+b}$$

$$\therefore\ AH=\frac{a+b}{2}\times\frac{2ab}{a+b}=ab=G^2$$

G is the geometric mean between A and H. Therefore, A, G, and H are in GP.

Again

$$A-G=\frac{a+b}{2}-\sqrt{ab}$$

$$=\frac{1}{2}(\sqrt{a}-\sqrt{b})^2$$

Since a and b are both real positive quantities, $\sqrt{a}$ and $\sqrt{b}$ are real; therefore, $(\sqrt{a}-\sqrt{b})^2$ is positive.

$$\therefore\ A-G>0 \Rightarrow A>G \qquad \text{(A)}$$

and

$$G-H=\sqrt{ab}-\frac{2ab}{a+b}$$

$$=\frac{\sqrt{ab}}{a+b}(\sqrt{a}-\sqrt{b})^2$$

Since a and b is positive, $\sqrt{ab}$ and $a+b$ are both positive, and therefore we have the following:

$$G-H>0$$

$$\therefore\ G>H \qquad \text{(B)}$$

Using Eqs. (A) and (B), we have $A > G > H$.

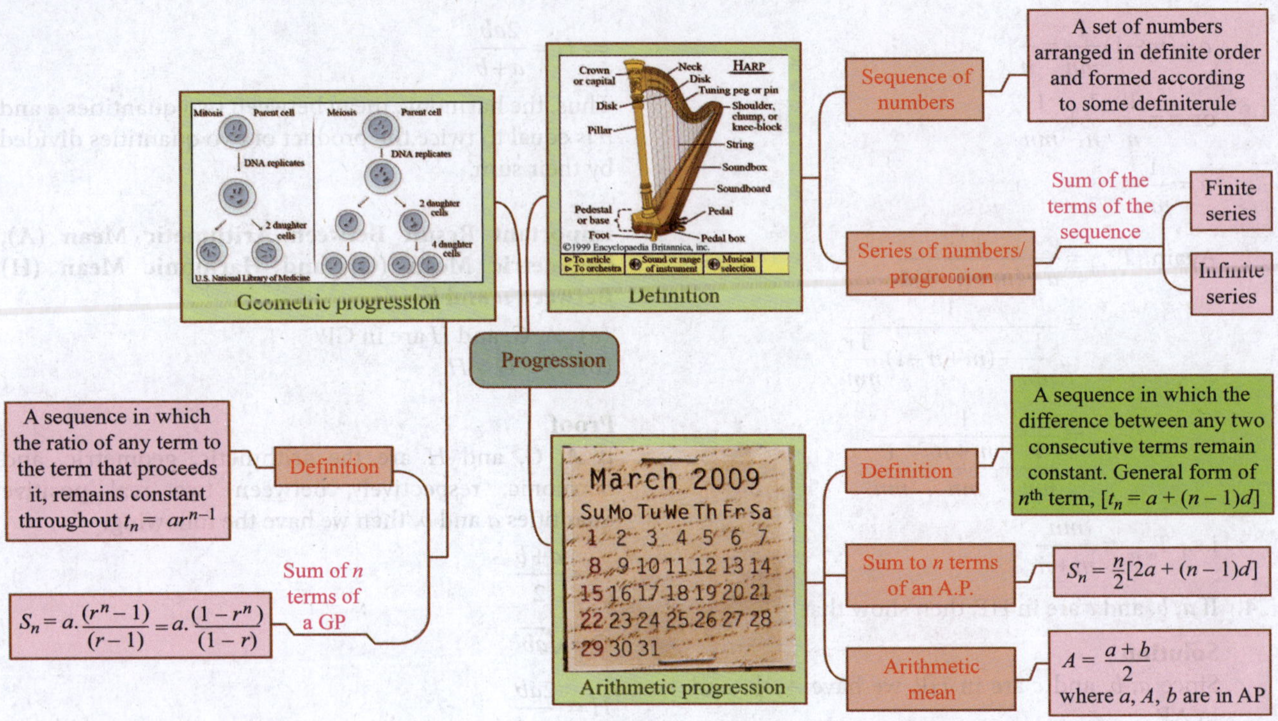

TEST YOURSELF

Level 1

1. Find the sum of the following sequences:
   (a) 2, 13/4, 9/2, … to 20 terms
   (b) 49, 44, 39, … to 17 terms
   (c) 3/4, 2/3, 7/12, …to 19 terms
   (d) 3.75, 3.5, 3.25, … to 16 terms
2. In an AP, the first term is 2, the last term is 29, and the sum is 155. Find the common difference.
3. Find the sum of the first n even numbers.
4. The sum of 15 terms of an AP is 600, and the common difference is 5. Find the first term.
5. Find the sum of 15 terms of the series whose nth term is $4n + 1$.
6. Find the sum of p terms of the series whose nth term is $(n/a) + b$.
7. Find the nth term and the term indicated against each sequence.
   (a) a, $a + 3d$, $a + 6d$, …, t_{15}
   (b) –5, –1, 3, …, t_{31}
   (c) 6, 11, …, t_{20}
8. How many terms are there in each of the following sequences?
   (a) 1, 4, 7, …, 49
   (b) 72, 70, 68, …, 40
   (c) 16, 11, 6, …, –19

Level 2

9. Find the first term and the number of terms in an AP if
   (a) the 4th, 16th, and last terms are 36, 0, and –42, respectively
   (b) the 7th, 15th, and last terms are –6, 6, and 15, respectively
   (c) the mth, nth, and last terms are n, m, and o, respectively.
10. Sum the following series:
    (a) 1 + 4 + 7 + ⋯ to 40 terms
    (b) – 4 + 4 + 12 + ⋯ to 20 terms
    (c) $(a - b)^2, (a^2 + b^2), (a + b)^2 + \cdots$ to n terms.
11. How many terms of the series 9 + 12 + 15 + ⋯ must be taken to make the sum 306?

12. If a, b, and c are in AP, show that $b + c$, $c + a$, and $a + b$ are also in AP.
13. Find three numbers in AP whose sum is 15 and the sum of the squares of the two extremes is 58.
14. Examine whether the following form a geometric progression. If so, find their sum.
 (a) $1/(1 + x^2), 1/(1 + x^2)^2, \ldots, 1/(1 + x^2)^n$
 (b) $1, x/(1 + x^2), x^2/(1 + x^2)^2, \ldots, x^n/(1 + x^2)^n$
15. Find the 8th term of the series $-1/3, 1/2, -3/4, \ldots$.
16. Find the 5th and 8th terms of the series 3, 6, 12, ….
17. Find the 10th and 16th terms of the series 256, 128, 64, ….
18. Which one of the following is in HP?
 (a) $1, \frac{1}{4}, \frac{1}{7}, \frac{1}{10}$
 (b) $\frac{1}{3}, \frac{1}{6}, \frac{1}{12}, \frac{1}{24}$
 (c) $2, \frac{1}{2}, \frac{1}{7}, \frac{1}{5}$
19. Write the nth term and the term indicated against each case:
 (a) $\frac{1}{a}, \frac{1}{a+3d}, \frac{1}{a+6d}, \ldots, t_{15}$
 (b) $\frac{4}{7}, \frac{12}{25}, \frac{12}{29}, \ldots, t_{31}$
 (c) $2, 1\frac{3}{4}, 1\frac{5}{9}, \ldots, t_{10}$
20. Find the nth term and the term indicated against each sequences:
 (a) $\sqrt{2}, \frac{1}{\sqrt{2}}, \frac{1}{2\sqrt{2}}, \ldots, t_{10}$
 (b) $0.3, 0.06, 0.012, \ldots, t_8$
 (c) $4, 2, 1, \ldots, t_{10}$
 (d) $1, 0.1, 0.01, 0.001, \ldots, t_{20}$
21. (a) Find the first term and the number of terms in a GP if the third, fifth, and last terms are $3\sqrt{3}$, $9\sqrt{3}$, and 729, respectively.
 (b) Find the first term and the number of terms in a GP if the fourth, seventh, and last terms are 10, 80, and 2560, respectively.
 (c) How many terms of the series 2 + 4 + 8 … make the sum 1022?
22. Find the sum of the following series:
 (a) $1 + \sqrt{3} + 3 + \cdots$ to 10 terms
 (b) $1 - 4 + 16 \ldots$ to 12 terms
 (c) $0.6 + 0.06 + 0.006 + \cdots$ to 16 terms
 (d) $1+\frac{1}{3}+\frac{1}{9}+\cdots+\frac{1}{3^{10}}$

Level 3

23. Find the sum of n terms of the series $(2a^2 - 1)/a$, $4a - 3/a$, $(6a^2 - 5)/a, \ldots$
24. Let a, b, and c be the pth, qth, and rth terms of an AP.
 Show that $(q - r)a + (r - p)b + (p - q)c = 0$.
25. Let a^2, b^2, and c^2 form an AP. Prove that the quantities $1/(b + c)$, $1/(c + a)$, and $1/(a + b)$ also form an AP.
26. In a GP, the first term is 7, the last term is 448, and the sum is 889. Find the common ratio.
27. The sum of three numbers in GP is 38, and their product is 1728. Find them.
28. If the pth, qth, and rth terms of a GP are a, b, and c, respectively, prove that $a^{q-r}b^{r-p}c^{p-q} = 1$.
29. Sum $1 + 2a + 3a^2 + 4a^3 + \cdots$ to n terms.
30. Sum $1 + 3x + 5x^2 + 7x^3 + \cdots$ to n terms.
31. Show that if a, b, c, and d are in GP, then $a + b$, $b + c$, and $c + d$ are also in GP.
32. If the mth term of an HP is n and the nth term is m, show that the rth term is $\frac{mn}{r}$.
33. Find the harmonic mean of 2 and 10.
34. Find the sum to n terms of the sequence $1^4 + 2^4 + 3^4 + 4^4 + \ldots n^4$.
35. Show that $1^2 + 3^2 + 5^2 + \cdots = \frac{n}{3}(4n^2 - 1)$.
36. Find the sum to n terms of the series 1.3.5 + 3.5.7 + 5.7.9 + ⋯
37. Find n such that $\frac{a^{n+1}+b^{n+1}}{a^n+b^n}$ be the arithmetic mean of a and b.
38. Find n such that $\frac{a^{n+1}+b^{n+1}}{a^n+b^n}$ be the geometric mean of a and b.
39. Find the sum of all natural numbers between 202 and 303 which are divisible by 6.
40. (a) If a, b, and c are in AP and x, y, and z are in GP, prove that $x^{b-c} \times y^{c-a} \times z^{a-b} = 1$.
 (b) Find the geometric mean of the following numbers:
  (i) 3 and 12
  (ii) a^2 and b^2
  (iii) $\sqrt{2}$ and $16\sqrt{8}$
41. Find x if
 (a) $\frac{1}{a-d}$, x, and $\frac{1}{a+d}$ are in HP
 (b) 3, x, and 9 are in HP
 (c) 2, x, and 6 are in HP

Multiple Choice Questions

42. The Fibonacci sequence is defined by $a_1 = a_2 = 1$, $a_n = a_{n-1} + a_{n-2}$, $n > 2$. Then the value of $a_5 - a_4 - a_3$ is

(A) 1 (B) 2
(C) 0 (D) −1

43. If for a sequence $<a_n>$, $S_n = 2n^2 + 9n$, where S_n is the sum of n terms, then value of a_{20} is

(A) 65 (B) 75
(C) 87 (D) 97

44. The first three terms of a sequence are 3, 3, 6. Each term after the second term is the sum of two terms preceding it. The 8th term of the sequence is

(A) 15 (B) 24
(C) 39 (D) 63

45. The common difference of the AP whose n^{th} term is given by $t_n = 5n + 4$ is

(A) 2 (B) 3
(C) 5 (D) 6

46. Which of the following can't be an AP (n^{th} term is given)?

(A) $3n + 2$ (B) $3n^2 + 2$
(C) $4n + 5$ (D) $7n + 2$

47. If mth term of an AP is $\frac{1}{n}$ and nth term is $\frac{1}{m}$ then mnth term of the AP is

(A) $\frac{1}{mn}$ (B) mn
(C) 1 (D) none of these

48. The sums of n terms of two AP's are in the ratio $(3n - 13) : (5n + 21)$. Find the ratio of their 24th terms.

(A) $\frac{1}{2}$ (B) $\frac{2}{3}$
(C) $\frac{2}{1}$ (D) $\frac{3}{2}$

49. How many terms of the series $20 + 19\frac{1}{3} + 18\frac{2}{3} + \ldots$ must be taken to make 300?

(A) 25 (B) 36
(C) both (A) and (B) (D) none of these

50. If N be the set of natural numbers is partitioned into groups $S_1 = \langle 1 \rangle, S_2 = \langle 2,3 \rangle, S_3 = \langle 4,5,6 \rangle, \ldots$ then the sum of the numbers in S_{50} is

(A) 62525 (B) 57567
(C) 62505 (D) none

51. If $(1 + 3 + 5 + \ldots + p) + (1 + 3 + 5 + \ldots + q) = (1 + 3 + 5 + \ldots + r)$, where each set of parentheses contains the sum of consecutive odd integers as shown then the smallest possible value of $p + q + r$ $(p > 6)$ is

(A) 18 (B) 19
(C) 20 (D) 21

52. Let α, β be roots of the equation $x^2 - 2x + A = 0$ and γ, δ be the roots of the equation $x^2 - 18x + B = 0$. If α, β, γ and δ $(\alpha < \beta < \gamma < \delta)$ are in arithmetic progression, then find the values of A and B.

(A) $A = -2, B = 77$ (B) $A = -3, B = 77$
(C) $A = 77, B = -3$ (D) none of these

53. If a, b, c and d are distinct integers which form an increasing AP such that $d = a^2 + b^2 + c^2$ then the value of $a + b + c + d$ is

(A) 1 (B) 2
(C) 3 (D) 0

54. The pth, q^{th} and r^{th} terms of an AP are a, b and c, respectively. The value of $a(q-r) + b(r-p) + c(p-q)$ is

(A) 1 (B) −1
(C) 0 (D) 1/2

55. Let S_n denotes the sum of n terms of an AP. If $S_{2n} = 3S_n$, then the ratio $\frac{S_{3n}}{S_n}$ is equal to

(A) 9 (B) 6
(C) 16 (D) 12

56. If $a_1, a_2, a_3, \ldots, a_n$ are in AP, where $a_i > 0$ for all i, then the value of

$$\frac{1}{\sqrt{a_1}+\sqrt{a_2}} + \frac{1}{\sqrt{a_2}+\sqrt{a_3}} + \ldots + \frac{1}{\sqrt{a_{n-1}}+\sqrt{a_n}}$$ is

(A) $\frac{1}{\sqrt{a_1}+\sqrt{a_n}}$ (B) $\frac{1}{\sqrt{a_1}-\sqrt{a_n}}$
(C) $\frac{n}{\sqrt{a_1}-\sqrt{a_n}}$ (D) $\frac{n-1}{\sqrt{a_1}+\sqrt{a_n}}$

57. If $(m + n)$th and $(m - n)$th terms of a GP are p and q, respectively, then the mth term of the GP is

(A) $p\left(\frac{q}{p}\right)^{\frac{m}{2n}}$ (B) $\sqrt{pq}$

(C) $\sqrt{p/q}$ (D) none of these

58. Find the sum up to n terms of the series $a + aa + aaa + aaaa + \ldots, \forall a \in N$ and $1 \le a \le p$.

(A) $\frac{a}{9}\left[\frac{10}{9}(10^n - 1) - n\right]$

(B) $a\left\{\frac{10}{9}(10^n - 1) - n\right\}$

(C) $\frac{1}{9}\left\{\frac{10}{9}(10^n - 1) - n\right\}$

(D) $\frac{a}{9}\left\{\frac{(10^n - 1)}{9} - n\right\}$

59. Find the sum up to n terms of the series $0.b + 0.bb + 0.bbb + 0.bbbb + \ldots, \forall b \in N$ and $1 \le b \le 9$.

(A) $b\left\{n - \frac{1}{9}\left(1 - \left(\frac{1}{10}\right)^n\right)\right\}$

(B) $\frac{b}{9}\left[n - \frac{1}{9}\left\{1 - \left(\frac{1}{10}\right)^n\right\}\right]$

(C) $\frac{1}{9}\left[n - \frac{1}{9}\left\{1 - \left(\frac{1}{10}\right)^n\right\}\right]$

(D) none of these

60. Let x_1, x_2 be the roots of the equation $x^2 - 3x + A = 0$ and let x_3, x_4 be the roots of the equation $x^2 - 12x + B = 0$. If the numbers x_1, x_2, x_3, x_4 (in that order) form an increasing GP, then values of A and B are

(A) $A = 2, B = 32$ (B) $A = -2, B = 32$

(C) $A = -2, B = -32$ (D) none of these

61. Let a, b, c be in AP and a^2, b^2, c^2 be in GP. If $a > b > c$ and $a + b + c = 3/2$ then find the values of a and c.

(A) $a = \frac{\sqrt{2}+2}{2\sqrt{2}}, b = \frac{\sqrt{2}-2}{2\sqrt{2}}$

(B) $a = \frac{2-\sqrt{2}}{2\sqrt{2}}, b = \frac{\sqrt{2}-2}{2\sqrt{2}}$

(C) $a = \frac{\sqrt{2}+2}{\sqrt{2}-2}, b = 1$

(D) $a = \frac{2+\sqrt{2}}{2\sqrt{2}}, b = \frac{2+\sqrt{2}}{2\sqrt{2}}$

62. Let a, b, c, d and e be five real numbers such that a, b, c are in AP; b, c, d are in GP; c, d, e are in HP. If $a = 2$ and $e = 18$ then find all possible values of b, c and d.

(A) $b = 6, c = 4, d = 3$

(B) $b = 4, c = 6, d = 9$

(C) $b = -6, c = -2, d = 9$

(D) none of these

63. Find n so that $\frac{a^{n+1} + b^{n+1}}{a^n + b^n}$ be the AM between a and b. (if $a \ne b$)

(A) 1 (B) 2

(C) 0 (D) –1

64. There are n AM's between 3 and 54 such that 8th mean is to $(n - 2)^{th}$ mean is as 3 to 5. Find n.

(A) $n = 12$ (B) $n = 13$

(C) $n = 16$ (D) $n = 15$

65. Find n, so that $\frac{a^{n+1} + b^{n+1}}{a^n + b^n}$ $(a \ne b)$ is the GM between a and b.

(A) 1/2 (B) –1/2

(C) 0 (D) none of these

66. Find n, so that $\frac{a^{n+1} + b^{n+1}}{a^n + b^n}$ $(a \ne b)$ be the HM between a and b.

(A) $n = 0$ (B) $n = 1$

(C) $n = -1/2$ (D) $n = -1$

67. If a, b and c are in AP then the value of $a^2(b + c) + b^2(c + a) + c^2(a + b)$ is

(A) $(a + b + c)^3$ (B) $4b^3$

(C) $6b^3$ (D) $10b^3$

68. Sum of the first n terms of the series $\frac{1}{2} + \frac{3}{4} + \frac{7}{8} + \frac{15}{16} + \ldots$ is equal to

(A) $n - 1 + \left(\frac{1}{2}\right)^n$ (B) $n + \left(\frac{1}{2}\right)^n$

(C) $n-1-\left(\frac{1}{2}\right)^n$ (D) none of these

OLYMPIAD AND NTSE LEVEL EXERCISE

69. If the pth term of an A.P. is q and qth term is p then its rth term will be
(A) $p-q+r$ (B) $p+q-r$
(C) $p+r-q$ (D) $p-q-r$

70. The sum of 24 terms of the following series $\sqrt{2}+\sqrt{8}+\sqrt{18}+\sqrt{32}+\ldots$ is
(A) 300 (B) $300\sqrt{2}$
(C) $200\sqrt{2}$ (D) none of these

71. If the sum of n terms of an A.P. is $2n^2+5n$ then the nth term will be
(A) $4n+3$ (B) $4n+5$
(C) $4n+6$ (D) $4n+7$

72. The sum of the first five terms of the series $3+4\frac{1}{2}+6\frac{3}{4}+\ldots$ will be
(A) $39\frac{9}{16}$ (B) $18\frac{3}{16}$
(C) $39\frac{7}{16}$ (D) $13\frac{9}{16}$

73. In an AP, of which a is the first term, if the sum of the first p terms is zero then the sum of the next q terms will be
(A) $-\frac{a(p+q)p}{q+1}$ (B) $\frac{a(p+q)p}{p+1}$
(C) $-\frac{a(p+q)q}{p-1}$ (D) none of these

74. If x, y and z are in G.P. and $a^x=b^y=c^z$ then
(A) $\log_b a=\log_a c$ (B) $\log_c b=\log_a c$
(C) $\log_b a=\log_c b$ (D) none of these

75. The value of $0.2^{\log_{\sqrt{5}}\left(\frac{1}{4}+\frac{1}{8}+\frac{1}{16}+\ldots\right)}$ is
(A) 4 (B) log 4
(C) log 2 (D) none of these

76. Match the following columns:

	Column I		Column II
(a)	If a, b and c are in GP, then $\log_a 10$, $\log_b 10$, $\log_c 10$ are in	(p)	AP
(b)	If $\frac{a+be^x}{a-be^x}=\frac{b+ce^x}{b-ce^x}=\frac{c+de^x}{c-de^x}$ then a, b, c and d are in	(q)	HP
(c)	If a, b, c are in AP; a, x, b are in GP and b, y, c are in GP, then x^2, b^2, y^2 are in	(r)	GP
(d)	If x, y and z are in GP, such that $a^x=b^y=c^z$, then log a, log b and log c are in	(s)	None of these

77. Match the following columns and then choose the correct option from amongst the given codes.

	Column I		Column II
(p)	If pth, qth, rth and sth terms of an AP are in GP then $p-q$, $q-r$, $r-s$	(1)	are all equal
(q)	If ln$x$, ln$y$ and ln$z$ ($x, y, z>1$) are in GP, then $2x+\ln(\ln x)$, $3x+\ln(\ln y)$, and $4x+\ln(\ln z)$	(2)	are in AP
(r)	If $n!$, $3\times n!$, and $(n+1)!$ are in GP, then $n!$, $5\times n!$ and $(n+1)!$	(3)	are in GP
(s)	If the arithmetic mean of $(b-c)^2$, $(c-a)^2$ and $(a-b)^2$ is the same as that of $(b+c-2a)^2$, $(c+a-2b)^2$ and $(a+b-2c)^2$ then a, b and c	(4)	are in HP

Codes:

	(p)	(q)	(r)	(s)
(A)	(1)	(2)	(4)	(3)
(B)	(2)	(1)	(3)	(4)
(C)	(3)	(2)	(2)	(1)
(D)	(2)	(1)	(4)	(3)

78. If $a, b, c, d \in R^+$ such that a, b, c and d are in H.P. then
(A) $a+d>b+c$ (B) $a+b>c+d$
(C) $a+c>b+d$ (D) none of these

GAME TIME

$\sum n = \frac{n(n+1)}{2}$ can be demonstrated as follows:

Start with a number of small square cards. Assume that the area of each square is 1 square unit. Arrange them as shown here.

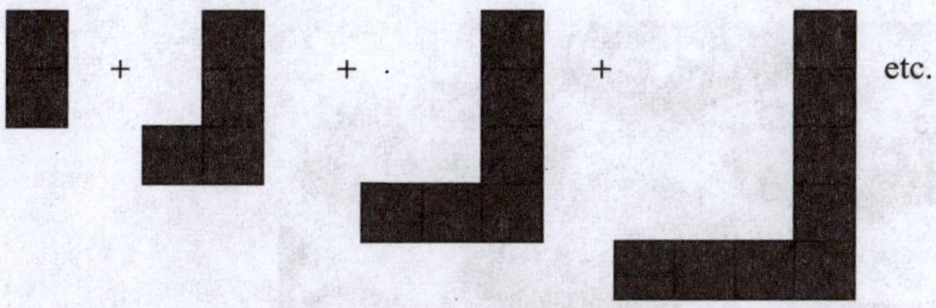

Total area = 2.1 + 2.2 + 2.3 + 2.4 + etc. = $2\Sigma n$

Now, remove all the '+' signs and bring all the arrangements together as shown below:

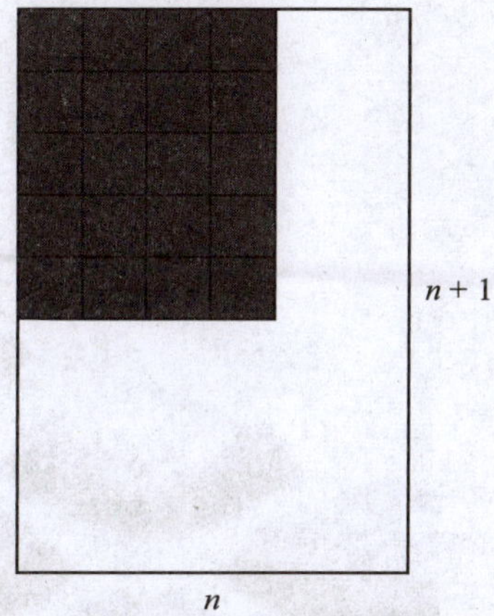

Total area = $n(n + 1)$.

Therefore, $2\,\Sigma n = n(n + 1)$ from which $\sum n = \frac{n(n+1)}{2}$.

9 Matrices

The only way to learn mathematics is to do mathematics. That tenet is the foundation of the do-it-yourself concept.

– Paul Halmos

OBJECTIVES

After reading this chapter you should be able to

- define matrix, represent a matrix, and identify types of matrices
- find the transpose of a matrix
- learn about equal matrices, symmetrical matrices
- perform operations on matrices
- represent linear equations in matrix form

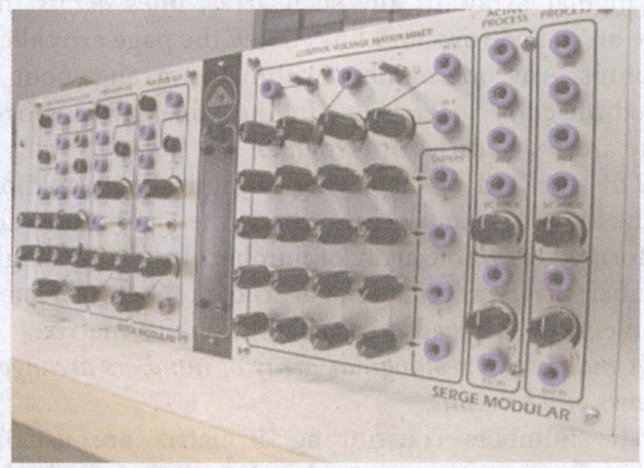

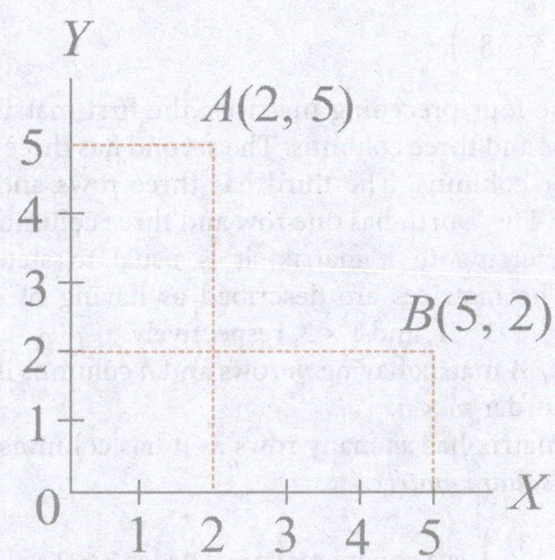

WHAT IS A MATRIX?

Sets of numerical information can often be presented in a rectangular arrangement. For example, newspapers often publish cricket league table as shown below.

	Played	Won	Drawn	Lost
Tamil Nadu	22	11	6	5
Karnataka	22	9	6	7
Bombay	21	7	8	6
Kerala	21	6	8	7
Delhi	21	6	6	9

The numbers which constitute the data form a rectangular array. In any such array, lines *across* the page are called rows and lines *down* the page are called columns. Any one number within the arrangement is called an *entry* or *element*. In the table given above, there are 5×4 entries or elements.

This method of displaying numerical information is the basis of an algebra where the operations are performed on whole array of numbers instead of single number; that is, a rectangular array of numbers is treated as an entity. This leads to the definition of a matrix:

A matrix is a rectangular array of numbers arranged in rows and columns.

The numbers constituting a matrix are usually enclosed in square or round brackets. For example the given below are all matrices.

$$\begin{pmatrix} 1 & 5 & 6 \\ 2 & 3 & 4 \end{pmatrix} \quad \begin{bmatrix} 1 & 2 & 3 & 10 \\ 4 & 5 & 6 & 11 \\ 7 & 8 & 9 & 12 \end{bmatrix} \quad \begin{bmatrix} 3 \\ 4 \\ 5 \end{bmatrix}$$

$$\begin{pmatrix} 1 & 7 & 8 \end{pmatrix}$$

In the four preceding matrices, the first matrix has two rows and three columns. The second has three rows and four columns. The third has three rows and one column. The fourth has one row and three columns.

In referring to a matrix, it is usual to state the order. The matrices are described as having order 2×3, 3×4, 3×1, and 1×3, respectively.

Thus, *A matrix having m rows and n columns is said to be of order $m \times n$.*

If a matrix has as many rows as it has columns, it is called a *square matrix*.

$\begin{pmatrix} 1 & 3 \\ 3 & 0 \end{pmatrix}$ is a square matrix of order 2×2.

$\begin{pmatrix} 1 & 2 & 3 \\ 4 & 5 & 6 \\ 7 & 8 & 9 \end{pmatrix}$ is a square matrix of order 3×3.

The order of a square matrix is sometimes given in abbreviated form. For example, the two matrices given above may be described as square matrices of order 2 and of order 3, respectively.

A matrix having only one row is called a *row matrix.*

A matrix having only one column is called a *column matrix.*

$(2 \quad 3)$ and $(2 \quad 3 \quad 5)$ are examples of row matrices.

$\begin{pmatrix} 2 \\ 3 \end{pmatrix}$ and $\begin{pmatrix} 2 \\ 3 \\ 5 \end{pmatrix}$ are examples of column matrices.

In coordinate geometry, we have seen ordered pairs and ordered triplets of numbers which fix the positions of points relative to two or three axes.

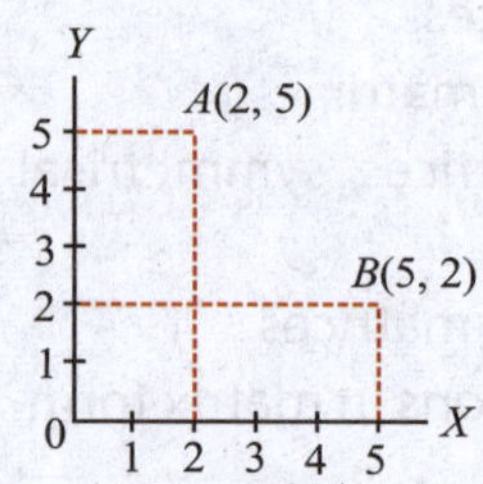

Fig. 9.1.

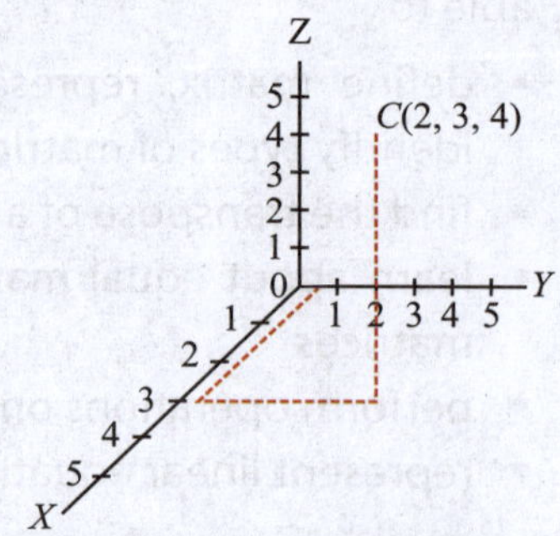

Fig. 9.2.

In Fig. 9.1, where *OX* and *OY* are perpendicular axes of reference, the ordered pair (2, 5) refers to *A*. The order in which the numbers appear is important. The same numbers in a different order, namely (5, 2), refer to a different point, namely *B*, in Fig. 9.1.

Similarly, in Fig. 9.2, where *OX*, *OY*, and *OZ* represent three mutually perpendicular axes, the ordered triplet (2, 3, 4) refers to the point *C*.

If the name *vector* is used for a straight line of given *length* in a *definite direction*, then it may be said that (2, 5) fixes a vector from *O* to *A* as in Fig. 9.3 and that (2, 3, 4) fixes the vector from O to *C* in Fig. 9.4.

Because ordered pairs and triplets can determine vectors, matrices such as (2, 5) and (2, 3, 4) are often called *row vectors*.

By analogy $(3 \quad 5 \quad 7 \quad 9)$ and $(1 \quad 2 \quad 3 \quad 0 \quad 7)$ for examples can also be called row vectors although they do not have any diagrammatic interpretation.

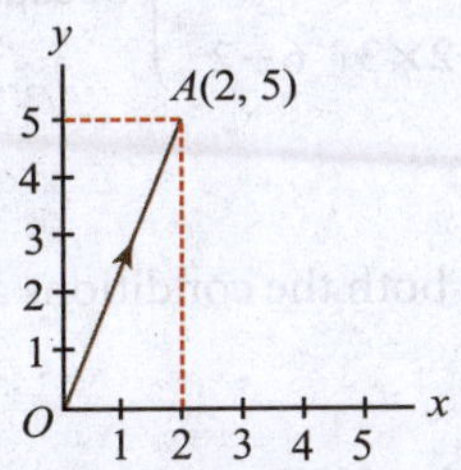

Fig. 9.3.

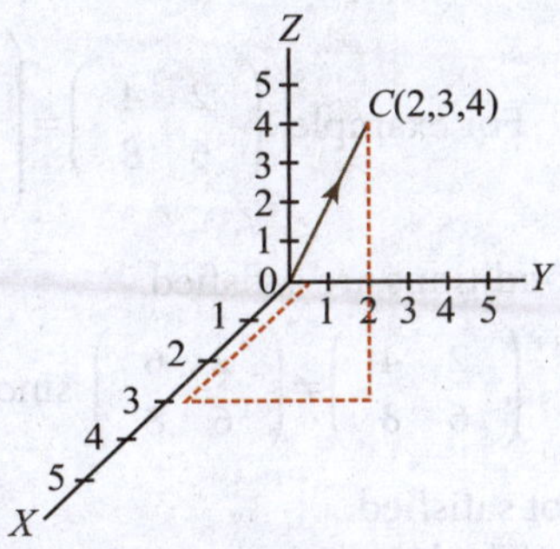

Fig. 9.4.

It is conventional that when dealing with coordinates of points, (2, 5) is written instead of $\begin{pmatrix} 2 \\ 5 \end{pmatrix}$ and (2 3 4) instead of $\begin{pmatrix} 2 \\ 3 \\ 4 \end{pmatrix}$. (2 5) and (2 3 4) are called *row vectors.*

$\begin{pmatrix} 2 \\ 5 \end{pmatrix}$ and $\begin{pmatrix} 2 \\ 3 \\ 4 \end{pmatrix}$ are called *column vectors.*

Notations

A matrix is often denoted by a capital letter and its entries are denoted by corresponding small letters appropriately suffixed.

For example, $A = \begin{pmatrix} a_{11} & a_{12} & a_{13} & a_{14} \\ a_{21} & a_{22} & a_{23} & a_{24} \\ a_{31} & a_{32} & a_{33} & a_{34} \end{pmatrix}$

represents in general terms a 3 by 4 matrix.

a_{11} represents the entry in row 1 and column 1.

a_{12} represents the entry in row 1 and column 2.

a_{23} represents the entry in row 2 and column 3 and so on.

In the same way, we write

$$B = \begin{pmatrix} b_{11} & b_{12} & b_{13} & b_{14} \\ b_{21} & b_{22} & b_{23} & b_{24} \\ b_{31} & b_{32} & b_{33} & b_{34} \end{pmatrix}.$$

This represents in general terms a 3 by 4 matrix.

In A above, let a_{ij} represent a typical element, namely the entry in the *i*th row and jth column. This leads us to a commonly accepted notation for a matrix *A*, namely $A = (a_{ij})_{3\times 4}$, read as 'A is the matrix of order 3 × 4 whose entries are of the form a_{ij}, where *i* takes the integral values from 1 to 3 and *j* integral values from 1 to 4'.

Types of Matrices

(a) **Rectangular matrix:** A matrix in which numbers of rows and columns are not equal is called a rectangular matrix.

(b) **Null matrix:** A matrix in which all elements are zero is called null matrix, normally denoted by $O_{m\times n}$. If $m = n$, then it is denoted by O_n.

(c) **Square matrix:** Square matrix is a matrix in which the number of rows is equal to the number of columns.

(d) **Triangular matrix:** A square matrix $A = [a_{ij}]$ is said to be *upper triangular* if every element below the principal diagonal is zero; i.e., $a_{ij} = 0$ when $i > j$.

A square matrix $A = [a_{ij}]$ is said to be *lower triangular* if every element above the principal diagonal is zero; i.e., $a_{ij} = 0$ when $i < j$.

A square matrix is said to be *triangular* if it is either upper or lower triangular.

Example:

$\begin{pmatrix} 3 & 5 \\ 0 & 2 \end{pmatrix}$ and $\begin{pmatrix} 1 & -3 & 4 \\ 0 & 3 & 5 \\ 0 & 0 & 4 \end{pmatrix}$ are upper triangular.

$\begin{pmatrix} -1 & 0 \\ 2 & 3 \end{pmatrix}$ and $\begin{pmatrix} 1 & 0 & 0 \\ 6 & 3 & 0 \\ 7 & 1 & 5 \end{pmatrix}$ are lower triangular.

(e) **Diagonal matrix:** A square matrix which is upper and lower triangular is called diagonal matrix, i.e., a square matrix in which all elements except those on its principal diagonal are zero.

Example

$\begin{pmatrix} 2 & 0 \\ 0 & 3 \end{pmatrix}$ and $\begin{pmatrix} 3 & 0 & 0 \\ 0 & 5 & 0 \\ 0 & 0 & 6 \end{pmatrix}$

(f) **Scalar matrix:** A diagonal matrix is said to be a scalar matrix when all the elements on the leading diagonal are same.

Example

$\begin{pmatrix} 2 & 0 \\ 0 & 2 \end{pmatrix}$ and $\begin{pmatrix} 3 & 0 & 0 \\ 0 & 3 & 0 \\ 0 & 0 & 3 \end{pmatrix}$

(g) **Unit matrix:** A diagonal matrix in which all leading diagonal elements are one is called a unit matrix.

Example:

$$\begin{pmatrix} 1 & 0 \\ 0 & 1 \end{pmatrix} \text{ and } \begin{pmatrix} 1 & 0 & 0 \\ 0 & 1 & 0 \\ 0 & 0 & 1 \end{pmatrix}$$

Transpose of a Matrix

Suppose writing papers are sold in various *sizes* and *quality* such that the cost matrix is as follows:

	Quality I	Quality II	Quality III
A_1	1500	1400	1300
A_2	1400	1300	1200
A_3	1300	1200	1100
A_4	1200	1100	1000

For example, A_3 size quality II costs ₹ 1200.

This information could be prepared in a different arrangement.

	A_1	A_2	A_3	A_4
Quality I	1500	1400	1300	1200
Quality II	1400	1300	1200	1100
Quality III	1300	1200	1100	1000

This is obtained by changing all the rows in the first matrix into columns so that the first row becomes first column the second row becomes second column etc. The second matrix is the *transpose* of the first. The notation for the transpose of A is A' or A^T.

$$\therefore \; A = \begin{pmatrix} 1 & 2 & 3 \\ 4 & 5 & 6 \end{pmatrix}$$

$$A' = \begin{pmatrix} 1 & 4 \\ 2 & 5 \\ 3 & 6 \end{pmatrix}$$

If A is of order $m \times n$ matrix, transpose of A will be $n \times m$ matrix.

Equal Matrices

Two matrices are said to be equal if they are of the same order and corresponding elements are equal.

For example, $\begin{pmatrix} 2 & 4 \\ 6 & 8 \end{pmatrix} = \begin{pmatrix} \frac{8}{4} & 2\times2 \\ 2\times3 & 6+2 \end{pmatrix}$ because conditions are satisfied.

$\begin{pmatrix} 2 & 4 \\ 6 & 8 \end{pmatrix} \neq \begin{pmatrix} 2 & 6 \\ 6 & 8 \end{pmatrix}$ since both the conditions are not satisfied.

The last two matrices are both of order 2, but all the corresponding entries are not equal.

This idea of equality of matrices makes possible the writing of sets of linear equations in matrix form. For example, $4x - 2y = 3$ and $3x + 5y = 4$ may be written as follows:

$$\begin{pmatrix} 4x-2y \\ 3x+5y \end{pmatrix} = \begin{pmatrix} 3 \\ 4 \end{pmatrix}$$

Note that both sides of the matrix statement are column matrices.

Similarly, $x + y + z = 5$, $2x - 3y - 2z = 7$ and $8x + 4y - 7z = 8$ may be written as follows:

$$\begin{pmatrix} x+y+z \\ 2x-3y-2z \\ 8x+4y-7z \end{pmatrix} = \begin{pmatrix} 5 \\ 7 \\ 8 \end{pmatrix}$$

Here, both sides of the matrix statement are column matrices.

Symmetric Matrices

Sometimes the transpose of a given matrix happens to be equal to the given matrix, as illustrated below:

(a) $A = \begin{pmatrix} x & y \\ y & z \end{pmatrix}$ $\quad A' = \begin{pmatrix} x & y \\ y & z \end{pmatrix}$

(b) $B = \begin{pmatrix} a & h & g \\ h & b & f \\ g & f & c \end{pmatrix}$ $\quad B' = \begin{pmatrix} a & h & g \\ h & b & f \\ g & f & c \end{pmatrix}$

Here, A and B are said to be symmetric. A square matrix A is said to be symmetric if $A' = A$.

Skew Symmetric Matrix

A matrix A is said to be *skew symmetric* if its transpose is negative of A, i.e., $A' = -A$.

Example

$$A = \begin{pmatrix} 0 & a & b \\ -a & 0 & c \\ -b & -c & 0 \end{pmatrix}$$

Note that in a skew-symmetric matrix diagonal elements are zero.

OPERATIONS ON MATRICES

Addition of Matrices

Suppose that on Monday a fruit seller sold 40 kg of apples and 30 kg of oranges, and on Tuesday 50 kg of apples and 40 kg of oranges. Agreeing on the order of apples and oranges, the sales could be shown as follows:

	Apples	Oranges
Monday	40	30
Tuesday	50	40

It is obvious that the total sales are (40 + 50) kg of apples and (30 + 40) kg of oranges. A matrix for the total sales could, therefore, be (90 70). Thus, using matrices it is possible to represent the above situation as (40 30)+ (50 40)= (90 70).

If column matrices are used, we have the following:

$$\begin{pmatrix} 40 \\ 30 \end{pmatrix} + \begin{pmatrix} 50 \\ 40 \end{pmatrix} = \begin{pmatrix} 90 \\ 70 \end{pmatrix}$$

In the above matrix statements, the entries in the final matrix are obtained by adding corresponding entries in the first two matrices.

Consider the attendance of students (boys and girls) in a month as shown in the following table:

	Class 9		Class 10	
	Boys	Girls	Boys	Girls
I week	520	400	310	250
II week	500	410	300	240
III week	525	425	340	230
IV week	450	450	305	205

The total number of boys and girls who attended in the same month week-wise will be as follows:

	Boys	Girls
I week	830	650
II week	800	650
III week	865	655
IV week	755	655

The above information may be presented in the following matrix form:

$$\begin{pmatrix} 520 & 400 \\ 500 & 410 \\ 525 & 425 \\ 450 & 450 \end{pmatrix} + \begin{pmatrix} 310 & 250 \\ 300 & 240 \\ 340 & 230 \\ 305 & 205 \end{pmatrix} = \begin{pmatrix} 830 & 650 \\ 800 & 650 \\ 865 & 655 \\ 755 & 655 \end{pmatrix}$$

This is a statement in the form $A + B = C$, where A, B, and C are matrices of the same order. Corresponding entries of first two matrices have been added to obtain the entries of the final matrix. This leads us to general definition of matrix addition:

If two matrices A and B are of the same order, their sum, A + B, is the matrix obtained by adding the elements of B to the corresponding elements of A.

Example 1

$$\begin{pmatrix} 1 \\ 3 \\ 4 \end{pmatrix} + \begin{pmatrix} 3 \\ 4 \\ 5 \end{pmatrix} = \begin{pmatrix} 1+3 \\ 3+4 \\ 4+5 \end{pmatrix} = \begin{pmatrix} 4 \\ 7 \\ 9 \end{pmatrix}$$

Example 2

$$\begin{pmatrix} 1 & 3 \\ 4 & 5 \end{pmatrix} + \begin{pmatrix} 1 & 0 \\ -1 & -2 \end{pmatrix} = \begin{pmatrix} 1+1 & 3+0 \\ 4-1 & 5-2 \end{pmatrix} = \begin{pmatrix} 2 & 3 \\ 3 & 3 \end{pmatrix}$$

Example 3

$$\begin{pmatrix} 3 & 2 & -5 \\ 1 & -7 & 2 \end{pmatrix} + \begin{pmatrix} -3 & -2 & +5 \\ -1 & +7 & -2 \end{pmatrix} = \begin{pmatrix} 0 & 0 & 0 \\ 0 & 0 & 0 \end{pmatrix}$$

Remember, then, that the only matrices of the same order can be added.

Is Matrix Addition Commutative?

The elements of a matrix will usually be real numbers. Any two numbers a and b satisfy the commutative law of addition, i.e., $a + b = b + a$.

Matrix addition is, therefore, commutative. In the same way, it can be shown that the matrix addition is associative, i.e., $A + (B + C) = (A + B) + C$ as we have $a + (b + c) = (a + b) + c$.

Repeated Addition

In the real number system, repeated addition such as $4 + 4 + 4 + 4 + 4 = 20$ may be symbolised as $5 \times 4 = 20$.

Repeated addition of matrices may be symbolised in a like manner.

Example 1

$$\begin{pmatrix} 4 \\ 5 \end{pmatrix}+\begin{pmatrix} 4 \\ 5 \end{pmatrix}+\begin{pmatrix} 4 \\ 5 \end{pmatrix}+\begin{pmatrix} 4 \\ 5 \end{pmatrix}=\begin{pmatrix} 16 \\ 20 \end{pmatrix}$$

By addition it may be written as follows:

$$4\begin{pmatrix} 4 \\ 5 \end{pmatrix}=\begin{pmatrix} 16 \\ 20 \end{pmatrix}$$

Example 2

$$\begin{pmatrix} 1 & 2 \\ 4 & 3 \end{pmatrix}+\begin{pmatrix} 1 & 2 \\ 4 & 3 \end{pmatrix}=\begin{pmatrix} 2 & 4 \\ 8 & 6 \end{pmatrix}$$

This may be written as follows:

$$2\begin{pmatrix} 1 & 2 \\ 4 & 3 \end{pmatrix}=\begin{pmatrix} 2 & 4 \\ 8 & 6 \end{pmatrix}$$

From such examples, it should be clear that multiplication of a given matrix by a positive integer k is affected by the multiplication of every element of the given matrix by k. Therefore, 'to multiply a given matrix by any real number k, multiply every entry of the matrix by k'.

This kind of multiplication is called scalar multiplication.

Identity Matrix for Addition

In the system of real numbers, zero is described as the identity element for the operation of addition, because for any real number a, $a + 0 = a = 0 + a$.

Suppose that A is any matrix of order 2×2. It should be clear that the result of adding to A a zero matrix of order 2×2 is a matrix identical with A.

$$\begin{pmatrix} a_{11} & a_{12} \\ a_{21} & a_{22} \end{pmatrix}+\begin{pmatrix} 0 & 0 \\ 0 & 0 \end{pmatrix}=\begin{pmatrix} a_{11} & a_{12} \\ a_{21} & a_{22} \end{pmatrix}$$

The zero matrix of order 2×2 may be described as the identity, under the operation of addition of matrices of order 2×2. Likewise, if A is any 2×3 matrix, then we have the following:

$$\begin{pmatrix} a_{11} & a_{12} & a_{13} \\ a_{21} & a_{22} & a_{23} \end{pmatrix}+\begin{pmatrix} 0 & 0 & 0 \\ 0 & 0 & 0 \end{pmatrix}=\begin{pmatrix} a_{11} & a_{12} & a_{13} \\ a_{21} & a_{22} & a_{23} \end{pmatrix}$$

The zero matrix of order 2×3 may be described as the identity under operation of addition for the matrices of order 2×3. Thus, zero matrix is the additive identity.

In the real number system, there is only one identity element under operation of addition, namely zero. For matrices under the operation of addition, there is an identity matrix for every type of matrix. The existence of so many matrices need not be confused because the order of the given matrix directs the attention to the order of the identity matrix required, i.e., the order of the identity matrix must be the same as the order of the given matrix.

Additive Inverse of a Matrix

If A and B are two matrices such that $A + B = O$ then B is said to be the additive inverse of A and is written as A so that $A + (-A) = O$.

Note that as the statement $A + B = O$ makes sense only if A, B, and O are matrices of the same order.

Consider the following:

$$A=\begin{pmatrix} a_{11} & a_{12} \\ a_{21} & a_{22} \end{pmatrix}$$

$$B=\begin{pmatrix} -a_{11} & -a_{12} \\ -a_{21} & -a_{22} \end{pmatrix}$$

On combining, we have the following:

$$A+B=\begin{pmatrix} a_{11} & a_{12} \\ a_{21} & a_{22} \end{pmatrix}+\begin{pmatrix} -a_{11} & -a_{12} \\ -a_{21} & -a_{22} \end{pmatrix}$$

$$=\begin{pmatrix} a_{11}+(-a_{11}) & a_{12}+(-a_{12}) \\ a_{21}+(-a_{21}) & a_{22}+(-a_{22}) \end{pmatrix}$$

$$=\begin{pmatrix} 0 & 0 \\ 0 & 0 \end{pmatrix}$$

The matrix $\begin{pmatrix} -a_{12} & -a_{12} \\ -a_{21} & -a_{22} \end{pmatrix}$ is the additive inverse of $\begin{pmatrix} a_{11} & a_{12} \\ a_{21} & a_{22} \end{pmatrix}$. Its entries are the negative of those in the matrix $\begin{pmatrix} a_{11} & a_{12} \\ a_{21} & a_{22} \end{pmatrix}$.

Subtraction of Matrices

In the field of real numbers, the difference between any real numbers a and b symbolised by $a - b$ has been defined as $a + (-b)$. A similar definition may be adopted for the difference of two matrices.

If A and B are matrices of the same order, then the difference symbolised by A – B is defined as A + (–B), i.e., the addition to A of the additive inverse of B.

Thus, if $A=\begin{pmatrix} a_{11} & a_{12} \\ a_{21} & a_{22} \end{pmatrix}$ and $B=\begin{pmatrix} b_{11} & b_{12} \\ b_{21} & b_{22} \end{pmatrix}$, then

$$A - B = A + (-B)$$

$$=\begin{pmatrix} a_{11} & a_{12} \\ a_{21} & a_{22} \end{pmatrix}+\begin{pmatrix} -b_{11} & -b_{12} \\ -b_{21} & -b_{22} \end{pmatrix}$$

$$=\begin{pmatrix} a_{11}+(-b_{11}) & b_{12}+(-b_{12}) \\ a_{21}+(-b_{21}) & a_{22}+(-b_{22}) \end{pmatrix}$$

$$=\begin{pmatrix} a_{11}-b_{11} & b_{12}-b_{12} \\ a_{21}-b_{21} & a_{22}-b_{22} \end{pmatrix}$$

As $a_{11} + (-b_{11})$, $a_{12} + (-b_{12})$, etc. are really subtractions of real numbers, the definition $A - B$ leads to the subtraction of corresponding entries in the given matrices. Hence, if $A=\begin{pmatrix} 4 & 5 & 6 \\ 7 & 8 & 9 \end{pmatrix}$ and $B=\begin{pmatrix} 1 & 1 & 4 \\ 2 & 0 & 1 \end{pmatrix}$, then $A - B$ may be written down immediately as $\begin{pmatrix} 3 & 4 & 2 \\ 5 & 8 & 8 \end{pmatrix}$.

Matrix Multiplication

Pencils are offered for sale in small size boxes of 100 and large size boxes of 500. A man buys five small size boxes and eight large size boxes. In matrix form, his purchase may be shown as (5 8) and contents of boxes as $\begin{pmatrix} 100 \\ 500 \end{pmatrix}$ column matrix being used to distinguish the content matrix from the purchase matrix.

Altogether, $5 \times 100 + 8 \times 500 = 4500$ pencils are bought. It is possible to regard this calculation as a matrix multiplication.

$$\begin{pmatrix} 5 & 8 \end{pmatrix}\begin{pmatrix} 100 \\ 500 \end{pmatrix} = (500+4000) = 4500$$

The entry in the final matrix, of order 1×1, is found by adding the product of the entries linked by arrow p to the product of the entries linked by arrow q.

Consider now a small hotel which offers three items, namely salad, veg rice, and ice cream, at prices ₹ 10, ₹ 30, ₹ 20 respectively.

Suppose that on Sunday numbers of salad, veg rice, and ice cream served were ₹ 40, 50, and 60, respectively. The total income in rupees from the items was $10 \times 40 + 50 \times 30 + 60 \times 20$.

∴ ₹ 400 + ₹ 1500 + ₹ 1200 = ₹ 3100

To represent this in matrix form, items served may be shown as (40 50 60) and the prices charged as $\begin{pmatrix} 10 \\ 30 \\ 20 \end{pmatrix}$, so that the calculation may be represented as the matrix multiplication.

$$\begin{pmatrix} 40 & 50 & 60 \end{pmatrix}\begin{pmatrix} 10 \\ 30 \\ 20 \end{pmatrix} = (3100)$$

As another illustration of a calculation presented as a multiplication of matrices, consider the following situation.

An export order for 100 'Car A' cars and 400 'Car B' cars is received by Maruthi manufacturer. This may be represented by (100 400). The production may be considered under headings material used and labour needed. Numbers of units required per car are given in the following table:

	Material	Labour
Car A	12	15
Car B	16	20

In matrix form, this may be written as follows:

$$\begin{pmatrix} 12 & 15 \\ 16 & 20 \end{pmatrix}$$

The material needed altogether is as follows:

$100 \times 12 + 400 \times 16$ units

$= 1200 + 6400 = 7600$ units

The labour required is as follows:
$100 \times 15 + 400 \times 20$ units
$= (1500 + 8000)$ units $= 9500$ units

Accordingly, the final requirements in material and labour are 7600 and 9500 units, respectively, which in matrix form may be written as (7600 9500).

This calculation may be presented as the following multiplication of matrices:

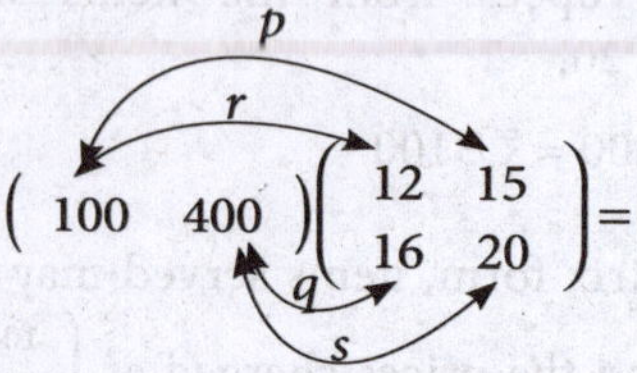

$$\begin{pmatrix} 100 & 400 \end{pmatrix}\begin{pmatrix} 12 & 15 \\ 16 & 20 \end{pmatrix} = \begin{pmatrix} 7600 & 9500 \end{pmatrix}$$

(a) The product of entries linked by arrow r is added to the product of entries linked by arrow q to obtain the first entry of the final matrix; this is the total number of units of material needed.

(b) The product of entries linked by arrow p is added to the product of entries linked by arrow s to obtain the second entry in the final matrix; this is the total number of units of labour needed.

If the cost of one unit of material is ₹ 600 and the cost of one unit of labour is ₹ 400, a further calculation to obtain the total cost of the cars may be shown in matrix form. As we already have a row matrix (7600 9500) indicating the total units of material and labour required, we represent the prices in a column matrix, namely $\begin{pmatrix} 600 \\ 400 \end{pmatrix}$. The calculation is then as follows:

$$\begin{pmatrix} 7600 & 9500 \end{pmatrix}\begin{pmatrix} 600 \\ 400 \end{pmatrix} = (7600 \times 600 + 9500 \times 400)$$

$$= (4560000 + 3800000) = (8360000)$$

The total cost of production of cars for the export order is thus ₹ 83,60,000. With the help of the examples illustrated above, we can say that for matrices A and B, the product AB can be performed only when the number of columns of A is equal to the number of rows in B. Matrices which satisfy the condition are said to be *conformable for multiplication.*

To form the product AB of two given matrices A and B, the useful procedure is as follows:

(a) Test whether multiplication is possible by looking at the number of columns of A and rows of B. (These should be equal.)

(b) If A and B are matrices of order $m \times p$ and $p \times n$, respectively, then the product AB is a matrix C of order $m \times n$. In C any element C_{ij} is obtained by adding the products of all the terms in the ith row of A with terms in corresponding position in the jth column of B.

Is Matrix Multiplication Commutative?

Having defined multiplication of two matrices A and B, it is important to know if multiplication is commutative, i.e. if $AB = BA$.

If $A = \begin{pmatrix} 4 & 1 \\ 2 & 5 \end{pmatrix}$ and $B = \begin{pmatrix} 2 & 3 \\ 1 & 6 \end{pmatrix}$, then we have the following:

$$AB = \begin{pmatrix} 4 & 1 \\ 2 & 5 \end{pmatrix}\begin{pmatrix} 2 & 3 \\ 1 & 6 \end{pmatrix} = \begin{pmatrix} 9 & 18 \\ 9 & 36 \end{pmatrix}$$

$$BA = \begin{pmatrix} 2 & 3 \\ 1 & 6 \end{pmatrix}\begin{pmatrix} 4 & 1 \\ 2 & 5 \end{pmatrix} = \begin{pmatrix} 14 & 17 \\ 16 & 31 \end{pmatrix}$$

So, in this case AB and BA are not equal.

Now, if $A = \begin{pmatrix} 1 & 2 \\ 2 & 3 \end{pmatrix}$ and $B = \begin{pmatrix} 3 \\ 5 \end{pmatrix}$, then we get the following:

$$AB = \begin{pmatrix} 1 & 2 \\ 2 & 3 \end{pmatrix}\begin{pmatrix} 3 \\ 5 \end{pmatrix} = \begin{pmatrix} 13 \\ 21 \end{pmatrix}$$

If the product BA is considered, the number of columns in B is not equal to the number of rows in A. It is, therefore, not possible to give any sensible meaning to BA so that again A and B are not commutative for multiplication.

In general, matrix multiplication is non-commutative. Exception to the rule exists.

For example, if $A = \begin{pmatrix} 1 & 0 \\ 0 & 1 \end{pmatrix}$ and $B = \begin{pmatrix} 3 & 5 \\ 1 & 2 \end{pmatrix}$, then we have the following:

$$AB = \begin{pmatrix} 1 & 0 \\ 0 & 1 \end{pmatrix}\begin{pmatrix} 3 & 5 \\ 1 & 2 \end{pmatrix} = \begin{pmatrix} 3 & 5 \\ 1 & 2 \end{pmatrix}$$

$$BA = \begin{pmatrix} 3 & 5 \\ 1 & 2 \end{pmatrix}\begin{pmatrix} 1 & 0 \\ 0 & 1 \end{pmatrix} = \begin{pmatrix} 3 & 5 \\ 1 & 2 \end{pmatrix}$$

In matrix multiplication, AB can be described as A – post-multiplied by B – and BA as A – pre-multiplied by B. AA is represented by A^2 but the product AA is defined only when A is a square matrix.

Zero Matrix as a Product

If $A=\begin{pmatrix}0 & 0\\ 0 & 1\end{pmatrix}$ and $B=\begin{pmatrix}1 & 0\\ 0 & 0\end{pmatrix}$, then

$AB=\begin{pmatrix}0 & 0\\ 0 & 0\end{pmatrix}$ and $BA=\begin{pmatrix}0 & 0\\ 0 & 0\end{pmatrix}$ This shows that the product of two matrices A and B can be a zero matrix without either A or B being zero matrix.

If $A=\begin{pmatrix}1 & 0 & 0\\ 2 & 0 & 0\\ 3 & 0 & 0\end{pmatrix}$ and $B=\begin{pmatrix}0 & 0 & 0\\ 1 & 2 & 3\\ 2 & 3 & 4\end{pmatrix}$, then we have the following:

$$AB=\begin{pmatrix}0 & 0 & 0\\ 0 & 0 & 0\\ 0 & 0 & 0\end{pmatrix}$$

$$BA=\begin{pmatrix}0 & 0 & 0\\ 1+4+6 & 0 & 0\\ 2+6+12 & 0 & 0\end{pmatrix}=\begin{pmatrix}0 & 0 & 0\\ 11 & 0 & 0\\ 20 & 0 & 0\end{pmatrix}$$

Here, neither A nor B is zero matrix and $AB = O$, but $BA \neq O$ unless, of course, for some values of the first column of A and the second and third rows of B are chosen so that the second and third elements of the first column of BA become zero.

Product Containing a Common Matrix

If $A = (1\ \ 2)$ and $B=\begin{pmatrix}1 & 5\\ 1 & 0\end{pmatrix}$, then $AB(3\ \ 5)$.

If $A = (1\ \ 2)$ and $C=\begin{pmatrix}2 & 3\\ \frac{1}{2} & 1\end{pmatrix}$, then $AC = (3\ \ 5)$.

Hence, AB may equal AC without A being a zero matrix or B being equal to C.

The last three sections lead to a comparison between the real number system and matrices.

Real number	Matrices
(a) $ab = ba$ always	$AB \neq BA$ in general
(b) If $ab = 0$, then either $a = 0$ or $b = 0$ or both a and b are equal to zero	If $AB = 0$, it does not necessarily follow that either $A = 0$ or $B = 0$ or both A and B are equal to zero.
(c) If $ab = 0$, then $ba = 0$	If $AB = 0$, then BA does not necessarily be equal to zero and in fact the product BA may not be defined
(d) If $ab = ac$ and $a \neq 0$, then $b = c$	If $AB = AC$ and $A \neq 0$, then B does not necessarily be equal to C

Sets of Linear Equations in Matrix Form

Consider the following:

$4x - 2y = 3$
$3x + 5y = 4$

The equations can be represented as the matrix equation:

$$\begin{pmatrix}4x-2y\\ 3x+5y\end{pmatrix}=\begin{pmatrix}3\\ 4\end{pmatrix}$$

where $4x - 2y$ and $3x + 5y$ are regarded as single entries in matrix of order 2×1 on the left.

In general, any two equations of this type may be written as follows:

$$\begin{aligned}a_1x + b_1y = c_1\\ a_2x + b_2y = c_2\end{aligned} \quad (1)$$

Matrix representation will be as follows:

$$\begin{pmatrix}a_1x+b_1y\\ a_2x+b_2y\end{pmatrix}=\begin{pmatrix}c_1\\ c_2\end{pmatrix} \quad (2)$$

Consider the matrix multiplication as follows:

$$\begin{pmatrix}a_1 & b_1\\ a_2 & b_2\end{pmatrix}\begin{pmatrix}x\\ y\end{pmatrix}$$

The product matrix is $\begin{pmatrix}a_1x+b_1y\\ a_2x+b_2y\end{pmatrix}$ which is exactly that matrix on the left-hand side of Eq. (2).

The set of Eq. (1) may, therefore, be presented as follows:

$$\begin{pmatrix} a_1 & b_1 \\ a_2 & b_2 \end{pmatrix}\begin{pmatrix} x \\ y \end{pmatrix} = \begin{pmatrix} c_1 \\ c_2 \end{pmatrix}$$

The matrix $\begin{pmatrix} a_1 & b_1 \\ a_2 & b_2 \end{pmatrix}$ is obtained from the coefficient of x and y in the set of Eq. (1). Call it as *coefficients matrix* and denote it by A. The matrix $\begin{pmatrix} x \\ y \end{pmatrix}$ is a column matrix containing the variables x and y in the given set of equations. Call it as the *variable matrix* and denote it by X.

The matrix $\begin{pmatrix} c_1 \\ c_2 \end{pmatrix}$ contains the constants c_1 and c_2 of the given equations. Call it as *constants matrix* and denote it by B. Then the set of Eq. (1) may be represented as $AX = B$ and can be solved using matrices, which is beyond the scope of this volume.

Identity Matrix for Multiplication

Examples of unit matrices are as follows:

$$\begin{pmatrix} 1 & 0 \\ 0 & 1 \end{pmatrix}$$

$$\begin{pmatrix} 1 & 0 & 0 \\ 0 & 1 & 0 \\ 0 & 0 & 1 \end{pmatrix}$$

$$\begin{pmatrix} 1 & 0 & 0 & 0 \\ 0 & 1 & 0 & 0 \\ 0 & 0 & 1 & 0 \\ 0 & 0 & 0 & 1 \end{pmatrix}$$

These types of unit matrices are called identity matrices for the operation of multiplication because when any given matrix is multiplied by one of them the given matrix has its identity preserved (see examples below). It is customary to denote such matrices by $I_{2\times 2}$, $I_{3\times 3}$, $I_{4\times 4}$ or I_2, I_3 or I_4, etc., although when its order is understood from its context, I alone is sufficient. As symmetric matrices are necessarily square, an identity matrix for the operation of multiplication is necessarily a square matrix.

(a) If, $A = \begin{pmatrix} p & q \\ r & s \end{pmatrix}$ and $I = \begin{pmatrix} 1 & 0 \\ 0 & 1 \end{pmatrix}$, then we have the following:

$$IA = \begin{pmatrix} 1 & 0 \\ 0 & 1 \end{pmatrix}\begin{pmatrix} p & q \\ r & s \end{pmatrix} = \begin{pmatrix} p & q \\ r & s \end{pmatrix}$$

$\therefore IA = A$

$$AI = \begin{pmatrix} p & q \\ r & s \end{pmatrix}\begin{pmatrix} 1 & 0 \\ 0 & 1 \end{pmatrix} = \begin{pmatrix} p & q \\ r & s \end{pmatrix}$$

$\therefore AI = A$

(b) If $A = \begin{pmatrix} p & q & r \\ s & t & u \\ v & w & x \end{pmatrix}$ and $I = \begin{pmatrix} 1 & 0 & 0 \\ 0 & 1 & 0 \\ 0 & 0 & 1 \end{pmatrix}$, then, we have the following:

$$IA = \begin{pmatrix} 1 & 0 & 0 \\ 0 & 1 & 0 \\ 0 & 0 & 1 \end{pmatrix}\begin{pmatrix} p & q & r \\ s & t & u \\ v & w & x \end{pmatrix}$$

$$= \begin{pmatrix} p & q & r \\ s & t & u \\ v & w & x \end{pmatrix}$$

Then $IA = A$

Now, if $A = \begin{pmatrix} a & b \\ c & d \\ e & f \end{pmatrix}$, find I such that $IA = A$. As A is of the order 3×2, I must have three columns for IA to be possible. Since identity matrix is always square, I must be of order 3 by 3. Then,

$$\begin{pmatrix} 1 & 0 & 0 \\ 0 & 1 & 0 \\ 0 & 0 & 1 \end{pmatrix}\begin{pmatrix} a & b \\ c & d \\ e & f \end{pmatrix} = \begin{pmatrix} a & b \\ c & d \\ e & f \end{pmatrix}$$

Using the above examples, we can say that in the matrix equation $IA = AI = A$, I's are of the same order if A is square but are of different orders if A is not a square matrix.

Look at the following to confirm the above statements:

$$\begin{pmatrix} 1 & 0 & 0 \\ 0 & 1 & 0 \\ 0 & 0 & 1 \end{pmatrix}\begin{pmatrix} a & b \\ c & d \\ e & f \end{pmatrix} = \begin{pmatrix} a & b \\ c & d \\ e & f \end{pmatrix}\begin{pmatrix} 1 & 0 \\ 0 & 1 \end{pmatrix}$$

$$= \begin{pmatrix} a & b \\ c & d \\ e & f \end{pmatrix}$$

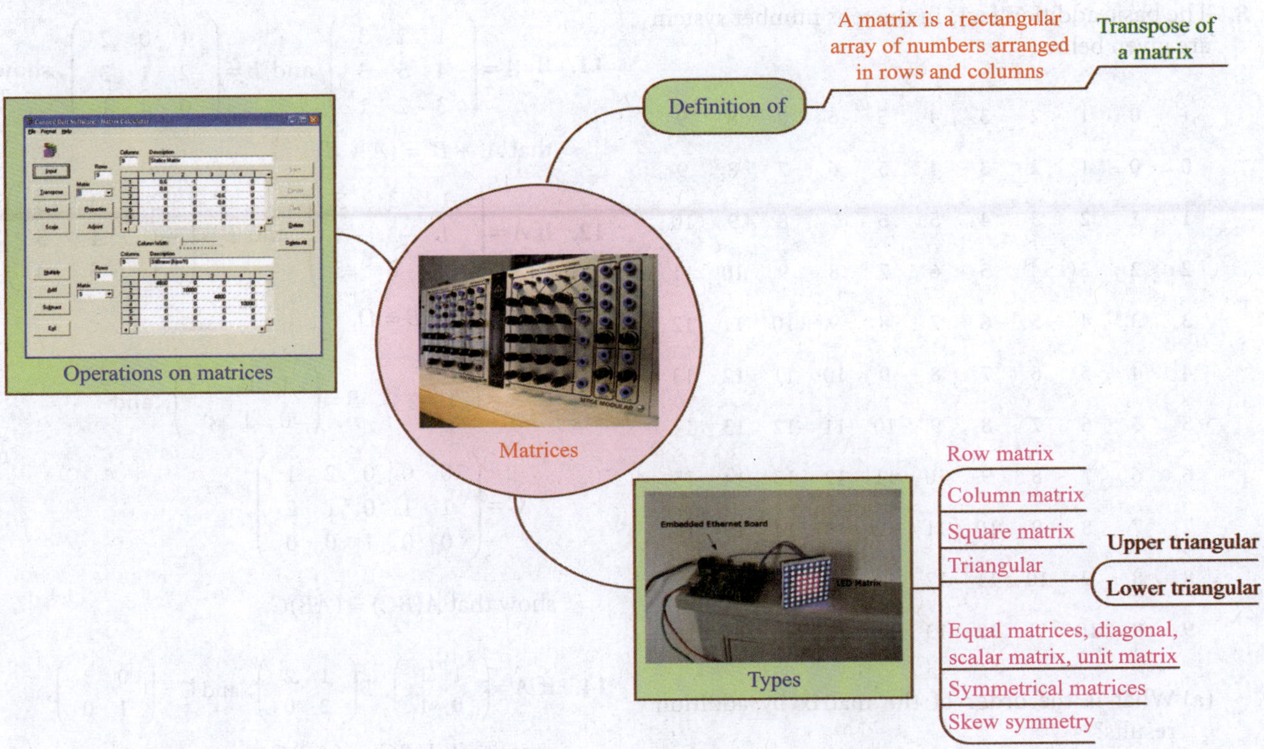

TEST YOURSELF

Level 1

1. From newspapers or magazines write some examples of information in matrix form.

2. State the order of the following matrices:

(a) $\begin{pmatrix} 1 & 2 & 3 \\ -4 & 6 & 7 \end{pmatrix}$ (b) $\begin{pmatrix} 7 \\ 8 \end{pmatrix}$

(c) $\begin{pmatrix} 3 & 7 & 8 & 9 \end{pmatrix}$ (d) $\begin{pmatrix} 1 & 2 \\ 3 & 4 \\ 5 & 6 \\ 7 & 9 \end{pmatrix}$

(e) $\begin{pmatrix} 1 & 2 & 3 \\ 4 & 5 & 6 \\ 7 & 8 & 9 \end{pmatrix}$ (f) $\begin{pmatrix} 0 & 1 & 0 & 1 \\ 1 & 0 & 1 & 1 \\ 1 & 0 & 0 & 1 \\ 0 & 0 & 0 & 1 \end{pmatrix}$

3. Which of the matrices in Q. 2 may be called row vectors or column vectors?

4. Show on a diagram, vectors which may be represented by the following matrices:

(a) $(0 \;\; 0)$ (b) $\begin{pmatrix} 1 \\ \sqrt{3} \end{pmatrix}$

(c) $(1 \;\; 1 \;\; 1)$ (d) $\begin{pmatrix} 2 \\ 3 \\ 4 \end{pmatrix}$

(e) $\begin{pmatrix} 2 \\ -2 \\ -2 \end{pmatrix}$

5. Write a 3×3 matrix in which element $a_{ij} = i + j$.

6. Write the transpose of all the matrices in Q. 2.

7. Write the full form of the following matrices:

$A = (a_{kj})_{3\times 2}$ $B = (b_{ij})_{3\times 3}$ $C = (c_{ij})_{5\times 1}$

$D = (d_{ij})_{1\times 5}$ $E = (f_{ij})_{4\times 4}$

8. The basic addition facts in the tens number system are given below:

+	0	1	2	3	4	5	6	7	8	9
0	0	1	2	3	4	5	6	7	8	9
1	1	2	3	4	5	6	7	8	9	10
2	2	3	4	5	6	7	8	9	10	11
3	3	4	5	6	7	8	9	10	11	12
4	4	5	6	7	8	9	10	11	12	13
5	5	6	7	8	9	10	11	12	13	14
6	6	7	8	9	10	11	12	13	14	15
7	7	8	9	10	11	12	13	14	15	16
8	8	9	10	11	12	13	14	15	16	17
9	9	10	11	12	13	14	15	16	17	18

(a) What is the order of the matrix by addition results?

(b) State entries in the leading diagonal.

(c) Is the matrix symmetric?

(d) If a typical element is denoted by a_{ij}, what are the following:
$a_{1,6}, a_{3,7}, a_{5,5}, a_{6,1}, a_{10,8}$

9. Write the following in matrix form:

(a) $2x + 3y = 5$
$4x + 8y = 9$

(b) $3x - 4y + 3z = 25$
$7x + 2y - 6z = 11$
$2x - 7y + 8z = 25$

(c) $x + y + z = 7$
$-y + z = 3$
$-x + 0y + z = 5$

Level 2

10. If $A = \begin{pmatrix} 1 & 2 \\ 3 & 4 \end{pmatrix}$, and $B = \begin{pmatrix} 5 & 6 \\ 4 & 3 \end{pmatrix}$, find the following:

(a) $A + B$ (b) $(A + B)'$ (c) $A' + B'$

Are the answers to (b) and (c) the same?

11. If $A = \begin{pmatrix} 1 & 2 & 3 \\ 4 & 5 & 3 \\ 3 & 2 & 1 \end{pmatrix}$, and $B = \begin{pmatrix} 1 & 0 & 2 \\ 2 & 1 & 3 \\ 0 & 4 & 1 \end{pmatrix}$, show that $A' + B' = (A + B)'$.

12. If $A = \begin{pmatrix} -1 & 3 & 5 \\ 1 & -3 & -5 \\ -1 & 3 & 5 \end{pmatrix}$, and $B = \begin{pmatrix} 2 & -3 & -5 \\ -1 & 4 & 5 \\ 1 & -3 & -4 \end{pmatrix}$, show that $AB = O$.

13. If $A = \begin{pmatrix} 1 & 2 \end{pmatrix}$, $B = \begin{pmatrix} 1 & 0 & 1 \\ 0 & 1 & 0 \end{pmatrix}$, and

$C = \begin{pmatrix} 0 & 0 & 0 & 2 & 1 \\ 1 & 1 & 0 & 1 & 2 \\ 0 & 0 & 1 & 0 & 0 \end{pmatrix}$,

show that $A(BC) = (AB)C$.

14. If $A = \begin{pmatrix} 1 & 1 \\ 0 & 1 \end{pmatrix}$, $B\begin{pmatrix} 1 & 2 \\ 2 & 0 \end{pmatrix}$, and $C = \begin{pmatrix} 0 & 3 \\ 1 & 0 \end{pmatrix}$, show that $A(BC) = (AB)C$.

15. If $A = \begin{pmatrix} 1 & 2 \end{pmatrix}$, $B = \begin{pmatrix} 3 \\ 1 \end{pmatrix}$, and $C = \begin{pmatrix} 5 \\ 2 \end{pmatrix}$, show that $A(B + C) = AB + AC$.

16. If $A = \begin{pmatrix} 1 & 0 \\ -3 & 4 \end{pmatrix}$, find the following:

(a) A^2 (b) A^3

(c) $A + A^2$ (d) $A + A^2 + A^3$

Level 3

17. If $A = \begin{bmatrix} 4 & 5 \\ 1 & 2 \end{bmatrix}$ and $B = \begin{bmatrix} 0 & 6 \\ 1 & 3 \end{bmatrix}$, find $3A - 2B$.

18. If $A + B = \begin{bmatrix} 12 & 8 \\ 8 & 4 \end{bmatrix}$ and $A - B = \begin{bmatrix} 4 & 6 \\ 6 & 2 \end{bmatrix}$, find the matrices A and B.

19. If $A = \begin{bmatrix} 3 & 1 \\ 2 & 5 \end{bmatrix}$, prove that $A^2 - 8A + 13I = 0$.

20. A warehouse deals with five items A, B, C, D, and E each of which being delivered in cartons of different sizes. The following matrix shows the number of cartons delivered by three vans in a day:

	A	B	C	D	E
Van I	4	6	2	8	8
Van II	10	12	0	2	4
Van III	3	4	5	10	1

If the weight of one carton of each type is A $\rightarrow$ 2 kg, B $\rightarrow$ 3.5 kg, C $\rightarrow$ 1 kg, D $\rightarrow$ 1.5 kg, and E $\rightarrow$ 0.75 kg, find the weight of cartons in each van. Also, if the volume in cc of one carton of each type A $\rightarrow$ 20, B $\rightarrow$ 30, C $\rightarrow$ 15, D $\rightarrow$ 20, and E $\rightarrow$ 45, find the total volume of goods delivered by all the vans put together.

Multiple Choice Questions

21. Which of the following matrices can be represented by the notation $B=\left[b_{ij}\right]_{3\times3}$?

(A) $\begin{bmatrix} b_{11} & b_{12} & b_{13} \\ b_{21} & b_{22} & b_{23} \\ b_{31} & b_{32} & b_{33} \end{bmatrix}$ (B) $\begin{bmatrix} b_{11} & b_{12} \\ b_{21} & b_{22} \\ b_{31} & b_{32} \end{bmatrix}$

(C) $\begin{bmatrix} b_{11} & b_{12} & b_{13} \\ b_{21} & b_{22} & b_{23} \end{bmatrix}$ (D) none of these

22. The short notation of the matrix $B=\begin{bmatrix} b_{11} & b_{12} & b_{13} \\ b_{21} & b_{22} & b_{23} \\ b_{31} & b_{32} & b_{33} \end{bmatrix}$ is

(A) $B=\left[b_{ij}\right]_{3\times2}$ (B) $B=\left[b_{ij}\right]_{2\times2}$

(C) $B=\left[b_{ij}\right]_{3\times3}$ (D) none of these

23. Which of the following is an upper triangular matrix?

(A) $\begin{bmatrix} 2 & 3 & 4 \\ 0 & 4 & 6 \\ 0 & 0 & 7 \end{bmatrix}$ (B) $\begin{bmatrix} 2 & 0 & 0 \\ 3 & 5 & 0 \\ 5 & 6 & 7 \end{bmatrix}$

(C) $\begin{bmatrix} 1 & 0 & 0 \\ 0 & 2 & 0 \\ 0 & 3 & 4 \end{bmatrix}$ (D) none of these

24. Which of the following is a diagonal matrix?

(A) $\begin{bmatrix} 2 & 3 & 4 \\ 5 & 0 & 0 \\ 0 & 6 & 7 \end{bmatrix}$ (B) $\begin{bmatrix} 0 & 2 & 3 \\ 4 & 0 & 3 \\ 6 & 7 & 0 \end{bmatrix}$

(C) $\begin{bmatrix} 2 & 0 & 0 \\ 0 & 4 & 0 \\ 0 & 0 & 8 \end{bmatrix}$ (D) none of these

25. What is the value of trace of the following matrix?

$$A=\begin{bmatrix} 2 & 3 & 4 \\ 7 & 8 & 10 \\ 0 & 0 & 3 \end{bmatrix}$$

(A) 10 (B) 8

(C) 3 (D) 13

26. A 2×3 matrix $A=\left[a_{ij}\right]$, whose elements are given by $a_{ij}=\dfrac{(i+2j)^2}{2}$ is

(A) $\begin{bmatrix} 2 & 4 & 5 \\ 7 & 8 & 9 \end{bmatrix}$ (B) $\begin{bmatrix} \frac{9}{2} & \frac{25}{2} & \frac{49}{2} \\ 8 & 18 & 32 \end{bmatrix}$

(C) $\begin{bmatrix} \frac{9}{2} & 8 \\ \frac{25}{2} & 18 \\ \frac{49}{2} & 32 \end{bmatrix}$ (D) none of these

27. Which of the following is represented by $c=\text{diag}\,(3,5,7)$?

(A) $c=[3\;5\;7]$ (B) $c=\begin{bmatrix} 3 & 0 & 0 \\ 0 & 5 & 0 \\ 0 & 0 & 7 \end{bmatrix}$

(C) $c=\begin{bmatrix} 3 \\ 5 \\ 7 \end{bmatrix}$ (D) none of these

28. Transpose of the matrix $A=\begin{bmatrix} 2 & 3 & 4 \\ 5 & 6 & 7 \end{bmatrix}_{2\times3}$ is

(A) $\begin{bmatrix} 2 & 3 \\ 4 & 5 \\ 6 & 7 \end{bmatrix}_{3\times2}$ (B) $\begin{bmatrix} 2 & 5 \\ 3 & 6 \\ 4 & 7 \end{bmatrix}_{3\times2}$

(C) $\begin{bmatrix} 2 & 3 \\ 5 & 6 \end{bmatrix}_{2\times2}$ (D) $\begin{bmatrix} 0 & 0 \\ 0 & 0 \end{bmatrix}_{2\times2}$

29. If $\begin{bmatrix} 2\alpha+1 & 3\beta \\ 0 & \beta^2-5\beta \end{bmatrix} = \begin{bmatrix} \alpha+3 & \beta^2+2 \\ 0 & -6 \end{bmatrix}$ then the values of α and β are

(A) $\alpha = 2, \beta = 3$ (B) $\alpha = 2, \beta = 2$

(C) $\alpha = 3, \beta = 3$ (D) none of these

30. Given

$$A = \begin{bmatrix} 1 & 3 & 5 \\ -2 & 0 & -2 \\ 0 & 4 & -3 \end{bmatrix} \quad B = \begin{bmatrix} 0 & 3 \\ -2 & 0 \\ 0 & 4 \end{bmatrix} \quad C = \begin{bmatrix} 4 & 1 & -2 \\ 3 & 2 & 1 \\ 2 & -1 & 7 \end{bmatrix}$$

Which of the following is defined?

(A) $A + B$ (B) $B + C$

(C) $A + C$ (D) none of these

31. If $A = \begin{bmatrix} 0 & 2 \\ 3 & -4 \end{bmatrix}$ and $kA = \begin{bmatrix} 0 & 3a \\ 2b & 24 \end{bmatrix}$ then find the value of $b - a - k$.

(A) 2 (B) 3

(C) 1 (D) 4

32. Given $A = \begin{bmatrix} 1 & 2 & -3 \\ 5 & 0 & 2 \\ 1 & -1 & 1 \end{bmatrix}$ and $B = \begin{bmatrix} 3 & -1 & 2 \\ 4 & 2 & 5 \\ 2 & 0 & 3 \end{bmatrix}$. The matrix C such that $A + 2C = B$ is

(A) $\begin{bmatrix} 1 & \frac{3}{2} & 1 \\ -\frac{1}{2} & \frac{7}{2} & \frac{3}{2} \\ \frac{1}{2} & \frac{8}{2} & 8 \end{bmatrix}$ (B) $\begin{bmatrix} \frac{5}{2} & \frac{3}{2} & 1 \\ -\frac{1}{2} & \frac{7}{2} & \frac{9}{2} \\ \frac{11}{2} & \frac{9}{2} & 8 \end{bmatrix}$

(C) $\begin{bmatrix} 1 & 2 & 4 \\ \frac{3}{2} & \frac{11}{2} & \frac{13}{2} \\ 4 & \frac{9}{2} & \frac{7}{2} \end{bmatrix}$ (D) none of these

33. The value of x and y for the equation

$2x - y = \begin{bmatrix} 3 & -3 & 0 \\ 3 & 3 & 2 \end{bmatrix}$ and $2y + x = \begin{bmatrix} 4 & 1 & 5 \\ -1 & 4 & -4 \end{bmatrix}$ are

(A) $x = \begin{bmatrix} 2 & -1 & 1 \\ 1 & 2 & 0 \end{bmatrix}$; $y = \begin{bmatrix} 1 & 1 & 2 \\ -1 & 1 & -2 \end{bmatrix}$

(B) $x = \begin{bmatrix} 1 & 1 & 2 \\ -1 & 1 & -2 \end{bmatrix}$; $y = \begin{bmatrix} 2 & -1 & 1 \\ 1 & 2 & 0 \end{bmatrix}$

(C) $x = \begin{bmatrix} \frac{1}{2} & \frac{1}{2} & 1 \\ -\frac{1}{2} & \frac{1}{2} & -1 \end{bmatrix}$; $y = \begin{bmatrix} 1 & -\frac{1}{2} & \frac{1}{2} \\ \frac{1}{2} & 1 & 0 \end{bmatrix}$

(D) none of these

34. If $A = \begin{bmatrix} 0 & 1 & 2 \\ 1 & 2 & 3 \\ 2 & 3 & 4 \end{bmatrix}$ and $B = \begin{bmatrix} 1 & -2 \\ -1 & 0 \\ 2 & -1 \end{bmatrix}$, then which of the following is defined?

(A) AB (B) BA

(C) both (A) and (B) (D) none of these

35. If $A = \begin{bmatrix} 0 & 1 & 2 \\ 1 & 2 & 3 \\ 2 & 3 & 4 \end{bmatrix}$ and $B = \begin{bmatrix} 1 & -2 \\ -1 & 0 \\ 2 & -1 \end{bmatrix}$, then AB is

(A) $\begin{bmatrix} 3 & -2 \\ 5 & -5 \\ 7 & -8 \end{bmatrix}$ (B) $\begin{bmatrix} 3 & 5 & 7 \\ -2 & -5 & -8 \end{bmatrix}$

(C) $\begin{bmatrix} 1 & 2 \\ 3 & 4 \\ 5 & 6 \end{bmatrix}$ (D) none

36. If $A = \begin{bmatrix} 1 & 2 \\ -2 & 3 \end{bmatrix}$, $B = \begin{bmatrix} 2 & 1 \\ 2 & 3 \end{bmatrix}$ and $C = \begin{bmatrix} -3 & 1 \\ 2 & 0 \end{bmatrix}$, then which of the following is true?

(A) $AB = BC$ (B) $(AB)C = A(BC)$

(C) $BA = AB$ (D) none of these

37. If $A = \begin{bmatrix} \alpha & 2 \\ 2 & \alpha \end{bmatrix}$ and $\det(A) = 0$, then the value of α is

(A) 2, –2 (B) 0

(C) –2, –2 (D) 2, 2

38. If $A = \begin{bmatrix} 1 & -1 \\ 2 & -1 \end{bmatrix}$, $B = \begin{bmatrix} a & 1 \\ b & -1 \end{bmatrix}$ and $(A+B)^2 = A^2 + B^2$, then the value of $a + b$ is

(A) 4 (B) 5

(C) 6 (D) 2

39. If $A=\begin{bmatrix}1 & 2\\ 2 & 3\end{bmatrix}$ and $A^2+\lambda A-I_2=0$, then λ is equal to

(A) -4 (B) -2
(C) 2 (D) 4

40. If $A\begin{bmatrix}\cos\theta & \sin\theta\\ -\sin\theta & \cos\theta\end{bmatrix}$, then $A^2=I$ is true for

(A) $\theta=0$ (B) $\theta=\frac{\pi}{4}$
(C) $\theta=\frac{\pi}{2}$ (D) $\theta=\frac{\pi}{3}$

41. If a, b; b, c and c, a are the roots of $x^2-4x+3=0$, $x^2-8x+15=0$ and $x^2-6x+5=0$, respectively, then the value of $\begin{bmatrix}a^2+c^2 & a^2+b^2\\ b^2+c^2 & a^2+c^2\end{bmatrix}$ is

(A) $\begin{bmatrix}10 & 22\\ 21 & 25\end{bmatrix}$ (B) $\begin{bmatrix}0 & 7\\ 8 & 9\end{bmatrix}$
(C) $\begin{bmatrix}26 & 10\\ 34 & 26\end{bmatrix}$ (D) $\begin{bmatrix}0 & 0\\ 0 & 0\end{bmatrix}$

42. If $A=\begin{bmatrix}\cos\theta & -\sin\theta\\ \sin\theta & \cos\theta\end{bmatrix}$, what is the value of θ satisfying the equation $A^T+A=I_2$?

(A) $\frac{\pi}{2}$ (B) $\frac{\pi}{3}$
(C) $\frac{\pi}{6}$ (D) π

43. Which of the following matrices is non-singular?

(A) $\begin{bmatrix}3 & 1\\ 9 & 3\end{bmatrix}$ (B) $\begin{bmatrix}2 & 2\\ 2 & 2\end{bmatrix}$
(C) $\begin{bmatrix}2 & 3\\ 4 & 5\end{bmatrix}$ (D) $\begin{bmatrix}0 & 0\\ 0 & 0\end{bmatrix}$

44. If $A=\begin{bmatrix}4 & x+2\\ 2x-3 & x+1\end{bmatrix}$ is symmetric, then x is equal to

(A) 2 (B) 3
(C) 4 (D) 5

45. The number of values of x for which the matrix $A=\begin{bmatrix}3-x & 2 & 2\\ 2 & 4-x & 1\\ -2 & -4 & -1-x\end{bmatrix}$ is singular, is

(A) 0 (B) 1
(C) 2 (D) 3

46. If matrix $A=\begin{bmatrix}2 & 3\\ 4 & 5\end{bmatrix}$ and matrix $B=\begin{bmatrix}0 & 2\\ 3 & 4\end{bmatrix}$, then the value of $3A-2B$ is

(A) $\begin{bmatrix}6 & 6\\ 6 & 6\end{bmatrix}$ (B) $\begin{bmatrix}6 & 5\\ 6 & 7\end{bmatrix}$
(C) $\begin{bmatrix}6 & 6\\ 5 & 7\end{bmatrix}$ (D) $\begin{bmatrix}5 & 7\\ 6 & 6\end{bmatrix}$

47. If $A=\begin{bmatrix}1 & 2 & 3\\ 4 & 5 & 6\\ 7 & 8 & 9\end{bmatrix}$ and $B=\begin{bmatrix}9 & 8 & 7\\ 6 & 5 & 4\\ 3 & 2 & 1\end{bmatrix}$ then the value of $\frac{1}{2}(A+B)$ is

(A) $\begin{bmatrix}4 & 6 & 7\\ 8 & 6 & 5\\ 4 & 4 & 4\end{bmatrix}$ (B) $\begin{bmatrix}5 & 5 & 5\\ 5 & 5 & 5\\ 5 & 5 & 5\end{bmatrix}$
(C) $\begin{bmatrix}0 & 0 & 0\\ 0 & 0 & 0\\ 0 & 0 & 0\end{bmatrix}$ (D) none of these

48. Consider the system of following equations.
$2x+3y=8$
$7x+8y=24$
This can be written in matrix form as

(A) $\begin{bmatrix}2 & 3\\ 7 & 8\end{bmatrix}\begin{bmatrix}y\\ x\end{bmatrix}=\begin{bmatrix}8\\ 24\end{bmatrix}$ (B) $\begin{bmatrix}2 & 3\\ 7 & 8\end{bmatrix}\begin{bmatrix}x\\ y\end{bmatrix}=\begin{bmatrix}8\\ 24\end{bmatrix}$
(C) $\begin{bmatrix}2 & 8\\ 7 & 24\end{bmatrix}\begin{bmatrix}x\\ y\end{bmatrix}=\begin{bmatrix}3\\ 8\end{bmatrix}$ (D) none of these

49. Consider the system of following equations.
$2x+3y+4z=0$
$5x+7y+8z=5$
$10x+11y+12z=12$
This can be written in matrix form as

(A) $\begin{bmatrix} 2 & 3 & 4 \\ 5 & 7 & 8 \\ 10 & 11 & 12 \end{bmatrix}\begin{bmatrix} x \\ y \\ z \end{bmatrix} = \begin{bmatrix} 0 \\ 5 \\ 12 \end{bmatrix}$

(B) $\begin{bmatrix} 2 & 3 & 0 \\ 5 & 7 & 9 \\ 10 & 11 & 12 \end{bmatrix}\begin{bmatrix} x \\ y \\ z \end{bmatrix} = \begin{bmatrix} 4 \\ 8 \\ 12 \end{bmatrix}$

(C) $\begin{bmatrix} 2 & 3 & 4 \\ 5 & 7 & 8 \\ 10 & 11 & 12 \end{bmatrix}\begin{bmatrix} 1 \\ 1 \\ 1 \end{bmatrix} = \begin{bmatrix} 0 \\ 5 \\ 12 \end{bmatrix}$

(D) none of these

50. If $A = \begin{bmatrix} 1 & 2 \\ 3 & 4 \end{bmatrix}$ and $B = \begin{bmatrix} 0 & 1 \\ 2 & 2 \end{bmatrix}$ then which of following is correct?

(A) $(A + B)^2 = A^2 + B^2 + 2AB$
(B) $(A + B)' = A' + B'$
(C) $(A + B)(A - B) = A^2 - B^2$
(D) none of these

OLYMPIAD AND NTSE LEVEL EXERCISES

51. If $A = \begin{bmatrix} 1 & 1 \\ 0 & 1 \end{bmatrix}$ then $A^n =$

(A) $\begin{bmatrix} 1 & n \\ 0 & 1 \end{bmatrix}$ (B) $\begin{bmatrix} n & n \\ 0 & n \end{bmatrix}$

(C) $\begin{bmatrix} n & 1 \\ 0 & n \end{bmatrix}$ (D) $\begin{bmatrix} 1 & 1 \\ 0 & n \end{bmatrix}$

52. If $A = \begin{bmatrix} ab & b^2 \\ -a^2 & -ab \end{bmatrix}$ and $A^n = O$ then the minimum value of n is

(A) 2 (B) 3
(C) 4 (D) 5

53. If $A = \begin{bmatrix} 2 & 0 & 0 \\ 0 & 2 & 0 \\ 0 & 0 & 2 \end{bmatrix}$ then $A^5 =$

(A) $5A$ (B) $10A$
(C) $16A$ (D) $32A$

54. If $A = \begin{bmatrix} 0 & 5 \\ 0 & 0 \end{bmatrix}$ and $f(x) = 1 + x + x^2 + ... + x^{16}$ then $f(a)$ is equal to

(A) 0 (B) $\begin{bmatrix} 1 & 5 \\ 0 & 1 \end{bmatrix}$

(C) $\begin{bmatrix} 1 & 5 \\ 0 & 0 \end{bmatrix}$ (D) $\begin{bmatrix} 0 & 5 \\ 1 & 1 \end{bmatrix}$

55. Let $A = \begin{bmatrix} 0 & 2y & z \\ x & y & -z \\ x & -y & z \end{bmatrix}$ and $A' \cdot A = I$. Then the value of $x^2 + y^2 + z^2$ is

(A) 1 (B) 2
(C) $\frac{1}{2}$ (D) None of these

56. If $2A + 3B = \begin{bmatrix} 2 & -1 & 4 \\ 3 & 2 & 5 \end{bmatrix}$ and $A + 2B = \begin{bmatrix} 5 & 0 & 3 \\ 1 & 6 & 2 \end{bmatrix}$ then $B =$

(A) $\begin{bmatrix} 8 & -1 & 2 \\ -1 & 10 & -1 \end{bmatrix}$ (B) $\begin{bmatrix} 8 & 1 & 2 \\ -1 & 10 & -1 \end{bmatrix}$

(C) $\begin{bmatrix} 8 & 1 & -2 \\ -1 & 10 & -1 \end{bmatrix}$ (D) $\begin{bmatrix} 8 & 1 & 2 \\ 1 & 10 & 1 \end{bmatrix}$

57. If $A_\alpha = \begin{bmatrix} \cos\alpha & \sin\alpha \\ -\sin\alpha & \cos\alpha \end{bmatrix}$ then $A_\alpha A_\beta$ is equal to

(A) $A_{\alpha\beta}$ (B) $A_{\alpha+\beta}$
(C) $A_{\alpha-\beta}$ (D) none of these

58. The equation, $[1\ x\ y]\begin{bmatrix} 1 & 3 & 1 \\ 0 & 2 & -1 \\ 0 & 0 & 1 \end{bmatrix}\begin{bmatrix} 1 \\ x \\ y \end{bmatrix} = [0]$ has for

(i) $y = 0$, (p) rational roots
(ii) $y = -1$ (q) irrational roots
(r) integral roots

then which of these is a correct match?

	(i)	(ii)
(A)	(p)	(r)
(B)	(q)	(p)

(C) (p) (q)

(D) (r) (p)

59. Consider a matrix $A = [a_{ij}]$ of order 3×3 such that $a_{ij} = (k)^{i+j}$ where $k \in I$.

Match Column I with Column II:

	Column I		Column II
(p)	All values of k if A is singular if	(1)	$k \in \{0\}$
(q)	A is null matrix if	(2)	$k \in \phi$
(r)	A is skew-symmetric which is not null matrix if	(3)	$k \in I$
(s)	$A^2 = 3A$, if	(4)	$k \in \{-1, 0, 1\}$

Now, choose the correct option from amongst the given codes.

Codes:

	(p)	(q)	(r)	(s)
(A)	(3)	(1)	(4)	(2)
(B)	(4)	(1)	(2)	(3)
(C)	(3)	(1)	(2)	(4)
(D)	(2)	(1)	(3)	(4)

60. The minimum number of zeros in a upper triangular matrix of order n is

(A) $\frac{n(n-1)}{2}$

(B) $\frac{n(n+1)}{2}$

(C) $\frac{2n(n-1)}{2}$

(D) none of these

10 Determinants

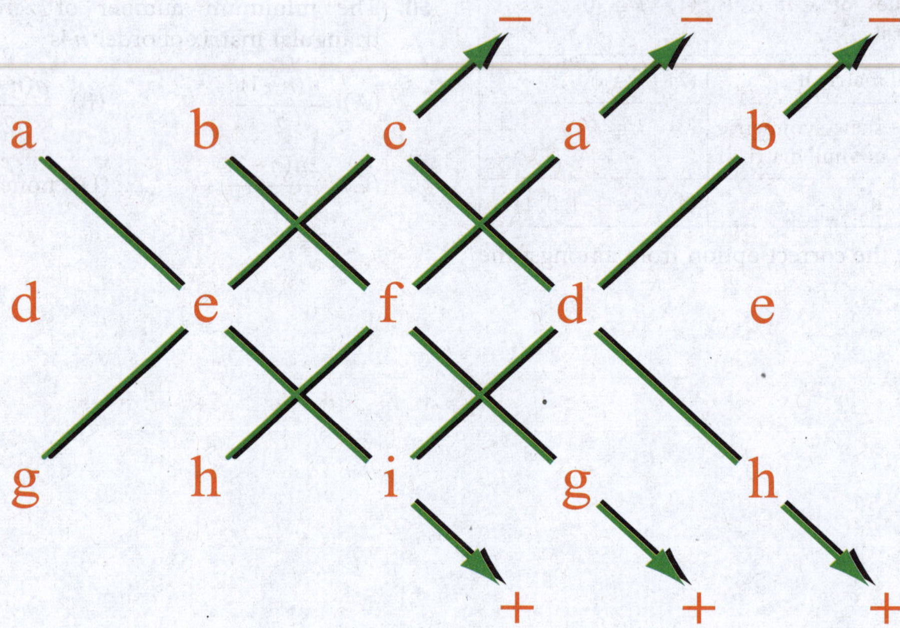

Value = aei + bfg + cdh – gec – hfa – idb

$$\begin{vmatrix} 1 & a & a^2 - bc \\ 1 & b & b^2 - ca \\ 1 & c & c^2 - ab \end{vmatrix}$$

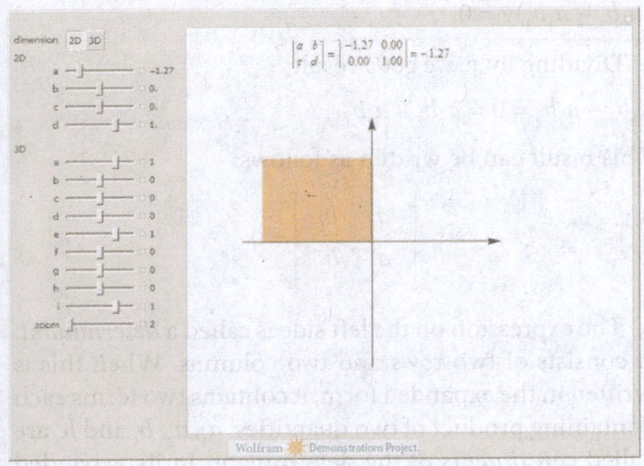

OBJECTIVES

After reading this chapter you should be able to

- understand the definition of determinant
- understand the general rule for expansion of determinants
- know about the application of determinants
- state and prove some properties of determinants

$$|A| = \begin{vmatrix} a_1 & a_2 & a_3 \\ b_1 & b_2 & b_3 \\ c_1 & c_2 & c_3 \end{vmatrix}$$

INTRODUCTION

Recall the method of solving homogeneous linear equations by elimination method.

$$a_1x + b_1y = 0 \qquad (1)$$

$$a_2x + b_2y = 0 \qquad (2)$$

Multiplying Eq. (1) by a_2, we have the following:

$$a_1a_2x + a_2b_1y = 0 \qquad (3)$$

Multiplying Eq. (2) by a_1, we have the following:

$$a_1a_2x + a_1b_2y = 0 \qquad (4)$$

Subtracting Eq. (3) from (4), we have the following:

$$(a_2b_1 - a_1b_2)y = 0$$

Dividing by y, we get a result

$$a_2b_1 - a_1b_2 = 0 = a_1b_2 - a_2b_1$$

This result can be written as follows:

$$\begin{vmatrix} a_1 & b_1 \\ a_2 & b_2 \end{vmatrix} = 0$$

The expression on the left side is called a *determinant.* It consists of two rows and two columns. When this is written in the expanded form, it contains two terms each containing product of two quantities. a_1, a_2, b_1 and b_2 are called *constituents* of the determinant. In its expanded form, the two terms a_1b_2 and a_2b_1 are called *elements* of the determinant.

Another way of looking at the determinants is as follows:

As we have already learnt matrices, the determinants have been related to square matrices.

With every square matrix $A = [a_{ij}]$, we associate a number called determinant of A denoted by det A or $|A|$. *Non-square matrices do not have determinants.*

Determinants are very useful tools for solving systems of linear equations.

In the determinant $\begin{vmatrix} a & b \\ c & d \end{vmatrix}$, the elements are arranged in two *rows a*, *b* and *c*, *d* and in two *columns a*, *c* and *b*, *d*. The term *ad* is the product of the two elements which lie on the leading diagonal (the diagonal that slopes downwards to the *right*).

Every term is the expansion of a determinant which contains one and only one element from each row and only one element from each column.

Worked Examples

1. Evaluate $\begin{vmatrix} -9 & 5 \\ -2 & 1 \end{vmatrix}$

Solution

The required value is as follows:

$(-9)1-(5)(-2) = 9 + 10 = 1$

2. Find the value of $\begin{vmatrix} \cos\theta & -\sin\theta \\ \sin\theta & \cos\theta \end{vmatrix}$.

Solution

The required value is as follows:

$\cos^2\theta - (-\sin^2\theta) = \sin^2\theta + \cos^2\theta = 1$

ORDER OF A DETERMINANT

A determinant having two rows and two columns is said to have order 2:

$$|A| = \begin{vmatrix} a & b \\ c & d \end{vmatrix} = ad - bc$$

A determinant of order 3 has three rows and three columns:

$$|A| = \begin{vmatrix} a_1 & a_2 & a_3 \\ b_1 & b_2 & b_3 \\ c_1 & c_2 & c_3 \end{vmatrix}$$

Determinants of higher order also exist. All properties of determinants can be learnt without loss of generality using determinants of order 2.

PROPERTIES OF DETERMINANTS

(a) The value of a determinant is unaltered if rows and columns are interchanged.

Proof

Let $D = \begin{vmatrix} a & b \\ c & d \end{vmatrix}$

$$D = \begin{vmatrix} a & b \\ c & d \end{vmatrix} = ad - bc$$

Interchanging rows with columns, we get the following:

$$\begin{vmatrix} a & c \\ b & d \end{vmatrix} = ad - bc$$

We observe that the values are the same.

(b) If two rows or columns are interchanged, the determinant is altered only in sign.

Let $D = \begin{vmatrix} a & b \\ c & d \end{vmatrix}$

If the columns be interchanged, we have the following:

$$D' = \begin{vmatrix} b & a \\ d & c \end{vmatrix}$$

$$D = \begin{vmatrix} a & b \\ c & d \end{vmatrix} = ad - bc$$

$$D' = \begin{vmatrix} b & a \\ d & c \end{vmatrix} = bc - ad$$

$$= -(ad - bc)$$

$$= -D$$

(c) If a determinant has two rows or columns identical, the value of the determinant is zero.

If $D = \begin{vmatrix} a & b \\ a & b \end{vmatrix}$ in which two rows are identical, the value is $ab - ab = 0$.

(d) If all the elements of one row or column are multiplied by the same number, the value of the determinant is multiplied by that number.

Let $D = \begin{vmatrix} a & b \\ c & d \end{vmatrix}$

If the elements of the first column are multiplied by k, then we have the following:

$$D' = \begin{vmatrix} ka & b \\ kc & d \end{vmatrix}$$

$$D' = kad - kcb$$

$$= k(ad - cb)$$

$$= k\begin{vmatrix} a & b \\ c & d \end{vmatrix}$$

$$= kd.$$

(e) If each element of any row or column is expressed as the sum of two terms, then the determinant can be expressed as the sum of two other determinants.

$$A = \begin{vmatrix} a+l & b \\ c+m & d \end{vmatrix}$$

$$= (a + l)d - (c + m)b$$

$$= ad + ld - cb - mb$$

$$= ad - cb + ld - mb$$

$$= \begin{vmatrix} a & b \\ c & d \end{vmatrix} + \begin{vmatrix} l & b \\ m & d \end{vmatrix}$$

(f) The value of a determinant is not changed by adding the elements of any row or column with the same multiples of the corresponding elements of any other row or column.

$$D = \begin{vmatrix} a+kb & b \\ c+kd & d \end{vmatrix}$$

$$= (a + kb)d - (c + kd)b$$

$$= ad + kbd - cb - kbd$$

$$= ad - cb + k(bd - bd)$$

$$= \begin{vmatrix} a & b \\ c & d \end{vmatrix} + k\begin{vmatrix} b & d \\ b & d \end{vmatrix}$$

$$= \begin{vmatrix} a & b \\ c & d \end{vmatrix} + k0$$

$$= D$$

APPLICATIONS

Let us solve the following simultaneous equations:

$$ax + by + c = 0 \quad (1)$$

$$a_1x + b_1y + c_1 = 0 \quad (2)$$

Multiplying Eq. (1) by b_1 and Eq. (2) by b, we get the following:

$$ab_1x + bb_1y + cb_1 = 0 \quad (3)$$

$$a_1bx + bb_1y + c_1b = 0 \quad (4)$$

Subtracting Eq. (4) from (3), we get the following:

$$(ab_1 - a_1b)x + cb_1 - c_1b = 0$$

$$\therefore (ab_1 - a_1b)x = bc_1 - b_1c$$

$$\therefore x = \frac{bc_1 - b_1c}{ab_1 - a_1b} \text{ if } ab_1 - a_1b \neq 0$$

Similarly, $y = \dfrac{ca_1 - c_1a}{ab_1 - a_1b}$ if $ab_1 - a_1b \neq 0$

Hence, we have the following:

$$\frac{x}{bc_1 - b_1c} = \frac{y}{ca_1 - c_1a} = \frac{1}{ab_1 - a_1b} \quad (A)$$

(This was learnt as rule of cross-multiplication under solutions to simultaneous linear equations.)

We can write A in determinant form as follows:

$$\frac{x}{\begin{vmatrix} b & c \\ b_1 & c_1 \end{vmatrix}} = \frac{y}{\begin{vmatrix} c & a \\ c_1 & a_1 \end{vmatrix}} = \frac{1}{\begin{vmatrix} a & b \\ a_1 & b_1 \end{vmatrix}}$$

Note 1

(a) The determinant $\begin{vmatrix} b_2 & c_2 \\ b_3 & c_3 \end{vmatrix}$ is called minor of a_1.

(b) The determinant $\begin{vmatrix} a_2 & c_2 \\ a_3 & c_3 \end{vmatrix}$ is called minor of b_1.

(c) The determinant $\begin{vmatrix} a_2 & b_2 \\ a_3 & b_3 \end{vmatrix}$ is called minor of c_1.

Note 2

To form the minor of an element of a determinant, omit the row and column in which the element occurs and write the determinant formed by the rest of the elements, keeping them in their respective positions.

Example

Write the minor of c in the given determinant:

$$\begin{vmatrix} a & h & g \\ h & b & f \\ g & f & c \end{vmatrix}$$

Omit the third row and third column in which c occurs. Write the rest of the element; i.e., the minor of c is

$$\begin{vmatrix} a & h \\ h & b \end{vmatrix}$$

Note 3

The value of $\begin{vmatrix} a_1 & b_1 & c_1 \\ a_2 & b_2 & c_2 \\ a_3 & b_3 & c_3 \end{vmatrix}$ is

$$a_1(b_2c_3 - b_3c_2) + b_1(c_1a_3 - c_3a_1) + c_1(a_2b_3 - a_3b_2)$$

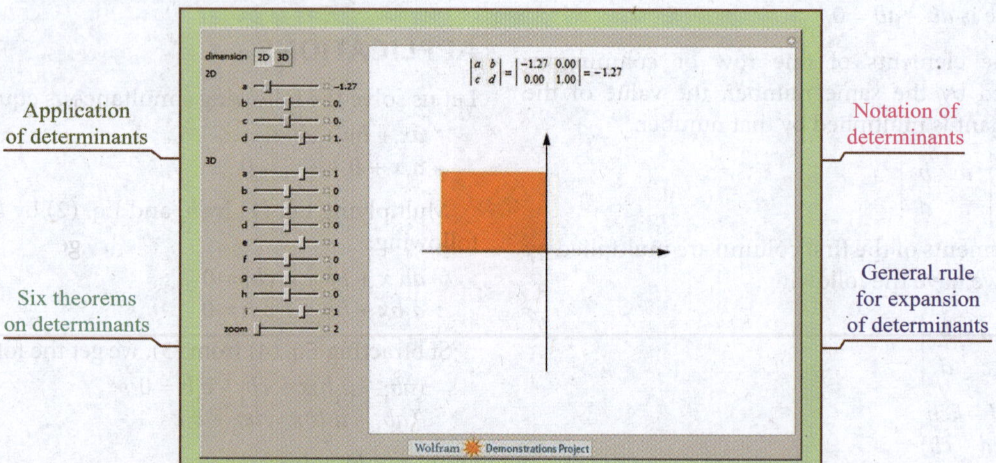

Determinants

TEST YOURSELF

Level 1

Evaluate the following determinants:

1. $\begin{vmatrix} 2 & 3 \\ 3 & 5 \end{vmatrix}$

2. $\begin{vmatrix} -4 & 5 \\ -5 & 6 \end{vmatrix}$

3. $\begin{vmatrix} 3 & 1 \\ -2 & -3 \end{vmatrix}$

4. $\begin{vmatrix} a & b \\ b & a \end{vmatrix}$

5. $\begin{vmatrix} 1 & \cos A \\ \cos A & 1 \end{vmatrix}$

6. $\begin{vmatrix} a-b & b-c \\ c-b & a-c \end{vmatrix}$

Level 2

Solve the following using determinants:

7. $19x - 4y = 13$
$23x - 13y = 19$

8. $2x + 5y - 20 = 0$
$3x + 4y - 25 = 0$

9. $5x + 2y = 45$
$\frac{2x}{7} + \frac{3y}{5} = 25$

10. $2x - 4y = -3$
$4x + 2y = 9$

11. $5x + 2y = -2$
$4x + 6y = -3$

Level 3

12. Write the minor of all the elements of the determinant $\begin{vmatrix} a & b & c \\ d & e & f \\ g & h & k \end{vmatrix}$.

13. Evaluate $\begin{vmatrix} 1 & 1 & 1 \\ 12 & 7 & 11 \\ 10 & 6 & 9 \end{vmatrix}$.

14. Evaluate $\begin{vmatrix} 5 & 20 & 25 \\ 8 & 72 & 56 \\ 1 & 6 & 6 \end{vmatrix}$.

15. Evaluate $\begin{vmatrix} 35 & 20 & 17 \\ 21 & 12 & 9 \\ 14 & 8 & 13 \end{vmatrix}$.

16. Evaluate $\begin{vmatrix} 15 & 9 & 13 \\ 1 & 0 & 0 \\ 8 & 6 & 3 \end{vmatrix}$.

17. Evaluate $\begin{vmatrix} 1 & a & b+c \\ 1 & b & c+a \\ 1 & c & a+b \end{vmatrix}$.

18. Evaluate $\begin{vmatrix} 1 & a & a^2-bc \\ 1 & b & b^2-ca \\ 1 & c & c^2-ab \end{vmatrix}$.

Multiple Choice Questions

19. Which of the following is independent from θ?

(A) $\begin{vmatrix} \sec\theta & \tan\theta \\ \tan\theta & \sec\theta \end{vmatrix}$ (B) $\begin{vmatrix} \sin\theta & \cos\theta \\ \cos\theta & \sin\theta \end{vmatrix}$

(C) $\begin{vmatrix} 1 & \sin\theta \\ \cos\theta & 1 \end{vmatrix}$ (D) $\begin{vmatrix} \cos\theta & 1 \\ \sin\theta & 1 \end{vmatrix}$

20. What is the value of the following determinant?

$\begin{vmatrix} 5 & 10 & 15 \\ 10 & 20 & 30 \\ 25 & 30 & 35 \end{vmatrix}$

(A) 0 (B) −10
(C) 10 (D) 100

21. The value of determinant $\begin{vmatrix} 5 & 20 & 25 \\ 8 & 72 & 56 \\ 1 & 6 & 6 \end{vmatrix}$ is

(A) 10 (B) 20
(C) 30 (D) 40

22. Evaluate the following determinant.

$\begin{vmatrix} 1 & a & b+c \\ 1 & b & c+a \\ 1 & c & a+b \end{vmatrix}$

(A) 1 (B) $a + b + c$
(C) $ab + bc + ca$ (D) 0

23. What is the value of the following determinant?

$\begin{vmatrix} 1 & a & a^2-bc \\ 1 & b & b^2-ca \\ 1 & c & c^2-ab \end{vmatrix}$

(A) 0
(B) $(a + b + c)^2$
(C) $(ab + bc + ca)(a + b + c)$
(D) $1 - abc$

24. What is the value of following determinant?

$$\begin{vmatrix} -1 & 2 & 1 \\ 3+2\sqrt{2} & 2+2\sqrt{2} & 1 \\ 3-2\sqrt{2} & 2-2\sqrt{2} & 1 \end{vmatrix}$$

(A) $-\sqrt{2}$ (B) $4\sqrt{2}$
(C) $-16\sqrt{2}$ (D) 1

25. Sum of real roots of the equation $\begin{vmatrix} 1 & 4 & 20 \\ 1 & -2 & 5 \\ 1 & 2x & 2x^2 \end{vmatrix} = 0$ is

(A) −2 (B) −1
(C) 0 (D) 1

26. If a, b and c are all different such that

$$\begin{vmatrix} a & a^2 & 1+a^3 \\ b & b^2 & 1+b^3 \\ c & c^2 & 1+c^3 \end{vmatrix} = 0,$$

then which of the following is true?

(A) $abc = -1$ (B) $abc = 0$
(C) $bc = 1$ (D) $abc = -1$

27. An equilateral triangle has each of its sides of length 4 cm. If (x_r, y_r) $(r = 1, 2, 3)$ are its vertices, the value of $\begin{vmatrix} x_1 & y_1 & 1 \\ x_2 & y_2 & 1 \\ x_3 & y_3 & 1 \end{vmatrix}^2$ is

(A) 192 (B) 768
(C) 1024 (D) 128

28. If $\begin{vmatrix} x & 3 & 6 \\ 3 & 6 & x \\ 6 & x & 3 \end{vmatrix} = \begin{vmatrix} 2 & x & 7 \\ x & 7 & 2 \\ 7 & 2 & x \end{vmatrix} = \begin{vmatrix} 4 & 5 & x \\ 5 & x & 4 \\ x & 4 & 5 \end{vmatrix} = 0$

then x is equal to

(A) 0 (B) −9
(C) 3 (D) 9

29. The value of $\begin{vmatrix} x & \sin\theta & \cos\theta \\ -\sin\theta & -x & 1 \\ \cos\theta & 1 & x \end{vmatrix}$ is

(A) $x^3 + 2x\cos\theta$ (B) $-x^3 + 2x$
(C) $-x^3\cos\theta + 2x$ (D) $-x^3$

30. If $\Delta = \begin{vmatrix} abc & b^2c & c^2b \\ abc & c^2a & ca^2 \\ abc & a^2b & b^2a \end{vmatrix} = 0$, such that a, b and c are all different non-zero real numbers, then which of the following is correct?

(A) $a + b + c = 0$ (B) $a = 2b + c$
(C) $a - b + c = 0$ (D) $a + b - c = 0$

31. Which of the following is not a root of the equation $\begin{vmatrix} x & -6 & -1 \\ 2 & -3x & x-3 \\ -3 & 2x & x+2 \end{vmatrix} = 0$?

(A) 2 (B) 0
(C) 1 (D) −3

32. What is the value of the determinant $\begin{vmatrix} 1 & 1 & 1 \\ x & y & z \\ x^3 & y^3 & z^3 \end{vmatrix}$?

(A) $(x-y)(y-z)(z-c)$
(B) $(x-y)(y-z)(z-x)(x+y+z)$
(C) $(x+y)(y+z)(z+x)$
(D) $(x+y+z)^3$

33. If $p + q + r = 0 = a + b + c$, then the value of the determinant $\begin{vmatrix} pa & qb & rc \\ qc & ra & pb \\ rb & pc & qa \end{vmatrix}$ is

(A) 0 (B) $pa + qb + rc$
(C) 1 (D) $6abcpqr$

34. Which of the following is the solution of the system of equations $x + y = 4$ and $3x - 2y = 9$?

(A) $\dfrac{x}{\begin{vmatrix} 4 & 1 \\ 9 & -2 \end{vmatrix}} = \dfrac{y}{\begin{vmatrix} 1 & 4 \\ 3 & 9 \end{vmatrix}} = \dfrac{1}{\begin{vmatrix} 1 & 1 \\ 3 & -2 \end{vmatrix}}$

(B) $\dfrac{x}{\begin{vmatrix} 1 & 4 \\ 3 & 9 \end{vmatrix}} = \dfrac{y}{\begin{vmatrix} 4 & 1 \\ 9 & -2 \end{vmatrix}} = \dfrac{1}{\begin{vmatrix} 1 & 1 \\ 3 & -2 \end{vmatrix}}$

(C) $\dfrac{x}{\begin{vmatrix} 1 & 1 \\ 3 & -2 \end{vmatrix}} = \dfrac{y}{\begin{vmatrix} 1 & 4 \\ 3 & 9 \end{vmatrix}} = \dfrac{1}{\begin{vmatrix} 4 & 1 \\ 9 & -2 \end{vmatrix}}$

(D) None of these

35. The value of determinant $\begin{vmatrix} b+c & c & b \\ c & c+a & a \\ b & a & a+b \end{vmatrix}$ is

(A) abc (B) $2abc$

(C) $3abc$ (D) $4abc$

OLYMPIAD AND NTSE LEVEL EXERCISES

36. The value of the determinant
$$\begin{vmatrix} 0 & b^3-a^3 & c^3-a^3 \\ a^3-b^3 & 0 & c^3-b^3 \\ a^3-c^3 & b^3-c^3 & 0 \end{vmatrix}$$ is equal to

(A) $a^3+b^3+c^3$ (B) $a^3-b^3-c^3$

(C) 0 (D) $-a^3+b^3+c^3$

37. $\begin{vmatrix} \sin^2 x & \cos^2 x & 1 \\ \cos^2 x & \sin^2 x & 1 \\ -10 & 12 & 2 \end{vmatrix} =$

(A) 0

(B) $12\cos^2 x - 10\sin^2 x$

(C) $12\sin^2 x - 10\cos^2 x - 2$

(D) $10\sin 2x$

38. If $a, b, c > 1, \Delta = \begin{vmatrix} \log_a(abc) & \log_a b & \log_a c \\ \log_b(abc) & 1 & \log_b c \\ \log_c(abc) & \log_c b & 1 \end{vmatrix}$ is

(A) 0

(B) $\log_a b + \log_b c + \log_c a$

(C) $\log_{abc}(a+b+c)$

(D) none of these

39. The roots of the equation $\begin{vmatrix} 1 & 4 & 20 \\ 1 & -2 & 5 \\ 1 & 2x & 5x^2 \end{vmatrix} = 0$ are

(A) –1, –2 (B) –1, 2

(C) 1, –2 (D) 1, 2

40. If $\Delta = \begin{vmatrix} abc & b^2c & c^2b \\ abc & c^2a & ca^2 \\ abc & a^2b & b^2a \end{vmatrix} = 0$, $(a, b, c \in R$ and are all different and non-zero) then

(A) $abc = 0$ (B) $ab + bc + ca = 0$

(C) $a + b + c = 0$ (D) $a^2 + b^2 + c^2 = 1$

41. If $f(x) = \begin{vmatrix} x-3 & 2x^2-18 & 3x^3-81 \\ x-5 & 2x^2-50 & 4x^3-500 \\ 1 & 2 & 3 \end{vmatrix}$

then $f(1) \cdot f(2) + f(3) \cdot f(5) + f(5) . f(1) =$

(A) $f(1)$ (B) $f(3)$

(C) $f(1) + f(3)$ (D) $f(1) + f(5)$

42. In triangle ABC,
$$\begin{vmatrix} 1 & 1 & 1 \\ \cot\frac{A}{2} & \cot\frac{B}{2} & \cot\frac{C}{2} \\ \tan\frac{B}{2}+\tan\frac{C}{2} & \tan\frac{C}{2}+\tan\frac{A}{2} & \tan\frac{A}{2}+\tan\frac{B}{2} \end{vmatrix} = 0$$
then triangle must be

(A) equilateral (B) isosceles

(C) obtuse angled (D) none of these

43. If $\begin{vmatrix} 1 & 1 & 1 \\ a & b & c \\ a^3 & b^3 & c^3 \end{vmatrix} = (a-b)(b-c)(c-a)(a+b+c)$,

where a, b, c are all different, then the determinant
$$\begin{vmatrix} 1 & 1 & 1 \\ (x-a)^2 & (x-b)^2 & (x-c)^2 \\ (x-b)(x-c) & (x-c)(x-a) & (x-a)(x-b) \end{vmatrix}$$
vanishes when

(A) $a + b + c = 0$ (B) $x = \frac{1}{3}(a+b+c)$

(C) $x = \frac{1}{2}(a+b+c)$ (D) $x = a + b + c$

44. If $a + b + c \neq 0$ and $\begin{vmatrix} a-x & c & b \\ c & b-x & a \\ b & a & c-x \end{vmatrix} = 0$ then

total number of different values of x is equal to

(A) 1 (B) 2

(C) 3 (D) None of these

45. Which of the following is not the root of the equation $\begin{vmatrix} x & -6 & -1 \\ 2 & -3x & x-3 \\ -3 & 2x & x+2 \end{vmatrix} = 0$?

(A) 2 (B) 0

(C) 1 (D) –3

PUZZLES AND GAMES

1. A 150-m-long steamer has changed its direction through 30 degrees while moving through a distance equal to twice its own length. What is the radius of the circle in which it has moved?

Solution

Let R be the radius of the circle.

$S = R\theta$

$$300 = R \times \frac{\pi}{6} \Rightarrow R = \frac{1800}{\pi} = 573 \text{ m}$$

2. A brick weighs 4 kg. What is the weight of a similar brick, four times smaller and made of the same material?

Solution

One would quickly come out with the answer, 1 kg, that is four times less. But this is wrong.

The length, breadth, and width of the smaller block are all one-fourth of that of the big block. Therefore, its volume and weight are $\frac{1}{4} \times \frac{1}{4} \times \frac{1}{4} = \frac{1}{64}$ times that of the big block.

The answer is $4000 \times 164 = 62.5$ g.

3. Can you divide the number 45 into four parts such that addition of 2 to the first part, subtraction of 2 from the second part, multiplication of 2 to the third part, and division by 2 of the fourth part result in the same number.

Solution

$A + B + C + D = 45$

$$A + 2 = B - 2 = 2C = \frac{D}{2}$$

$$A = B - 4; \quad C = \frac{B-2}{2}; \quad D = 2(B-2)$$

$B - 4 + B - 2 + 2(B - 2) = 45$

$B = 12$, $A = 8$, $C = 5$, and $D = 20$

Thus, $12 - 2 = 8 + 2 = 5 \times 2 = \frac{20}{2} = 10$

The numbers are 8, 12, 5, and 20.

4. A cylindrical container has a radius of 8 inches and a height of 3 inches. How many inches should be added to either the radius or the height to give the same increase in the volume?

Solution

Suppose the change in the radius or the height be x.

Then $(8)^2 (3 + x) = (8 + x)^2 (3)$

$$x = \frac{16}{3}$$

5. A bottle and its cork together cost ₹1.10 and the bottle costs ₹ 1 more than its cork. What is the cost of the bottle?

Solution

No, the answer is not ₹ 1. If the cost of the bottle is ₹ 1 and that of the cork is 10 paise, then the bottle will cost only 90 paise more than the cork. Considering it more carefully, we come out with the correct answer: 105 paise.

Exercise

1. Find the two rectangles which have dimensions in such a way that they are integers and their area and perimeter are equal.
2. Here is a multiplication. $7451 \times 3982 = 29\_69882$. One digit is missing from the product. Can you find the missing number without performing the multiplication?
3. Here is an octagon. Can you dissect it into five sections so that it can be rearranged to form a square?
4. One day while driving through the countryside, I saw a farmer tending his pigs and ducks in his yard. I was curious to know how many of each he has. I stopped the car and inquired. He replied, 'They have altogether 60 eyes and 86 feet between them'. I drove off trying to calculate in my mind the exact number of ducks and pigs he has? What do you think the answer is?
5. A heavy tree trunk can be sawn into 12 ft long piece in 1 minute. How long will it take to saw it into 12 equal pieces?

11 Rational Expressions

Four 4's Puzzle

$$0 = 44 - 44$$

$$1 = \frac{44}{44}$$

$$2 = \frac{4 \times 4}{4 + 4}$$

$$3 = \frac{4 + 4 + 4}{4}$$

$$4 = 4 + (4 - 4) \times 4$$

$$5 = \frac{4 \times 4 + 4}{4}$$

$$6 = \frac{4 + 4}{4} + 4$$

$$7 = 4 + 4 - \frac{4}{4}$$

$$8 = 4 \times 4 - (4 + 4)$$

$$9 = 4 + 4 + \frac{4}{4}$$

$$10 = \frac{44 - 4}{4}$$

> The advancement and perfection of mathematics are intimately connected with the prosperity of the state.
>
> – Napoleon

$$\begin{vmatrix} 1 & a & a^2-bc \\ 1 & b & b^2-ca \\ 1 & c & c^2-ab \end{vmatrix}$$

OBJECTIVES

After reading this chapter you should be able to

- express polynomials as rational expression and vice versa
- identify the types of rational expressions
- reduce the rational expression to the lowest form
- perform the fundamental operations on the set of rational expressions

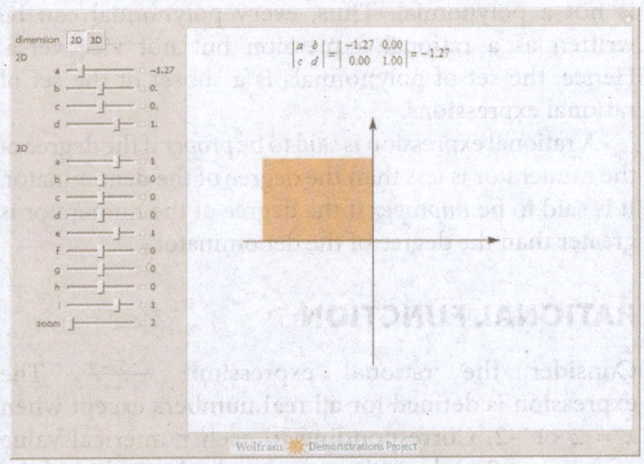

$$|A| = \begin{vmatrix} a_1 & a_2 & a_3 \\ b_1 & b_2 & b_3 \\ c_1 & c_2 & c_3 \end{vmatrix}$$

INTRODUCTION

You know that $\frac{p}{q}$, where p and q are integers and $q \neq 0$, is a rational number. This idea can easily be extended to rational expressions by taking polynomials P and Q instead of integers p and q.

Hence, a rational expression is an expression of the form $\frac{P}{Q}$, where P and Q are polynomials and $Q \neq 0$.

Every polynomial can be expressed as a rational expression by taking one in the denominator.

e.g., $P \Leftrightarrow \frac{P}{1}$

Let us see if every rational expression is a polynomial.

Consider a rational expression $\frac{x^2+1}{x} \Leftrightarrow x + \frac{1}{x}$ which is not a polynomial. Thus, every polynomial can be written as a rational expression but not vice versa. Hence, the set of polynomials is a subset of the set of rational expressions.

A rational expression is said to be *proper* if the degree of the numerator is less than the degree of the denominator. It is said to be *improper* if the degree of the numerator is greater than the degree of the denominator.

RATIONAL FUNCTION

Consider the rational expression $\frac{x^2-5x+4}{x^2-4}$. The expression is defined for all real numbers except when $x = 2$ or -2. Corresponding to each numerical value of x except 2 and -2, there corresponds a value of the rational expression.

Example

$$x = 0 \Rightarrow \frac{x^2-5x+4}{x^2-4} = -1$$

$$x = 4 \Rightarrow \frac{x^2-5x+4}{x^2-4} = 0$$

Thus, the rational expression $\frac{x^2-5x+4}{x^2-4}$ is a function of one variable x for all real numbers except 2 and -2.

A rational function of x is a function which is defined by an expression of the form $\frac{P(x)}{Q(x)}$, where $P(x)$ and $Q(x)$ are polynomials and $Q(x)$ is not a zero polynomial.

Reduction to the Lowest Form

A rational expression is said to be in its lowest form when the numerator and the denominator have no common factor other than one. Thus, simplification of a rational expression consists of the following:

(a) Factorisation of the numerator and denominator

(b) Cancelling out the common factors from the numerator and denominator

Hence, if P, Q, and R are polynomials so that $Q \neq 0$ and $R \neq 0$, then $\frac{P \times R}{Q \times R} = \frac{P}{Q}$.

Worked Examples

1. Reduce the following to the lowest term:

(a) $\dfrac{15x^3y^2}{35xy^5}$, for all $x \neq 0$ and $y \neq 0$

(b) $\dfrac{x^2-x-6}{x^2-5x+6}$, for all $x \neq 2$ and $x \neq 3$

Solution

(a) $\dfrac{15x^3y^2}{35xy^5} = \dfrac{5xy^2}{5xy^2} \times \dfrac{3x^2}{7y^3} = \dfrac{3x^2}{7y^3}$

(b) $\dfrac{x^2-x-6}{x^2-5x+6} = \dfrac{(x-3)(x+2)}{(x-3)(x-2)} = \dfrac{x+2}{x-2}$

2. Simplify the following rational expressions indicating the values of the variable which are not admissible:

(a) $\dfrac{2x^2-3x-44}{x^2-16}$ (b) $\dfrac{x^4+4y^4}{x^2+2y^2-2xy}$

Solution

(a) $\dfrac{2x^2-3x-44}{x^2-16} = \dfrac{2x^2-3x-44}{(x+4)(x-4)}$

The given rational expression is not defined if $x + 4 = 0$ and $x - 4 = 0$ because in either case the denominator would be zero and division by zero is not possible.

Therefore, for $x \neq 4$ and $x \neq -4$, we have the following:

$$\frac{2x^2-3x-44}{x^2-16} = \frac{(2x-11)(x+4)}{(x+4)(x-4)} = \frac{2x-11}{x-4}$$

(b) $\dfrac{x^4+4y^4}{x^2+2y^2-2xy}$

$x^2 + 2y^2 - 2xy \neq 0$

$$\frac{x^4+4y^4}{x^2+2y^2-2xy}$$

$$= \frac{(x^2)^2+(2y^2)^2+4x^2y^2-4x^2y^2}{(x^2+2y^2-2xy)}$$

$$= \frac{(x^2+2y^2+2xy)(x^2+2y^2-2xy)}{(x^2+2y^2-2xy)}$$

$$= x^2+2y^2+2xy$$

Four Fundamental Operations on the Set of Rational Expressions

Recall the four fundamental operations on rational numbers. The procedure is same for rational expressions.

(a) **Rational expressions having the same denominator:** In the case of rational numbers, in order to add rational numbers having the same denominators, we divide the sum of the numerators by the common denominator. Similarly, in order to add rational expressions having the same denominators, we add the polynomials in the numerators and divide the sum by the common denominator.

If P, Q, and R are three polynomials so that $Q \neq 0$, then we have the following:

$$\frac{P}{Q}+\frac{R}{Q}=\frac{P+R}{Q}$$

(b) **Rational expressions having different denominators:** In this case it is not possible to apply the above rule/method. In order to add rational expressions having different denominators, first we bring them to have the same denominators, then add the numerators, and then divide the sum by the common denominator. Thus, if P, Q, R, and S are four polynomials so that $Q \neq 0$ and $S \neq 0$, then we have the following:

$$\frac{P}{Q}+\frac{R}{S}=\frac{PS}{QS}+\frac{RQ}{QS}=\frac{PS+RQ}{QS}$$

Steps for finding the sum of rational expressions are as follows:

(i) Reduce the rational expressions to equivalent rational expressions having the lowest common denominator

(ii) Find the sum of the numerators of these rational expressions and divide the sum of numerators by the common denominator.

(iii) Reduce the resulting rational expression to the lowest terms if possible.

(c) **Subtraction of rational expressions:** Subtracting a rational expression from another rational expression is the same as adding the additive inverse of the second rational expression to the first rational expression. Thus, if P, Q, R, and S are four polynomials so that $Q \neq 0$ and $S \neq 0$, then we have the following:

$$\frac{P}{Q}-\frac{R}{S}=\frac{PS}{QS}-\frac{RQ}{QS}=\frac{PS-RQ}{QS}$$

(d) **Multiplication of rational expressions:** If P, Q, R, and S are four polynomials so that $Q \neq 0$ and $S \neq 0$, then it can be expressed as follows:

$$\frac{P}{Q}\times\frac{R}{S}=\frac{PR}{QS}$$

The product of two rational expressions is also a rational expression whose numerator is the product of the numerators of the expressions and the denominator is the product of the denominators of the given expressions.

(e) **Division of rational expressions:** To divide a rational expression by another rational expression, multiply the first rational expression by reciprocal of the other rational expression. If P, Q, R, and S are polynomials so that $Q \neq 0$, $R \neq 0$, and $S \neq 0$, then we have the following:

$$\frac{\frac{P}{Q}}{\frac{R}{S}}=\frac{P}{Q}\div\frac{R}{S}=\frac{P}{Q}\times\frac{S}{R}=\frac{PS}{QR}$$

Worked Examples

Addition and Subtraction

1. Simplify $\frac{11+x}{x}+\frac{x-10}{x}$.

Solution

$$\frac{11+x}{x}+\frac{x-10}{x}$$

$$=\frac{11+x+x-10}{x}=\frac{2x+1}{x} \text{ for all } x\neq 0$$

2. Simplify $\frac{x+10}{y-2}-\frac{8}{y-2}+\frac{x-2}{y-2}$.

Solution

$$\frac{x+10}{y-2}-\frac{8}{y-2}+\frac{x-2}{y-2}$$

$$=\frac{(x+10)-8+(x-2)}{y-2}=\frac{2x+0}{y-2}$$

$$=\frac{3x}{y-2} \text{ for all } y\neq 2$$

3. Simplify $\frac{2p+1}{p^2-5}+\frac{p+11}{p^2-5}-\frac{p+1}{p^2-5}$.

Solution

$$\frac{2p+1}{p^2-5}+\frac{p+11}{p^2-5}-\frac{p+1}{p^2-5}$$

$$= \frac{(2p+1)+(p+11)-(p+1)}{p^2-5}$$

$$= \frac{(2p+p-p)+1+11-1}{p^2-5}$$

$$= \frac{2p+11}{p^2-5} \text{ for all } p^2-5 \neq 0$$

4. Simplify $\frac{5}{(x+1)(x+2)} + \frac{11}{x+1}$.

Solution

$$\frac{5}{(x+1)(x+2)} + \frac{11}{x+1}$$

$$= \frac{5}{(x+1)(x+2)} + \frac{11(x+2)}{(x+1)(x+2)}$$

$$= \frac{5+11(x+2)}{(x+1)(x+2)} = \frac{5+11x+22}{(x+1)(x+2)}$$

$$= \frac{11x+27}{(x+1)(x+2)} \text{ for all } x \neq -1 \text{ and } x \neq -2$$

5. Simplify $\frac{6}{(x+2)} + \frac{4}{x(x+2)} - \frac{4}{x}$.

Solution

$$\frac{6}{(x+2)} + \frac{4}{x(x+2)} - \frac{4}{x}$$

$$= \frac{6x}{x(x+2)} + \frac{4}{x(x+2)} - \frac{4(x+2)}{x(x+2)}$$

$$= \frac{6x+4-4x-8}{x(x+2)}$$

$$= \frac{6x+4-4x-8}{x(x+2)} = \frac{2x-4}{x(x+2)}$$

6. Simplify $\frac{x+4}{x^2-1} + \frac{x-1}{x^2-3x-4} - \frac{x+1}{x^2-5x+4}$.

Solution

$$\frac{x+4}{x^2-1} + \frac{x-1}{x^2-3x-4} - \frac{x+1}{x^2-5x+4}$$

$$= \frac{x+4}{(x+1)(x-1)} + \frac{(x-1)}{(x-4)(x+1)} - \frac{x+1}{(x-4)(x-1)}$$

$$= \frac{(x+4)(x-4)}{(x+1)(x-1)(x-4)} + \frac{(x-1)(x-1)}{(x-4)(x+1)(x-1)} - \frac{(x+1)(x+1)}{(x-4)(x-1)(x+1)}$$

$$= \frac{(x+4)(x-4)+(x-1)^2-(x+1)^2}{(x+1)(x-1)(x-4)}$$

$$= \frac{x^2-16+x^2-2x+1-x^2-2x-1}{(x+1)(x-1)(x-4)}$$

$$= \frac{x^2-4x-16}{(x+1)(x-1)(x-4)}$$

$$x \neq -1, x \neq 1 \text{ and } x \neq 4$$

Multiplication

7. Multiply $\frac{2x^2y}{5x}$ by $\frac{15x^3}{6y}$ if $x \neq 0$ and $y \neq 0$.

Solution

$$\frac{2x^2y}{5x} \times \frac{15x^3}{6y} = \frac{2x^2y \times 15x^3}{5x \times 6y} = x^4$$

8. Multiply $\frac{11x}{5x-10}$ by $\frac{(x-2)^3}{5(2-x)}$ for all $x \neq 2$.

Solution

$$= \frac{11x}{5(x-2)} \times \frac{(x-2)^3}{5(2-x)}$$

$$= \frac{11x(x-2)^3}{5(x-2)(-5)(x-2)} = \frac{11x(x-2)}{-25} = \frac{11x(2-x)}{25}$$

9. Assuming that no denominator has zero polynomial, simplify the following expression:

$$\frac{x^2+x-6}{x^2-5x+6} \times \frac{x^3-27}{x^2+2xy+2y^2} \times \frac{x^4+4y^4}{x^2-9}$$

Solution

Factorising numerators and denominators, we have the following:

$$\frac{(x+3)(x-2)}{(x-3)(x-2)} \times \frac{(x-3)(x^2+3x+9)}{x^2+2xy+2y^2}$$

$$\times \frac{(x^2+2xy+2y^2)(x^2-2xy+2y^2)}{(x+3)(x-3)}$$

$$= \frac{(x^2+3x+9)(x^2-2xy+2y^2)}{(x-3)}$$

Division

10. Divide $\frac{x^2-5x}{x^2-25}$ by $\frac{x^2-8x+15}{x-5}$ specifying the restrictions on the values of x.

Solution

$$\frac{x^2-5x}{x^2-25} \div \frac{x^2-8x+15}{x-5}$$

Factorising numerators and denominators, we get the following:

$$\frac{x(x-5)}{(x+5)(x-5)} \times \frac{(x-5)}{(x-5)(x-3)}$$

Therefore, the given expression is not defined when $x + 5 = 0$, $x - 5 = 0$, and $x - 3 = 0$.

$$\frac{x(x-5)(x-5)}{(x+5)(x-5)(x-5)(x-3)}$$

for all $x \neq -5$, $x \neq 5$, and $x \neq 3$

$$\frac{x}{(x+5)(x-3)} = \frac{x}{x^2+2x-15}$$

11. Simplify the following:

$$\frac{x^4-8x}{2x^2+5x-3} \times \frac{2x-1}{x^2+2x+4} \div \frac{x^2-2x}{x+3}$$

Solution

Assuming that no denominator has 0 value,

$$\frac{x^4-8x}{2x^2+5x-3} \times \frac{2x-1}{x^2+2x+4} \div \frac{x^2-2x}{x+3}$$

$$= \frac{x(x^3-8)}{(x+3)(2x-1)} \times \frac{2x-1}{(x^2+2x+4)} \times \frac{x+3}{x(x-2)}$$

$$= \frac{x(x-2)(x^2+2x+4)(2x-1)(x+3)}{(x+3)(2x-1)(x^2+2x+4)x(x-2)} = 1$$

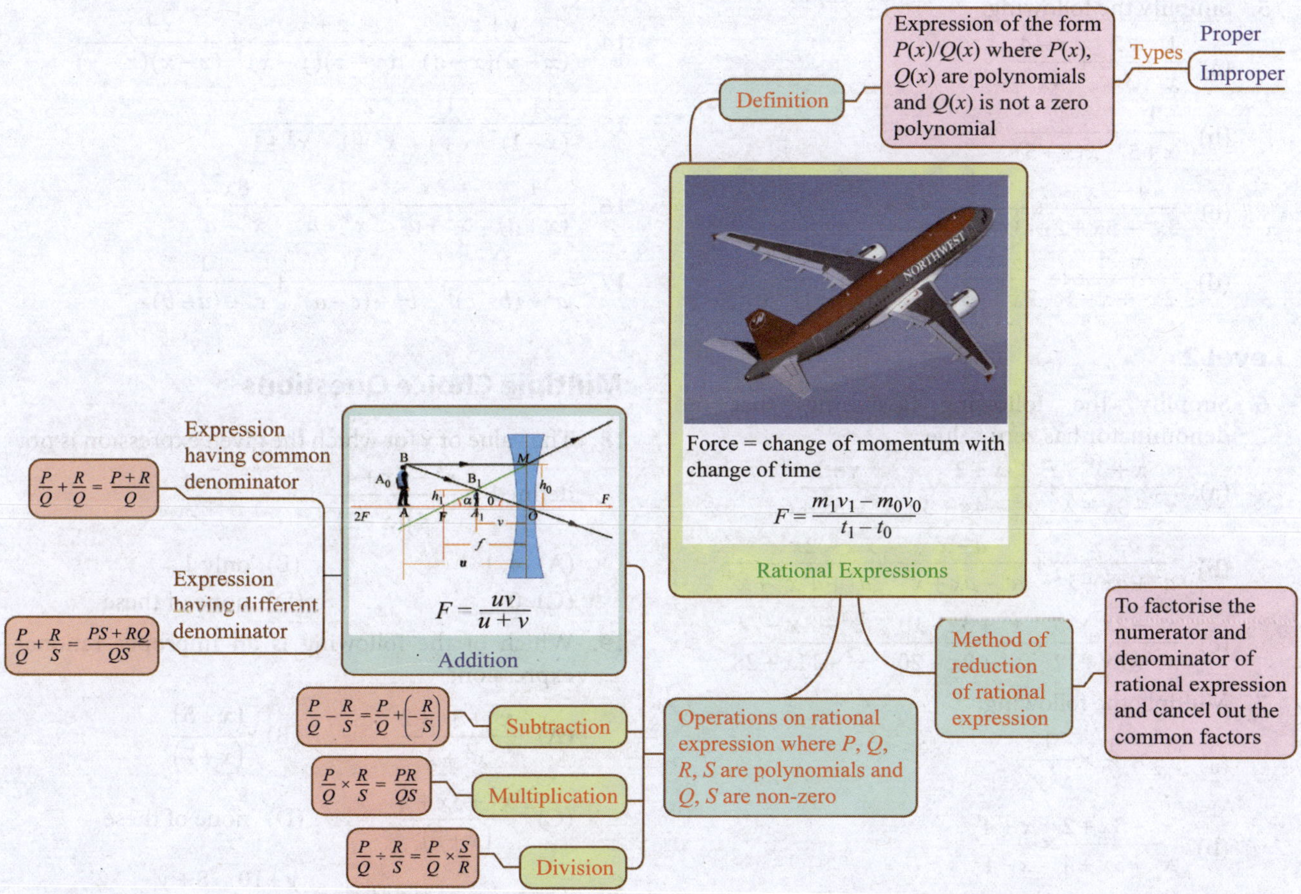

TEST YOURSELF

Level 1

1. Reduce the following expressions to the lowest common denominator:

$$\frac{x}{yz}, \frac{y}{zx}, \frac{z}{xy}$$

2. $\dfrac{x+y}{x-y}, \dfrac{x^2+xy+y^2}{x+y}, \dfrac{x^2+y^2}{x^2-y^2}$

3. $\dfrac{x+4}{x^2-3x-4}, \dfrac{x}{x^3+1}, \dfrac{x-4}{x^3-x^2+x}$

4. Reduce the following to their lowest terms indicating the restriction:

(a) $\dfrac{2x+10}{5x+25}$

(b) $\dfrac{15x^2y^3z^5}{3xy^4z^2}$

(c) $\dfrac{x^4-x^2y^2}{x^2+3xy+y^2}$

(d) $\dfrac{x^2-3x-10}{x^2-11x+30}$

5. Simplify the following:

(a) $\dfrac{1}{x}+\dfrac{2}{3x^2}+\dfrac{x-4}{4x^3}$

(b) $\dfrac{1}{x+5}-\dfrac{2}{x(x+5)}$

(c) $\dfrac{4-7x}{3x^2-5x+2}+\dfrac{2}{3x-2}-\dfrac{3}{x-1}$

(d) $\dfrac{x-1}{2x^2+x-1}-\dfrac{x+1}{2x^2-3x+1}+\dfrac{4}{2(x^2-1)}$

Level 2

6. Simplify the following assuming that no denominator has zero value:

(a) $\dfrac{x+3}{x^2-3x+2}+\dfrac{x+2}{x^2-4x+3}+\dfrac{x+1}{x^2-5x+6}$

(b) $\dfrac{a+x}{a^2+ax+x^2}+\dfrac{a-x}{a^2-ax+x^2}+\dfrac{2x^3}{a^4+a^2x^2+x^4}$

(c) $\dfrac{x^2-2x-15}{x^2+7x+12}+\dfrac{x^2+3x-10}{x^2+9x-20}-\dfrac{x^2-4x-77}{x^2+11x+28}$

7. Multiply the following:

(a) $\dfrac{x^2}{y^2}\times\dfrac{y}{x^3}\times\dfrac{1}{x^2y^2}$

(b) $\dfrac{x^2+3x+2}{x^2+5x+4}\times\dfrac{x+4}{x-1}$

(c) $\dfrac{x^4-8x}{2x^2+5x-3}\times\dfrac{2x-1}{x^2+2x+4}\times\dfrac{x+3}{x^3-2x}$

Level 3

8. Prove that $\dfrac{1}{1-x}-\dfrac{1}{1+x}-\dfrac{2x}{1+x^2}-\dfrac{4x^3}{1+x^4}=\dfrac{8x^7}{1-x^8}$

Simplify the following:

9. $\dfrac{a^4+4b^4}{a^2+2ab+2b^2}$

10. $\dfrac{2x^3-4x^2-6x}{3x^5-6x^4-6x^3-9x^2}$

11. $\dfrac{x^6+y^6}{x^4+4y^4}\times\dfrac{x^2+2xy+y^2}{x^2+y^2}\times\dfrac{x^2+2xy+2y^2}{x^2-x^2y^2+y^4}$

12. $\dfrac{a}{(a-b)(a-c)}+\dfrac{b}{(b-c)(c-a)}+\dfrac{c}{(c-a)(c-b)}$

13. $\dfrac{a+b-c}{(b-c)(a-c)}+\dfrac{b+c-a}{(a-b)(c-a)}+\dfrac{c+a-b}{(b-a)(b-c)}$

14. $\dfrac{y+z}{(x-y)(x-z)}+\dfrac{z+x}{(y-z)(y-x)}+\dfrac{x+y}{(z-x)(z-y)}$

15. $\dfrac{1}{(x-1)}-\dfrac{1}{x+1}-\dfrac{2}{x^2+1}-\dfrac{4}{x^4+1}$

16. $\dfrac{1}{(x+a)}+\dfrac{2x}{x^2+a^2}+\dfrac{4x^3}{x^4+a^4}-\dfrac{8x^7}{x^8-a^8}$

17. $\dfrac{1}{a^2-(b-c)^2}+\dfrac{1}{b^2-(c-a)^2}+\dfrac{1}{c^2-(a-b)^2}$

Multiple Choice Questions

18. The value of x for which the given expression is not defined $\dfrac{x^2-5x+4}{(x-4)(x-1)^2}$ is

(A) 4, 1 (B) only 1
(C) 0 (D) none of these

19. Which of the following is an improper rational expression?

(A) $\dfrac{x^2+3x+2}{x^2+2}$ (B) $\dfrac{(x+5)}{(x+7)}$

(C) $\dfrac{x^2+3x+2}{(x+5)}$ (D) none of these

20. The value of expression $\dfrac{y+10}{x-2}-\dfrac{8+y}{x-2}-\dfrac{2}{x-2}$

(A) is 0 (B) depends on x
(C) depends on y (D) is 1

21. The expression $\dfrac{2x+1}{x^2-5}+\dfrac{x+11}{x^2-5}+\dfrac{x+1}{x^2-5}$ is equivalent to

(A) $\dfrac{2x+1}{x^2-5}$ (B) $\dfrac{3x+14}{x^2-5}$

(C) $\frac{2x}{x^2-5}$ (D) $\frac{2x+11}{x^2-5}$

22. Simplified form of the expression $\frac{10}{(x+1)(x+2)}+\frac{11}{x+1}-\frac{8}{x+2}$ is

(A) $\frac{(x+8)}{(x+1)(x+2)}$ (B) $\frac{3(x+8)}{(x+1)(x+2)}$

(C) $\frac{3}{(x+2)}$ (D) $\frac{(x+8)}{(x+1)}$

23. Simplified form of the rational expression $\frac{y+4}{y^2-1}+\frac{y-1}{y^2-3y-4}-\frac{y+1}{y^2-5y+4}$ is

(A) $\frac{y^2+4y-16}{(y-4)(y^2-1)}$ (B) $\frac{y^2-4y-16}{(y-4)(y^2-1)}$

(C) $\frac{y^2-4y+16}{(y-4)(y^2-1)}$ (D) None of these

24. Divide $\frac{x^2-5x}{x^2-25}$ by $\frac{x^2-8x+15}{x-5}$ and indicate the values of the variable which are not admissible.

(A) $\frac{x}{(x+5)(x-3)}, x \neq -5, 5, 3$

(B) $\frac{3x}{(x+5)(x+3)}, x \neq -5, -3$

(C) $\frac{2x+5}{(x+5)(x+3)}, x \neq -5, -3$

(D) None of these

25. The simplified form of the expression $\frac{x^2+x-6}{x^2-5x+6}\times\frac{x^3-27}{x^2-2xy+2y^2}\times\frac{x^4+4y^4}{x^2-9}$ is

(A) $\frac{(x^2+3x+9)(x^2+2xy+2y^2)}{x-3}$

(B) $\frac{(x^2-3x+9)(x^2-2xy+2y^2)}{x+3}$

(C) $\frac{(x^2+3x+9)(x^2-2xy+2y^2)}{x+3}$

(D) None of these

26. Find the value of x for the following expression:

$$\frac{2}{(x+1)(x-3)}-\frac{8(x+4)}{(x+1)(x+2)}+\frac{4x}{(x-3)(x+2)}=0$$

(A) $\frac{-1\pm\sqrt{201}}{2}$ (B) $\frac{-1\pm\sqrt{401}}{4}$

(C) $\frac{1\pm\sqrt{401}}{4}$ (D) $\frac{1\pm\sqrt{201}}{2}$

27. Find the value of x for the following expression:

$$\frac{x+3}{x^2-3x+2}+\frac{x+2}{x^2-4x+3}+\frac{x+1}{x^2-5x+6}=0$$

(A) $\pm\frac{1}{\sqrt{3}}$ (B) $\pm\frac{14}{\sqrt{3}}$

(C) $\pm\sqrt{\frac{14}{3}}$ (D) None of these

28. Find the value of x for the following expression:

$$\frac{1}{x}+\frac{2}{3x^2}+\frac{x-4}{4x^3}=0$$

(A) $\frac{11\pm\sqrt{697}}{12}$ (B) $\frac{11\pm\sqrt{697}}{24}$

(C) $\frac{-11\pm\sqrt{697}}{24}$ (D) None of these

29. Find the value of x for the following expression:

$$\frac{x-1}{2x^2+x-1}-\frac{x+1}{2x^2-3x+1}+\frac{4}{2(x^2-1)}=0$$

(A) $x = 4, 0$ (B) no value of x

(C) $x = 2, 3$ (D) $x = ½, 1$

30. Find the value of x for which the given equation is satisfied:

$$\frac{x+5}{(x-2)(x+3)}+\frac{x+8}{(x+3)(x-7)}=0$$

(A) $1\pm\frac{\sqrt{106}}{2}$ (B) $\frac{2\pm\sqrt{106}}{2}$

(C) $\frac{1\pm\sqrt{106}}{2}$ (D) $-1\pm\frac{\sqrt{106}}{2}$

31. $\frac{a}{(a-b)(a-c)}+\frac{b}{(b-c)(c-a)}+\frac{c}{(c-a)(c-b)}$ is equivalent to

(A) $\frac{a}{(b-c)(c-a)}$ (B) $\frac{-b}{(a-b)(a-c)}$

(C) $\dfrac{b}{(a-b)(a-c)}$ (D) None of these

32. $\dfrac{x^2-2x-15}{x^2+7x+12}+\dfrac{x^2+3x-10}{x^2+9x+20}-\dfrac{x^2-4x-77}{x^2+11x+28}$ is equivalent to

(A) 0 (B) 1
(C) 2 (D) 3

33. $\dfrac{1}{a^2-(b-c)^2}+\dfrac{1}{b^2-(c-a)^2}+\dfrac{1}{c^2-(a-b)^2}$ is equivalent to

(A) $\dfrac{1}{(a+b-c)(a-b+c)(-a+b+c)}$

(B) $\dfrac{-a-b-c}{(a+b-c)(a-b+c)(-a+b+c)}$

(C) $(a+b+c)$

(D) $\dfrac{a+b+c}{(a+b-c)(a-b+c)(-a+b+c)}$

34. $\dfrac{2x^4-4x^3-6x^2}{3x^4-6x^3-6x^2+12x}$ is equivalent to

(A) $\dfrac{2x(x-3)(x+1)}{3(x-2)(x^2-2)}$ (B) $\dfrac{x(x-3)(x+1)}{(x-2)(x^2-2)}$

(C) $\dfrac{(2x+1)(x-1)}{(x-2)(x^2-2)}$ (D) None of these

35. If $\dfrac{4-x}{x}+\dfrac{3}{2}=-\dfrac{4}{x}$ then $x=$

(A) –6 (B) 16
(C) –16 (D) 0

36. If $\dfrac{x}{x+3}+\dfrac{5}{x-7}=\dfrac{30}{x^2-4x-21}$, then $x=$

(A) 3, – 5 (B) 5
(C) 3, –3 (D) 5, –5

37. If $\dfrac{2}{x-1}+\dfrac{3}{4}=\dfrac{5}{x-1}$ then $x=$

(A) 5 (B) –5
(C) 0 (D) 2

38. If $\dfrac{2}{a+1}+\dfrac{5}{a-1}=\dfrac{10}{a^2-1}$ then $a=$

(A) 1 (B) –1
(C) 2 (D) no value of a

39. If $\dfrac{x}{x-2}-\dfrac{4}{x+2}=\dfrac{8}{x^2-4}$ then $x=$

(A) 0, 2 (B) 2
(C) 0 (D) 0, –2

40. If $\dfrac{-3x}{x+1}+\dfrac{4x+1}{x}=\dfrac{-5}{x^2+x}$ then $x=$

(A) –1 (B) –1, –4
(C) –2, –3 (D) None of these

41. The expression $\dfrac{x^2-5x}{x^2-2x-15}\times\dfrac{x^2-9}{x^2-3x}$ is equivalent to

(A) 1 (B) 0
(C) 2 (D) None

42. The expression $\dfrac{a^2-8a+16}{a}\times\dfrac{a^3}{4-a}$ is equivalent to

(A) $a^2(a-4)$ (B) $a^2(4-a)$
(C) $a^2(a-1)$ (D) $-a^2(a\ 1)$

43. $\dfrac{x^2-4x-21}{x^2-49}\div\dfrac{x^2+8x+15}{x^2+2x-35}=$

(A) $\dfrac{x-5}{x+7}$ (B) $\dfrac{x+5}{x-5}$

(C) $\dfrac{x+5}{x-3}$ (D) $\dfrac{x-5}{x+5}$

OLYMPIAD AND NTSE LEVEL EXERCISES

44. If $a + b + c = 0$, then the value of $\dfrac{(a+b)^2}{ab}+\dfrac{(b+c)^2}{bc}+\dfrac{(c+a)^2}{ca}$ is

(A) 0 (B) 1
(C) 2 (D) 3

45. The simplified form of $\dfrac{a}{x+a}+\dfrac{b}{x-a}-\dfrac{c}{x^2-a^2}$ is

(A) $\dfrac{a(x-a)+b(x+a)-c}{x^2-a^2}$

(B) $\dfrac{b(x-a)+a(x+a)-c}{(x-a)^2}$

(C) $\dfrac{b(x-a)+a(x+a)-c}{x^2-a^2}$

(D) $\dfrac{a(x-a)+b(x+a)+c}{x^2-a^2}$

46. Simplify

$$\frac{1}{(x+1)(x+2)}+\frac{1}{(x+2)(x+3)}+\frac{1}{(x+3)(x+1)}.$$

(A) $\dfrac{1}{(x+1)(x+3)}$ (B) $\dfrac{2}{(x+1)(x+3)}$

(C) $\dfrac{3}{(x+2)(x+3)}$ (D) $\dfrac{3}{(x+1)(x+3)}$

47. Simplify $\dfrac{x^3-3x^2}{9x^2-x^4}$.

(A) $\dfrac{1}{x+3}$ (B) $-\dfrac{1}{x+3}$

(C) $\dfrac{1}{3x+3}$ (D) $\dfrac{1}{3x-3}$

48. Simplify $\dfrac{x^2-y^2}{(x+y)^2}$.

(A) $\dfrac{x+y}{x-y}$ (B) $\dfrac{(x+y)^2}{x-y}$

(C) $\dfrac{x+y}{x^2-y^2}$ (D) $\dfrac{x-y}{x+y}$

49. Write $\dfrac{a^2-b^2}{a^2-3ab+2b^2}$ in the simplest form.

(A) $\dfrac{(a+b)^2}{(a-2b)}$ (B) $\dfrac{a+b}{a-2b}$

(C) $\dfrac{(a-b)}{(a-2b)}$ (D) $\dfrac{(a+b)}{(a+2b)}$

50. Simplify $\dfrac{x^2+2x-8}{x^2-x-20}$.

(A) $\dfrac{x-2}{x-5}$ (B) $\dfrac{x+2}{x-5}$

(C) $\dfrac{x+4}{x+5}$ (D) $\dfrac{x-4}{x-5}$

51. $\dfrac{x-5}{x+2}-\dfrac{3}{2-x}-\dfrac{14-x}{x^2-4}=$

(A) $\dfrac{x+2}{x-1}$ (B) $\dfrac{x-1}{x+2}$

(C) $\dfrac{x+1}{x-2}$ (D) $\dfrac{-x-1}{-2-x}$

52. $\dfrac{2(m+2)}{4m+m^2-12}-\dfrac{1}{m-2}=$

(A) $\dfrac{1}{m-2}$ (B) $\dfrac{m-2}{m+6}$

(C) $\dfrac{1}{m+6}$ (D) $\dfrac{m+6}{m-2}$

53. Let a, b and x be non-zero real numbers such that $\dfrac{5}{x}=\dfrac{2}{a}+\dfrac{3}{b}$. Then x is equal to

(A) $\dfrac{-5ab}{2b+3a}$ (B) $\dfrac{5ab}{2b+3a}$

(C) $\dfrac{b}{2b-3a}$ (D) None of these

GAME TIME

1. A certain number of balls are arranged in rows to form an equilateral triangle. The first row consists of one ball, the second row of two balls, and so on. If 669 more balls are added, then all the balls can be arranged in the shape of a square and each of the sides contains eight balls less than each side of the triangle. Find the initial number of balls.
2. A man passed 1/6 of his life in childhood, 1/12 in youth, and 1/7 in bachelorhood. A son was born to him 5 years after he was married. The son died 4 years before his father at half the age at which the father died. At what age did the man die?
3. Two trains of equal length of 120 m move at different speeds in the same direction. The faster train overtakes the slower train completely in 15 seconds. But if the speed of the slower train is reduced to half its original speed, the faster train can overtake it in 10 seconds. At what speeds are the two trains moving (in metres per second)?
4. A herd of cattle, of which 10 are lame and have only three feet, rushes on to a street and suddenly people and cattle are found running along the street. A quiet count yields 120 heads and 300 feet. How many cattle and how many people are there?
5. Seventy-five people went to a party. The party had 25 teachers, 20 doctors, 18 engineers, and 12 bank employees. All the participants together spent ₹ 1330. It was found that 5 teachers spent as

much as 4 doctors, and 12 doctors spent as much as 9 engineers, and 6 engineers spent as much as 8 bank employees. How much did each professional group spend?

6. Take the ten digits 0, 1, 2, 3, 4, 5, 6, 7, 8, 9 and make up an example in addition using three of the digits in the top line and another set of three digits in the second line. Then, add the two numbers to get an answer, whose four digits should be those which are not used in the addends. Thus, every digit is used once and once only in this addition problem.

	x	x	x	addend
	x	x	x	addend
x	x	x	x	sum

7. There are three switches outside a cabin. One of these switches is connected to a lamp which is fixed inside the cabin. Explain how you identify this switch.

Rules:
(a) You can operate any two switches only.
(b) You can enter the cabin only once.

8. A word I know, six letters it contains, subtract just one and twelve remains. What is the word?
9. Two boys were discussing and one said, 'Give me one of your pencils, and I'll have as many as you have'. The other replied, 'Yes, but if you give me one of your pencils, I'll have twice as many as you have'. How many pencils did each boy have in the beginning?
10. Each of the Verma brothers has as many brothers as he has sisters. But each of the Verma sisters has twice as many brothers as she has sisters. How many brothers and sisters are there in the Verma family?
11. Two erratic clocks A and B are set at some point of time to show 3 O'clock. A loses 15 minutes every hour and B gains 20 minutes every hour. At the end of a certain time, both A and B show exact hours. What is the time in hours read on the clocks A and B?
12. A person on vacation for 'd' days observed the following:
(i) It rained seven times, morning or afternoon.
(ii) When it rained in the afternoon, it was clear in the morning.
(iii) There were five clear afternoons.
(iv) There were six clear mornings.
Then, 'd' equals

(A) 7 (B) 9 (C) 10 (D) 11

13. There are seven numbers 5, 7, 11, 13, 17, 19, and 23. Arrange the given prime numbers in the seven circles so that the rows and the diagonals add up to give the same prime number.

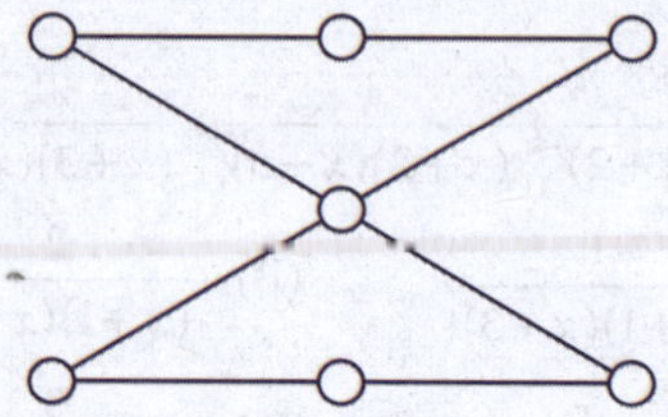

14. A train starts every day from Kashmir to Kanyakumari, while another starts daily from Kanyakumari to Kashmir and the journey lasts 7 days. How many trains will a traveller meet during the journey from Kashmir to Kanyakumari?
15. When a screen is placed 10 m from a projector, the picture occupies 10 m^2. How large will the picture be when the projector is 15 m from the screen?
16. The sales persons in 'warehouse' receive a salary of ₹ 300 a week, plus 1 % of all sales over ₹ 10,000 a week. If one person earned a total of ₹ 350, in one week, what was the amount of total sales for that week?
17. Your bed and light switch are 12 feet apart. How can you switch off the light and get into bed before the room is dark?

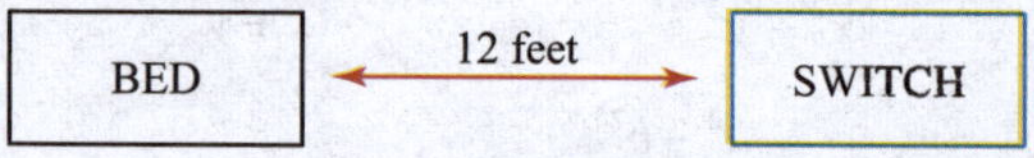

18. With one continuous line, not lifting your pencil from the paper, cross over all of the four dots, using only three straight lines. You may not retrace over any part of the line drawn earlier.

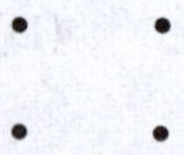

19. Change the arrangement of coins from (1)(1)(1)(2)(2) to (1)(2)(1)(2)(1).

Rules:
(a) Two adjacent coins are to be shifted at a time.
(b) Minimum number of shifts is to be employed.
(c) Order of the coins should not change.

20. A man goes to the well with a 5-litre jar and a 3-litre jar, which is 8 litre in all. However, he wants 7 litres of water. How can he get 7 litres without using a measuring cup, using only the 5-litre jar and the 3-litre jar?

12 Transformations

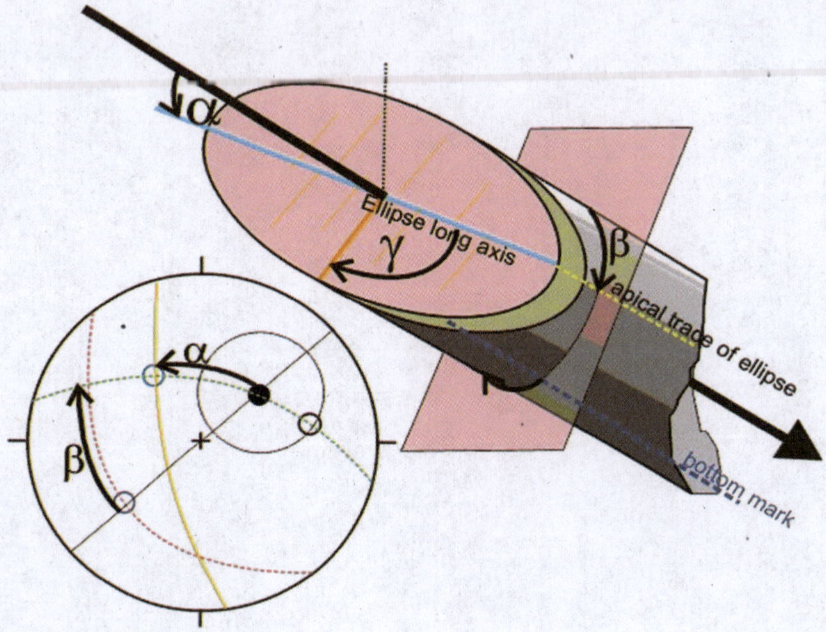

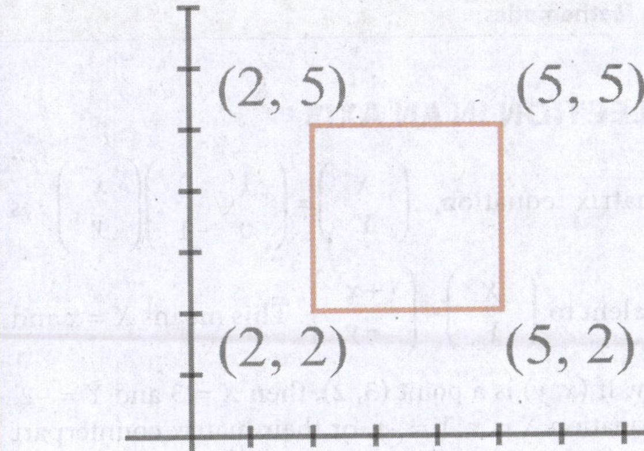

OBJECTIVES

After reading this chapter you should be able to

- learn about reflection on *x*- and *y*-axes
- learn about reflection on any line through the origin and in the origin

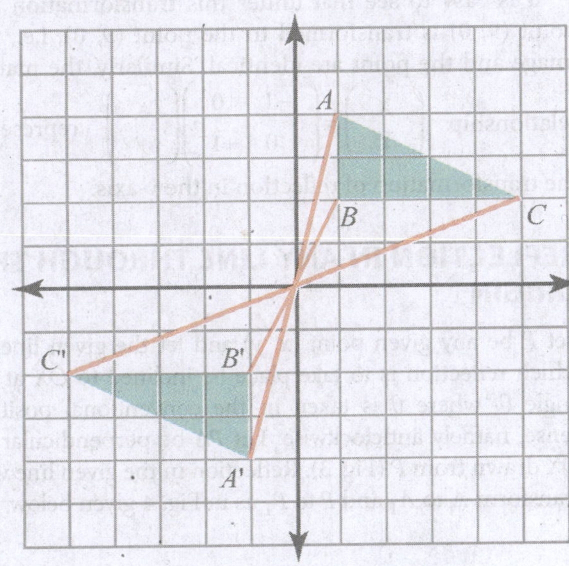

REFLECTION IN AN AXIS

By matrix equation, $\begin{pmatrix} X \\ Y \end{pmatrix} = \begin{pmatrix} 1 & 0 \\ 0 & -1 \end{pmatrix} \begin{pmatrix} x \\ y \end{pmatrix}$ is equivalent to $\begin{pmatrix} X \\ Y \end{pmatrix} = \begin{pmatrix} +x \\ -y \end{pmatrix}$. This means $X = x$ and $Y = -y$. If (x, y) is a point (3, 2), then $X = 3$ and $Y = -2$. The equation $X = x$, $Y = -y$, or their matrix counterpart may be said to have changed, mapped, or transformed the given point (3, 2) to the point (3, 2). The point (3, –2) is a reflection of (3, 2) in the x-axis as shown in Fig. 1.

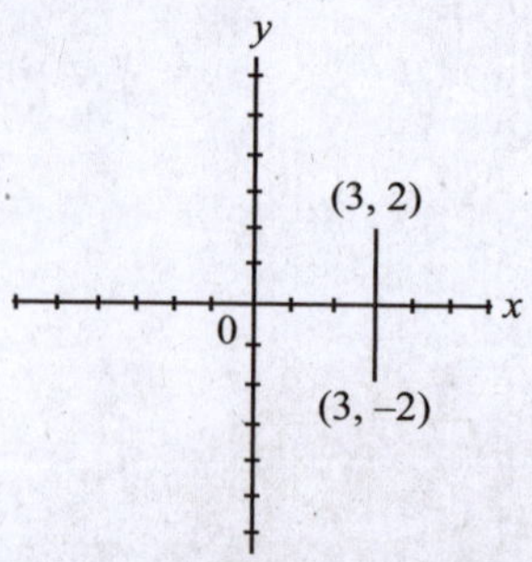

Fig. 1

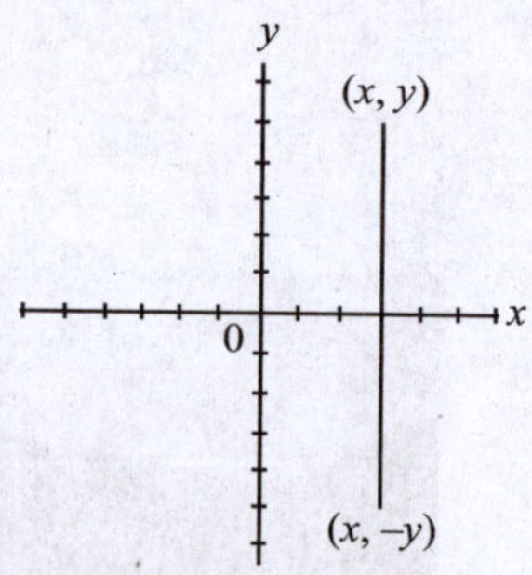

Fig. 2

The general situation is shown in fig. 2, when the given point (x, y) is transformed to $(x, -y)$. This too is obviously a reflection of (x, y) in the x-axis which may be regarded as a mirror. The matrix relationship $\begin{pmatrix} X \\ Y \end{pmatrix} = \begin{pmatrix} 1 & 0 \\ 0 & -1 \end{pmatrix} \begin{pmatrix} x \\ y \end{pmatrix}$ is seen as an algebraic representation of reflection in the x-axis.

It is easy to see that under this transformation the point (9, 0) is transformed to the point (9, 0), i.e., the image and the point are identical. Similarly, the matrix relationship $\begin{pmatrix} X \\ Y \end{pmatrix} = \begin{pmatrix} -1 & 0 \\ 0 & +1 \end{pmatrix} \begin{pmatrix} x \\ y \end{pmatrix}$ represents the transformation of reflection in the y-axis.

REFLECTION IN ANY LINE THROUGH THE ORIGIN

Let P be any given point (x, y) and let the given line in which reflection is to take place be inclined to OX at an angle θ, where θ is taken in the conventional positive sense, namely anticlockwise. Let PA be perpendicular to OX drawn from P (Fig. 3). Reflection in the given line will transform A to A_1 and P to P_1 as in Fig. 4 given below. To apply trigonometry to Fig. 5, draw A_1B, P_1C perpendicular to OX and P_1D perpendicular to A_1B as in Fig. 5. Note $A_1\hat{O}B$ is 2θ so that $O\hat{A}_1B$ is the complement of 2θ. Thus, as DA_1P_1 is a right triangle, $D\hat{A}_1P_1$ is 2θ. It is easily seen from the diagram that any point on the given line reflects onto itself.

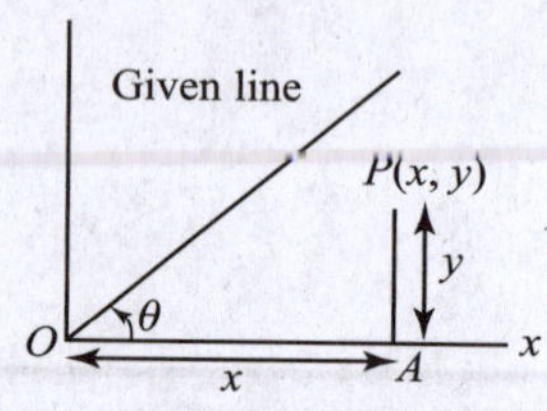

Fig. 3

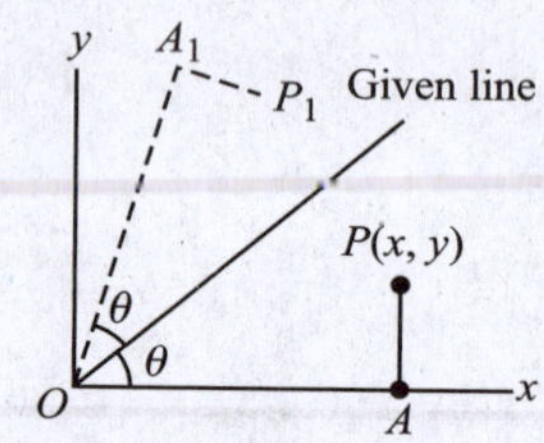

Fig. 4

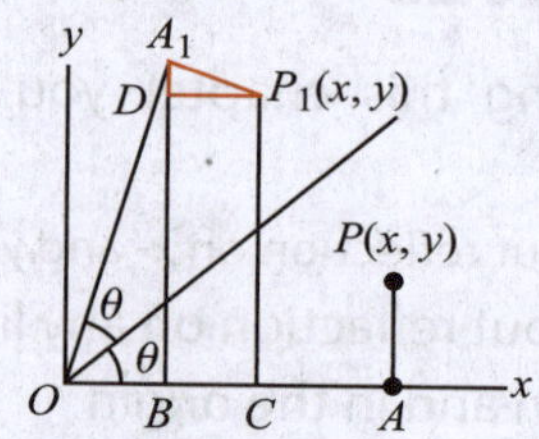

Fig. 5

(X, Y) the coordinates of P_1 are required

$$X = OC$$
$$= OB + BC$$
$$= OB + DP_1$$
$$= OA_1 \cos 2\theta + A_1P_1 \sin 2\theta$$
$$= x \cos 2\theta + y \sin 2\theta$$
$$Y = CP_1$$
$$= BD$$
$$= BA_1\ DA_1$$
$$= OA_1 \sin 2\theta\ A_1P_1 \cos 2\theta$$
$$= x \sin 2\theta\ y \cos 2\theta$$

Thus, the equations $X = x \cos 2\theta + y \sin 2\theta$ and $Y = x \sin 2\theta - y \cos 2\theta$ map the point P_1, namely (x, y) onto the point P_1, namely (X, Y). In matrix form we have the following:

$$\begin{pmatrix} X \\ Y \end{pmatrix} = \begin{pmatrix} \cos 2\theta & \sin 2\theta \\ \sin 2\theta & -\cos 2\theta \end{pmatrix} \begin{pmatrix} x \\ y \end{pmatrix}$$

This matrix relationship represents a reflection on a line making a positive angle θ with the x-axis. When $x = 0$ and $y = 0$, it can be easily seen from the matrix that $X = 0$ and $Y = 0$. Thus, under the transformation of reflection, any line through the origin, whatever the value of the angle θ, the origin maps into itself.

When $\theta = 0$, $\cos 2\theta = 1$, and $\sin 2\theta = 0$, the matrix relationship becomes

$$\begin{pmatrix} X \\ Y \end{pmatrix} = \begin{pmatrix} 1 & 0 \\ 0 & -1 \end{pmatrix} \begin{pmatrix} x \\ y \end{pmatrix}$$

When $\theta = 90°$, the relationship becomes

$$\begin{bmatrix} X \\ Y \end{bmatrix} = \begin{pmatrix} -1 & 0 \\ 0 & 1 \end{pmatrix} \begin{bmatrix} x \\ y \end{bmatrix}$$

REFLECTION IN THE ORIGIN

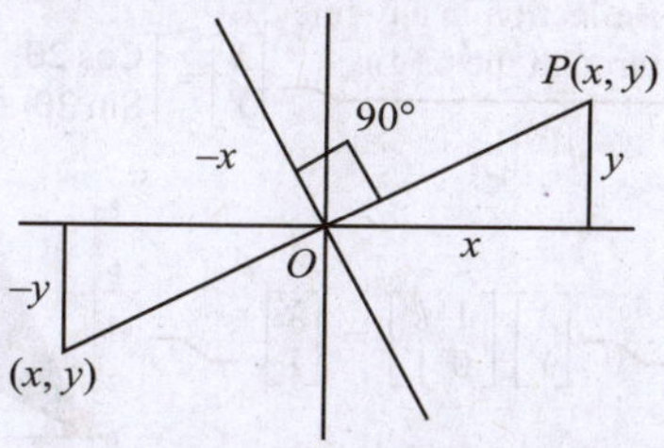

In the figure, point $P(x, y)$ is joined to the origin O. A line is drawn through O perpendicular to OP. If P is reflected in this line, the point P_1 is obtained, where $OP = OP_1$. Let the coordinates of P_1 be (X, Y). By congruency of triangles, it is easy to show that X is of equal length to x and y of equal length to y so that the coordinate of P_1 is actually (x, y).

The transformation which maps (x, y) onto $(-x, -y)$ is called reflection in the origin.

In matrix form the transformation is $\begin{pmatrix} X \\ Y \end{pmatrix} = \begin{pmatrix} -1 & 0 \\ 0 & -1 \end{pmatrix} \begin{pmatrix} x \\ y \end{pmatrix}$. There are also other transformations such as shear, stretch, compression, and enlargement.

Worked Examples

1. Use $\begin{pmatrix} X \\ Y \end{pmatrix} = \begin{pmatrix} -1 & 0 \\ 0 & 1 \end{pmatrix} \begin{pmatrix} x \\ y \end{pmatrix}$ to find the coordinates of images of the following points when reflected on the y-axis:

   (i) (1, 1) (ii) (2, 3)

Solution

(i) $\begin{pmatrix} X \\ Y \end{pmatrix} = \begin{pmatrix} -1 & 0 \\ 0 & 1 \end{pmatrix} \begin{pmatrix} 1 \\ 1 \end{pmatrix} = \begin{pmatrix} -1+0 & 0+1 \end{pmatrix}$

$= (-1, 1)$

The reflected point is (–1, 1).

(ii) $\begin{pmatrix} X \\ Y \end{pmatrix} = \begin{pmatrix} -1 & 0 \\ 0 & 1 \end{pmatrix} \begin{pmatrix} -2 \\ 3 \end{pmatrix}$

$= \begin{pmatrix} (-1)(-2)+0 & (0)\times(-2)+3 \end{pmatrix}$

$= (2 \quad 3)$

The reflected point is (2, 3).

2.

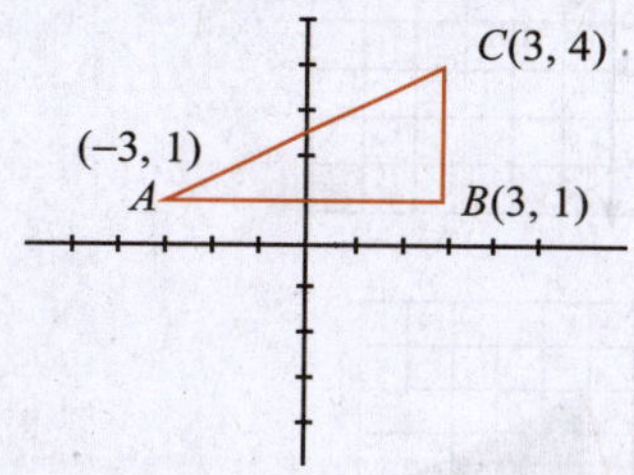

By the transformation matrix, find the image of the triangle shown in the above given figure in the x-axis.

Solution

We know that the matrix relationship of reflection in the x-axis is as follows:

$$\begin{pmatrix} X \\ Y \end{pmatrix} = \begin{pmatrix} 1 & 0 \\ 0 & -1 \end{pmatrix} \begin{pmatrix} x \\ y \end{pmatrix}$$

Therefore, the image of the point C is as follows:

$$\begin{pmatrix} X \\ Y \end{pmatrix} = \begin{pmatrix} 1 & 0 \\ 0 & -1 \end{pmatrix} \begin{pmatrix} x \\ y \end{pmatrix}$$

$$= \begin{pmatrix} 1\times3+0\times4 & 0\times3+-1\times4 \end{pmatrix} = (3, -4)$$

The image of B is as follows:

$$\begin{pmatrix} X \\ Y \end{pmatrix} = \begin{pmatrix} 1 & 0 \\ 0 & -1 \end{pmatrix} \begin{pmatrix} 3 \\ 1 \end{pmatrix}$$

$$= \begin{pmatrix} 1\times3+(0)(4) & 0\times3+(-1)(-1) \end{pmatrix} = \begin{pmatrix} 3 & -1 \end{pmatrix}$$

The image of A is as follows:

$$\begin{pmatrix} X \\ Y \end{pmatrix} = \begin{pmatrix} 1 & 0 \\ 0 & -1 \end{pmatrix} \begin{pmatrix} -3 \\ 1 \end{pmatrix}$$

$$= \begin{pmatrix} +1\times(-3)+0\times1 & 0\times(-3)+(-1)(1) \end{pmatrix}$$

$$= \begin{pmatrix} -3, & -1 \end{pmatrix}$$

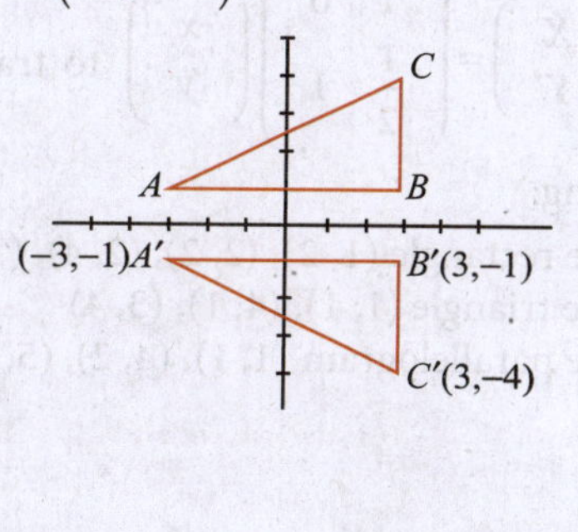

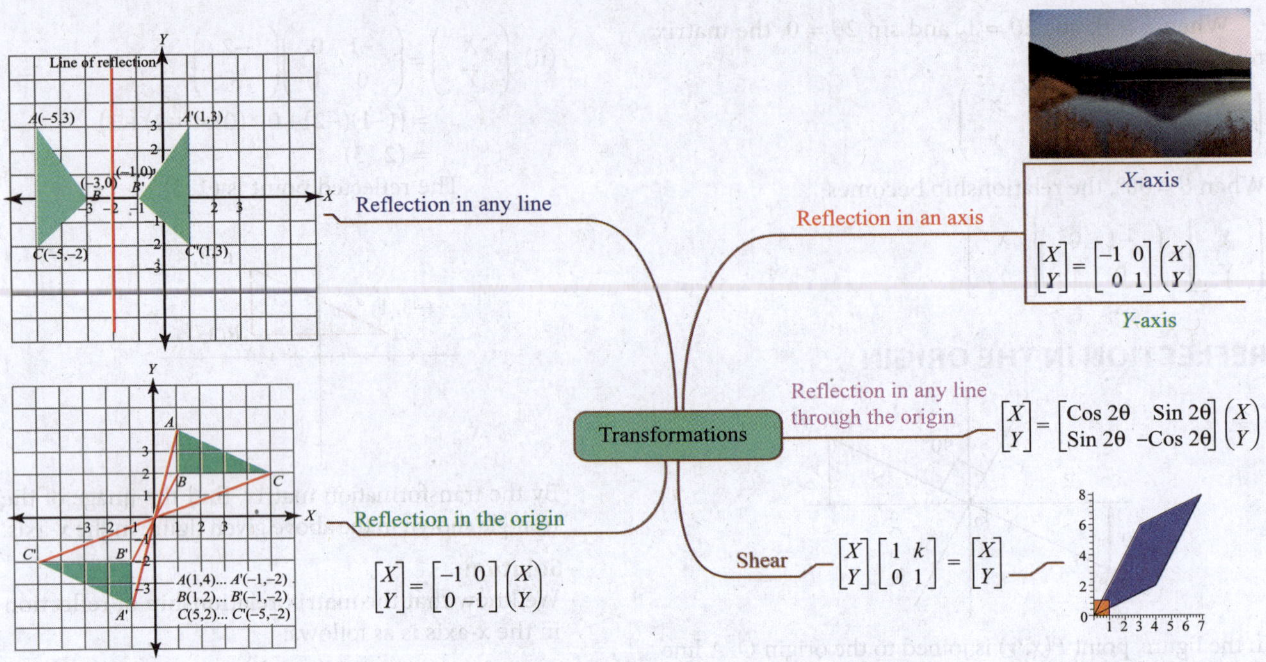

TEST YOURSELF

Level 1

1. Find the reflection of the point $P(3, 4)$ in the following using the transformation of matrices:

   (a) x-axis (b) y-axis (c) through the origin

2. Find the reflection of the segment formed by the points using matrix transformation:

   (a) $A(1, 3)$ and $B(2, 5)$ on x-axis
   (b) $C(3, 4)$ and $D(4, 6)$ on y-axis.

3. Find the reflection of the triangle using matrix transformation formed by the points $A(1, 1)$, $B(3, 5)$, and $C(2, 7)$ on y-axis.

4. Use $\begin{pmatrix} X \\ Y \end{pmatrix} = \begin{pmatrix} 1 & 0 \\ \frac{1}{2} & 1 \end{pmatrix}\begin{pmatrix} x \\ y \end{pmatrix}$ to transform the following:

   (a) The rectangle (1, 2), (2, 2), (2, 4), (1, 4)
   (b) The triangle (1, 1), (4, 1), (3, 4)
   (c) The parallelogram (1, 1), (4, 2), (5, 4), (2, 3)

5. Use $\begin{pmatrix} X \\ Y \end{pmatrix} = \begin{pmatrix} 1 & \frac{1}{2} \\ 0 & 1 \end{pmatrix}\begin{pmatrix} x \\ y \end{pmatrix}$ to transform the figures given the above questions.

Level 2

6. Using $\begin{pmatrix} X \\ Y \end{pmatrix} = \begin{pmatrix} 3 & 0 \\ 0 & 3 \end{pmatrix}\begin{pmatrix} x \\ y \end{pmatrix}$, transform the rectangle whose vertices are (1, 1), (3, 1), (3, 2), (1, 2).

7. Using $\begin{pmatrix} X \\ Y \end{pmatrix} = \begin{pmatrix} -\frac{1}{3} & 0 \\ 0 & -\frac{1}{3} \end{pmatrix}\begin{pmatrix} x \\ y \end{pmatrix}$, transform the triangle whose vertices are (3, 3), (9, 3), (3, 12).

8. If a transformation is given by the matrix form, $\begin{pmatrix} X \\ Y \end{pmatrix} = \begin{pmatrix} \cos\theta & \sin\theta \\ -\sin\theta & \cos\theta \end{pmatrix}\begin{pmatrix} x \\ y \end{pmatrix}$, then find the transformation matrix, where $\theta = 30°$, $\theta = 45°$, and $\theta = 60°$.

9. Find the reflection of the triangle using matrix transformation, whose coordinates are (3, 4), (3, 8), and (–2, 6) on x-axis and y-axis.

10. Find the reflection of the figure formed by joining the coordinates of the points (0, –7), (1, –5), (4, –3), (2, 0), (3, 3), (0, 4), (4, 6), using the matrix transformation on y-axis.

11.

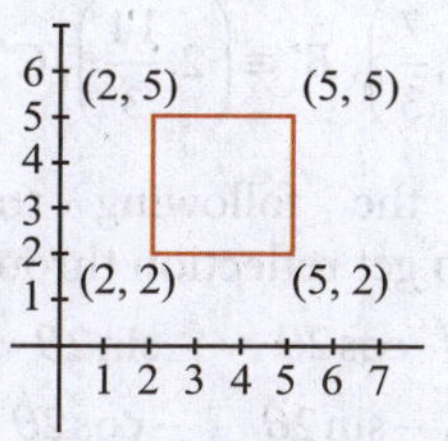

Using the above figure, find the coordinates of the images of the vertices of the square:

(a) Reflection in the x-axis

(b) Reflection in the y-axis

(c) Reflection in the origin

Multiple Choice Questions

12. The coordinates of images of the point (2, 3) when reflected on the y-axis are

(A) (2, 3) (B) (2, –3)

(C) (–2, –3) (D) (–2, 3)

13. When the point $(x + y, x - y)$ is reflected through x-axis, the point (3, 5) is obtained. The values of x and y are

(A) $x = -1$ and $y = 4$

(B) $x = 1$ and $y = 4$

(C) $x = 1$ and $y = -4$

(D) $x = -1$ and $y = -4$

14. If the point (3, 2) is reflected through x-axis, the point $(2\alpha + 3\beta, 3\alpha - 2\beta)$ is obtained. The values of α and β are

(A) $\alpha = 1, \beta = 0$ (B) $\alpha = 0, \beta = 1$

(C) $\alpha = 1, \beta = 1$ (D) none of these

15. The coordinates of the point which is obtained by taking reflection of point (3, 5) through origin are

(A) (–3, 5) (B) (3, 5)

(C) (–3, –5) (D) (3, –5)

16. Triangle ABC is reflected through x-axis as shown in the figure. The coordinates of A', B' and C' are

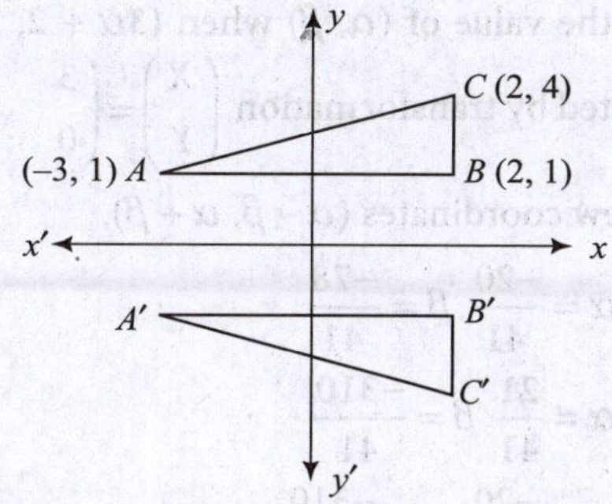

(A) $A' \equiv (-3, -1)$; $B' \equiv (2, -1)$; $C' \equiv (2, -4)$

(B) $A' \equiv (-3, -1)$; $B' \equiv (2, -4)$; $C' \equiv (2, -1)$

(C) $A' \equiv (-3, 1)$; $B' \equiv (2, -1)$; $C' \equiv (2, -4)$

(D) none of these

17. The line segment shown in the figure is reflected through x-axis. The equation of reflected line segment will be

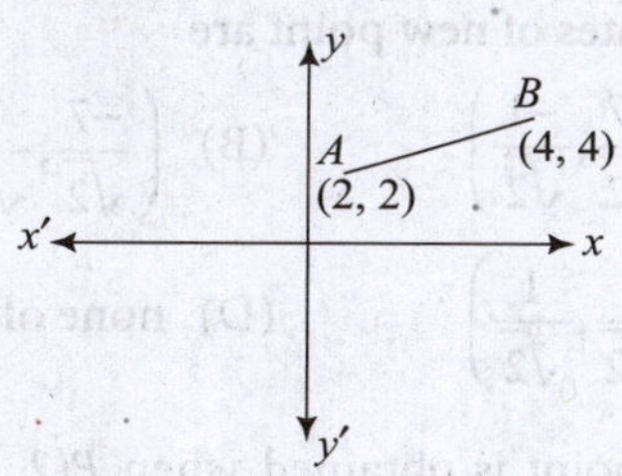

(A) $y = x$ (B) $y = x + 2$

(C) $y = -x$ (D) none of these

18. Find the coordinates of new square when $ABCD$ is reflected through origin.

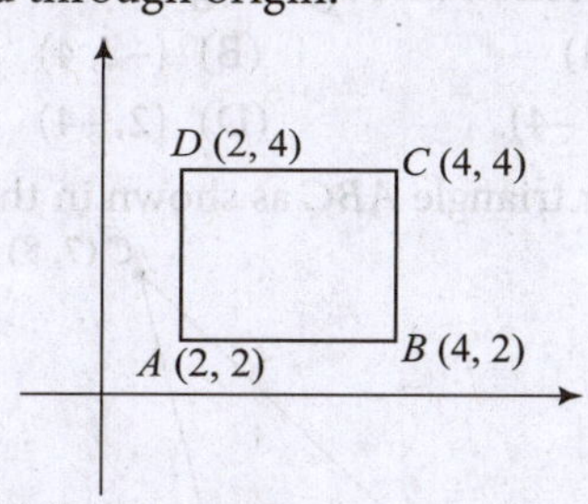

(A) $A' \equiv (-2, -2)$; $B' \equiv (-4, -2)$; $C' \equiv (-4, -4)$; $D' \equiv (-2, -4)$

(B) $A' \equiv (2, 2)$; $B' \equiv (4, 2)$; $C' \equiv (-4, -4)$; $D' \equiv (-2, -4)$

(C) $A' \equiv (-2, -2)$; $B' \equiv (-4, -2)$; $C' \equiv (4, 4)$; $D' \equiv (2, 4)$

(D) $A' \equiv (2, -2)$; $B' \equiv (4, -2)$; $C' \equiv (4, -4)$; $D' \equiv (2, -4)$

19. Find the value of (α, β) when $(3\alpha + 2, 3\beta + 2)$ is reflected by transformation $\begin{pmatrix} X \\ Y \end{pmatrix} = \begin{pmatrix} 3 & 0 \\ 0 & 2 \end{pmatrix}\begin{pmatrix} x \\ y \end{pmatrix}$ to get new coordinates $(\alpha - \beta, \alpha + \beta)$.

(A) $\alpha = \frac{-20}{41}, \beta = \frac{-78}{41}$

(B) $\alpha = \frac{21}{41}, \beta = \frac{-310}{41}$

(C) $\alpha = \frac{-20}{41}, \beta = \frac{-310}{41}$

(D) $\alpha = \frac{-20}{41}, \beta = \frac{78}{41}$

20. The point $P(3, 4)$ is transformed using $\begin{pmatrix} X \\ Y \end{pmatrix} = \begin{pmatrix} \cos\theta & \sin\theta \\ -\sin\theta & \cos\theta \end{pmatrix}\begin{pmatrix} x \\ y \end{pmatrix}$, where $\theta = 45°$. The coordinates of new point are

(A) $\left(\frac{-7}{\sqrt{2}}, \frac{-1}{\sqrt{2}}\right)$ (B) $\left(\frac{-7}{\sqrt{2}}, \frac{1}{\sqrt{2}}\right)$

(C) $\left(\frac{7}{\sqrt{2}}, \frac{1}{\sqrt{2}}\right)$ (D) none of these

21. Which point is obtained when $P(2, 4)$ undergoes the following transformation?

(I) Reflection in the x-axis, after that

(I) Reflection in the y-axis, and then

(III) Reflection through origin

(A) (2, 4) (B) (–2, 4)

(C) (–2, –4) (D) (2, –4)

22. Consider triangle ABC as shown in the figure.

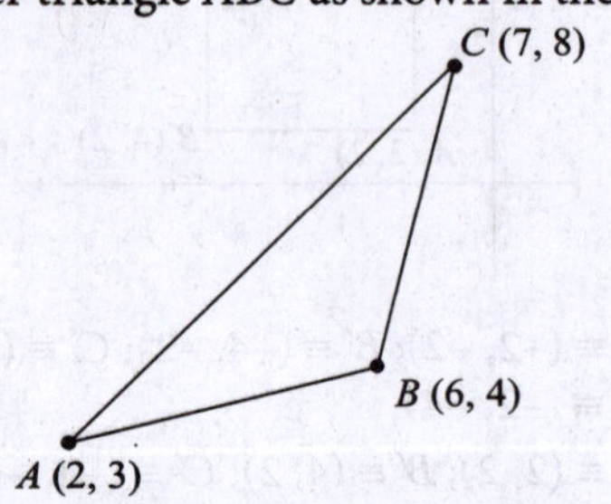

The new vertices of triangle when it undergoes the transformation $\begin{pmatrix} X \\ Y \end{pmatrix} = \begin{pmatrix} \frac{-1}{3} & 1 \\ 1 & \frac{-1}{3} \end{pmatrix}\begin{pmatrix} x \\ y \end{pmatrix}$ will be

(A) $A' \equiv \left(\frac{7}{3}, 1\right), B' \equiv \left(2, \frac{14}{3}\right), C' \equiv \left(\frac{17}{3}, \frac{13}{3}\right)$

(B) $A' \equiv \left(\frac{7}{3}, 1\right), B' \equiv \left(\frac{14}{3}, 2\right), C' \equiv \left(\frac{17}{3}, \frac{13}{3}\right)$

(C) $A' \equiv \left(\frac{7}{3}, 1\right), B' \equiv \left(2, \frac{14}{3}\right), C' \equiv \left(\frac{13}{3}, \frac{13}{3}\right)$

(D) $A' \equiv \left(1, \frac{7}{3}\right), B' \equiv \left(2, \frac{14}{3}\right), C' \equiv \left(\frac{17}{3}, \frac{13}{3}\right)$

23. Which of the following transformations is equivalent to get reflection through x-axis?

(A) $\begin{pmatrix} X \\ Y \end{pmatrix} = \begin{pmatrix} \cos 2\theta & \sin 2\theta \\ -\sin 2\theta & -\cos 2\theta \end{pmatrix}\begin{pmatrix} x \\ y \end{pmatrix}, \theta = 45°$

(B) $\begin{pmatrix} X \\ Y \end{pmatrix} = \begin{pmatrix} \cos\theta & \sin 2\theta \\ -\sin 2\theta & -\cos\theta \end{pmatrix}\begin{pmatrix} x \\ y \end{pmatrix}, \theta = 0°$

(C) $\begin{pmatrix} X \\ Y \end{pmatrix} = \begin{pmatrix} \sin 2\theta & \cos\theta \\ \cos 2\theta & -\sin 2\theta \end{pmatrix}\begin{pmatrix} x \\ y \end{pmatrix}, \theta = 90°$

(D) none of these

OLYMPIAD AND NTSE LEVEL EXERCISES

24. If the vertices of a right-angled triangle are reflected through x-axis, the new vertices so obtained will form

(A) an equilateral triangle

(B) an isosceles triangle

(C) a right-angled triangle

(D) a scalene triangle

25. If the point (–3, 4) is first reflected through origin and then the new point undergoes the transformation $\begin{pmatrix} X \\ Y \end{pmatrix} = \begin{pmatrix} \frac{1}{2} & 1 \\ 0 & 1 \end{pmatrix}\begin{pmatrix} x \\ y \end{pmatrix}$ then the final point will be

(A) (3, 4) (B) (11/2, 4)

(C) (11/2, –4) (D) (5/2, 4)

26. If the point (2, –5) is first reflected through x-axis and then the new point undergoes the transformation $\begin{pmatrix} X \\ Y \end{pmatrix} = \begin{pmatrix} 0 & 1 \\ 0 & 1 \end{pmatrix}\begin{pmatrix} x \\ y \end{pmatrix}$ then the final point will be

(A) (2, 5) (B) (2, 2)

(C) (5, 5) (D) (5, 2)

27. If the point (–1, –3) is first reflected through y-axis and then the new point undergoes the transformation $\begin{pmatrix} X \\ Y \end{pmatrix} = \begin{pmatrix} 1 & -1 \\ 0 & -1 \end{pmatrix}\begin{pmatrix} x \\ y \end{pmatrix}$, then the final point will be
(A) (4, 3) (B) (1, –3)
(C) (3, 4) (D) (0, 4)

28. If a point is reflected through any of coordinate axes, what will be the effect on its distance from origin?
(A) It will become half
(B) It will become twice
(C) It will remain the same
(D) None of these

29. A point located in second quadrant is reflected through the origin, what will be the location of the new point?
(A) First Quadrant
(B) Second Quadrant
(C) Third Quadrant
(D) Fourth Quadrant

30. If a point $A(4, 6)$ is reflected through x-axis to get point B and then B is transformed using $\begin{pmatrix} X \\ Y \end{pmatrix} = \begin{pmatrix} \frac{1}{2} & 1 \\ -1 & -\frac{1}{2} \end{pmatrix}\begin{pmatrix} x \\ y \end{pmatrix}$ to get point C, then the area of triangle ABC is
(A) 104 sq. unit (B) 52 sq. unit
(C) 50 sq. unit (D) 100 sq. unit

31. Which of the following transformations of point (x, y), where $x \neq 0, y \neq 0$, will give the points forming a right-angled triangle?
(A) Reflection through origin and then reflection through x-axis
(B) Reflection through y-axis and then reflection through origin
(C) Reflection through x-axis and then reflection through y-axis
(D) All of these

32. A point (–6, 4) undergoes two consecutive transformations $\begin{pmatrix} X \\ Y \end{pmatrix} = \begin{pmatrix} \frac{2}{3} & -1 \\ 0 & -\frac{1}{4} \end{pmatrix}\begin{pmatrix} x \\ y \end{pmatrix}$ and $\begin{pmatrix} X \\ Y \end{pmatrix} = \begin{pmatrix} 1 & \frac{-1}{2} \\ 0 & -1 \end{pmatrix}\begin{pmatrix} x \\ y \end{pmatrix}$. The final point will be
(A) $\left(-\frac{17}{2}, 1\right)$ (B) $\left(-\frac{15}{2}, -1\right)$
(C) $\left(-\frac{17}{2}, -1\right)$ (D) $\left(-\frac{15}{2}, 1\right)$

33. If all the vertices of a triangle undergo the same transformation $\begin{pmatrix} X \\ Y \end{pmatrix} = \begin{pmatrix} \frac{1}{2} & \frac{-1}{2} \\ -1 & 1 \end{pmatrix}\begin{pmatrix} x \\ y \end{pmatrix}$, what will be the effect on the area of the triangle?
(A) It will remain unchanged
(B) It will become half
(C) It will become 0
(D) It will get double

13 Circles and Cyclic Quadrilaterals

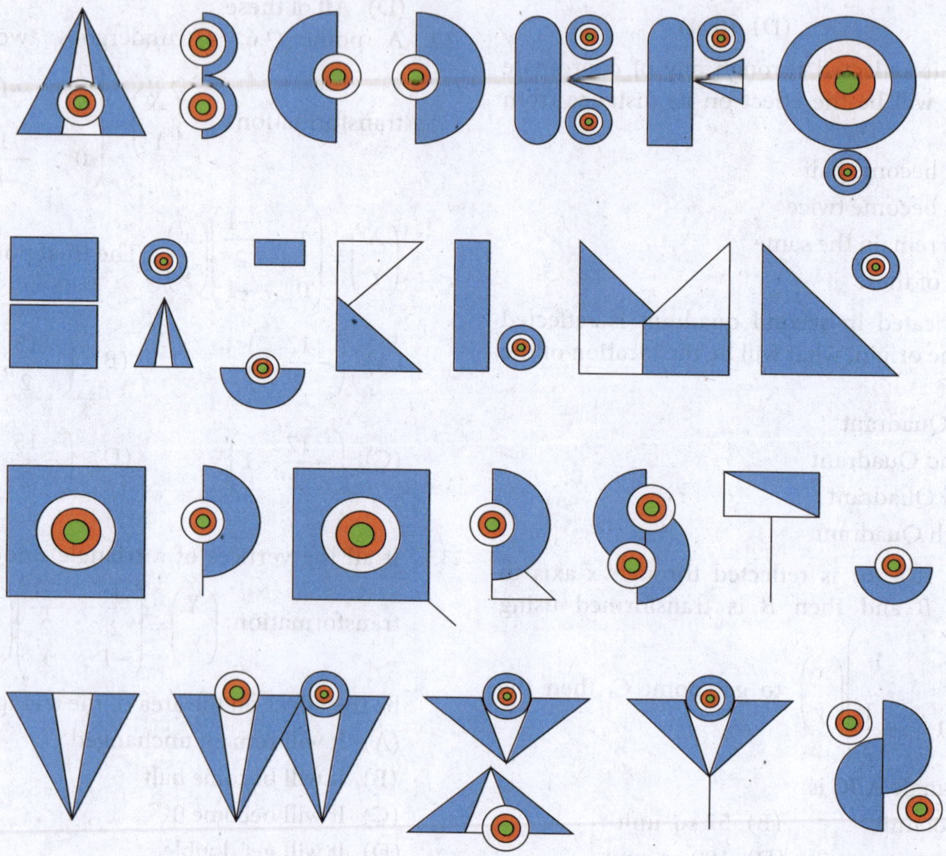

> Where there is matter, there is geometry.
>
> – Johannes Kepler

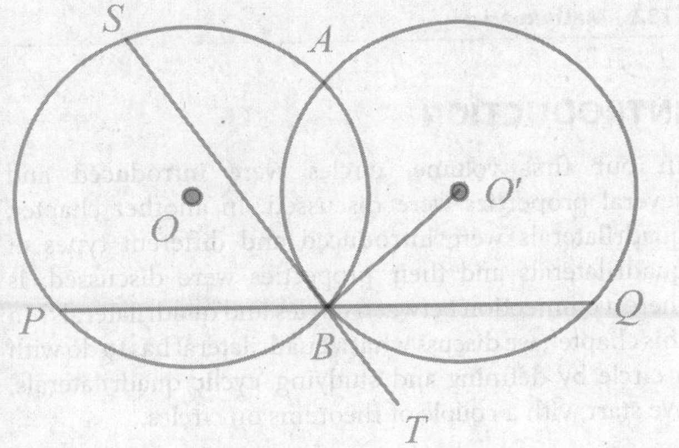

OBJECTIVES

After reading this chapter you should be able to

- state and prove the theorems on circles
- define cyclic quadrilateral
- state and prove the theorems on cyclic quadrilaterals
- solve the related riders

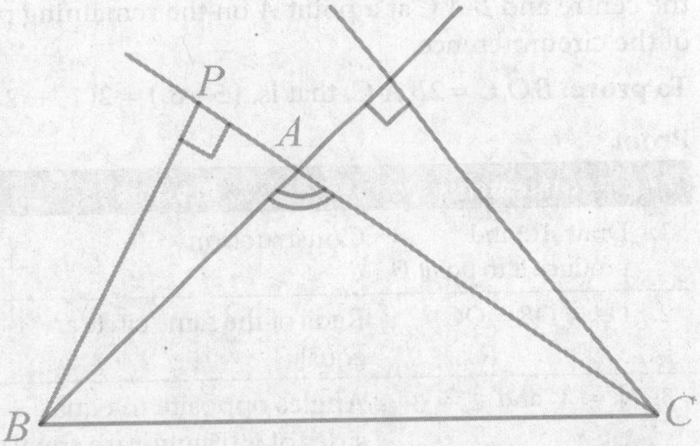

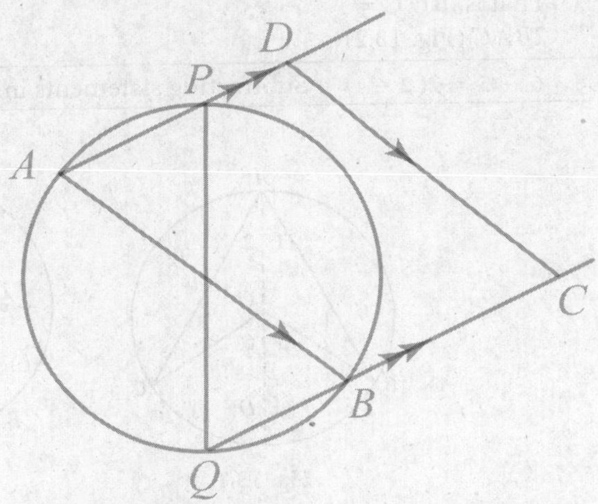

INTRODUCTION

In our first volume, circles were introduced and several properties were discussed. In another chapter, quadrilaterals were introduced and different types of quadrilaterals and their properties were discussed. Is there a connection between circles and quadrilaterals? In this chapter, we discuss what a quadrilateral has to do with a circle by defining and studying 'cyclic quadrilaterals'. We start with a couple of theorems on circles.

Theorem

The angle subtended by the arc of a circle at the centre is double the angle subtended at any point on the circumference.

Given: A circle with centre *O*. Arc *BC* subtends $B\hat{O}C$ at the centre and $B\hat{A}C$ at a point *A* on the remaining part of the circumference.

To prove: $B\hat{O}C = 2B\hat{A}C$, that is, $(\hat{5}+\hat{6}) = 2(\hat{1}+\hat{2})$.

Proof

Statements	Reasons
1. Draw *AO* and produce it to point *D*	Construction
2. $OA = OB = OC$	Radii of the same circle are equal
3. $\hat{1} = \hat{4}$ and $\hat{2} = \hat{3}$	Angles opposite to equal sides of a triangle are equal
4. $\hat{5} = \hat{4}+\hat{1}$ and $\hat{6} = \hat{2}+\hat{3}$ (Fig. 13.1)	The exterior angle of a triangle is equal to the sum of the two interior opposite angles
5. $\hat{5} = 2(\hat{1})$ and $\hat{6} = 2(\hat{2})$	From statements (3) and (4)
6. $\hat{5}+\hat{6} = 2(\hat{1}+\hat{2})$	Adding equations in (5)
7. That is, $B\hat{O}C = 2B\hat{A}C$ (Fig. 13.2)	
8. $\hat{6}-\hat{5} = 2(\hat{2}-\hat{1})$	Subtracting statements in (5)
9. $B\hat{O}C = 2B\hat{A}C$ (Fig. 13.3)	$\hat{6}-\hat{5} = B\hat{O}C$, $\hat{2}-\hat{1} = B\hat{A}C$
10. $\hat{5} = 2\times\hat{1}$	From statement (5)
11. $B\hat{O}C = 2\hat{A}$	From statement (10)

Theorem

Angles in the same segment of a circle are equal.

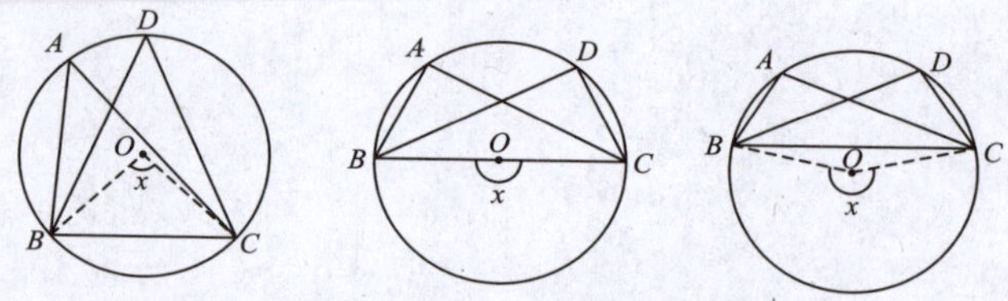

Given: A circle with centre *O*. Segment *BADC* is part of the circle. Draw any two angles *A* and *D* in the segment *BADC*.

To prove: $\hat{A} = \hat{D}$

Proof

Statements	Reasons
1. Draw $OB = OC$	Construction
2. $\hat{A} = \frac{x}{2}$, $\hat{D} = \frac{x}{2}$	The angle at the circumference is equal to half the angle at the centre
3. $\hat{A} = \hat{D}$	Each angle $= \frac{x}{2}$

Aliter

Statements	Reasons
1. $\hat{A} = \frac{BC}{2}$	The angle at the circumference is equal to half the angle subtended at the centre
2. $\hat{D} = \frac{BC}{2}$	-do-
3. $\hat{A} = \hat{D}$	Each angle $= \frac{BC}{2}$

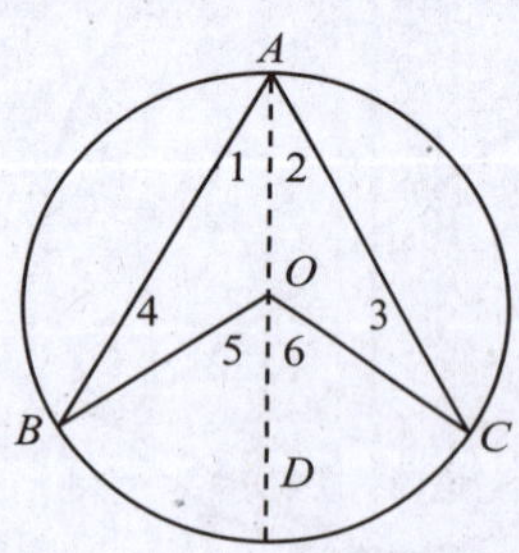

Fig. 13.1.

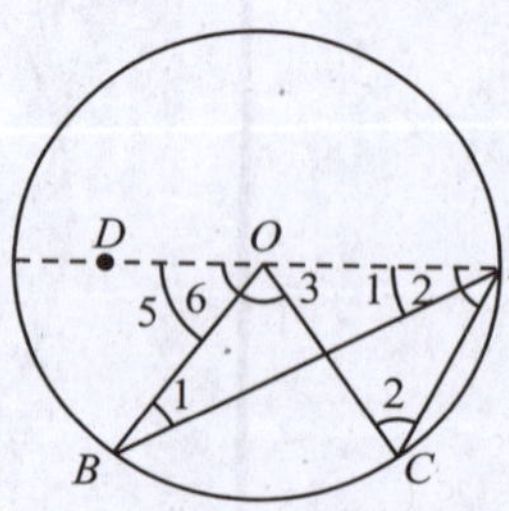

Fig. 13.2.

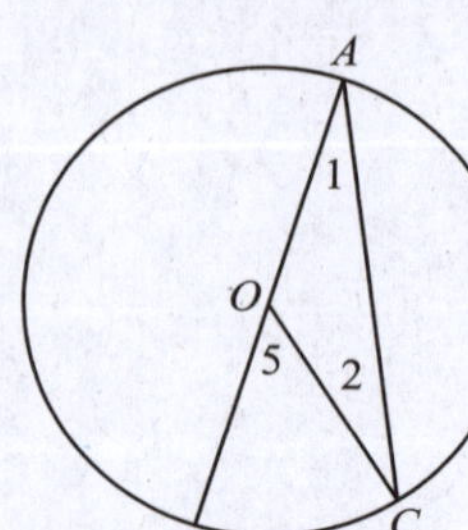

Fig. 13.3.

Corollary

An angle in a semicircle is a right angle.

Given: A circle with centre *O*. *BOC* is the diameter.

To prove: $B\hat{A}C = 90°$

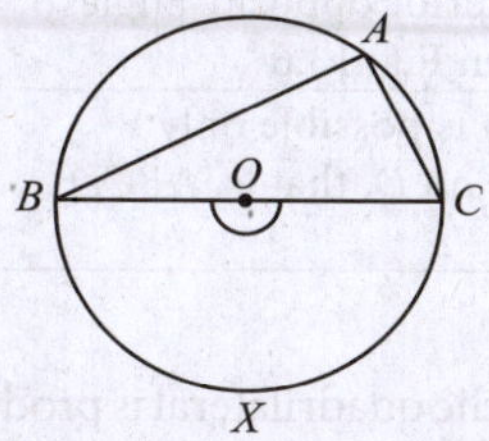

Proof

Statements		Reasons
1.	Arc $B\hat{A}C$ is a semicircle	Diameter bisects a circle
2.	$B\hat{A}C = \frac{1}{2}B\hat{O}C$	The angle at the circumference is equal to half the angle at the centre
3.	$B\hat{A}C = \frac{1}{2}(180°) = 90°$	$B\hat{O}C$ = straight line = 180°

Aliter 1

Statements		Reasons
1.	$B\hat{A}C = \frac{1}{2}BXC$	The angle at the circumference is equal to half the arc intercepted by the angle
2.	$B\hat{A}C = \frac{1}{2}(180°) = 90°$	A semicircle subtends a straight angle at the centre

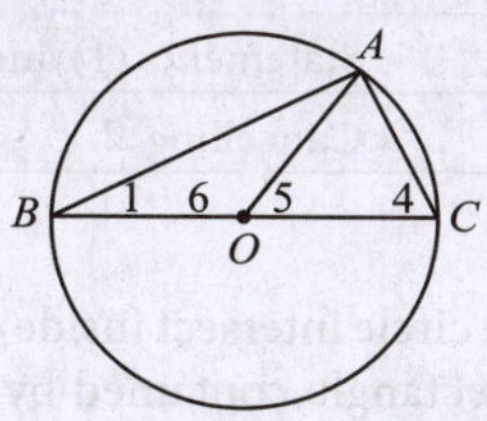

Aliter 2

Statements		Reasons
1.	Draw *AO*	Construction
2.	*OA* = *OB* = *OC*	Radii of the same circle are equal
3.	$\hat{1} = \hat{2}$ and $\hat{3} = \hat{4}$	Angles opposite to the equal sides are equal
4.	$\hat{5} = \hat{1} + \hat{2} = 2 \times \hat{2}$	The exterior angle of a triangle is equal to the sum of the interior opposite angles and statement (3)
5.	$\hat{6} = \hat{3} + \hat{4} = 2 \times \hat{3}$	The exterior angle of a triangle is equal to the sum of the interior opposite angles and statement (3)
6.	$\hat{5} + \hat{6} = 2(\hat{2} + \hat{3})$	Adding statements (4) and (5)
7.	$B\hat{A}C = \frac{180°}{2} = 90°$	*BOC* is a straight line

CYCLIC QUADRILATERAL

A quadrilateral is said to be cyclic if all its vertices lie on a circle. In other words, a quadrilateral is said to be cyclic if it can be inscribed in a circle.

Concyclic Points

Points are said to be concyclic if they lie on a circle.

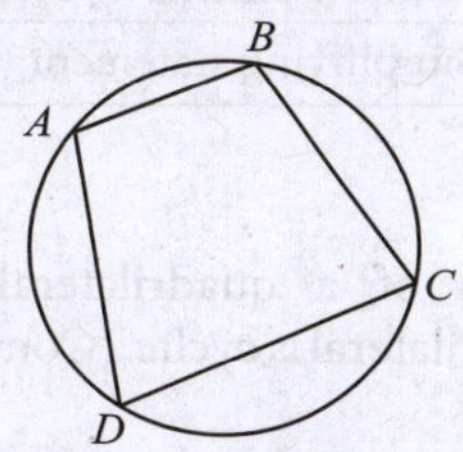

Fig. 13.4.

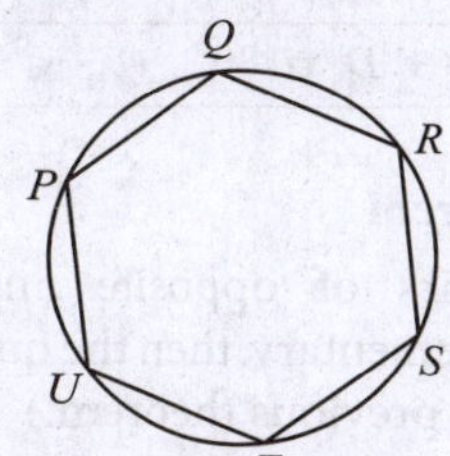

Fig. 13.5.

In Figs. 13.4 and 13.5, points *A*, *B*, *C*, and *D* and *P*, *Q*, *R*, *S*, *T*, and *U*, respectively, are concyclic, that is, lie on a circle. Therefore, quadrilateral *ABCD* is cyclic, and *PQRSTU* is a cyclic hexagon.

Theorem

The opposite angles of a cyclic quadrilateral are supplementary.

Given: A quadrilateral *ABCD* inscribed in a circle with centre *O*.

To prove: (a) $\hat{A} + \hat{C} = 180$ and (b) $\hat{B} + \hat{D} = 180°$

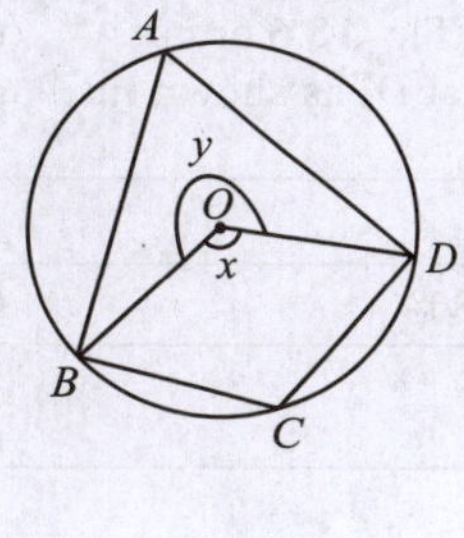

Proof

Statements	Reasons
1. Draw BO and DO	Construction
2. $\hat{A} = \frac{x}{2}$	The angle at the circumference is equal to half the angle at the centre
3. $\hat{C} = \frac{y}{2}$	The angle at the circumference is equal to half the angle at the centre
4. $\hat{A}+\hat{C} = \frac{x+y}{2}$	Adding statements (2) and (3)
5. $\hat{A}+\hat{C} = \frac{360^\circ}{2} = 180^\circ$	The angle at a point is equal to 360°
6. $\hat{A} + \hat{B} + \hat{C} + \hat{D} = 360^\circ$	The sum of all the angles of a quadrilateral equal to 360°
7. $\hat{B} + \hat{D} + \hat{A} + \hat{C} = 360^\circ$	Rearranging statement (6)
8. $\hat{B} + \hat{D} + 180^\circ = 360^\circ$	From statements (5) and (7)
9. $\hat{B} + \hat{D} = 180^\circ$	Simplifying statement (8)

Theorem

If pairs of opposite angles of a quadrilateral are supplementary, then the quadrilateral is cyclic. (Converse of the previous theorem.)

Given: A quadrilateral $ABCD$ in which $\hat{B}+\hat{D}=180^\circ$

To prove: $ABCD$ is cyclic.

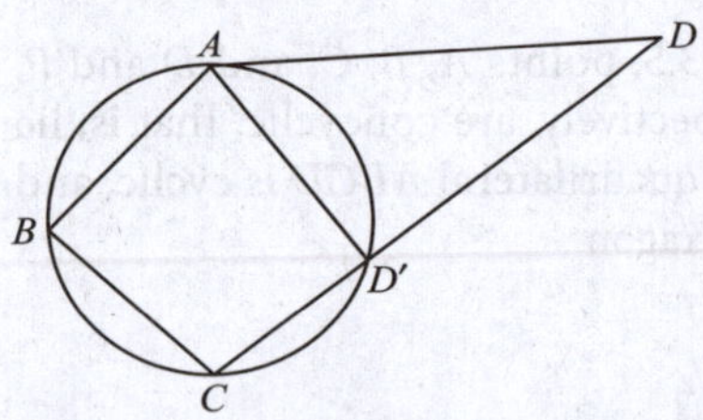

Fig. 13.6.

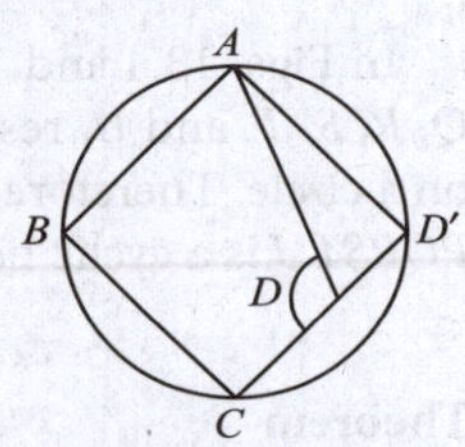

Fig. 13.7.

Proof

Statements	Reasons
1. If $ABCD$ is not cyclic, let the circle passing through ABC cut DC at D' as in Fig. 13.6 or cut CD produced at D', as shown in Fig. 13.7	There is a unique circle passing through three given points
2. $\hat{B} + D' = 180^\circ$	$ABCD'$ is cyclic
3. But $\hat{B} + \hat{D} = 180^\circ$	Given
4. $D = D'$	Using statements (2) and (3)
5. This is not possible because the exterior angle $\hat{D}$ is equal to one of the interior opposite angles of the triangle ADD' in Fig. 13.6 The exterior angle D' is equal to one of the interior opposite angles D, as shown in Fig. 13.6	
6. Statement (5) is possible only when D′ falls on D, that is, $ABCD$ is cyclic	

Theorem

If one side of a cyclic quadrilateral is produced, the exterior angle so formed is equal to the interior opposite angle.

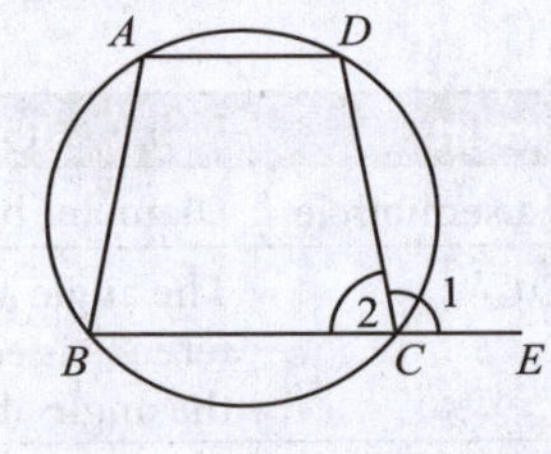

Given: A cyclic quadrilateral $ABCD$. BC is produced to E so that the exterior angle $D\hat{C}E = (\hat{1})$ is formed.

To prove: $\hat{1} = \hat{A}$

Proof

Statements	Reasons
1. $\hat{A}+\hat{2} = 180^\circ$	The opposite angles of a cyclic quadrilateral are supplementary
2. $\hat{2}+\hat{1} = 180^\circ$	L.P.A. are supplementary
3. $\hat{A}+\hat{2} = \hat{2}+\hat{1}$	Statements (1) and (2)
4. $\hat{A} = \hat{1}$	Cancelling $\hat{2}$

Theorem

If two chords of a circle intersect inside or outside (when produced), the rectangle contained by the segments of one is equal to the rectangle contained by the segments of the other.

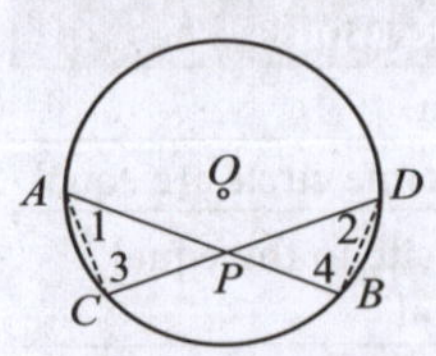

Fig. 13.8.

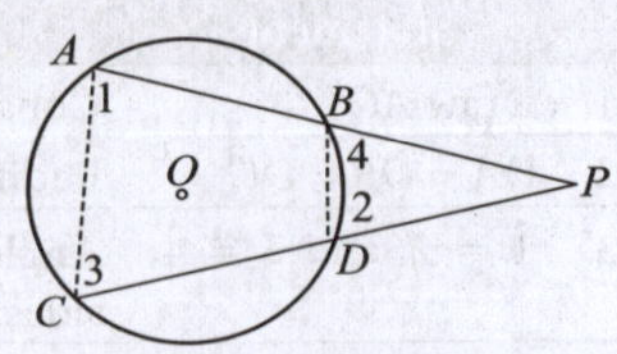

Fig. 13.9.

Given: Two chords AB and CD of a circle with centre O, intersecting at P inside or outside when produced.

To prove: $PA \times PB = PC \times PD$

Proof

Statements	Reasons
1. Draw AC and BD	Construction
2. In triangles PAC and PDB $\hat{1}=\hat{2}$ and $\hat{3}=\hat{4}$ (Fig. 13.8)	Angles in the same segment are equal
3. $\Delta PAC \sim \Delta PDB$	AA criterion
4. $\frac{PA}{PD}=\frac{PC}{PB}$	Sides of similar triangles are proportional
5. $PA \times PB = PC \times PD$	From statement (4) by cross-multiplication
6. In Fig. 13.9, $\hat{1}=\hat{2}$ and $\hat{3}=\hat{4}$	$ABCD$ is a cyclic quadrilateral and the exterior angle is equal to the interior opposite angle
7. $\Delta PAC \sim \Delta PDB$	AA criterion
8. $\frac{PA}{PD}=\frac{PC}{PB} \Rightarrow PA \times PB = PC \times PD$	The corresponding sides of similar triangles are proportional

Worked Examples

1. In the given figure, AB and CD are two chords of a circle intersecting at O. Prove that $AO \times OB = DO \times OC$.

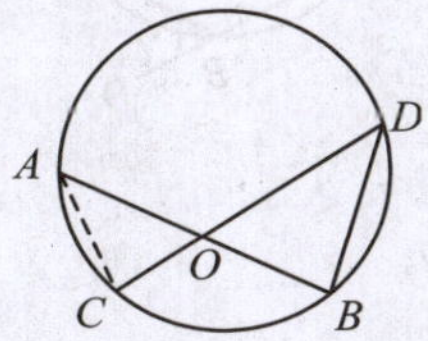

Proof

Statements	Reasons
1. Draw AC and DB	Construction
2. In triangles AOC and DOB (i) $\hat{A}=\hat{D}$ and $\hat{C}=\hat{B}$	Angles in the same segment are equal
3. $\Delta AOC \sim \Delta DOB$	AA criterion
4. $\frac{AO}{DO}=\frac{OC}{OB}=\frac{AC}{DB}$	Sides of similar triangles are proportional
5. $AO \times OB = DO \times OC$	From statement (4), taking first two ratios

2. In the figure, radius OC is perpendicular to chord AB at D. Prove that $AC = CB$.

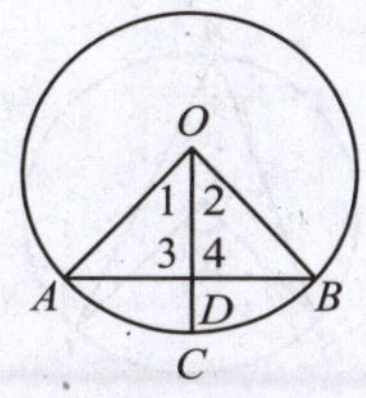

Proof

Statements	Reasons
1. $OC \perp AB$	Given
2. $OA = OB$	Radii of the same circle are equal
3. $OD = OD$	Common
4. $\hat{3}=\hat{4}$	Each is a right angle and statement (1)
5. $\Delta OAD \cong \Delta OBD$	R.H.S.
6. $\hat{1}=\hat{2}$	
7. Therefore, $AC = BC$	In the same or equal circles, equal central angles have equal arcs

3. In the given figure, chord BC = chord BD in a circle with centre O. AB bisects $C\hat{B}D$. Prove that $AC = AD$.

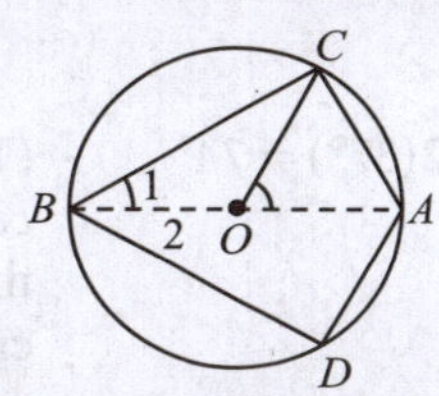

Proof

Statements	Reasons
1. AB bisects $C\hat{B}D$	Given
2. $\hat{1}=\hat{2}$	AB bisects $C\hat{B}D$
3. $BC = BD$	Given
4. $AB = AB$	Identity
5. $\Delta ABC \cong \Delta ABD$	Using SAS axiom on triangle congruence of triangles
6. $AC = AD$	
7. $A\hat{O}C = 2(\hat{1})$	The exterior angle is equal to the sum of the interior opposite angles
8. $A\hat{O}D = 2(\hat{2})$	The exterior angle (extended) is equal to the sum of interior opposite angles
9. $A\hat{O}C = A\hat{O}D$	$\hat{1}=\hat{2}$ (given)
10. $AC = AD$	In the same or equal circles, equal central angles have equal arcs

4. (a) In the given figure, ABC is a circle with centre O. If $B\hat{O}C = 80°$, then $\hat{x} = ?$

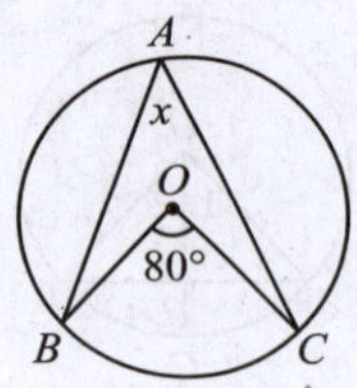

Solution

$\hat{x} = \dfrac{B\hat{O}C}{2} = \dfrac{80°}{2} = 40°$ (The angle at the circumference is equal to half the angle at the centre)

(b) In the given figure, ABC is a circle with centre O. If $A\hat{C}B = 37°$, then $\hat{x} =$?

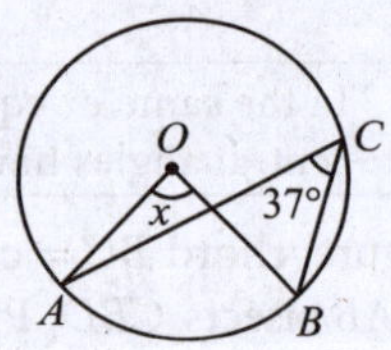

Solution

$\hat{x} = 2A\hat{C}B = 2(37°) = 74°$ (The angle at the centre is double the angle at the circumference)

(c) In the given figure, AOB is a diameter of circle with centre O. Find $\hat{x}$.

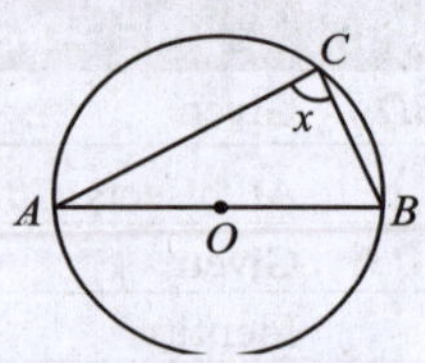

Solution

$\hat{x}=90°$ (The angle in a semicircle is a right angle)

(d) In the given figure, O is the centre of the circle. $\hat{p}$, $\hat{q}$, and $\hat{r}$ stand on the same minor arc ACB. If $\hat{m} = 100°$, then

(i) Find $\hat{p}$ = ______, $\hat{q}$ = ______, $\hat{r}$ = ______

(ii) What is the relation between $\hat{p}$, $\hat{q}$, *and* $\hat{r}$?

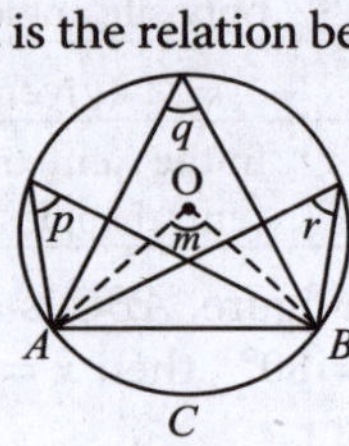

Solution

(i) $\hat{p} = \frac{m}{2}, = \frac{100°}{2} = 50°$

$\hat{q} = \frac{m}{2} = \frac{100°}{2} = 50°$

$\hat{r} = \frac{m}{2} = \frac{100°}{2} = 50°$ (The angle at the circumference is half the angle at the centre)

(ii) Angles in the same segment are equal.

5. In the given figure, if $C\hat{B}D = 110°$, then find x.

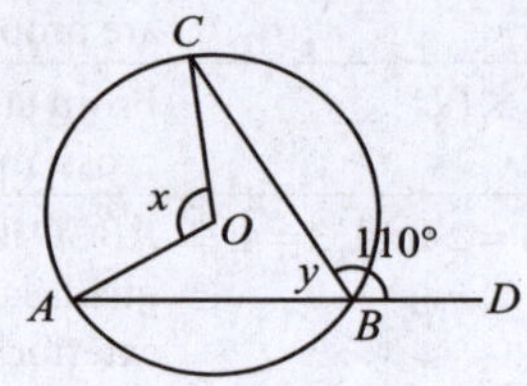

Solution

$\hat{y} = 180° - 110° = 70°$

$\hat{x} = 2\,\hat{y} = 2(70°)$ (The angle at the centre is double the angle on the remaining arc)

$= 140°$

6. In the given figure, $A\hat{O}C = 120°$. Find $C\hat{B}D$.

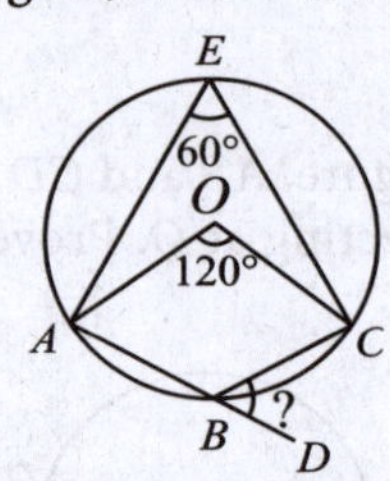

Solution

$A\hat{O}C = 120°$ (Given)

Reflex angle $AOC = 360° - 120° = 240°$.

$A\hat{B}C = \frac{1}{2}$ reflex angle $A\hat{O}C$ (The angle at the centre is equal to double the angle at the remaining arc)

$= \frac{1}{2}(240°) = 120°$

$C\hat{B}D = 180° - A\hat{B}C = 180° - 120° = 60°$

$C\hat{B}D$ and $A\hat{B}C$ are linear pair of angles.

Take a point E and join EA and EC.

$A\hat{E}C = \frac{1}{2} \times 120° = 60°$

$\therefore C\hat{B}D = A\hat{E}C$ (The exterior angle of a cyclic quadrilateral is equal to the interior opposite angle)

$= 60°$

7. In the given figure, find β.

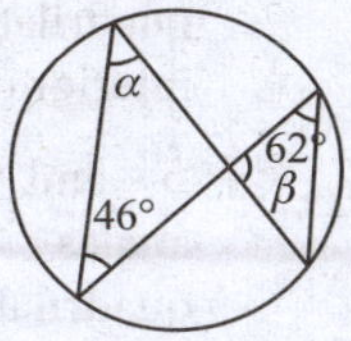

Solution

$a = 62°$ (Angles in the same segment)

$\beta = 180° - (62° + 46°) = 180° - 108° = 72°$

(The sum of the three angles of a triangle = 180°)

8. In the given figure, if $C\hat{A}B = 30°$ and $A\hat{D}B = 40°$, then find $A\hat{B}C$.

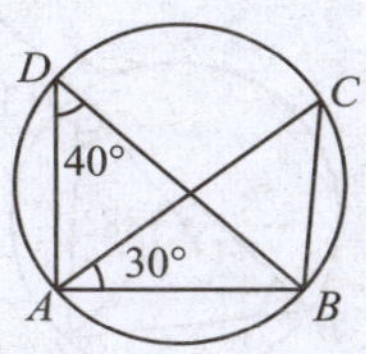

Solution

$\hat{C} = \hat{D} = 40°$ (Angles in the same segment)

$A\hat{B}C = 180° - (B\hat{A}C + \hat{C})$

$A\hat{B}C = 180° - (30° + 40°) = 110°$

9. In the given figure, if $A\hat{D}C = 84°$ and $AB = BC$, then find $B\hat{A}C$.

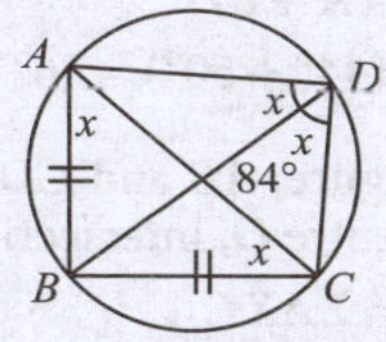

Solution

Let $B\hat{A}C = x$

$A\hat{C}B = B\hat{A}C = x$ (Angles opposite to equal sides are equal)

$A\hat{D}C = 84°$ (Given)

$C\hat{D}B = B\hat{A}C = x$ (Angles in the same segment)

$B\hat{D}A = B\hat{C}A = x$

$B\hat{A}C = \frac{1}{2} A\hat{D}C = 42°$

This problem can be solved using the property of cyclic quadrilateral.

$A\hat{D}C + A\hat{B}C = 180°$ (Opposite angles of a cyclic quadrilateral are supplementary)

$A\hat{B}C = 180° - 84° = 96°$

$$B\hat{A}C = B\hat{C}A = \frac{180° - 96°}{2} = 42°$$

10. In the given figure, AOB is the diameter and $B\hat{A}C = 37°$. Find $A\hat{D}C$.

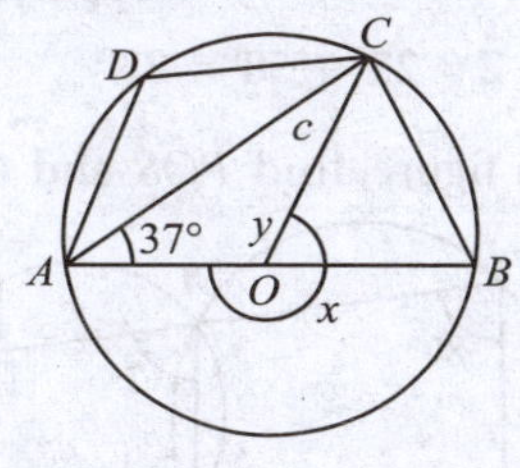

Solution

Join O and C (Construction)

$OA = OC$ (Radii of the same circle)

$C = O\hat{A}C = 37°$ (Angles opposite to equal sides are equal in a triangle)

$\hat{y} = 180° - (37° + 37°) = 106°$

$\hat{x} = 360° - 106° = 254°$ (The angle at a point = 360°)

$A\hat{D}C = \frac{\hat{x}}{2} = \frac{254°}{2} = 127°$ (The angle at the circumference is equal to half the angle at the centre)

Aliter:

$A\hat{C}B = 90°$ The angle in a semicircle is a right angle)

$\hat{B} = 90° - 37° = 53°$ (Two acute angles in a right-angled triangle are complementary)

$A\hat{D}C + C\hat{B}A = 180°$ (Opposite angles of cyclic quadrilateral $ABCD$ are supplementary)

Therefore, $A\hat{D}C = 180° - 53° = 127°$.

From the previous statements, $C\hat{B}A = 53°$.

11. In the given figure, $Q\hat{X}R = 25°$ and $Q\hat{R}X = 33°$. Find XYZ and PZQ.

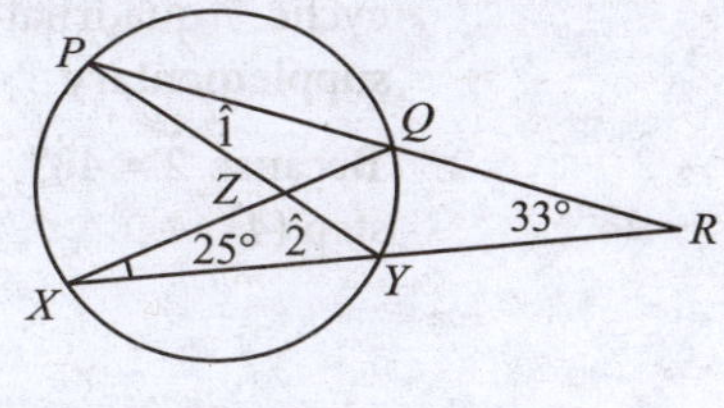

Solution

$\hat{1} = \hat{x} = 25°$ (Angles in the same segment are equal)

$\hat{2} = \hat{1} + R = 25° + 33° = 58°$ (The exterior angle of a triangle is equal to the sum of the two interior opposite angles)

$P\hat{Z}Q = \hat{X} + \hat{2} = 25° + 58° = 83°$

12. In the given figure, find $P\hat{Q}S$ and $Q\hat{S}R$.

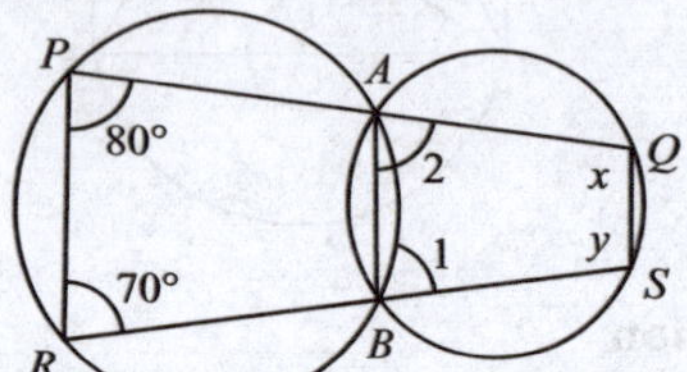

Solution

$\hat{1} = \hat{P} = 80°$ (The exterior angle of a cyclic quadrilateral is equal to the interior opposite angle)

$\hat{2} = \hat{R} = 70°$

$\hat{1} + \hat{x} = 180° \Rightarrow x = 180° - 80° = 100°$

$2 + \hat{y} = 180° \Rightarrow y = 180° - \hat{2} = 180° - 70° = 110°$

13. In the given figure, $CD = CE$. Find (i) $A\hat{B}D$ (ii) $B\hat{F}D$

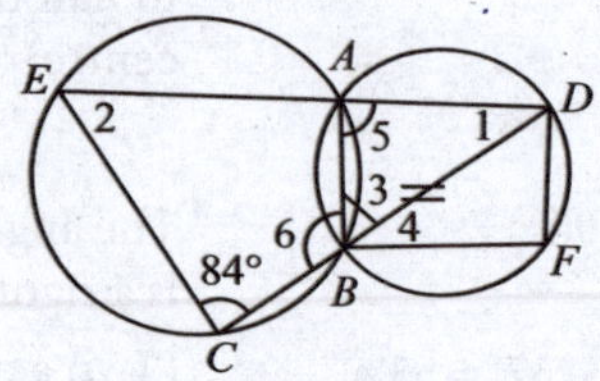

Solution

1. $CE = CD$ — Given
2. $\hat{C} = 84°$ — Given
3. $\hat{1} = \hat{2}$ — Angles opposite to equal sides of a triangle are equal.
4. $\hat{1} = \hat{2} = \dfrac{180° - 84°}{2} = 48°$ — The sum of the three angles of a triangle is 180°
5. $\hat{6} + \hat{2} = 180°$ — Opposite angles of a cyclic quadrilateral are supplementary
6. $\hat{6} = 180° - \hat{2} = 180° - 48° = 132°$ — Because $\hat{2} = 48°$ (proved) step (4)
7. $\hat{5} = E\hat{C}B = 84°$ — The exterior angle of a cyclic quadrilateral is equal to the interior opposite angle
8. $B\hat{F}D = 180° - \hat{5} = 180° - 84° = 96°$ — $\hat{5}$ and $\hat{F}$ are opposite angles of cyclic quadrilateral $ABFD$
9. $\hat{3} = 180° - (\hat{5} + \hat{1}) = 180° - (84° + 48) = 48° = A\hat{B}D$

14. In the given figure, O is the centre of the circle ABC. Calculate the measure of each angle of ΔABC.

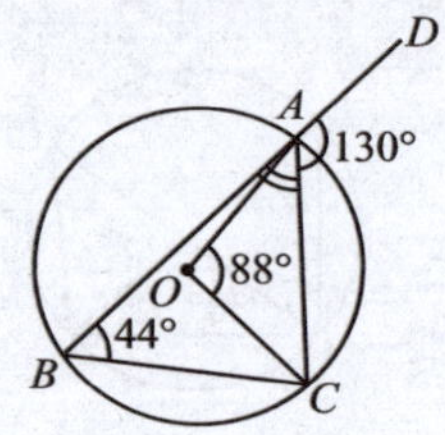

Solution

$A\hat{B}C = \frac{1}{2} A\hat{O}C$

$= \frac{1}{2}(88°) = 44°$ (The angle at the circumference is equal to half the angle at the centre)

$B\hat{A}C = 180° - C\hat{A}D = 180° - 130° = 50°$

$B\hat{C}A = 180° - (A + B)$

$= 180° - (44° + 50°) = 86°$

15. In the given figure, AB and CD are two chords of a circle with centre O, intersecting at X. Prove that $A\hat{O}C + B\hat{O}D = 2A\hat{X}C$.

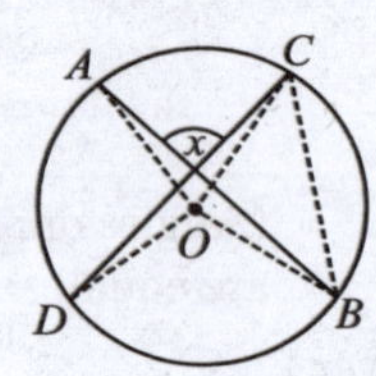

Proof

Statements	Reasons
1. Draw BC	Construction
2. $A\hat{X}C = X\hat{C}B + C\hat{B}X$	The exterior angle of the triangle XCB is equal to the sum of the two interior opposite angles

3.	$A\hat{X}C = \frac{B\hat{O}D}{2} + \frac{A\hat{O}C}{2}$	The angle at the circumference is equal to half the angle at the centre
4.	$2A\hat{X}C = B\hat{O}D + A\hat{O}C$	Multiplying statement (3) by 2

16. In the given figure, two circles intersect at points X and Y. A straight line PXQ is drawn cutting the circles at P and Q. Prove that $A\hat{X}B = P\hat{Y}Q$.

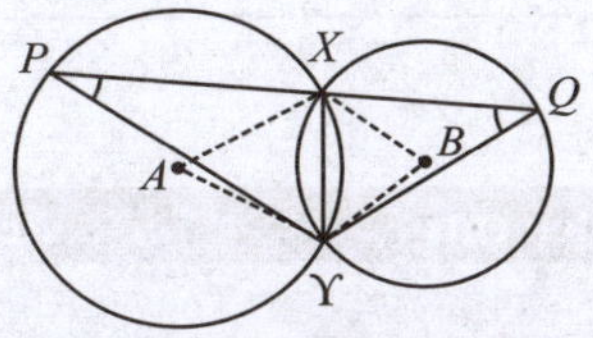

Proof

	Statements	Reasons
1.	Draw AX, AY, BX, BY, and XY	Construction
2.	$\hat{P} = \frac{\hat{A}}{2}$ and $\hat{Q} = \frac{\hat{B}}{2}$	The angle at the circumference is equal to half the angle at the centre
3.	$A\hat{X}Y = A\hat{Y}X$ and $B\hat{X}Y = B\hat{Y}X$	AX = AY and BX = BY (Angles opposite to the equal sides are equal in a triangle)
4.	$\begin{cases}\hat{A} + 2A\hat{X}Y = 180° \\ \hat{B} + 2B\hat{X}Y = 180°\end{cases}$	Sum of the angles of a triangle
5.	$A\hat{X}Y = 90° - \frac{\hat{A}}{2}$ and $B\hat{X}Y = 90° - \frac{B}{2}$	Using statement (4)
6.	$A\hat{X}B = \left(90° - \frac{\hat{A}}{2}\right) + \left(90° - \frac{\hat{B}}{2}\right)$	
7.	$P\hat{Y}Q = 180° - (\hat{P} + \hat{Q})$	Using sum of three angles of a triangle = 180°
8.	$P\hat{Y}Q = 180° - \left(\frac{A}{2} + \frac{B}{2}\right)$	From statements (2) and (7)
9.	$A\hat{X}B = P\hat{Y}Q$	Statements (5) and (8)

17. In the given figure, $ABCD$ is a cyclic quadrilateral in which the diagonals AC and BD intersect at right angles at M. If MX is drawn perpendicular to BC and produced to cut AD at Y, then prove that Y is the midpoint of AD.

Given: $ABCD$ is a cyclic quadrilateral. BD cuts AC at M perpendicularly. MX is perpendicular to BC and produced to meet AD at Y.

To prove: $AY = YD$; let 1, 2, 3, 4, 5, 6, 7, 8, and 9 be the angles, as shown in the figure.

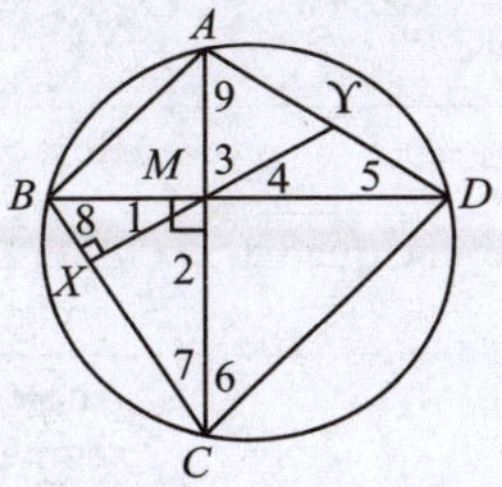

Proof

	Statements	Reasons
1.	$\hat{1} = \hat{7}$	$B\hat{M}X = M\hat{C}X$
2.	$\hat{7} = \hat{5}$	Angles in the same segment
3.	$\hat{1} = \hat{4}$	Vertically opposite angles
4.	$\hat{4} = \hat{5}$	Using statements (1), (2), and (3)
5.	YD = YM Similarly, we can prove	Sides opposite to equal angles of a triangle
6.	$YM = YA$	Similarly, we can prove
7.	$AY = YD$	Using statements (5) and (6)

18. Prove that the altitudes of a triangle are concurrent.

Given: ΔABC with BE and CF as altitudes meeting at O.

To prove: Altitude AD passes through O.

Let BE and CF be perpendiculars to AC and AB meeting at O.

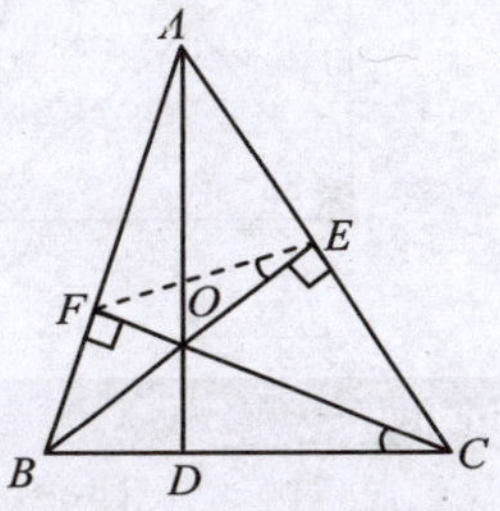

Proof

	Statements	Reasons
1.	(a) Join EF (b) Join AO and produce it to meet BC at D.	Construction
2.	In quadrilateral $AFOE$, (a) $A\hat{F}O = A\hat{E}O = 90°$ (b) Quadrilateral $AFOE$ is cyclic	 (a) $BE \perp AC$, $CF \perp AB$ (b) Opposite angles are supplementary

3.	$A\hat{E}F = A\hat{O}F$	Angles in the same segment are equal
4.	$A\hat{O}F = D\hat{O}C$	V.O.A. are equal
5.	$B\hat{F}C = B\hat{E}C = 90°$	Reason 2 (i)
6.	$BFEC$ is a cyclic quadrilateral	B, F, E, and C on the circle
7.	$F\hat{E}O = F\hat{C}B$	Angles in the same segment are equal
8.	$F\hat{E}O = O\hat{C}D$	Same as above
9.	$D\hat{O}C + O\hat{C}D = A\hat{E}F + F\hat{E}O = A\hat{E}O = 90°$	
10.	$O\hat{D}C = 90°$	If two angles of a triangle are complementary, then the remaining angle is a right angle
11.	$AD \perp BC$	Perpendicular lines contain a right angle
12.	But AD passes through O	By construction
13.	Altitudes are concurrent	By data and statement (11)

19. In the given figure, prove that $ABCD$ is a cyclic quadrilateral. Prove that $B\hat{A}D + B\hat{C}D = 180°$, that is, quadrilateral $ABCD$ is cyclic.

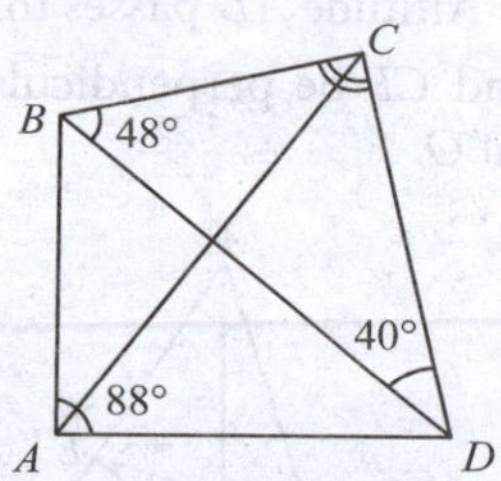

Proof

Statements	Reasons
1. In ΔBCD, $B\hat{C}D = 180° - (C\hat{B}D + C\hat{D}B)$	The sum of the three angles of a triangle = 180°
2. $B\hat{C}D = 180° - (48° + 40°) = 92°$	$C\hat{B}D = 48°$ and $C\hat{D}B = 40°$ (given)
3. $B\hat{A}D = 88°$	Given
4. $B\hat{C}D + B\hat{A}D = 92° + 88° = 180°$	Adding statements (2) and (3)
5. Quadrilateral $ABCD$ is cyclic	The opposite angles of a quadrilateral are supplementary

20. In the given figure, prove that $BCEF$ is a cyclic quadrilateral.

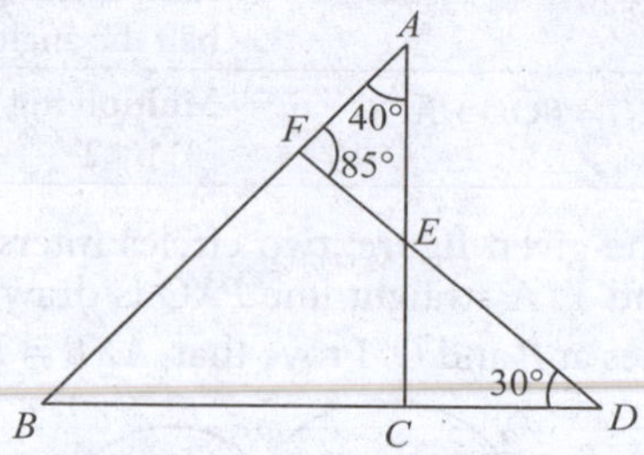

Proof

Statements	Reasons
1. $B\hat{F}E = 180° - 85° = 95°$	They are linear pair
2. $F\hat{E}A = 180° - (85° + 40°) = 55°$	The sum of the three angles of a triangle = 180°
3. $F\hat{B}D = A\hat{F}E - F\hat{D}B = 85° - 30° = 55°$	The exterior angle of a triangle is equal to the sum of the two interior opposite angles (ΔFDB)
4. $F\hat{B}D = F\hat{E}A = 55°$	From statements (2) and (3)
5. Quadrilateral $BCEF$ is cyclic	The exterior angle of a quadrilateral is equal to the interior opposite angle

21. In the given figure, $ABCD$ is a parallelogram. A circle through A and B cuts DA and CB produced at P and Q, respectively. Prove that D, C, Q, and P are concyclic.

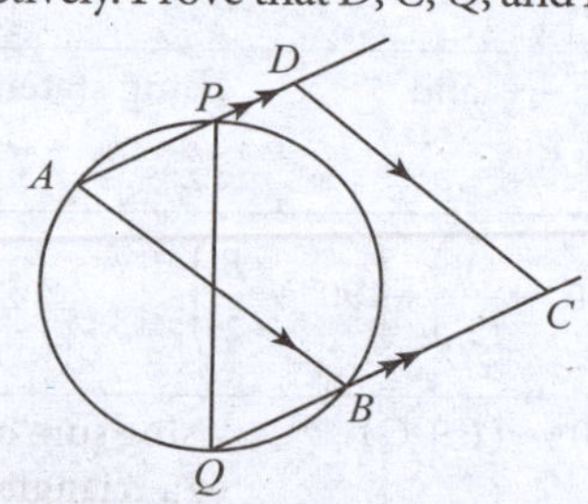

Proof

Statements	Reasons
1. $ABCD$ is a parallelogram	Given
2. $\hat{A} + \hat{D} = 180°$	The sum of the two adjacent angles = 180°
3. $\hat{A} = \hat{Q}$	Angles in the same segment are equal
4. $\hat{D} + \hat{Q} = 180°$	From statements (2) and (3)
5. P, Q, C, and D are concyclic points	The opposite angles of quadrilateral are supplementary

22. Prove that a parallelogram inscribed in a circle is a rectangle.
Given: A parallelogram *ABCD* inscribed in a circle with centre *O*.
To prove: *ABCD* is a rectangle ($\hat{A}=\hat{B}=\hat{C}=\hat{D}=90°$).

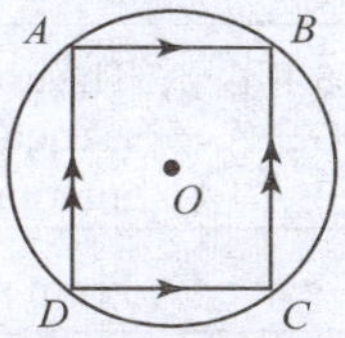

Proof

Statements	Reasons
1. *ABCD* is a parallelogram	Given
2. $\hat{A}=\hat{C}$ and $\hat{B}=\hat{D}$	The opposite angles of a parallelogram are equal
3. $\hat{A}+\hat{C}=180°$ and $\hat{B}+\hat{D}=180°$	The opposite angles of a cyclic quadrilateral are supplementary
4. $2\hat{A}=2\hat{C}=180°$ and $2\hat{B}=2\hat{D}=180°$	From the statements (2) and (3)
5. $\hat{A}=\hat{C}=90°$ and $\hat{B}=\hat{D}=90°$	From the statement 4
6. *ABCD* is a rectangle	All angles are right angles

23. Prove that a rhombus inscribed in a circle is a square.
Given: A rhombus *ABCD* inscribed in a circle.
To prove: *ABCD* is a square ($A = B = C = D = 90°$).

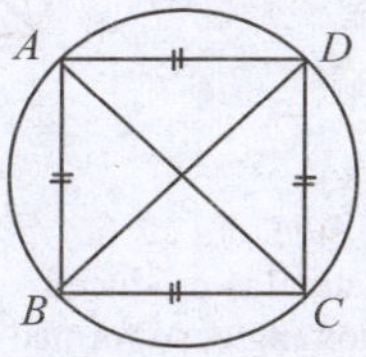

Proof

Statements	Reasons
1. *ABCD* is a rhombus inscribed in the circle	Given
2. $AB = BC = CD = DA$	All sides of a rhombus are equal
3. $\hat{A}=\hat{C}$ and $\hat{B}=\hat{D}$	The opposite angles of a parallelogram are equal
4. $\hat{A}+\hat{C}=180°$ and $\hat{B}+\hat{D}=180°$	The opposite angles of a cyclic quadrilateral are supplementary
5. $2\hat{A}=2\hat{C}=180°$ and $2\hat{B}=2\hat{D}=180°$	From statements (3) and (4)
6. $\hat{A}=\hat{C}=90$ and $\hat{B}=\hat{D}=90°$	From statements (5)
7. *ABCD* is a square	From statements (2) and (6)

24. In the given figure, *AOB* and *COD* are two perpendicular diameters. Chords *CPQ* and *CSR* cut *AB* at *P* and *S*, respectively. Prove that the points *P*, *Q*, *R*, and *S* are concyclic.
Let 1, 2, 3, 4, and 5 be the angles, as shown in the figure.

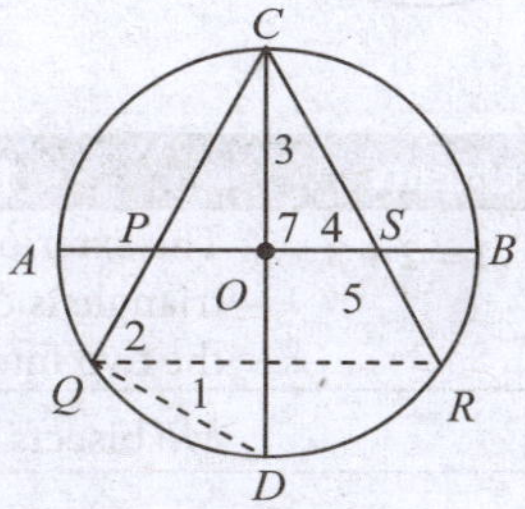

Proof

Statements	Reasons
1. Draw *QR* and *QD*	Construction
2. $\hat{1}=\hat{3}$	Angles in the same segment are equal
3. $\hat{1}+\hat{2}=90°$	The angle in a semicircle is a right angle
4. $\hat{3}+\hat{4}=90°$	In any right angled triangle, two acute angles are complementary
5. $\hat{3}+\hat{4}=\hat{1}+\hat{2}$	From statements (2) and (4)
6. $\hat{2}=\hat{4}$	From statements (2) and (5)
7. $\hat{4}+\hat{5}=180°=\hat{2}+\hat{5}$	Linear pair $\hat{4}$, $\hat{5}$
8. Quadrilateral *PQRS* is cyclic	The sum of the opposite angles is 180°
9. The points *P*, *Q*, *R*, and *S* are concyclic	From statement (7)

25. Prove that the angle bisectors of the angles formed by producing the sides of a cyclic quadrilateral (provided they are not parallel) intersect at right angles.
Given: A cyclic quadrilateral *ABCD* in which sides *BC* and *AD* are produced to meet at *E* and sides *AB* and *DC* are produced to meet at *F*. The bisectors of *E* and *F* meet at *L*. $\hat{3}=\hat{4}$ and $\hat{9}=\hat{10}$.

To prove: $\hat{7} = \hat{8} = 90°$

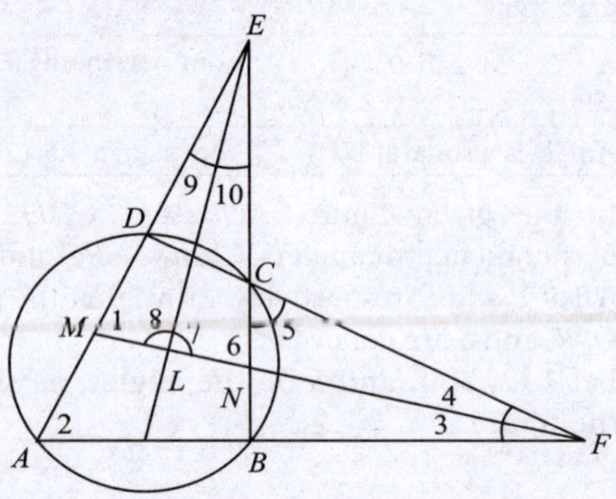

Proof

Statements	Reasons
1. In ΔAMF, $\hat{1} = \hat{2} + \hat{3}$	The exterior angle of a triangle is equal to the sum of the two interior opposite $\hat{S}$
2. $\hat{3} = \hat{4}$,	*MF* bisects $A\hat{F}D$
3. $\hat{1} = \hat{2} + \hat{4}$	From statements (1) and (2)
4. In cyclic quadrilateral *ABCD*, $\hat{5} = \hat{2}$	The exterior angle of a cyclic quadrilateral is equal to interior opposite angle
5. $\hat{1} = \hat{5} + \hat{4}$	From statements (2) and (4)
6. In ΔCNF, $\hat{6} = \hat{5} + \hat{4}$	The exterior angle of a triangle is equal to the sum of the two interior opposite angles
7. $\hat{6} = \hat{1}$	From statements (5) and (6)
8. In triangles *EML* and *ELN*, $\hat{1} = \hat{6}$, $\hat{9} = \hat{10}$, and $\hat{8} = \hat{7}$	From statement (7). If two angles of two triangles are equal, then the third angle will be equal
9. $\hat{8} + \hat{7} = 180°$	They are L.P.A.
10. $\hat{8} = \hat{7} = 90°$	From statements (8) and (9)

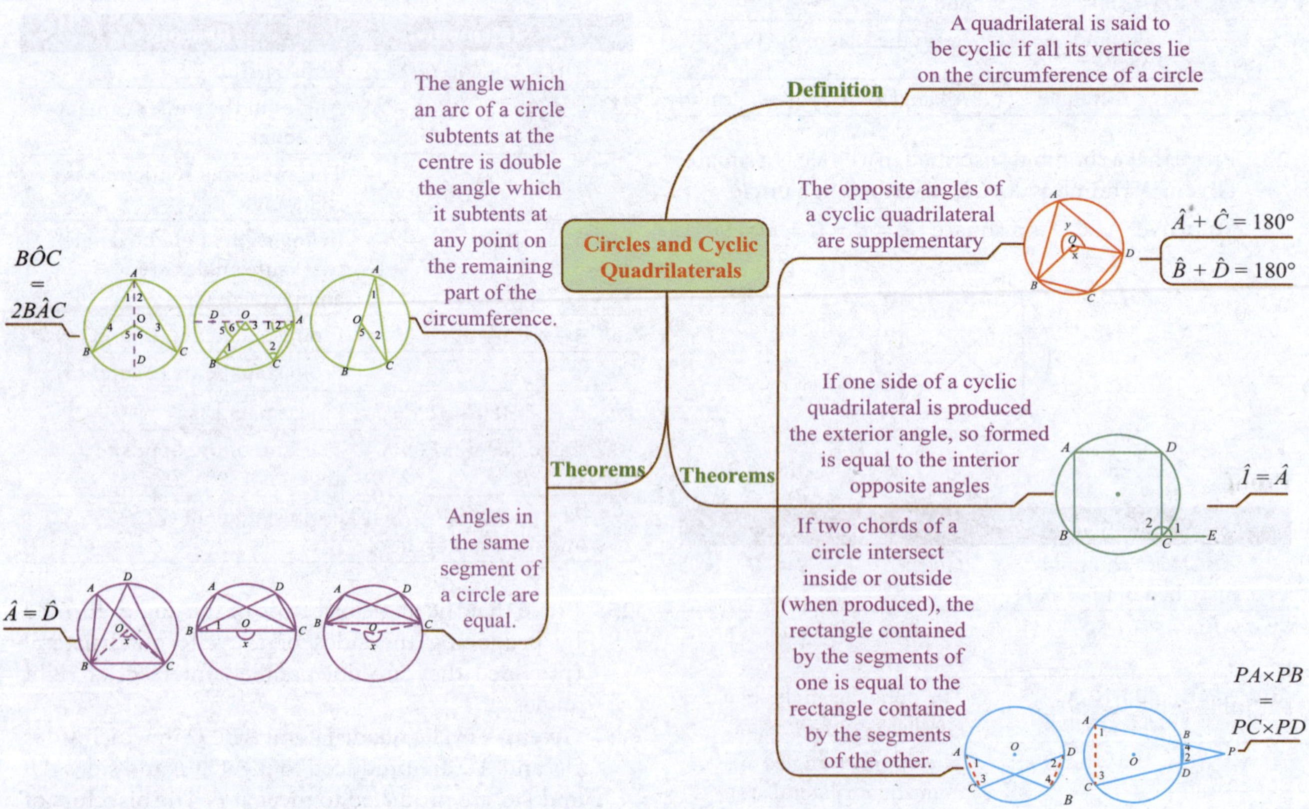

TEST YOURSELF

Level 1

1. O is a point outside a circle. Two equal lines OP and OQ are drawn touching the circle at point Q. Prove that the bisector of the angle POQ will pass through the centre of the circle.
2. A quadrilateral $ABCD$ is inscribed in a circle and $AB = CD$, prove that $AC = BD$.
3. In a circle equal chords PQ, QR, RS, and ST are drawn. Prove that $PR = QS = RT$.
4. In the given figure, AD is divided into three equal parts and all the arcs are semicircles. Show that four curved lines that connect A with D are of equal length.

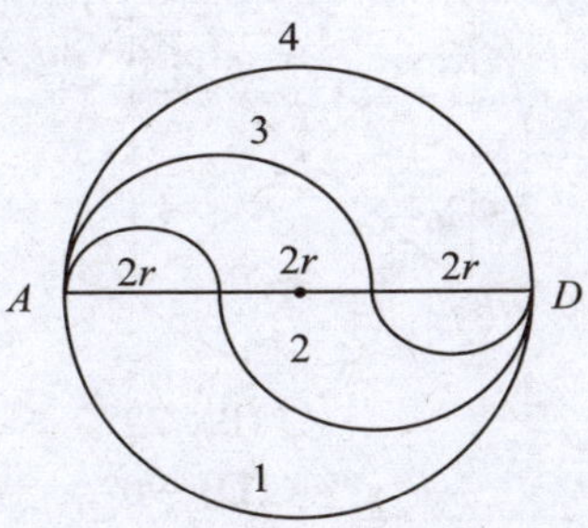

5. $ABCDEF$ is an irregular hexagon inscribed about a circle.
Prove that $AB + CD + EF = BC + DE + FA$.

Level 2

6. AD is perpendicular to the base BC of ΔABC. AE is a diameter of the circumcircle. Show that angles of the ΔABD and ΔAEC are equal (i.e., they are equiangular), and also ΔACD and ΔAEB are equiangular.
7. BE and CF are altitudes of a triangle ABC. Prove that $BFEC$ are concyclic.
8. $ABCD$ is a quadrilateral inscribed in a circle and $\hat{A} = 60°$, O is the centre of the circle. Prove that $O\hat{B}D + O\hat{D}B = C\hat{B}D + C\hat{D}B$.
9. $ABCD$ is a parallelogram. A circle drawn through A, B cuts AD, BC (produced if necessary) at and F, respectively. Prove that E, F, C, and D are concyclic.
10. In the given figure, $PQRS$ is a cyclic quadrilateral. $R\hat{Q}N = 50°$ and $a = 3\beta$. Find the value of β.

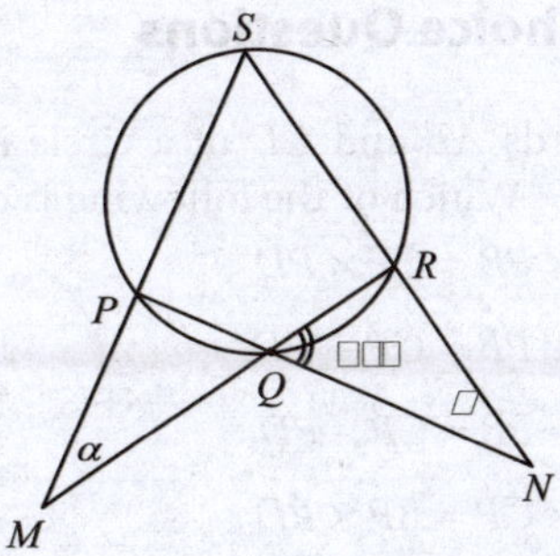

11. In the figure, angle $A > 90°$. PB is perpendicular to AC and QC is perpendicular to AB. Prove that $BA \times AQ = CA \times AP$.

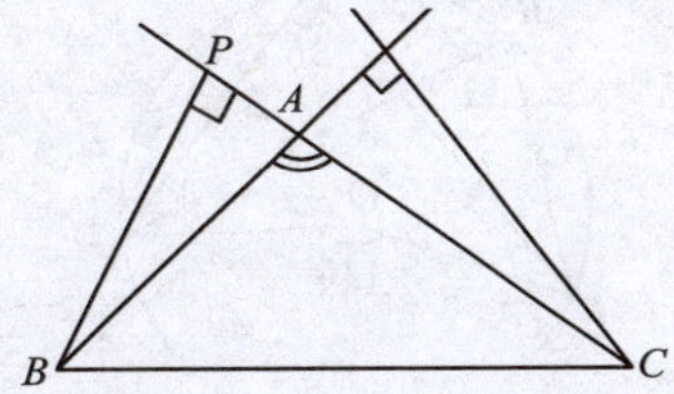

12. In the given figure, O and OO' are centres of the two congruent circles (of radius r) cutting each other at A and B. PBQ is a line segment through B, and SBT is tangent to the circle with centre OO', intersecting the other circle at S. Prove that $SP = BQ$.

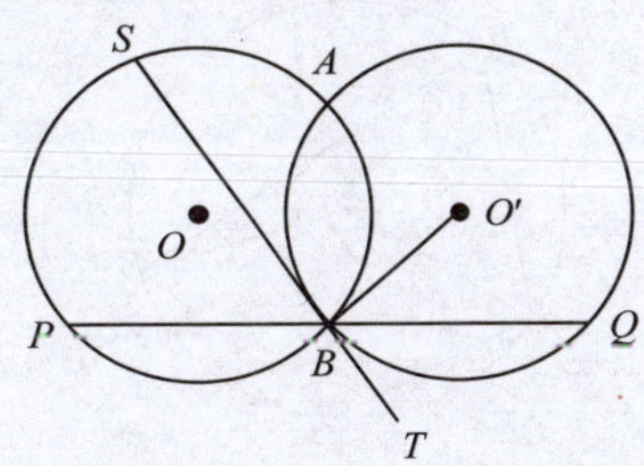

13. Prove that circles on any two sides of a triangle as diameters intersect on the third side.
14. If a pair of opposite sides of a cyclic quadrilateral are equal, then prove that the other two sides are parallel.
15. In the following figure, O is the centre of a circle, show that $B\hat{O}C = 2[A\hat{B}C + A\hat{C}B]$.

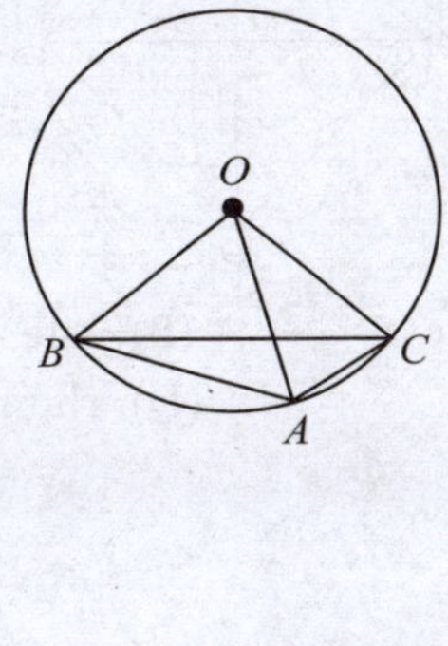

Multiple Choice Questions

16. Two chords AB and CD of a circle intersect each other at P. Which of the following is correct?

(A) $AP \times PB = CP \times PD$

(B) $AP + PB = CP + PD$

(C) $AP - PB = CP - PD$

(D) $AP \times CP = BP \times PD$

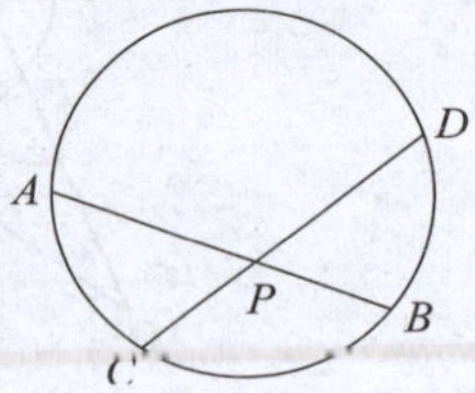

17. In the given figure, which of the following is correct (O is the centre of circle)?

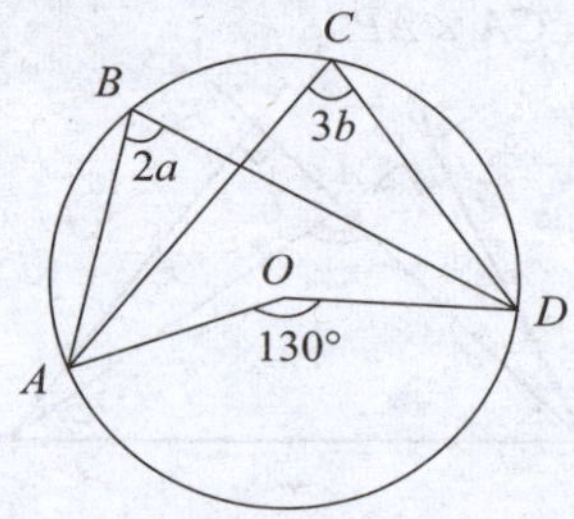

(A) $a = b$ (B) $3b = 2a$

(C) $2b = 3a$ (D) None of these

18. In the given figure, value of y (O is the centre of the circle) is

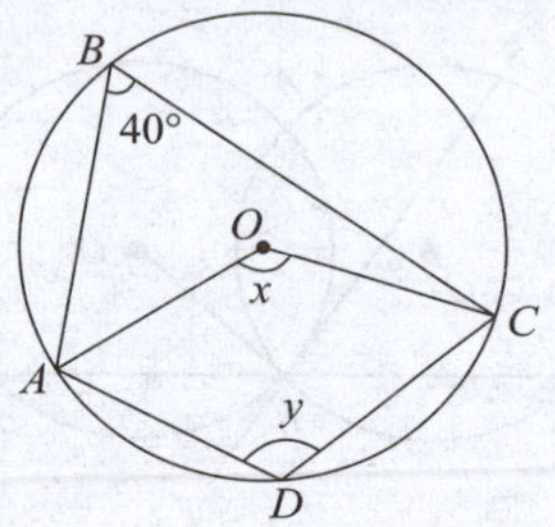

(A) 100° (B) 120°

(C) 130° (D) 140°

19. If the given figure, O is centre of the circle and $AP = PB$ then $APBQ$ is a

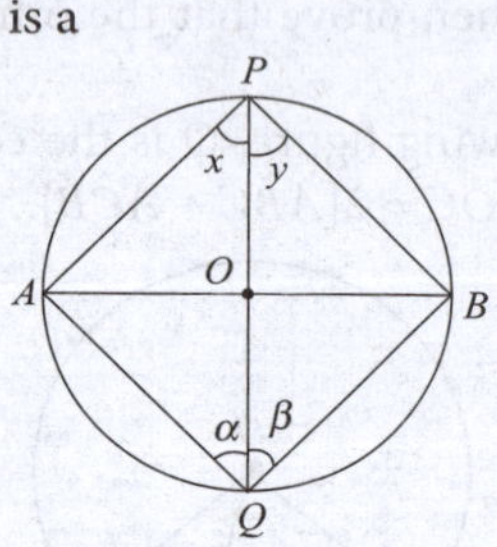

(A) rectangle (B) kite

(C) square (D) rhombus

20. In the given figure, find the value of x, if it is given that O is the centre of circle.

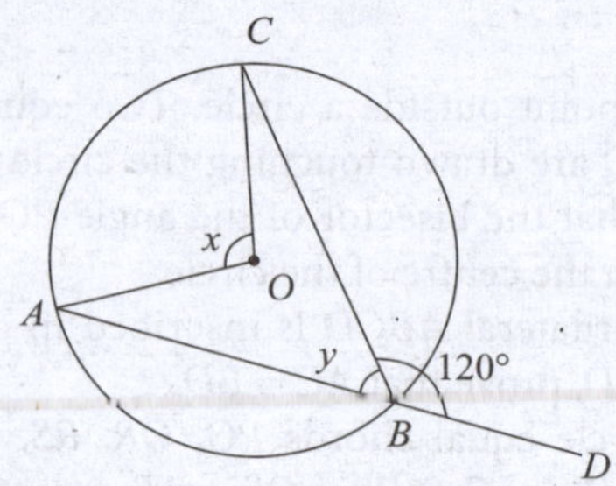

(A) 120° (B) 60°

(C) 100° (D) 75°

21. In the given figure, the value of β is

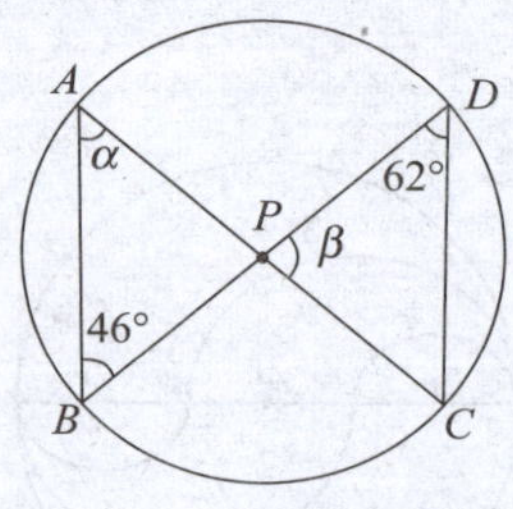

(A) 36° (B) 72°

(C) 54° (D) 90°

22. In the given figure, the value of $\angle ABC$ is

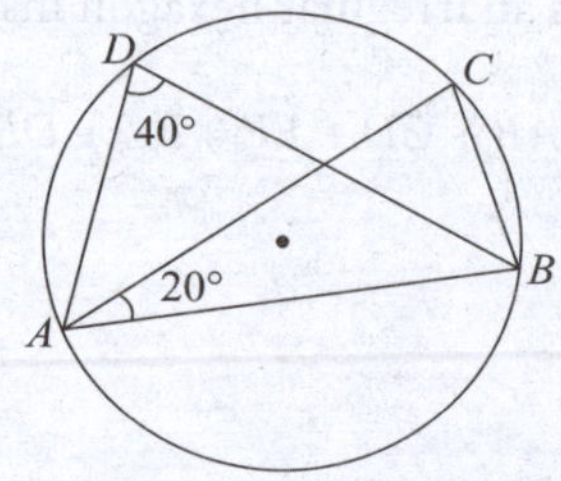

(A) 60° (B) 80°

(C) 100° (D) 120°

23. In the figure, $\angle ACD = 130°$. Then $\angle COE =$

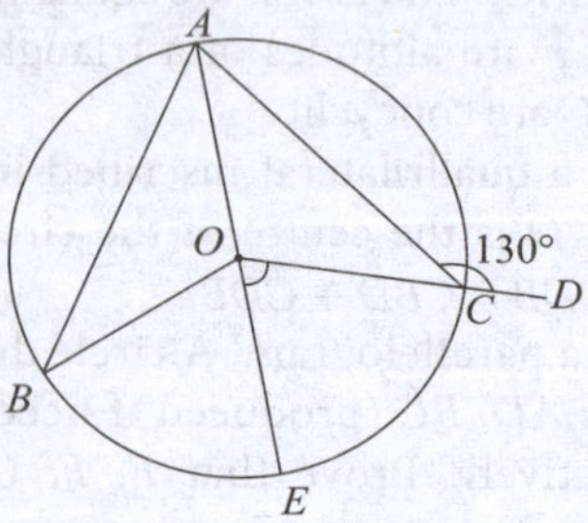

(A) 80° (B) 50°

(C) 70° (D) 100°

24. In the given figure, O is the centre of circle. What is the measure of $\angle ADB$?

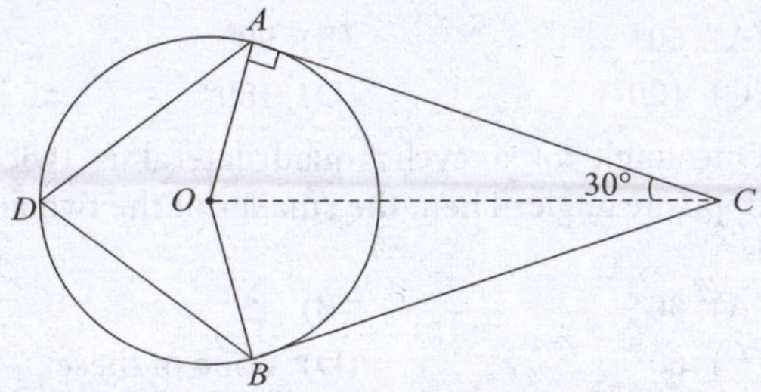

(A) 60° (B) 120°
(C) 30° (D) 45°

25. If O is the centre of the circle then find $\angle ADC$.

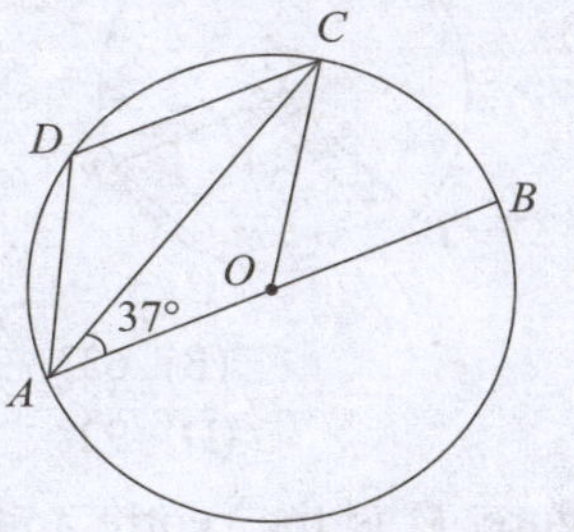

(A) 125° (B) 127°
(C) 90° (D) 137°

26. In the given figure, O is the centre of the circle. Then $\angle AED$ measures

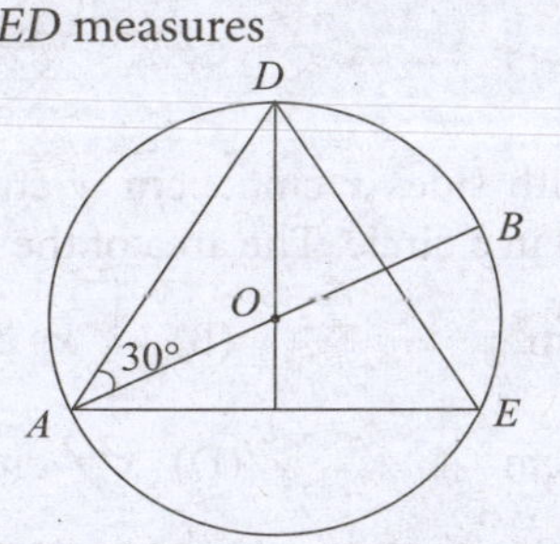

(A) 30° (B) 45°
(C) 60° (D) 90°

27. In the given figure, the measure of $\angle ADB$ is

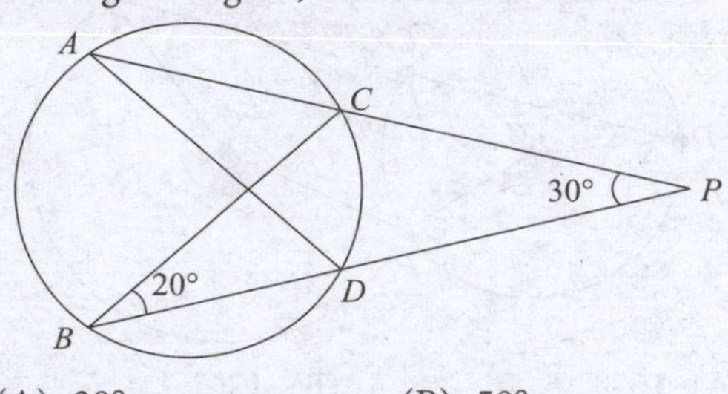

(A) 30° (B) 50°
(C) 80° (D) 70°

28. In the given figure (if O is the centre of circle), $\angle COE$ measures

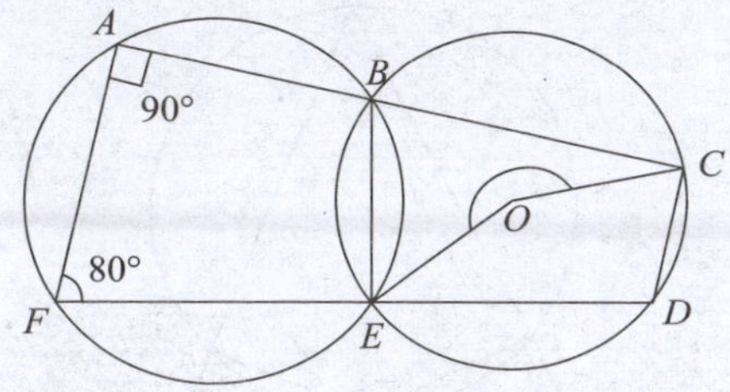

(A) 100° (B) 200°
(C) 150° (D) 50°

29. What is the angle $\angle AED$ if O is the centre of the circle?

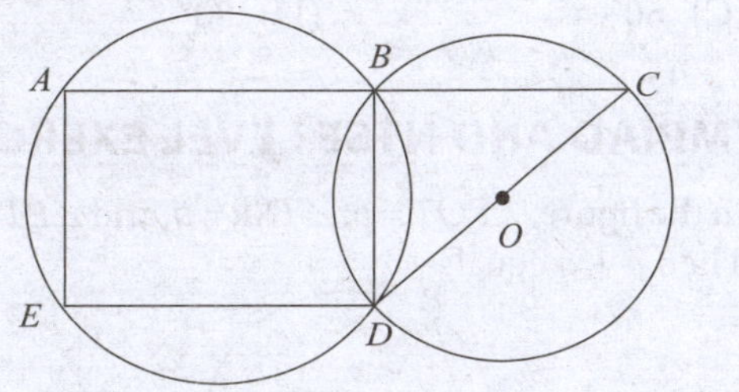

(A) 90° (B) 45°
(C) 120° (D) None of these

30. If O is the centre of circle and $\angle ADE = 60°$ then find $\angle AOC$.

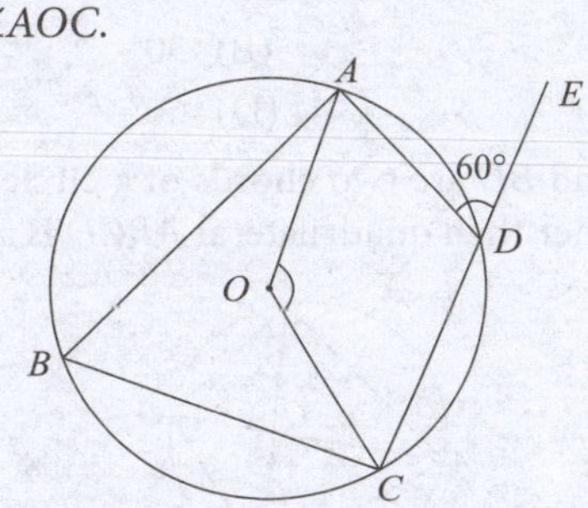

(A) 60° (B) 100°
(C) 120° (D) None of these

31. In the given figure, $BCEF$ is a

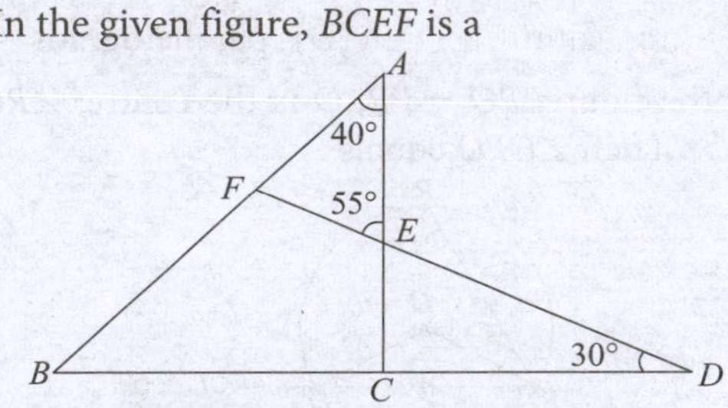

(A) cyclic quadrilateral (B) rhombus
(C) rectangle (D) None

32. If $\angle RQN = 50°$ and $\alpha = 3\beta$ then the value of β is

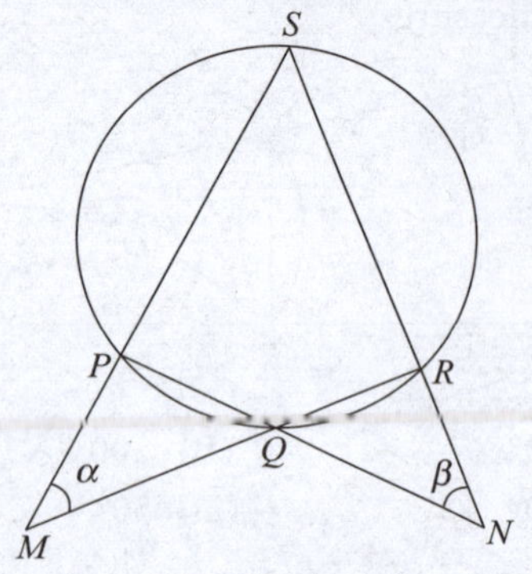

(A) 40° (B) 20°
(C) 60° (D) 80°

OLYMPIAD AND NTSE LEVEL EXERCISES

33. In the figure, $\angle PQT = p$, $\angle TSR = q$, and $\angle PTQ = 60°$. Then $p + q$ equals

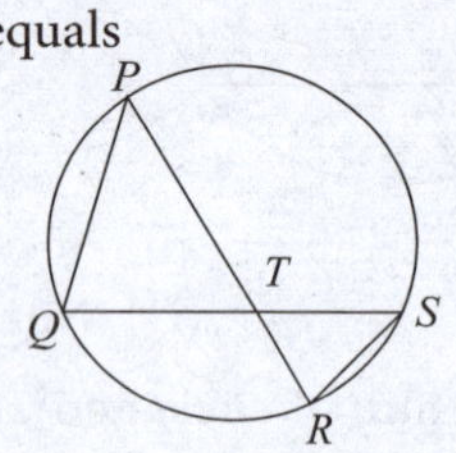

(A) 300° (B) 30°
(C) 120° (D) 60°

34. If AC and BD are two chords of a circle that bisect each other then quadrilateral $ABCD$ is a

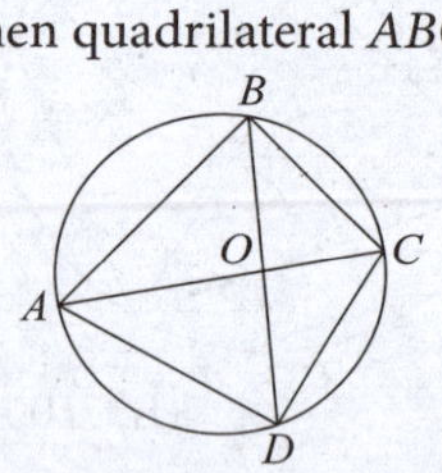

(A) rectangle (B) square
(C) trapezium (D) parallelogram

35. In the figure, $PQ = PR$, O is the centre, $\angle RQT$ = [illegible]°. Then $\angle OPQ$ equals

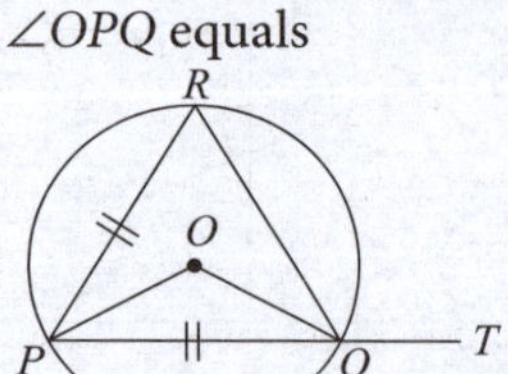

(A) 70° (B) 140°
(C) 40° (D) 20°

36. If an equilateral triangle ABC is inscribed in a circle with centre O then the measure of $\angle AOB$ is

(A) 60° (B) 90°
(C) 120° (D) 180°

37. One angle of a cyclic quadrilateral is twice its opposite angle. Then, the smaller of the two angles is

(A) 30° (B) 45°
(C) 60° (D) none of these

38. In the figure, O is the centre of the circle. $\angle PQR$ is equal to

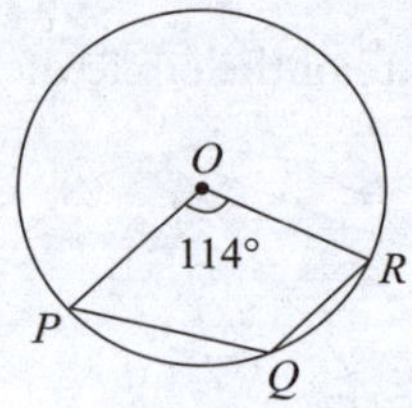

(A) 57° (B) 63°
(C) 123° (D) 92°

39. In the figure, O is the centre and $\angle ACB = 20°$. $\angle AOD$ is equal to

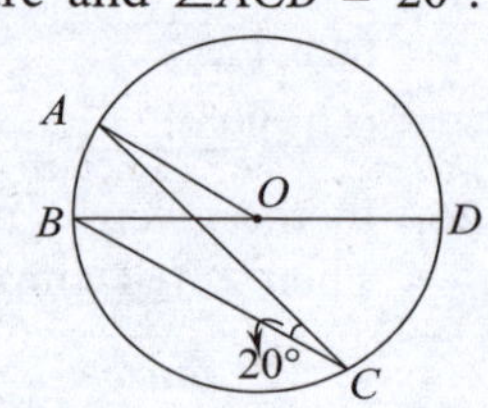

(A) 140°
(B) 160°
(C) 120°
(D) 130°

40. A kite with sides x cm, x cm, y cm, and y cm is inscribed in a circle. The area of the kite is

(A) xy cm^2 (B) $\frac{1}{2}xy$ cm^2
(C) $2xy$ cm^2 (D) x^2y^2 cm^2

41. In the figure, A and C are the centres of the circles. DEB is a straight line. If reflex angle $BAF = 208°$ then obtuse angle DCF =

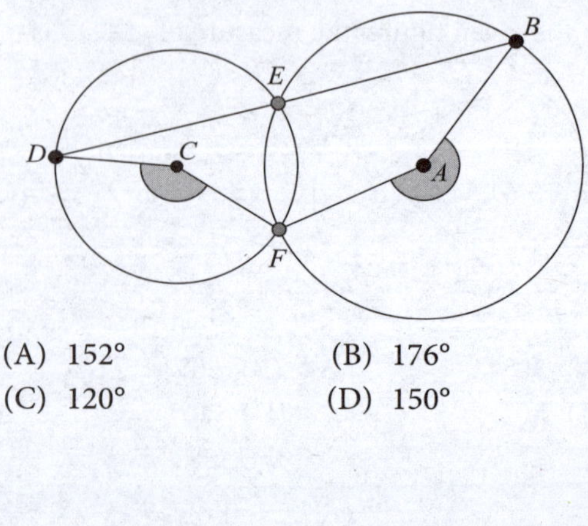

(A) 152° (B) 176°
(C) 120° (D) 150°

42. Two concentric circles have their radii as a and b ($a > b$). PQ is the diameter of the larger circle. From Q, a tangent is drawn to smaller circle which touches it at R. Length of PR is

(A) $\sqrt{b^2+a^2}$ (B) $\sqrt{3b^2+a^2}$

(C) $2\sqrt{a^2-b^2}$ (D) $\sqrt{a^2-b^2}$

GAME TIME

You know that it is not possible to find the exact value of π. But many mathematicians and other enthusiasts have calculated the approximate value of π correct to a number of decimal places. The following simple msethod known as Buffon's needle experiment is an experimental method based on the theory of probability.

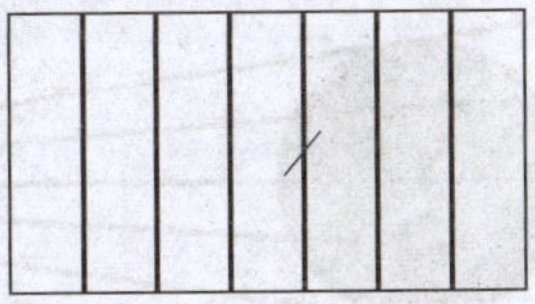

This is a pair activity.

Materials required:

A large sheet of paper and a needle of length 5 cm.

Procedure:

On a drawing sheet (full size) draw equidistant (5 cm gap) parallel lines till the page is filled. Place the sheet on the floor. One student will drop a needle from a height of about a metre on the sheet and see whether it crosses a line or falls in between the lines without touching any line. The other student should note down on a sheet of paper the number of trials and the number of times the needle crosses or touches a line. The needle has to be dropped a minimum of 100 times.

Tabulate the results as follows:

Trial No.	Crossed?
1	Yes
2	Yes
3	No
etc.	

Calculate the quotient = Q

$$= \frac{\text{Number of trials}}{\text{Number of times needle crossed}}$$

Approximate value of $\pi = 2Q$.

14 Tangents to Circles

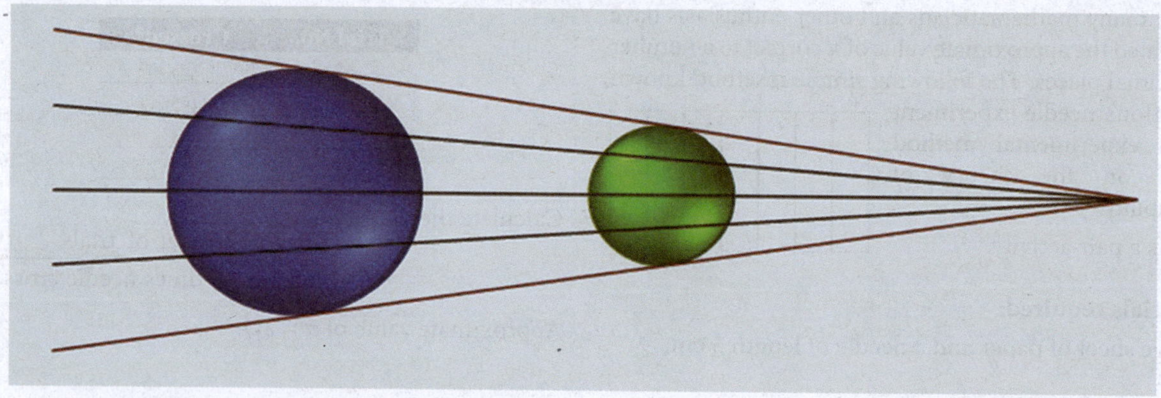

What is the figure formed by all the tangents common to two spheres of different sizes?

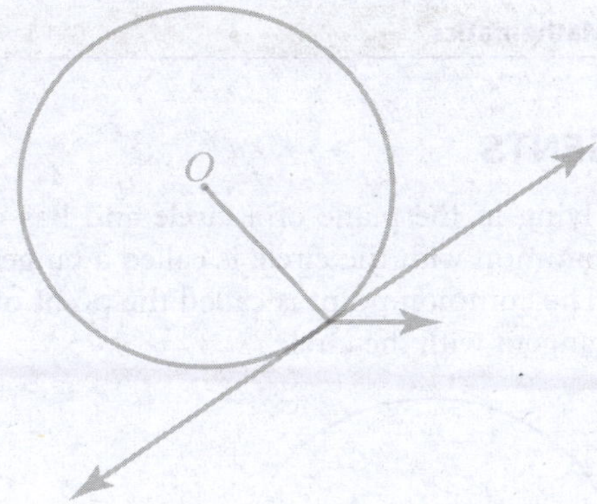

OBJECTIVES

After reading this chapter you should be able to

- learn about tangents to circles, touching circles
- learn about common tangents to two circles
- learn about sector and segment of a circle
- state and prove theorems on tangents to circles
- solve riders based on theorems on tangents to circles

TANGENTS

A line lying in the plane of a circle and has only one point common with the circle is called a tangent to the circle. The common point is called the point of contact of the tangent with the circle.

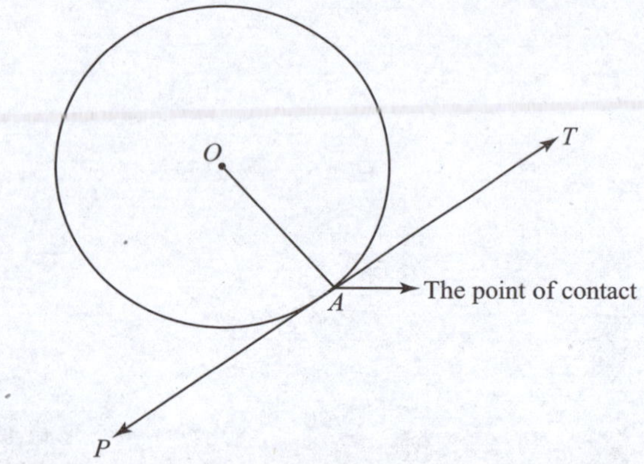

TOUCHING CIRCLES

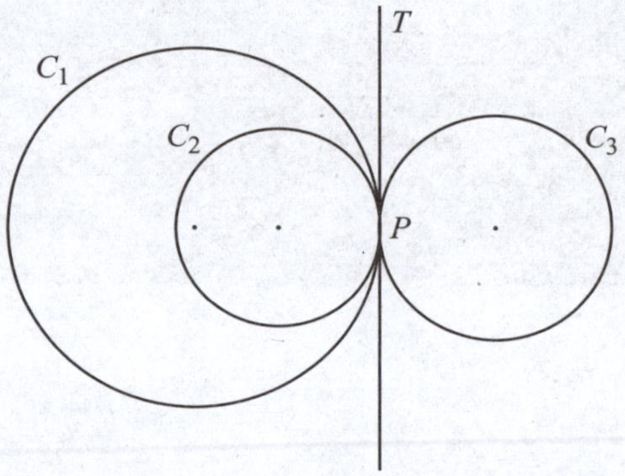

Two circles are said to be touching circles if they are coplanar and have one line as tangent.

Circles C_1 and C_2 are said to touch internally if their centres lie on the same side of the common tangent PT. Circles C_1 and C_3 are said to touch externally if their centres lie on opposite sides of the common tangent.

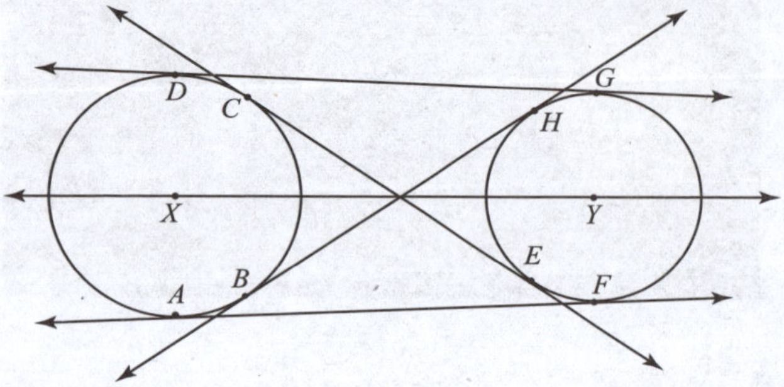

Euclid

The Greek mathematician Euclid (300 BC) wrote the *Elements*, a collection of geometrical theorems. The oldest major mathematical work in the Western world, it set a standard for logical exposition for over 2000 years.

Virtually nothing is known of Euclid's personality. It is not even known for certain whether he was a creative mathematician himself or was simply good at compiling the work of others. Most of the information about Euclid comes from Proclus, a Greek scholar of the 5th century AD. Because Archimedes refers to Euclid and Archimedes lived immediately after the time of Ptolemy I, King of Egypt (306–283 BC), Proclus concluded that they were contemporaries. Euclid's mathematical education may well have been obtained from Plato's pupils in Athens, since it was there that most of the earlier mathematicians upon whose work the *Elements* is based had studied and taught.

Ptolemy

Little is known about Ptolemy's life and education, other than that he lived and worked in Alexandria, one of the primary centres of Greek culture in late antiquity, and that he has no genealogical relationship whatsoever to the Pharaoh dynasty bearing the same name (as believed by many in medieval times). He wrote on many topics, including geography, astrology, musical theory, optics, physics, and of course astronomy. He died most probably between AD 141 and 151. His landmark astronomical work is his mathematical compilation (or Syntaxis, as Prolemy himself referred to it), better known under the name Amalgest, given to it by its later Arabic translators and commentators. By making extensive use of the geometrical constructions known as epicycles and equants, Ptolemy constructed a mathematical model of planetary motion that did far better at predicting planetary positions than anything else produced in antiquity. It is the fusion of Ptolemy's model with the cosmology and physics of Aristotle that was to be adopted in the late antiquity and medieval Western world, and literally defines mankind's view of the universe for over a millennium.

Ptolemy was also an accomplished geographer. His maps of Asia and Africa are said to have inspired Christopher Columbus, many centuries later, in his westward expedition to India across the Atlantic.

The above figure illustrates the circles and several lines, all in one plane. The line that passes through the centres of two circles is called the line of centres (*XY* is the line of centres). A common tangent line such as *BH* or *CE* that intersects the *line of centres* at a point between the centres is a common internal tangent or *transverse common tangent* (T.C.T.). A common tangent line such as *AF* or *DG* that does not intersect the line of centres between the centres of the circles is a common external tangent or *direct common tangent* (D.C.T.)

SECTOR AND SEGMENT OF A CIRCLE

A sector of a circle is the part of the interior of a circle bounded by two radii and an arc. *AOB* is a sector of the circle with centre *O* (Fig. 14.1).

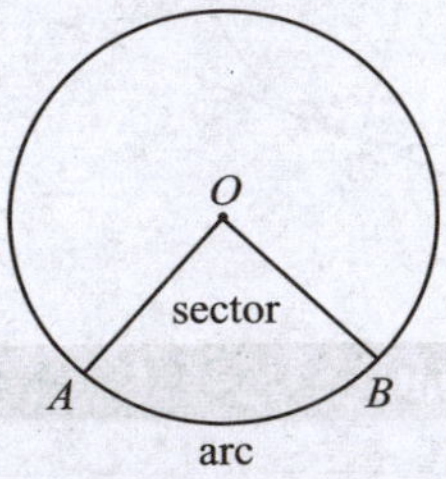

Fig. 14.1.

A segment of a circle is a part of a circle bounded by a chord and its arc.

In Fig. 14.2, *AXB* is the segment of the circle with centre *O*.

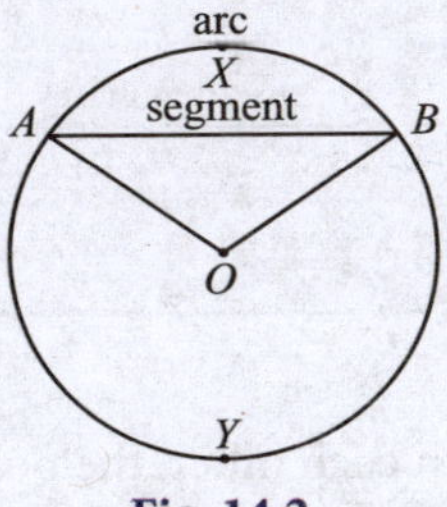

Fig. 14.2.

Major Segment

A segment of a circle which contains the centre is called a major segment. In Fig. 14.2, *AYB* is the major segment (greater part of the circle).

Minor Segment

A segment of a circle which does not contain the centre is a minor segment. In Fig. 14.2, *AXB* is the minor segment.

Alternate Segment

A straight line *APB* touches a circle at *P* and *PQ* is a chord through the point of contact *P*.

With respect to $B\hat{P}Q$, the segment *PMQ* which lies on the other side of the chord *PQ* is called the alternate segment. Similarly, with respect to $A\hat{P}Q$, the segment *PNQ* is called the alternate segment.

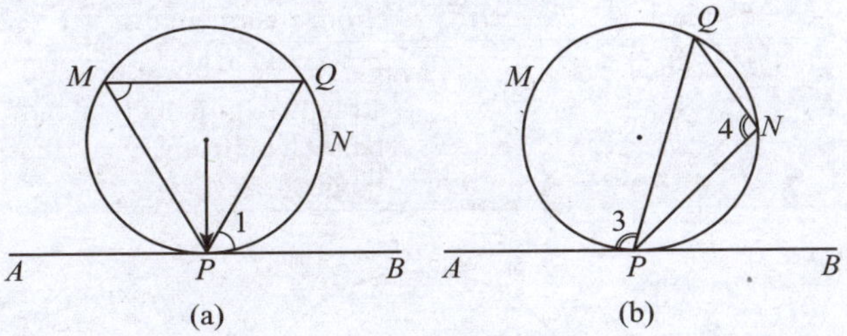

Fig. 14.3.

A straight line *AB* touches a circle at *P* and *PQ* is a chord through *P*.

Then $B\hat{P}Q$ and $Q\hat{P}A$ are the two angles formed by the chord *PQ* with the tangent at *P* (one end point of chord *PQ*).

In Fig. 14.3(a), $Q\hat{M}P$ is an angle in the alternate segment *PMQ* corresponding to $Q\hat{P}B$.

In Fig. 14.3(b), $Q\hat{N}P$ is an angle in the alternate segment *PNQ* corresponding to $Q\hat{P}A$.

By measurement, we find that $Q\hat{P}B = Q\hat{M}P$ in Fig.14.3(a) and $Q\hat{P}A = Q\hat{N}P$ in Fig. 14.3(b).

THEOREMS ON TANGENTS TO CIRCLES

Theorem 1

The tangent at any point of a circle and the radius through the point of contact are perpendicular to one another.

Given: A circle with centre *O*. *APB* is a tangent at *P*. *OP* is the radius through the point of contact *P*.

To prove: $OP \perp AB$.

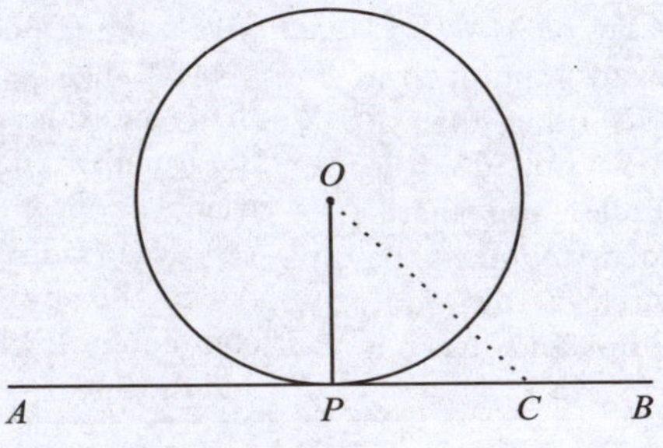

Proof

Statements	Reasons
1. Choose any point C on AB other than P and draw OC	Construction
2. APB is a tangent to the circle	Given and definition of the tangent
3. C lies outside the circle	Always $OC > OP$
4. $OP < OC$	From statement 3
5. $OP \perp AB$ at P	OP is the shortest of all straight segments that can be drawn from O to the tangent AB

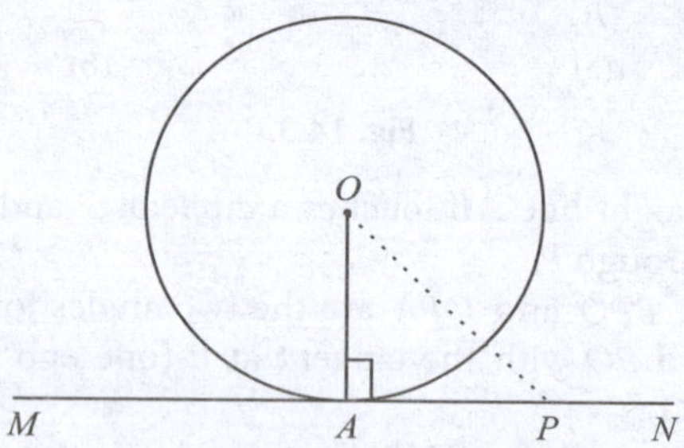

Theorem 2

The line drawn perpendicular to a radius of a circle at its extremity is a tangent to the circle.

Given: A circle with centre O. OA is the radius. Line MN is perpendicular to OA at A.

To prove: MAN is a tangent.

Proof

Statements	Reasons
1. Choose any point P on MN and join OP	Construction
2. In DOAP (a) $O\hat{A}P = 90°$ (b) $O\hat{A}P > O\hat{P}A$ (c) $OP > OA$ (d) P lies outside the circle (e) A lies on MAN (f) Every point on MN other than A lies outside the circle, and therefore MN meets the circle at only one point and hence a tangent at A	 (a) $OA \perp MN$ (b) In any right-angled triangle, the two acute angles are complementary (c) The side opposite the greater angle is greater than the side opposite the smaller angle in a triangle. (d) $OP > OA$ (radius) (e) A is the only point common to OA, circle, and the line

Theorem 3

If a pair of tangents is drawn to a circle from an external point,

(a) they are equal in length,
(b) they subtend equal angles at the centre, and
(c) they are equally inclined to the line joining the centre and to the external point.

Given: A circle with centre O. P is a point outside the circle. PA and PB are the pair of tangents drawn from P.

To prove: $PA = PB$, $\hat{1} = \hat{2}$, and $\hat{3} = \hat{4}$.

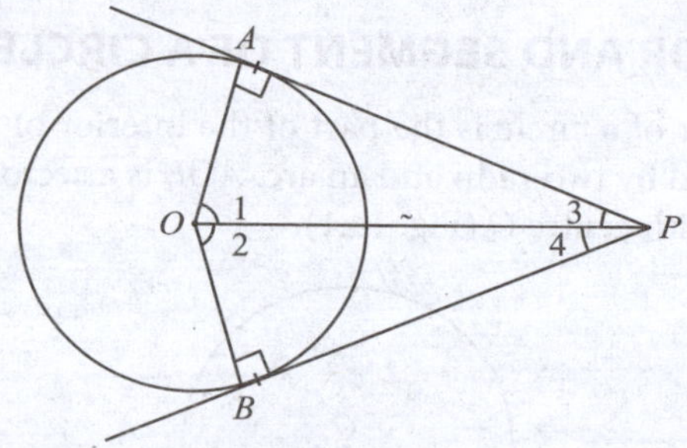

Proof

Statements	Reasons
1. OA, OB, and OP are drawn In ΔOAP and DOBP (a) $\hat{A} = \hat{B} = 90°$ (b) $OA = OB$ (c) $OP = OP$	Construction (i) Tangent and radius are perpendicular to each other at the point of contact (ii) Radii of the same circle are equal (iii) Identity
2. $\Delta OAP \cong \Delta OBP$	RHS
3. $PA = PB$, $\hat{1} = \hat{2}$, and $\hat{3} = \hat{4}$	

Theorem 4

If two circles touch each other, the point of contact and the centres of the circles are collinear.

Given: Two circles with centres A and B touching each other at P externally (Fig. 14.4) and internally (Fig. 14.5).

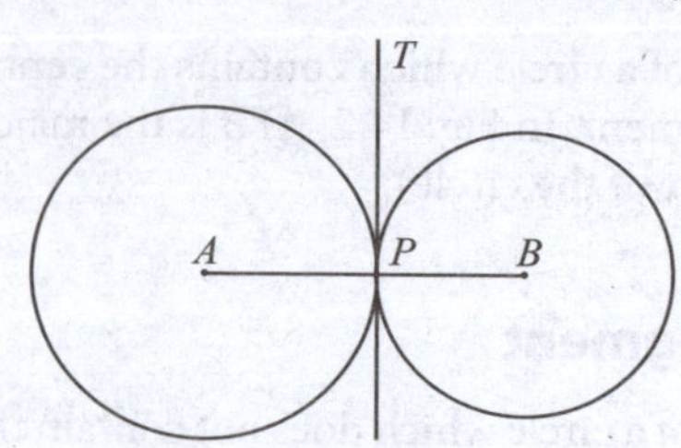

Fig. 14.4.

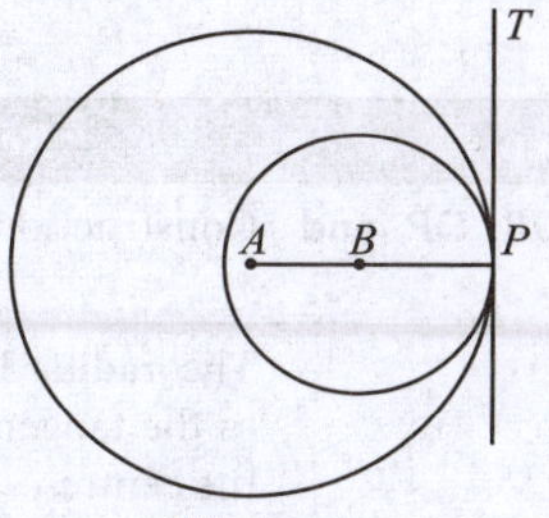

Fig. 14.5.

To prove: A, P, and B are collinear.

Proof

Statements	Reasons
1. Join *PA* and *PB*; draw a common tangent *PT* to the circle at *P*	Construction
2. $A\hat{P}T = 90°$ and $B\hat{P}T = 90°$	Radius perpendicular to the tangent at the point of contact
3. In Fig. 14.4, $A\hat{P}T + B\hat{P}T = 180° = A\hat{P}B$	Adding in statement 2
4. *APB* is a straight line	A straight line contains 180°
5. For Fig. 14.5, $T\hat{P}A = T\hat{P}B$	Each one is a right angle
6. *PA* and *PB* are perpendicular; therefore, *PA* must coincide with *PB*	Only one perpendicular can be drawn from a point on a line to the line
7. *APB* is collinear	From statement 6

Theorem 5

If a line touches a circle and a chord is drawn from the point of contact, the angle which the chord makes with the tangent is equal to the angle in the alternate segment.

Given: A circle with centre O. QPT is a tangent to the circle at P, and PB is any chord through the point of contact P.

Let $B\hat{P}T = \hat{1}$, $B\hat{P}C = \hat{2}$, $B\hat{C}P = \hat{3}$, $B\hat{D}P = \hat{4}$, $B\hat{P}Q = \hat{5}$, and $B\hat{A}P = \hat{6}$.

To prove: $\hat{1} = \hat{4}$ and $\hat{5} = \hat{6}$.

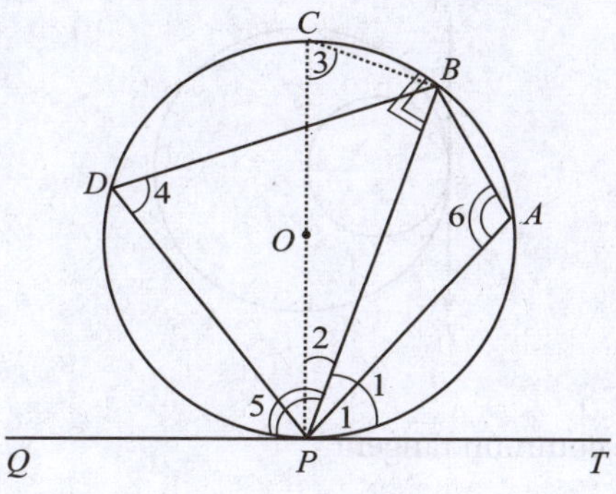

Proof

Statements	Reasons
1. Join *PO* and produce it to meet the circumference at *C* and join *B* and *C*	Construction
2. $C\hat{P}T = 90°$	Radius and tangent are perpendicular to each other at the point of contact
3. $\hat{1} + \hat{2} = C\hat{P}T = 90°$	
4. $P\hat{B}C = 90°$	The angle in a semicircle is a right angle (*POB* is a diameter)
5. $\hat{2} + \hat{3} = 90°$	The two acute angles in a right-angled triangle are complementary
6. $\hat{2} + \hat{3} = \hat{1} + \hat{2} \Rightarrow \hat{1} = \hat{3}$	From statements 3 and 5
7. $\hat{4} = \hat{3}$	Angles in the same segment are equal
8. $\hat{1} = \hat{4}$	From statements 6 and 7
9. *PABD* is a cyclic quadrilateral	*P*, *A*, *B*, and *D* lie on circumference of the circle
10. $\hat{4} + \hat{6} = 180°$	Opposite angles of a cyclic quadrilateral are supplementary
11. $\hat{1} + \hat{5} = 180°$	They are *LPA*
12. $\hat{5} = \hat{6}$	From statements 8, 10, and 11

Note: This theorem can be simply stated as follows: 'angles in the alternate segments are equal'.

Theorem 6

If through an extremity of a chord a circle on a straight line is drawn, making with the chord an angle equal to the angle in the alternate segment, then the line touches the circle.

Given: Chord PQ of a circle with centre O; straight line APB through P such that $Q\hat{P}B = Q\hat{R}P$ in the alternate segment. $Q\hat{P}B = Q\hat{R}P$, i.e., $\hat{1} = \hat{2}$.

To prove: APB is a tangent to the circle at P.

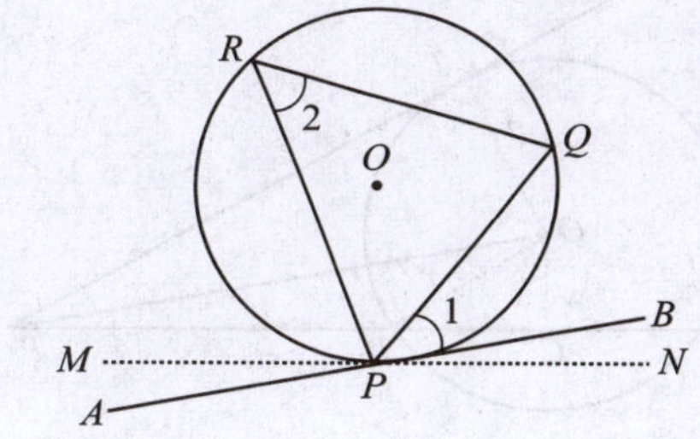

Proof

If APB is not a tangent, let MPN be a tangent to the circle.

$Q\hat{P}N = \hat{2}$ (angles in the alternate segment are equal), but $B\hat{P}Q = \hat{2}$ (given).

Therefore, $Q\hat{P}N = B\hat{P}Q$, which is impossible unless PN falls on PB. Hence, PB or APB is a tangent to the circle.

Theorem 7

If a secant and a tangent are drawn to a circle from an external point, the rectangle contained by the whole secant and the part of it outside the circle are equal to the square on the tangent.

Given: A circle with centre O. The tangent PT and the secant PBA at P are drawn to the circle.

To prove: $PT^2 = PA \times PB$.

Let $\hat{1}$, $\hat{2}$, and $\hat{3}$ be the angles as shown in the figure.

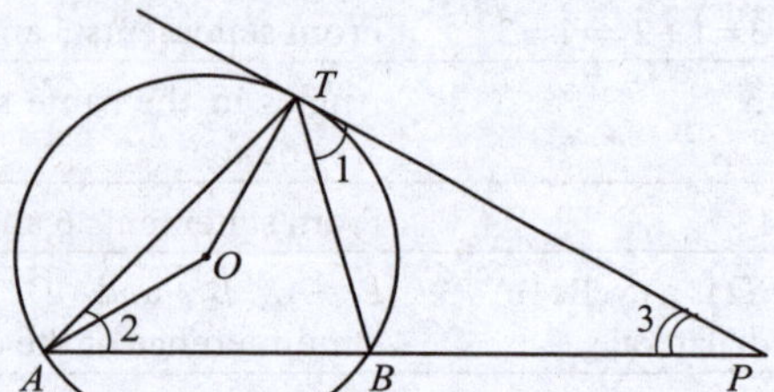

Proof

Statements	Reasons
1. Draw TA and TB	Construction
2. In ΔPTB and ΔPAT	
(a) $\hat{1} = \hat{2}$ (b) $\hat{3} = \hat{3}$	(a) Angles in the alternate segment are equal (b) Identity
3. $\Delta PTB \sim \Delta PAT$	AA criterion
4. $\frac{PT}{PA} = \frac{PB}{PT} \Rightarrow PT^2 = PA \times PB$	Corresponding sides of similar triangles are proportional

Aliter

(using Pythagoras theorem)

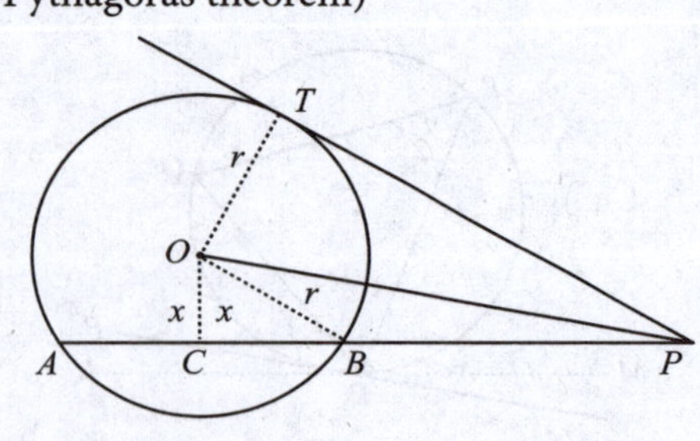

Proof

Statements	Reasons
1. Draw OT, OB, OP, and $OC \perp AB$	Construction
2. $O\hat{T}B = 90°$	The radius is perpendicular to the tangent at the point of the centre
3. $AC = CB = x$	Perpendicular to chord from the centre of the circle bisects the chord
4. $PT^2 = OP^2 - r^2$	From the right-angled ΔOTP
5. $PT^2 = OC^2 + CP^2 - r^2$	From right-angled ΔOCP, $OP^2 = OC^2 + CP^2$
6. $PT^2 = r^2 - x^2 + CP^2 - r^2 = CP^2 - x^2$	Form the right-angled ΔOCB
7. $PT^2 = (CP + x)(CP - x)$	Using $A^2 - B^2 = (A + B)(A - B)$
8. $PT^2 = PA \times PB$	$CP + x = CP + CA = PA$ and $CP - x = CP - CB = PB$

COMMON TANGENTS TO TWO CIRCLES

Note 1

(a) One circle lies inside the other

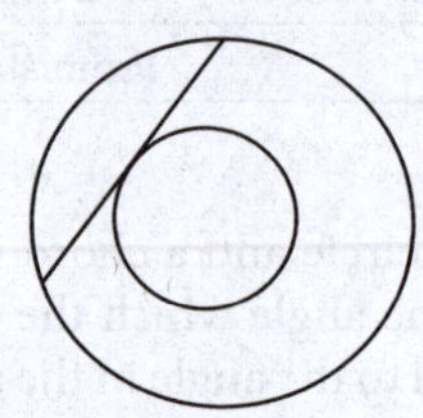

No common tangent

(b) Two circles touch internally

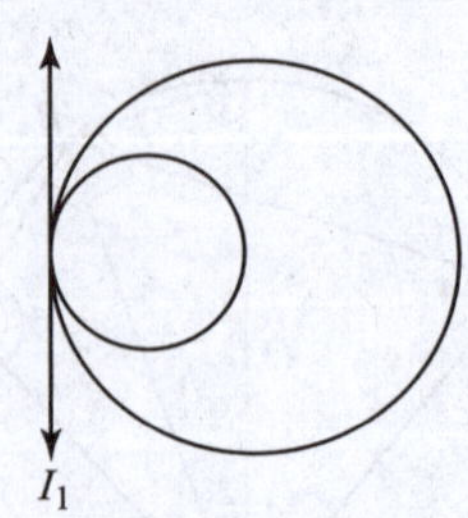

One common tangent

(c) Two circles intersect at two distinct points

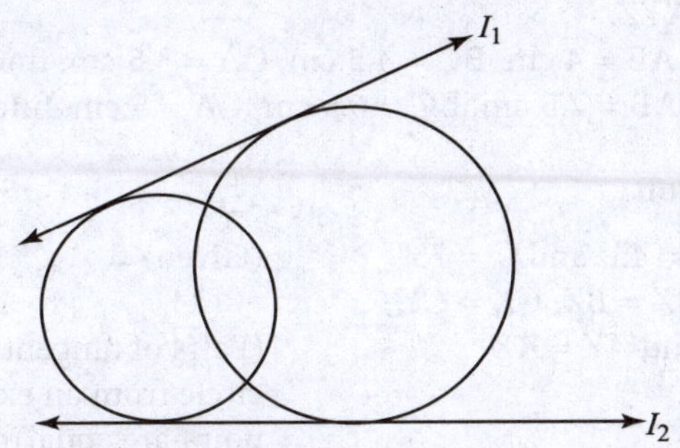

Two common tangents

(d) Two circles intersect at one distinct point

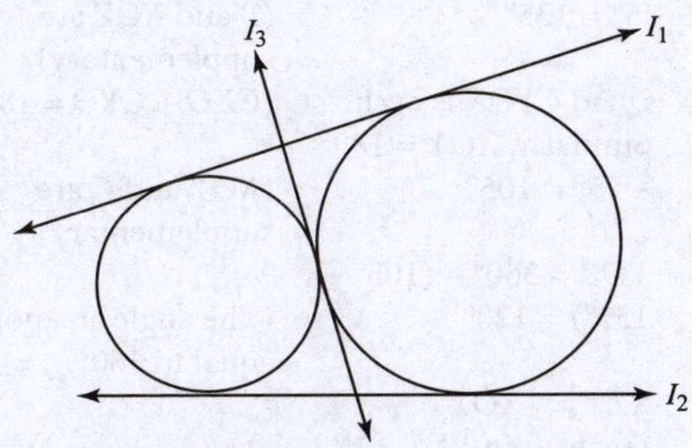

Three common tangents

(e) One circle lies entirely outside the other

(i) m_1 and m_2 are called direct (exterior) common tangents

(ii) m_3 and m_4 are called transverse (interior) common tangents

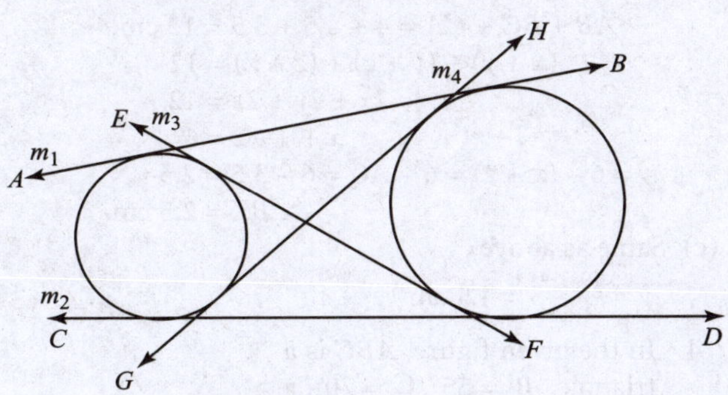

Four common tangents

Note 2

(a) Centres of similitude – the points of intersection of the direct common tangents, transverse common tangents, and the centres of the two circles (O, O', and A and A') – all are collinear.

(b) The points of intersection of the direct common tangent and the transverse common tangent to the two circles divide the segment joining the two centres externally and internally, respectively, in the ratio of their radii.

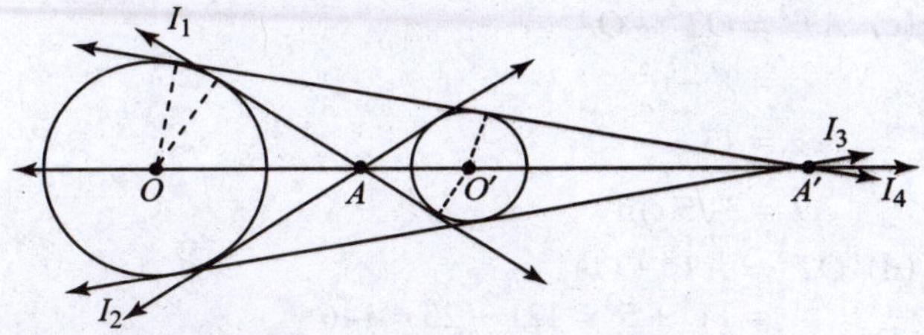

Worked Examples

1. In the figure, TAC is a tangent at A to the circle with centre O.
   (a) If $D\hat{A}C = 30°$, find $O\hat{A}D$ and $A\hat{O}D$.
   (b) If $A\hat{O}D = 60°$, find $D\hat{A}C$.
   (c) If $OA = 6$ cm and $OT = 9$ cm, find TA.
   (d) If $TA = 11$ cm and $OA = 5$ cm, find OT.

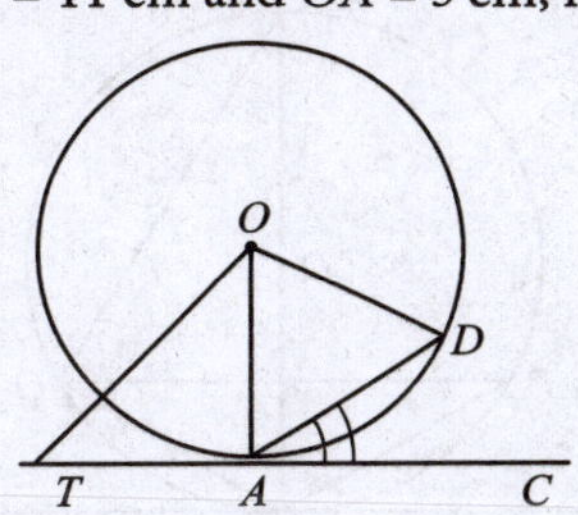

Solution

(a) $O\hat{A}C = 90°$ (The radius and the tangent at the point of contact are perpendicular)

$O\hat{A}D = 90° - D\hat{A}C$
$= 90° - 30° = 60°$

$OA = OD$ (Radii of the same circle are equal)

$O\hat{D}A = O\hat{A}D = 60°$ (Angles opposite equal sides are equal in a triangle)

$AOD = 180° - (60° + 60°)$
$= 60°$ (Using the sum of three angles of a triangle is equal to 180°)

(b) $A\hat{O}D = 60°$ (Given)

$OA = OD$ (Radii of the same circle are equal)

$O\hat{A}D = O\hat{D}A$ (The sum of three angles of a triangle is equal to 180°)

$= \dfrac{180° - 60°}{2} = 60°$

$O\hat{A}C = 90°$ (The tangent is perpendicular to the radius at the point of contact)

$D\hat{A}C = 90° - 60° = 30°$ (Complementary angles)

(c) $AT^2 = OT^2 - OA^2$

$= 9^2 - 6^2$

$= 45$

$AT = 3\sqrt{5}$ cm

(d) $OT^2 = AT^2 + OA^2$

$= 11^2 + 5^2 = 121 + 25 = 146$

$\therefore OT = \sqrt{146}$

2. In the figure, AC is a diameter of the circle with centre O. TAB is a tangent.

(a) If $Q\hat{A}B = 40°$, find $Q\hat{A}C$.

(b) If $A\hat{P}Q = 45°$, find $Q\hat{A}B$.

(c) If $Q\hat{A}T = 140°$, find the angle in the minor segment AQ.

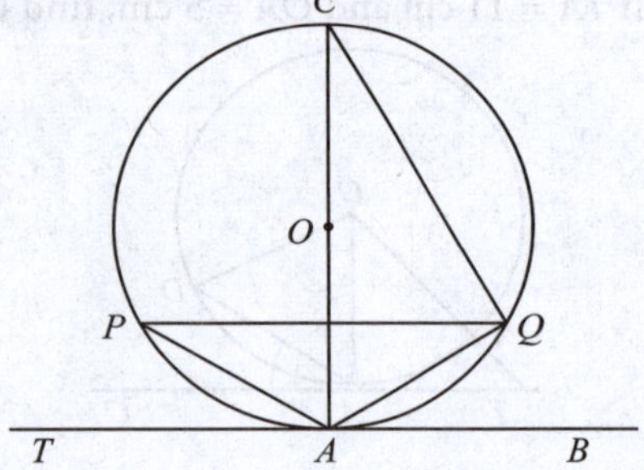

Solution

(a) $Q\hat{A}B = Q\hat{C}A = 40°$ (Angles in the alternate segment are equal)

$Q\hat{A}C = C\hat{A}B - Q\hat{A}B$ (CA perpendicular to tangent TA at A)

$= 90° - 40° = 50°$

(b) $A\hat{Q}B = A\hat{P}Q = 45°$ (Angles in the alternate segment)

(c) $Q\hat{A}T =$ any angle in the minor segment $AQ = 140°$

3. In the figure, with a circle, centre O touches the sides of ΔABC at X, Y, and Z.

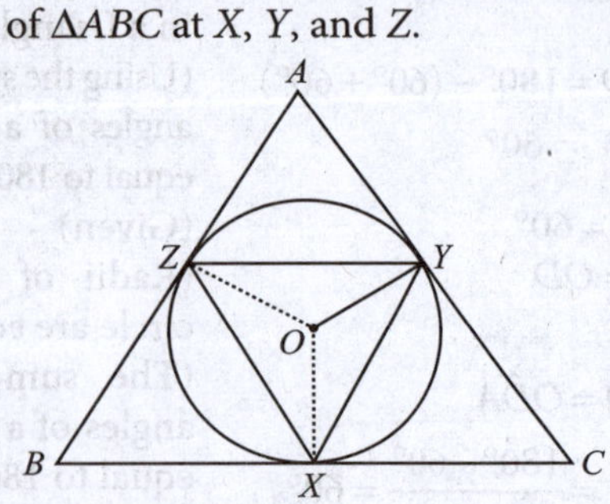

(a) If $\hat{B} = 45°$ and $\hat{C} = 75°$, find $X\hat{O}Z$ and the angles of ΔXYZ.

(b) If AB = 4 cm, BC = 4.5 cm, CA = 3.5 cm, find BX

(c) If AB = 7.5 cm, BC = 6.5 cm, CA = 5 cm, find BX

Solution

(a) $\hat{B} = 45°$ and $\hat{C} = 75°$ (Given)

$BX = BZ$, $CX = CY$, and $AY = AZ$ (Pairs of tangents to a circle from an external point are equal)

Quad $BXOZ$ is cyclic ($B\hat{X}O + B\hat{Z}O = 90° + 90° = 180°$)

$Z\hat{O}X = 180° - \hat{B} = 180° - 45° = 135°$ ($\hat{B}$ and $X\hat{O}Z$ are supplementary)

Quad $CYOX$ is cyclic ($C\hat{X}O + C\hat{Y}O = 180°$)

Similarly, $X\hat{O}Y = 180° - 75° = 105°$ ($X\hat{O}Y$ and $\hat{C}$ are supplementary)

$Y\hat{O}Z = 360° - (105° + 135°) = 120°$ (The angle at a point is equal to 360°)

$X\hat{Z}Y = \frac{1}{2}X\hat{O}Y = \frac{1}{2}(105°) = 52\frac{1}{2}°$ (The angle at circumference is half the angle at the centre)

$Z\hat{X}Y = \frac{1}{2} = Y\hat{O}Z = \frac{1}{2}(120°) = 60°$

$X\hat{Y}Z = \frac{1}{2}X\hat{O}Z = \frac{1}{2}(135°) = 67\frac{1}{2}°$

(b) Let $AZ = AY = x$, $BX = BZ = y$, and $CX = CY = z$

$AB + BC + CA = 4 + 4.5 + 3.5 = 12$ cm

$(x + y) + (y + z) + (z + x) = 12$

$2x + 2y + 2z = 12$

$x + y + z = 6$

$y = 6 - (x + z) = 6 - AC = 6 - 3.5 = 2.5$

$\therefore BX = 2.5$ cm

(c) Same as above

$y = \frac{19}{2} - 5 = 4.5$ cm

4. In the given figure, ABC is a triangle, $\hat{B} = 55°$, $\hat{C} = 70°$, a circle touches BC, AC produced, AB produced at P, Q, and R, respectively. Find $Q\hat{P}R$.

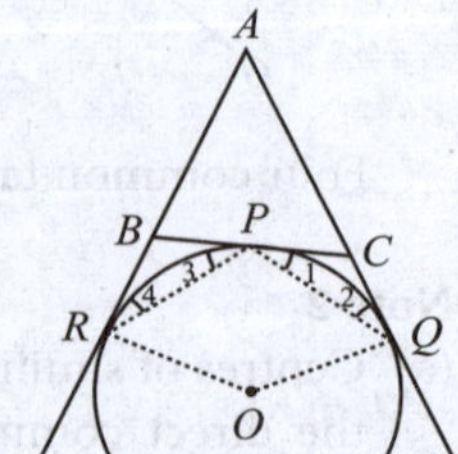

Solution

Joint RP, RO, QO, and QP.

$\hat{B} = 55°$ and $\hat{C} = 70°$ (Given)

$\hat{A} = 180° - (\hat{B} + \hat{C}) =$

$180° - (70° + 55°) = 55°$ (The sum of three angles of a triangle is equal to 180°)

$\hat{1} = \hat{2} = \frac{70°}{2} = 35°$ ($CP = CQ$; the exterior angle of a triangle is equal to the sum of the interior opposite angles)

$\hat{3} = \hat{4} = \frac{55°}{2} = 27\frac{1°}{2}$ ($CP = CQ$; the exterior angle of a triangle is equal to the sum of the interior opposite angles)

$Q\hat{P}R = 180° - (\hat{1} + \hat{3})$

$= 180 - \left(62\frac{1}{2}\right)° = 117°30'$

Observe that $Q\hat{P}R = \frac{1}{2}[P\hat{B}R + P\hat{C}Q] = \frac{1}{2}(125 + 110)°$

$= \frac{235°}{2} = 117°30'$

5. In the figure, $ABCD$ is a quadrilateral circumscribing a circle; DCY, ABY, ADX, and BCX are straight lines. Prove that $XB + BY = XD + DY$.

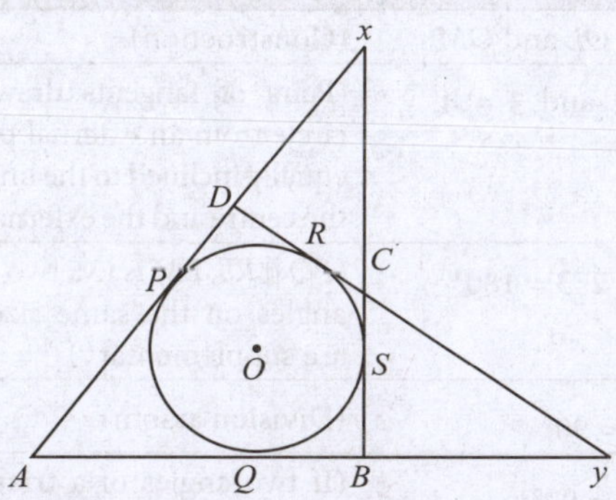

Proof

Statements	Reasons
1. Sides AB, BC, CD, and DA touch the circle at Q, S, R, and P, respectively	(The tangent touches the circle at one and only one point)
2. $XB + BY = XS + SB + YQ - QB$	
3. $XB + BY = XP + YR$	($SB = QB$; the tangent to a circle from an external
$= XD + DP + YD - DR$	point is equal cancelling DP and DR which are equal)
$= XD + YD$	

6. A pair of tangents drawn to a circle from an external point is perpendicular to each other. Prove that the length of each tangent is equal to the radius of the circle.

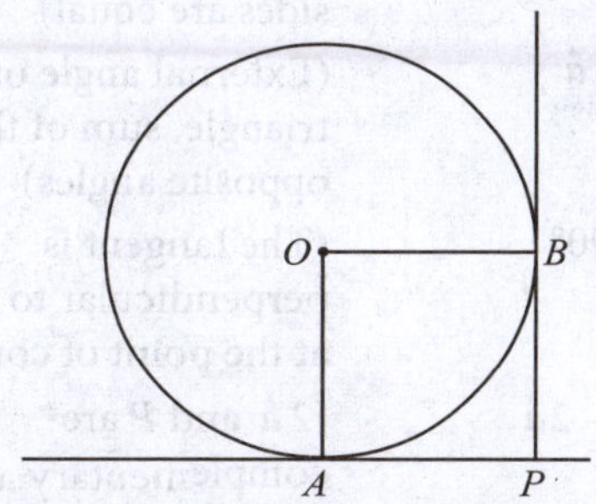

Given: Let PA and PB a pair of tangents drawn to the circle with centre O from are external point P $A\hat{P}B = 90°$

To prove: $AP = PB = OA$ (radius).

Proof

Statements	Reasons
1. $\hat{P} = 90°$	(Given)
2. $\hat{A} = \hat{B} = 90°$	(The tangent is perpendicular to the radius at the point of contact)
3. $\hat{O} = 90°$	(If three angles of a quadrilateral are each a right angle, thus the remaining angle is a right angle)
4. $OA = OB$	(Radii of the same circle are equal)
5. $AP = BP$	(Tangents to a circle from an external point are equal)
6. $OAPB$ is a rectangle	(Each angle is a right angle)
7. $OAPB$ is a square	(If a pair of adjacent sides of a rectangle is equal, then it is a square)
8. $PA = PB =$ radius $=$ OA or OB	(All the sides of a square are equal)

7. In the figure, AP is a tangent at A to the circle, centre O. Find in terms of a.

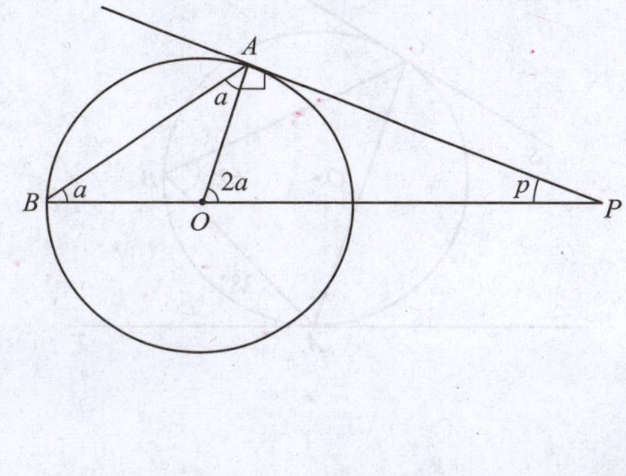

Solution

(a) $BA = OB$ (Radii of the same circle are equal)

(b) $\hat{B} = \hat{a}$ (Angles opposite equal sides are equal)

(c) $A\hat{O}P = 2\hat{a}$ (External angle of a triangle, sum of the interior opposite angles)

(d) $P\hat{A}O = 90°$ (The tangent is perpendicular to the radius at the point of contact)

(e) $\hat{P} = 90° - 2\hat{a}$ ($2\hat{a}$ and P are complementary angles)

8. In the figure, I is the in the centre of ΔABC. The circle touches the sides at X, Y, and Z. Find the angles of ΔABC.

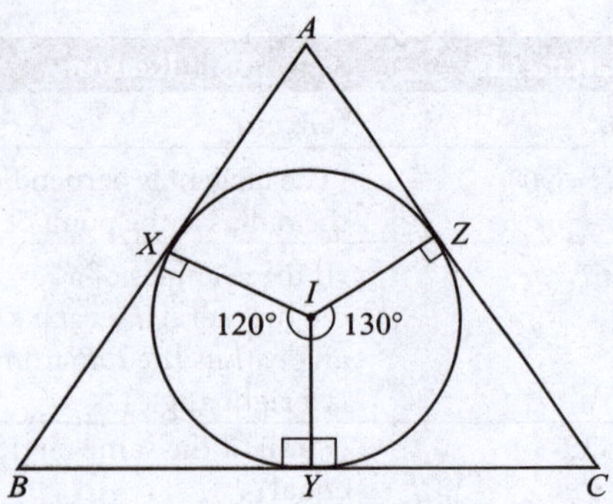

Solution

The quadrilaterals $IXAZ$, $IXBY$, and $IYCZ$ are cyclic.
$\hat{B} = 180° - 120° = 60°$ and $\hat{C} = 180° - 130° = 50°$ (opposite angles of a cyclic quadrilateral are supplementary)
$\hat{A} = 180° - (60° + 50°) = 70°$ (the sum of three angles of a triangle is equal to 180°)

9. In the figure, AT and CS are tangents to the circle. Find $B\hat{C}S$.

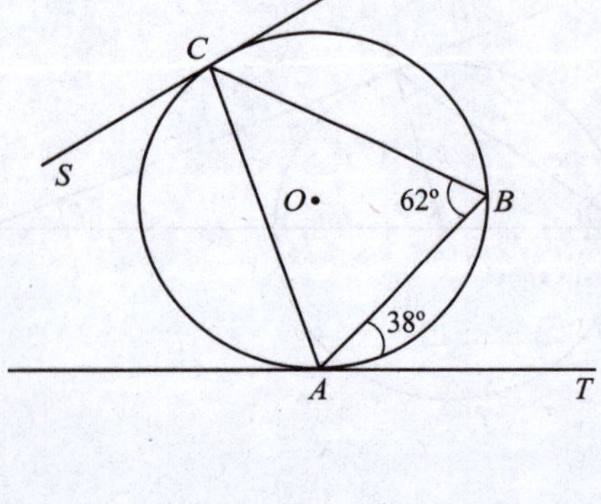

Solution

Draw AC.
$A\hat{C}B = 38°$ (The angle in the alternate segment)
$S\hat{C}A = 62°$ (The angle in the alternate segment)
$B\hat{C}S = A\hat{C}B + S\hat{C}A = 38° + 62° = 100°$

10. Prove that the intercept of a tangent between two parallel tangents to a circle subtends a right angle at the centre.

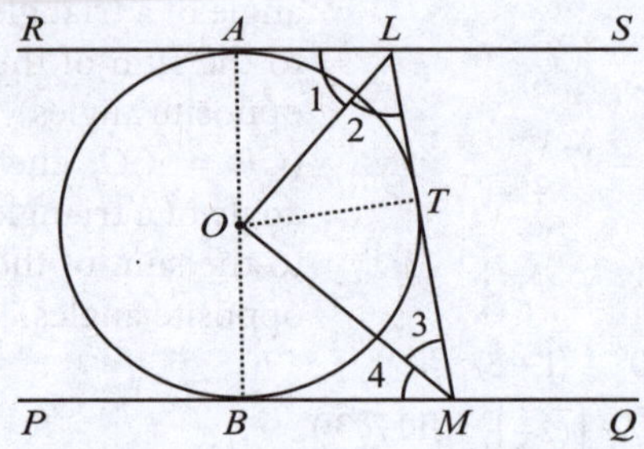

Given: A circle with centre O. Tangents PQ and RS at B and A, respectively, are parallel.
LM is a tangent at T intersecting RS at L and PQ at M.
To prove: $L\hat{O}M = 90°$.

Proof

Statements	Reasons
1. Draw OL and OM	(Construction)
2. $\hat{1} = \hat{2}$ and $\hat{3} = \hat{4}$	(Pairs of tangents drawn to a circle from an external point are equally inclined to the line join of the centre and the external point)
3. $2.\hat{2} + 2.\hat{3} = 180°$	($PQ \parallel RS$; LM is t.v.; two interior angles on the same size of t.v. are supplementary)
4. $\hat{2} + \hat{3} = 90°$	(Division axiom)
5. $L\hat{O}M = 90°$	(If two angles of a triangle are complementary, then the third angle is a right angle)

11. If ΔABC is isosceles with $AB = AC$, prove that the tangent at A to the circle is parallel to BC.

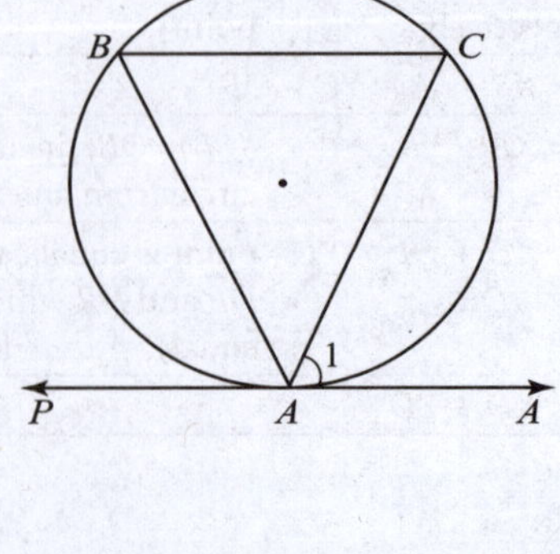

Proof

Statements	Reasons
1. $AB = AC$	(Given)
2. $\hat{B} = \hat{C}$	(Angles opposite equal sides are equal)
3. $\hat{1} = \hat{B}$	(Angles in alternate segments)
4. $\hat{C} = \hat{1}$	(From statements 2 and 3)
5. PQ || BC	(Alternate angles are equal)

12. In the figure, TA touches circle ABE at A. PQT bisects $A\hat{T}B$. Prove that $AP = AQ$.

Given: TA is a tangent to the circle circumscribing the ΔABC. TP bisects $A\hat{T}B$ cutting AC at Q.

To prove: $AP = AQ$.

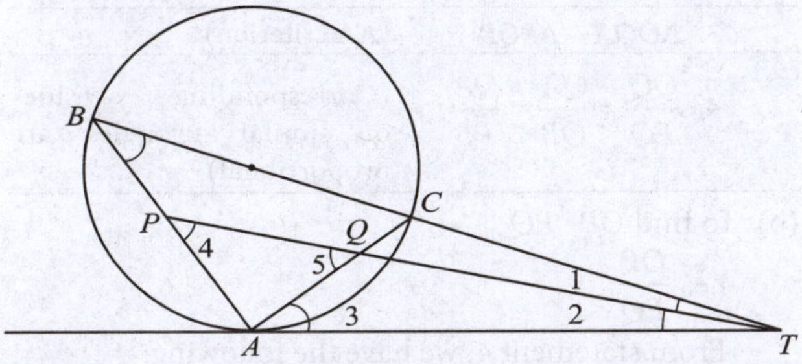

Proof

Statements	Reasons
1. $\hat{B} = \hat{3}$	(Angles in the alternate segment)
2. $\hat{1} = \hat{2}$	(TP bisects $A\hat{T}B$)
3. $\hat{4} = \hat{1} + \hat{B}$	(The exterior angle of a triangle is equal to the sum of interior opposite angles in DPBT)
4. $\hat{5} = \hat{2} + \hat{3}$	(The exterior angle of a triangle is equal to the sum of interior opposite angles in DAQT)
5. $\hat{5} = \hat{1} + \hat{B}$	(Using statements 1 and 2)
6. $\hat{4} = \hat{5}$	(From statements 3 and 5)
7. AP = AQ	(Sides opposite equal angles are equal)

13. In the figure, two circles touch each other at P externally; APB and CPD are straight lines. Prove that $CA \parallel BD$.

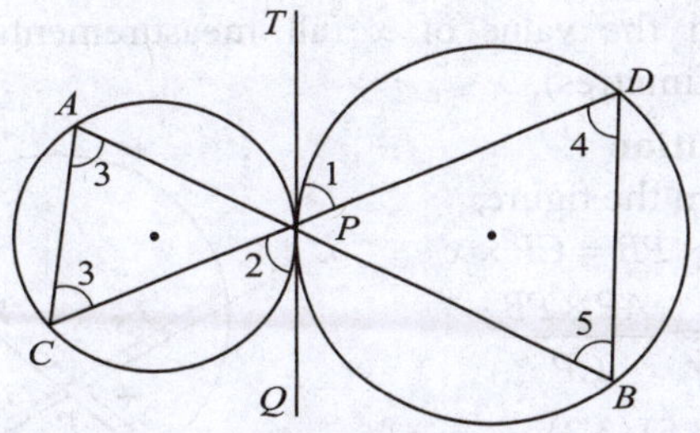

Proof

Statements	Reasons
1. Draw a common tangent TPQ at P	(Construction)
2. $T\hat{P}D = P\hat{B}D$, i.e., $\hat{1} = \hat{5}$	(Angles in the alternate segment are equal)
3. $C\hat{P}Q = T\hat{P}D$, i.e., $\hat{2} = \hat{1}$	(V.O.A. is equal)
4. $C\hat{A}P = C\hat{P}Q$, i.e., $\hat{3} = \hat{2}$	(Angles in the alternate segment are equal)
5. $\hat{3} = \hat{5}$	(From statements 2, 3, and 4)
6. $CA \parallel DB$	(Alternate angles are equal)

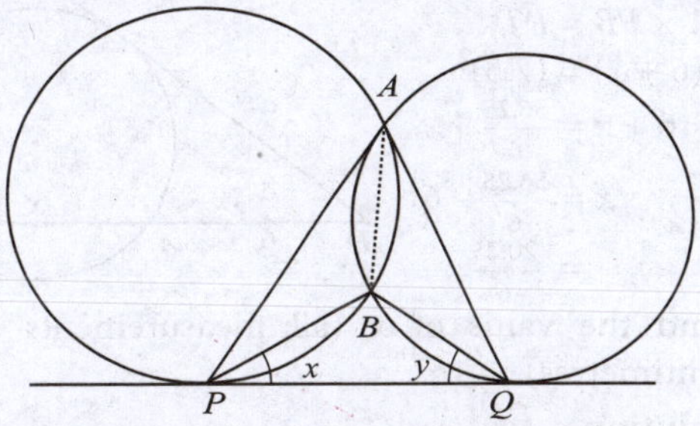

14. In the figure, two circles intersect at A and B. PQ is a common tangent to the circles. Prove that $P\hat{A}Q + P\hat{B}Q =$ two right angles.

Proof

Statements	Reasons
1. Draw AB	(Construction)
2. $P\hat{B}Q = 180° - (x + y)$	(The sum of the angles of a triangle = 180°)
3. $P\hat{A}B = \hat{x}$ and $B\hat{A}Q = y$	(Angles in the all segments are equal)
4. $P\hat{A}Q + P\hat{B}Q = (x + y) + 180° - (x + y)$	(Adding statements 2 and 3)
5. $P\hat{A}Q + P\hat{B}Q = 180°$ = two right angles in problems 16 to 19.	

15. Find the value of x (all measurements are in centimetres).

Solution

From the figure,

$AP \times PB = CP \times x$

$\therefore x = \dfrac{AP \times PB}{CP}$

$= \dfrac{(3.5)(3.2)}{2.8} = 4$ cm

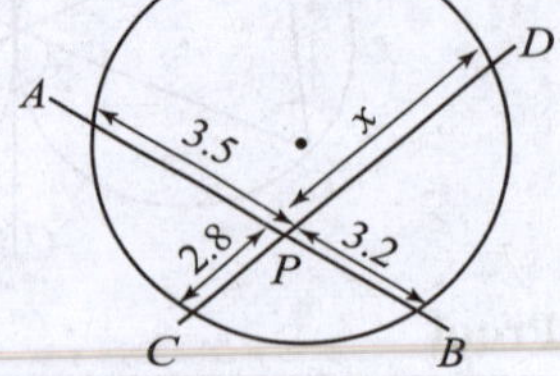

16. Find the value of x (all measurements are in centimetres).

Solution

From the figure,

$CP \times PD = AP \times PB$

$x(9) = (8.1)(5)$

$x = \dfrac{(8.1)(5)}{9} = 4.5$ cm

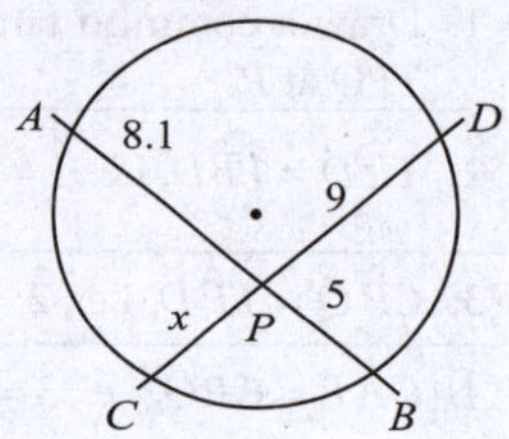

17. Find the value of x (all measurements are in centimetres).

Solution

From the figure,

$PA \times PB = PT^2$

$6(6 + x) = (7.5)^2$

$6 + x = \dfrac{56.25}{6}$

$x = \dfrac{56.25}{6} - 6$

$= \dfrac{20.25}{6}$

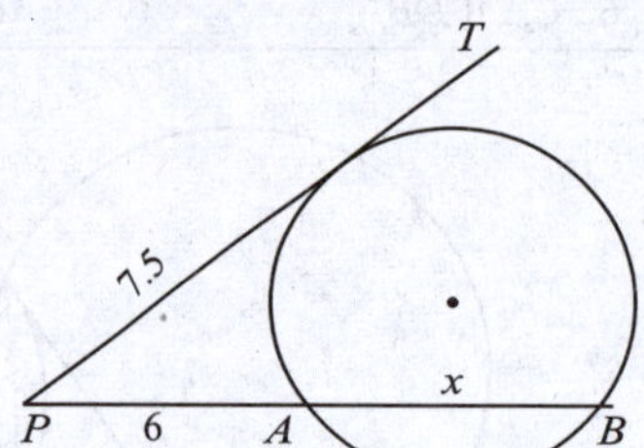

18. Find the value of x (all measurements are in centimetres).

Solution

From the figure,

$PA \times PB = PT^2$

$x(x + 14) = 24^2$

$x^2 + 14x - 576 = 0$

$x^2 + 32x - 18x - 576 = 0$

$x(x + 32) - 18(x + 32) = 0$

$(x - 18)(x + 32) = 0$

$x - 18 = 0$ or $x + 32 = 0$

$x = 18$

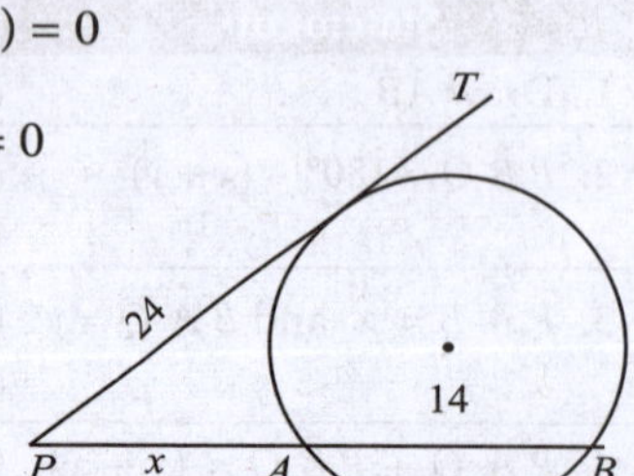

The other value is negative and hence rejected.

19. AB is a direct common tangent drawn to two circles with centres O and P, and radii 8 cm and 5 cm, respectively. AB and OP are produced to meet at Q.

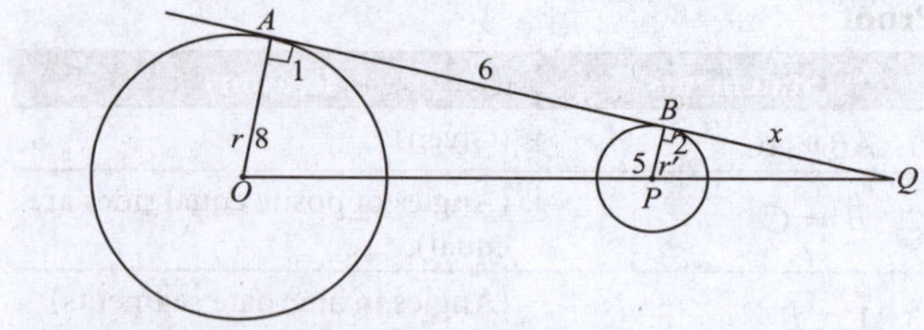

(a) Prove $\Delta BPQ \sim \Delta AOQ$.

(b) Find $OP : PQ$.

(c) If $AB = 6$ cm, find AQ.

Proof

Statements	Reasons
(a) 1. Draw OA and PB	(Construction)
2. In ΔOQA and ΔPQB	
(i) $\hat{1} = \hat{2}$ (ii) $\hat{Q} = \hat{Q}$	(The tangent is perpendicular to the radius at the point of contact) (Identity)
3. $\Delta OQA \sim \Delta PQB$	(AA criterion)
4. $\dfrac{OQ}{PQ} = \dfrac{QA}{QB} = \dfrac{OA}{PB}$	(Corresponding sides of similar triangles are proportional)

(b) To find $OP : PQ$

i.e., $\dfrac{OP}{PQ}$

From statement 4, we have the following:

$\dfrac{OQ}{PQ} = \dfrac{OA}{PB} = \dfrac{8}{5} \rightarrow$ (A)

Subtracting 1 on either side of (A), we have the following:

$\dfrac{OQ}{PQ} - 1 = \dfrac{8}{5} - 1$

$\dfrac{OQ - PQ}{PQ} = \dfrac{8-5}{5} = \dfrac{3}{5}$

$\dfrac{OP}{PQ} = \dfrac{3}{5}$

i.e., $OP : PQ = 3 : 5$

(c) Taking BQ as x,

$\dfrac{AQ}{QB} = \dfrac{8}{5}$

$\dfrac{AB + BQ}{QB} = \dfrac{8}{5}$

$\dfrac{6 + x}{x} = \dfrac{8}{5}$

$30 + 5x = 8x$

$3x = 30$

$x = 10$

$\therefore AQ = AB + BQ = 6 + 10 = 16$ cm

Seven theorems on tangents to circles

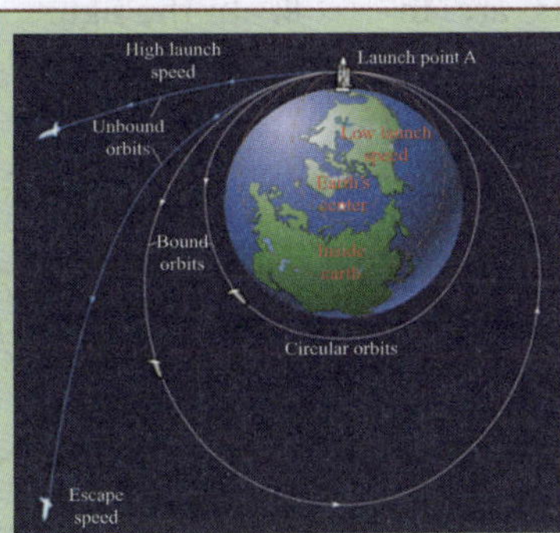

Tangents to circles

Segment: Part of the interior of a circle bounded by a chord and its arc

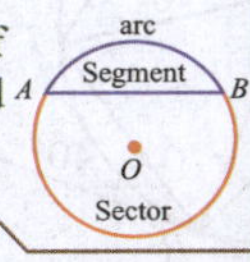

Major sector

Minor sector

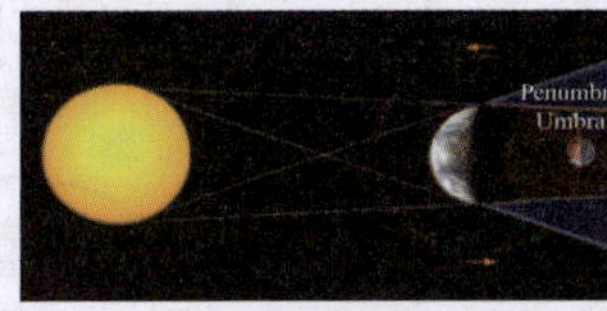

Centres of similitude

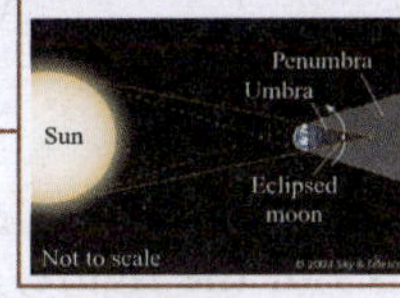

A line in the plane of a circle meeting the circle in one point is called a tangent to the circle.

Part of the interior of a circle bounded by two radii and an arc

Major sector

Minor sector

TEST YOURSELF

Level 1

1. Refer to the figure for the problem below in which AT and BT are tangent from T to circle centre O.
   (a) If $A\hat{T}B = 65°$, find $A\hat{O}T$.
   (b) If $A\hat{T}B = 64°$, find $A\hat{O}B$.
   (c) If $A\hat{T}B = 40°$, find $T\hat{A}B$.

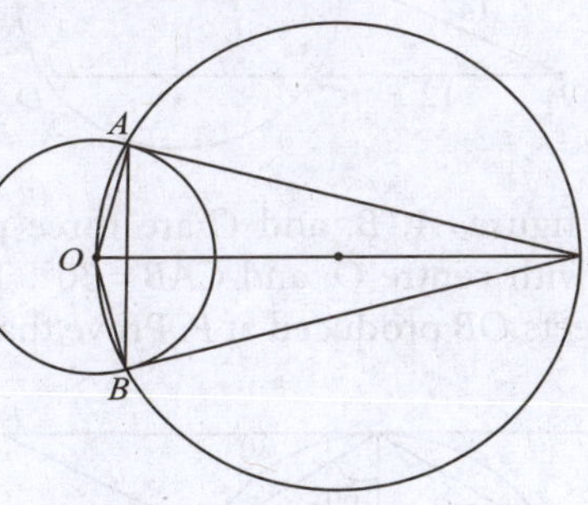

2. In ΔABC, $BC = 3$ cm, $CA = 6$ cm, $AB = 7$ cm; a circle is drawn to touch AB produced at R, AC produced at Q, and BC at P. Find the length of AR and CP.

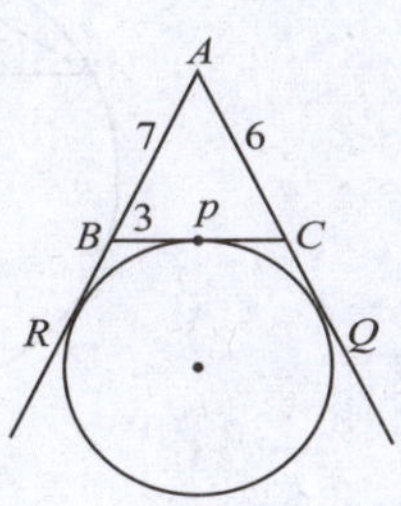

3. Two circles touch each other at the point P (internally and externally), and from a point T, tangents TQ and TR are drawn to two circles. Prove that $TQ = TR$.

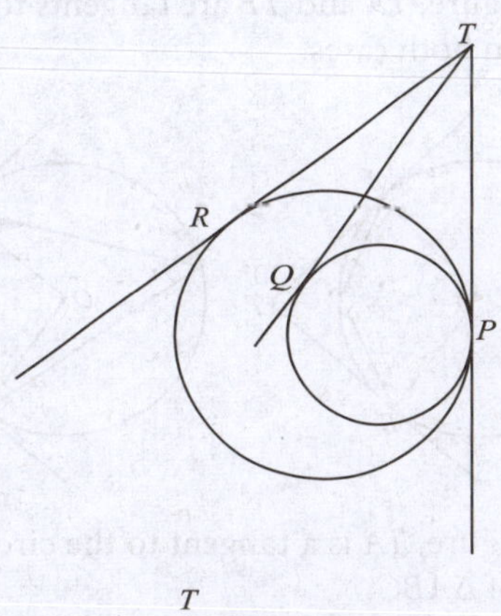

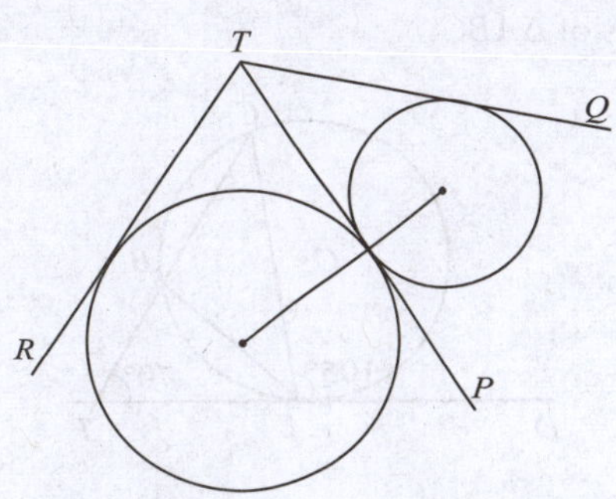

4. Find the value of x (all measurements are in centimetres).

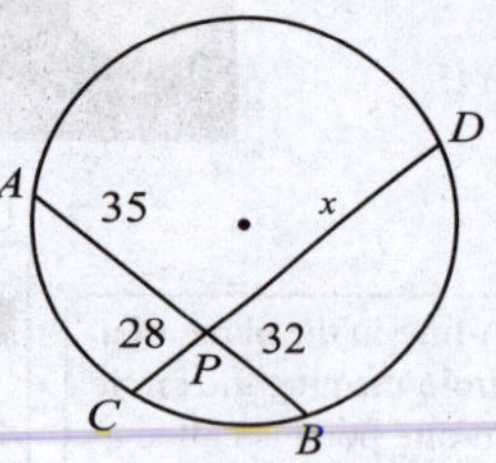

5. In the figure, PQ is a common chord and ST is a common tangent. Prove that PQ produced bisects ST at R.

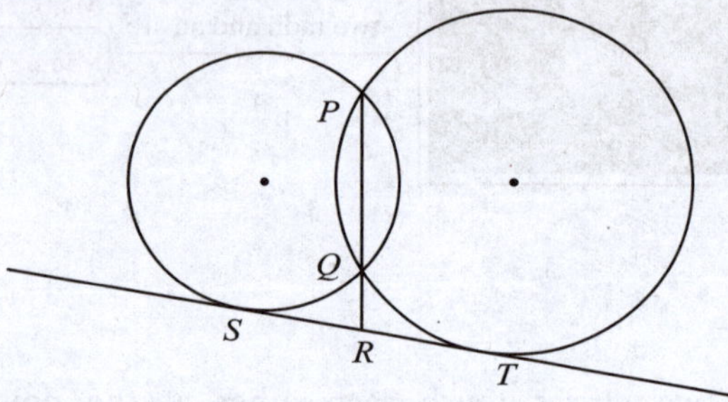

Level 2

6. In the figure, TA and TB are tangents to the circles. Find x in both cases.

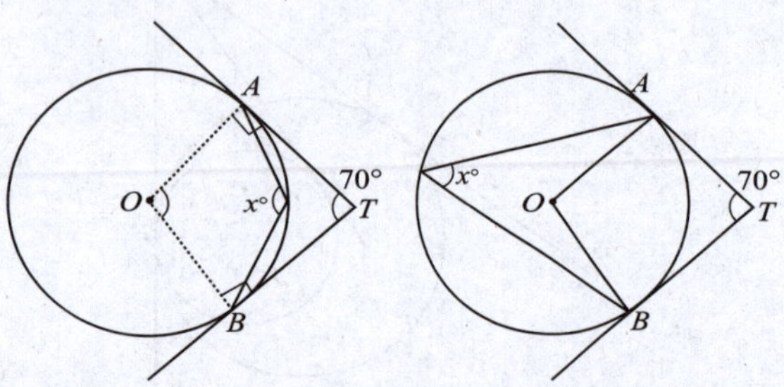

7. In the figure, TA is a tangent to the circle. Find the angles of ΔABC.

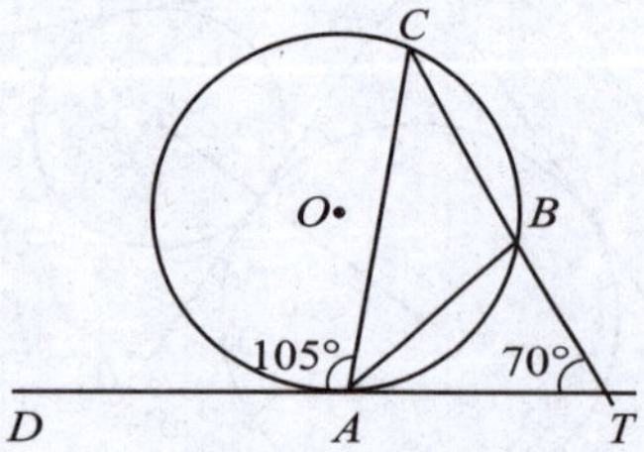

8. In the figure, PQ and PR are tangents to the circle with centre O. Find $R\hat{S}Q$, $R\hat{T}Q$, and $R\hat{P}Q$.

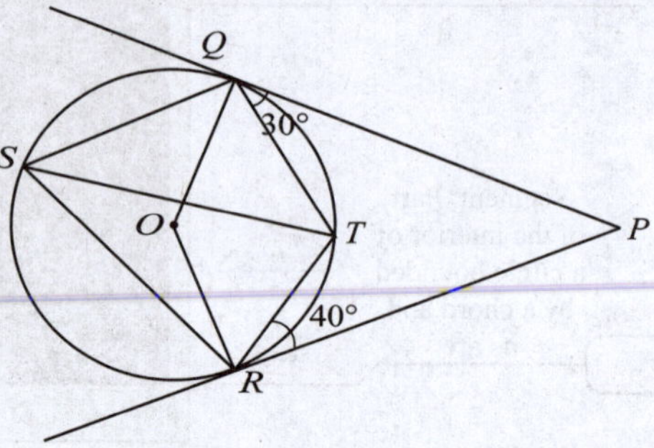

9. In a right-angled ΔABC right angled at B, a circle with the side AB as diameter is drawn to intersect the hypotenuse AC in P. Prove that the tangent to the circle at P bisects side BC.

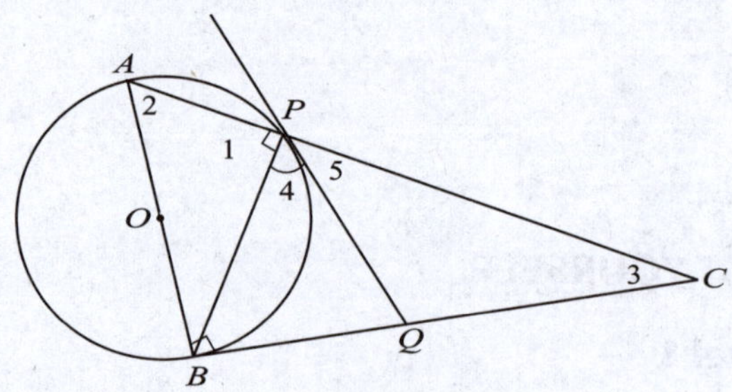

10. Find the value of x in the following figure.

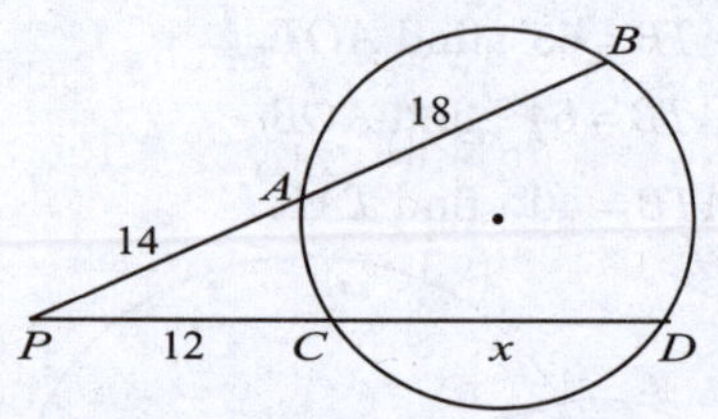

11. In the figure, A, B, and C are three points on a circle, with centre O, and $C\hat{A}B = 30°$. The tangent at C meets OB produced at P. Prove that $OB = BP$.

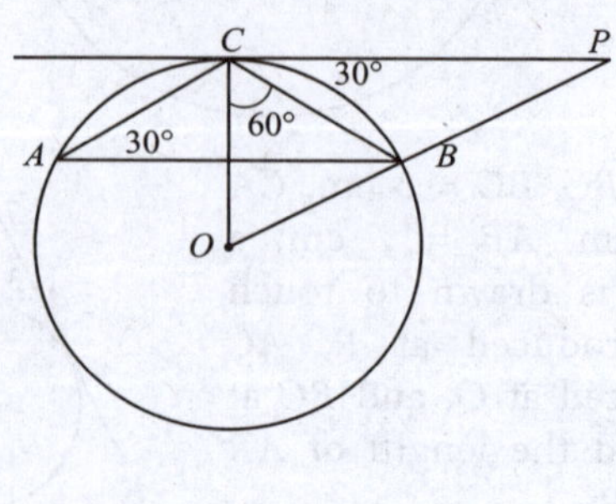

Level 3

12. In the figure, O is the centre of the circle. $A\hat{O}T$ is a right angle and TP is a tangent. Prove that $TP = TQ$.

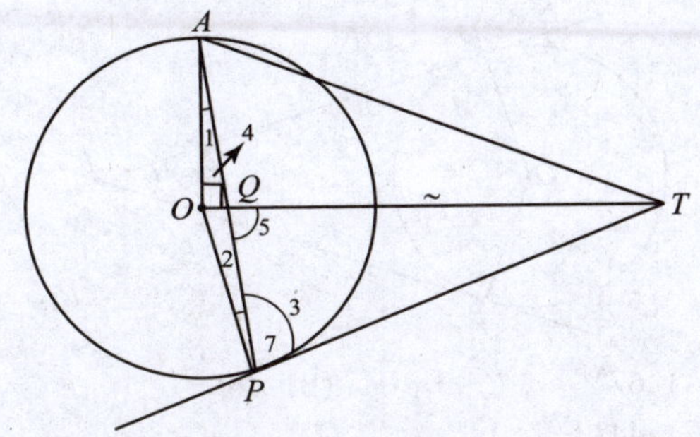

13. In the figure, $ABCD$ is a cyclic quadrilateral; AX and DY are tangents; $X\hat{A}B = 44°$, $Y\hat{D}C = 57°$, and $B\hat{D}C = 26°$. Find the angles of quadrilateral $ABCD$.

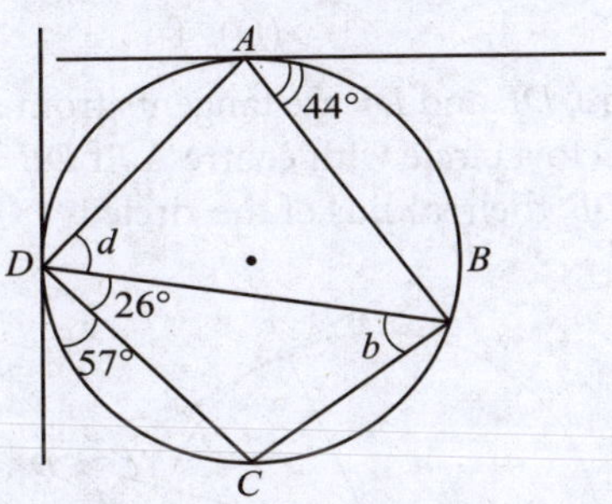

14. Two circles with centres O and O' and radii r and r' touch each other at P (internally or externally). A line is drawn to pass through P intersecting two circles at Q and R respectively. Prove that $OQ \parallel O'R$.

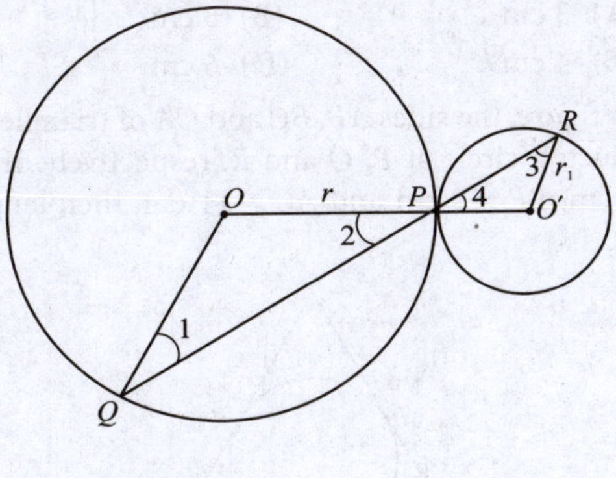

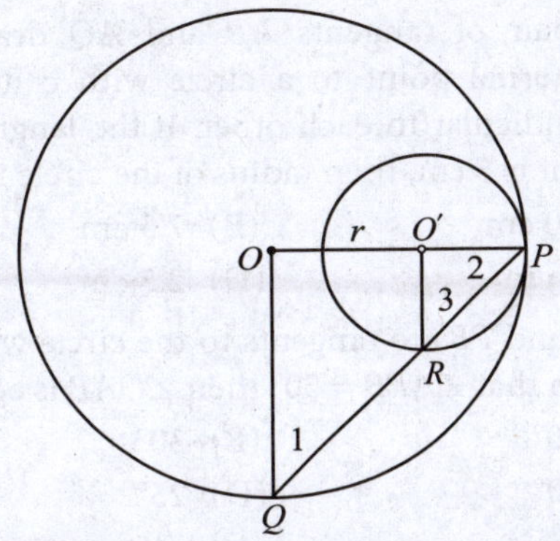

15. In the figure, PQ is a diameter of a circle with centre O. QR is a tangent to the circle equal to OQ; OT is perpendicular to PR. QT is produced to cut the circle at S. Prove that $SP = ST$.

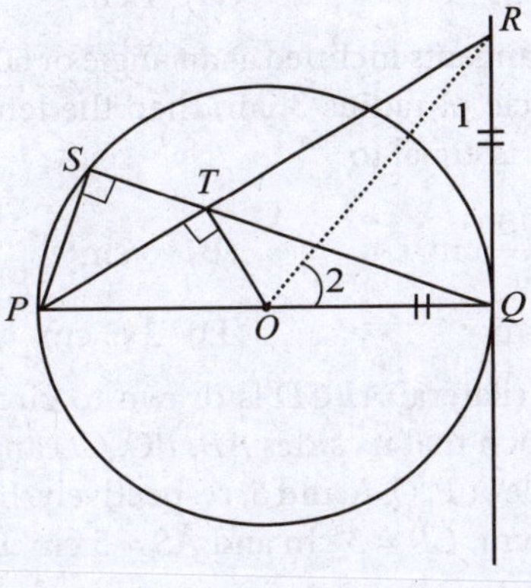

16. Two spherical balls of radii 10 cm and 5 cm lie on a plane touching each other. Find the distance between the points where the balls touch the plane.

Multiple Choice Questions

17. If ABC is tangent to the circle at point B whose centre is O, PQ is a chord parallel to AC and $\angle QBC = 60°$, then $\angle QBP$ is equal to

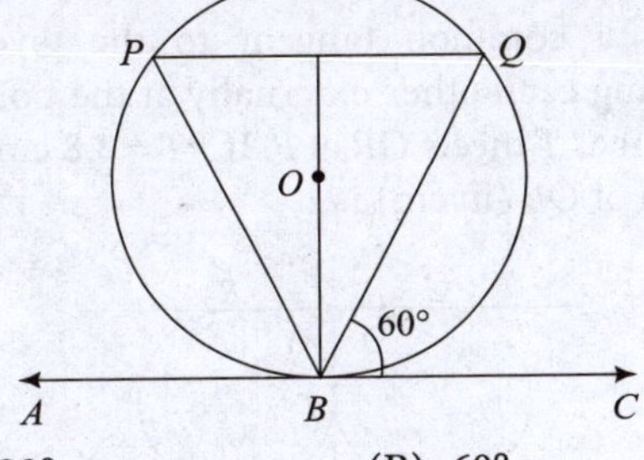

(A) 30° (B) 60°
(C) 90° (D) 80°

18. The pair of tangents AP and AQ drawn from an external point to a circle with centre O are perpendicular to each other. If the length of each tangent is 5 cm, then radius of the circle is

(A) 10 cm (B) 7.5 cm
(C) 5 cm (D) 2.5 cm

19. If PA and PB are tangents to the circle with centre O such that $\angle APB = 50°$ then $\angle OAB$ is equal to

(A) 50° (B) 30°
(C) 40° (D) 25°

20. If radii of two concentric circles are 4 cm and 5 cm. Then the length of the chord of one circle which is tangent to the other circle is

(A) 3 cm (B) 6 cm
(C) 9 cm (D) 1 cm

21. If two tangents inclined at an angle of 60° are drawn to a circle of radius 3 cm, then the length of each tangent is equal to

(A) $\frac{3\sqrt{3}}{2}$ cm (B) 6 cm
(C) 3 cm (D) $3\sqrt{3}$ cm

22. A quadrilateral $ABCD$ is drawn to circumscribe a circle such that its sides AB, BC, CD and DA touch the circle at P, Q, R and S, respectively. If $AB = x$ cm, $BC = 7$ cm, $CR = 3$ cm and $AS = 5$ cm, then x = ?

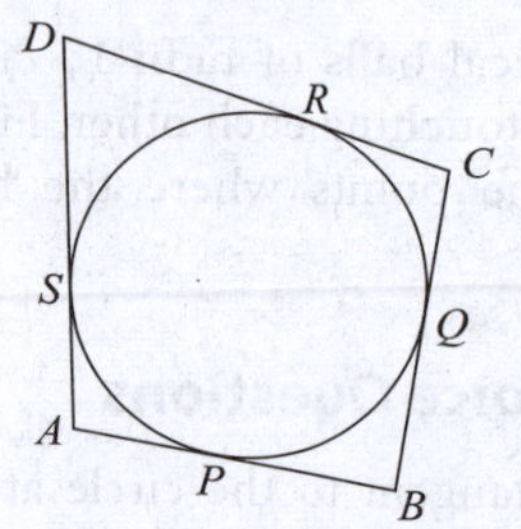

(A) 10 cm (B) 8 cm
(C) 9 cm (D) 7 cm

23. QR is a common tangent to the given circles touching each other externally at the point T. The tangent at T meets QR at P. If $PT = 3.8$ cm, then the length of QR (in cm) is

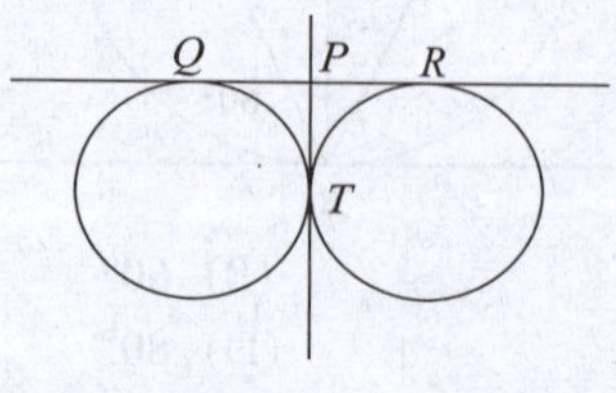

(A) 3.8 (B) 7.6
(C) 5.7 (D) 1.9

24. If PQ and PR are two tangents to a circle with centre O. If $\angle QPR = 46°$, then $\angle QOR$ is equal to

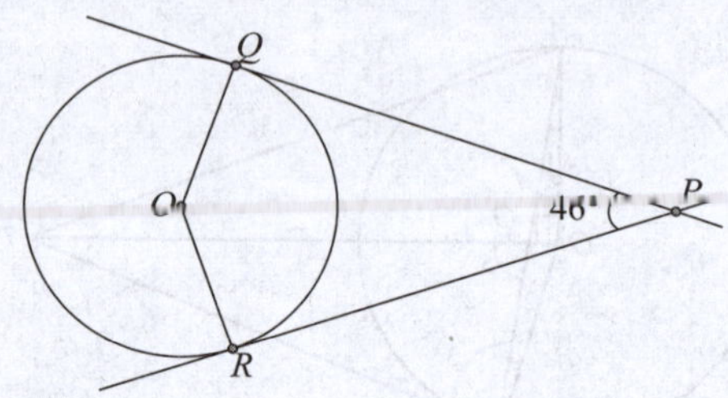

(A) 67° (B) 46°
(C) 44° (D) 134°

25. In a right triangle ABC, right angled at B, $BC = 12$ cm and $AB = 5$ cm. The radius of the circle inscribed in the triangle (in cm) is

(A) 4 (B) 3
(C) 2 (D) 1

26. In figure, DE and DF are tangents from an external point D to a circle with centre A. If $DE = 5$ cm and $DE \perp DF$, then radius of the circle is

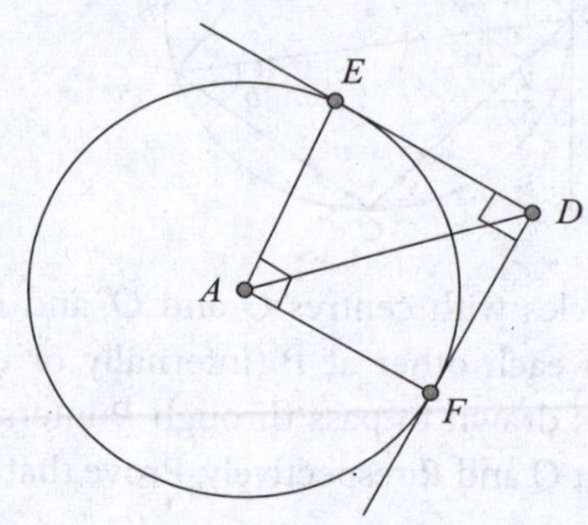

(A) 3 cm (B) 5 cm
(C) 4 cm (D) 6 cm

27. In figure, the sides AB, BC and CA of triangle ABC, touch a circle at P, Q and R, respectively. If $PA = 4$ cm, $BP = 3$ cm and $AC = 11$ cm, then length of BC is

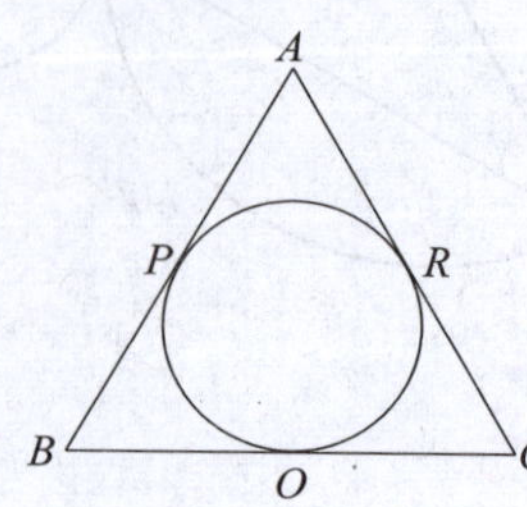

(A) 11 cm (B) 15 cm
(C) 14 cm (D) 10 cm

28. In the figure, if $AB = 8$ cm and $PE = 3$ cm then $AE =$

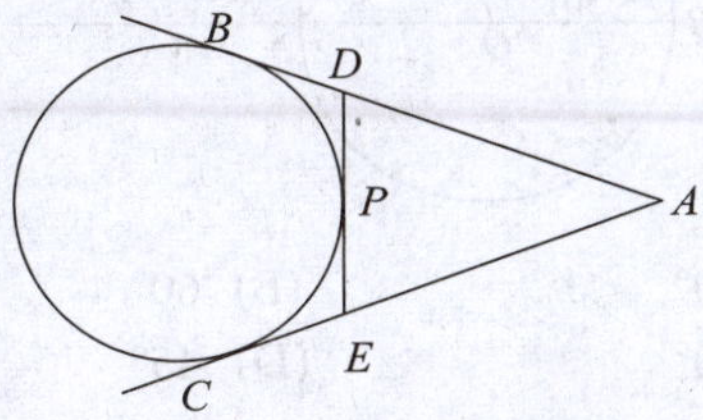

(A) 11 cm (B) 7 cm
(C) 5 cm (D) 3 cm

29. In figure, PQ and PR are the tangents drawn from P to a circle with centre O. If $\angle OPQ = 35°$ then

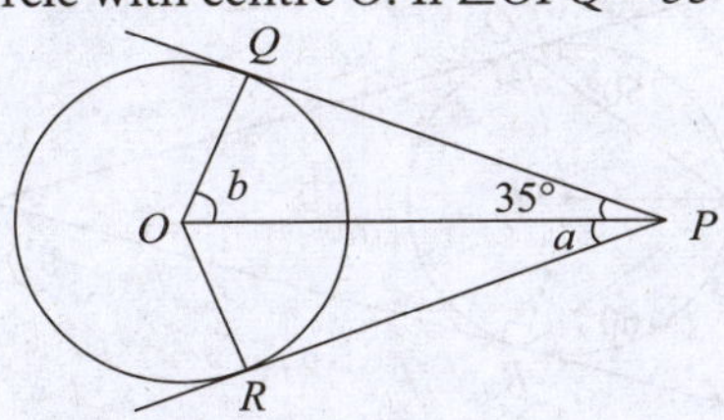

(A) $a = 30°, b = 60°$ (B) $a = 35°, b = 55°$
(C) $a = 40°, b = 50°$ (D) $a = 45°, b = 45°$

30. In figure, if TP and TQ are tangents drawn from an external point T to a circle with centre O such that $\angle TQP = 60°$, then $\angle OPQ =$

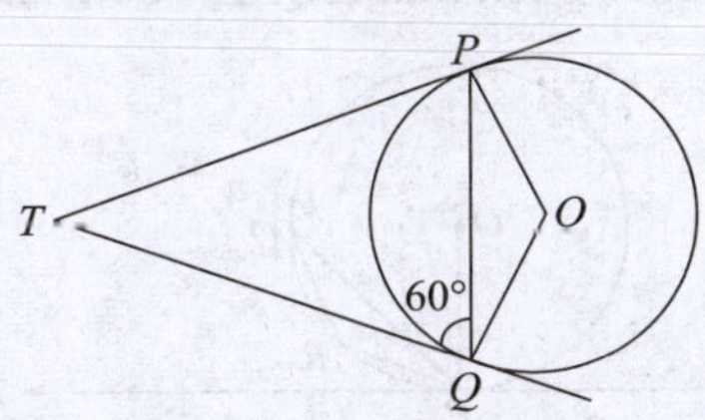

(A) 25° (B) 60°
(C) 40° (D) 30°

31. There are two concentric circles with centre O. PR and PQ are tangents to the inner circle from point P lying on the outer circle. If $PR = 7.5$ cm, then PS is equal to

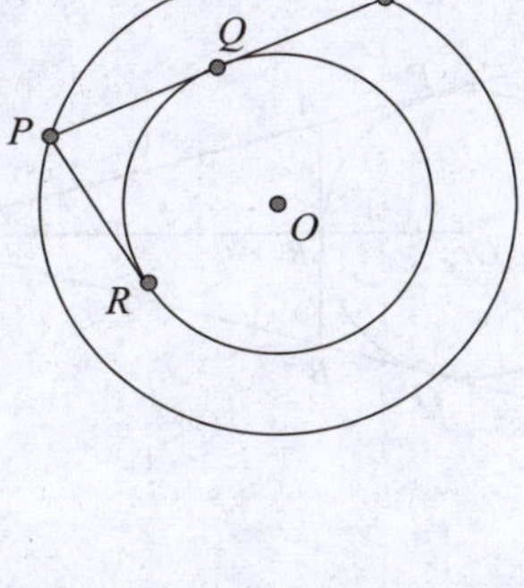

(A) 10 cm (B) 12 cm
(C) 15 cm (D) 18 cm

32. In figure, $PR = ?$

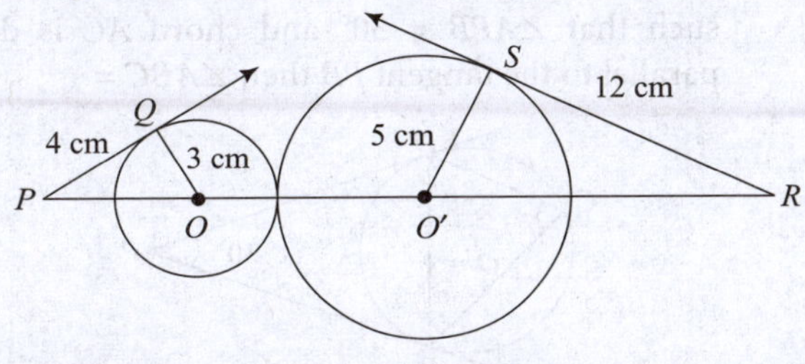

(A) 20 cm (B) 22 cm
(C) 24 cm (D) 26 cm

33. In figure, PR is tangent to the circle at P, and O is the centre of the circle. The value of $\angle POQ$ is

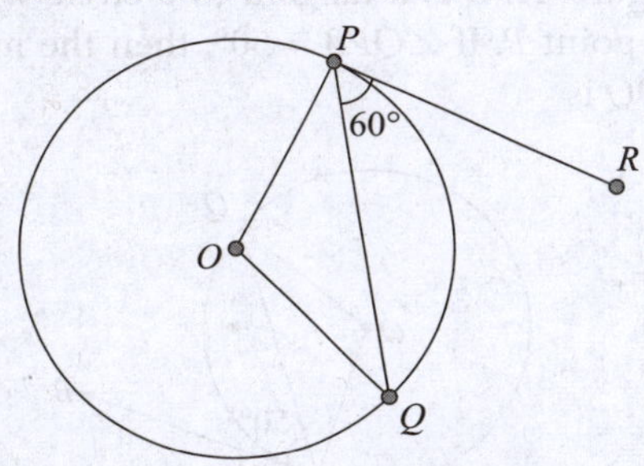

(A) 110° (B) 100°
(C) 120° (D) 90°

34. In figure, if $AP = 10$ cm, then $BP =$

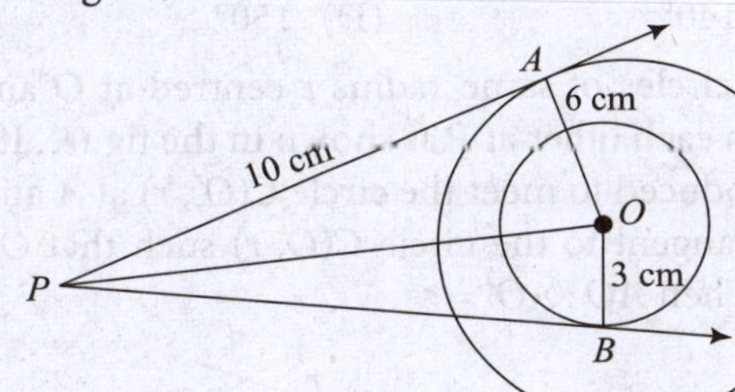

(A) $\sqrt{91}$ cm (B) $\sqrt{127}$ cm
(C) $\sqrt{119}$ cm (D) $\sqrt{109}$ cm

35. In figure, if $AB = 12$ cm, $BC = 8$ cm and $AC = 10$ cm, then AD is

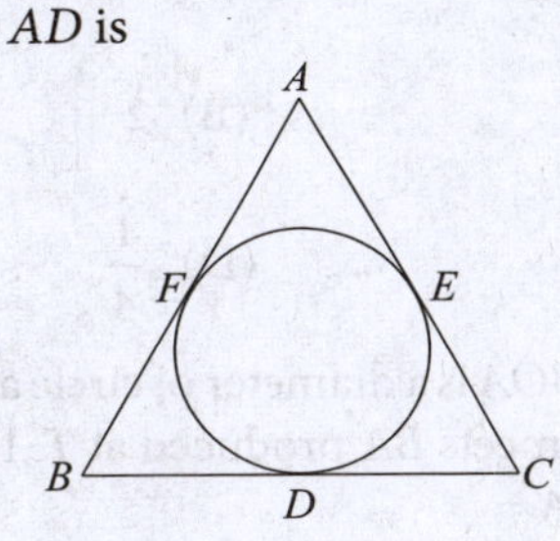

(A) 5 cm (B) 4 cm

(C) 6 cm (D) 7 cm

36. In figure, tangents PA and PB are drawn to a circle such that $\angle APB = 30°$ and chord AC is drawn parallel to the tangent PB then $\angle ABC =$

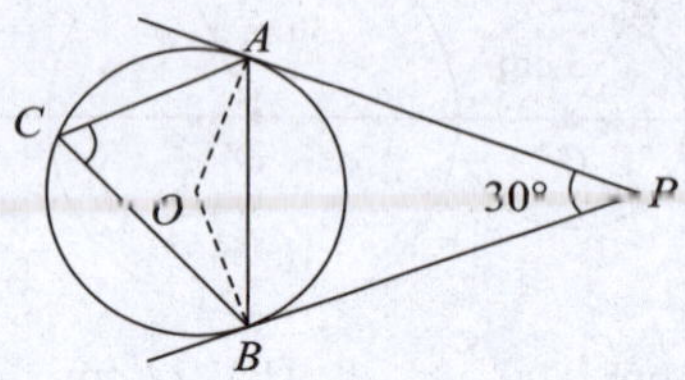

(A) 60° (B) 50°

(C) 30° (D) none of these

37. In figure, APB is a tangent to a circle with centre O at point P. If $\angle QPB = 50°$, then the measure of $\angle POQ$ is

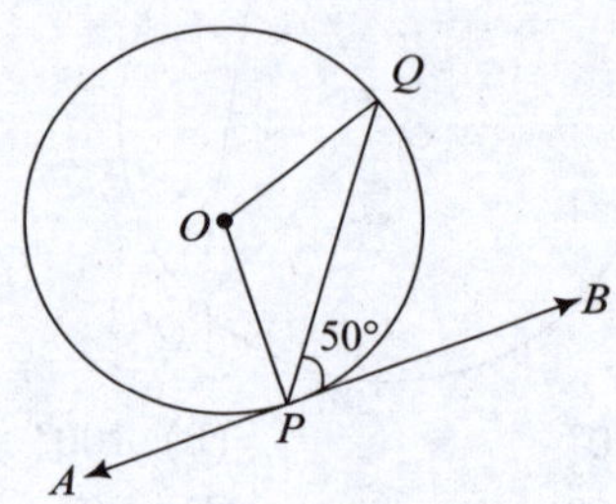

(A) 100° (B) 120°

(C) 140° (D) 150°

38. Two circles of same radius r centred at O and O' touch each other at P as shown in the figure. If OO' is produced to meet the circle $C(O', r)$ at A and AT is a tangent to the circle $C(O, r)$ such that $O'Q \perp AT$. Then $AO : AO' =$

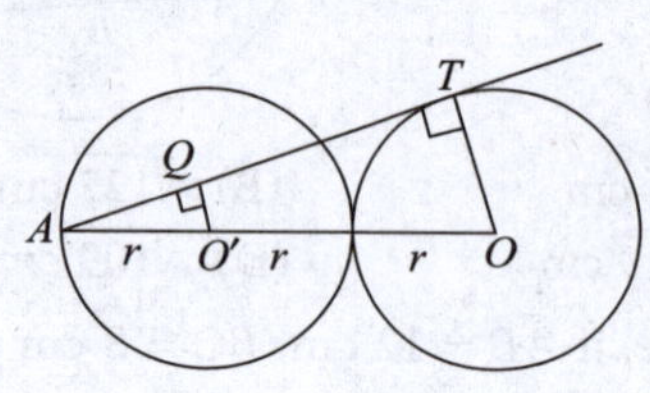

(A) $\frac{3}{2}$ (B) 2

(C) 3 (D) $\frac{1}{4}$

39. In figure, BOA is a diameter of circle and the tangent at point P meets BA produced at T. If $\angle PBO = 30°$, then $\angle PTA =$

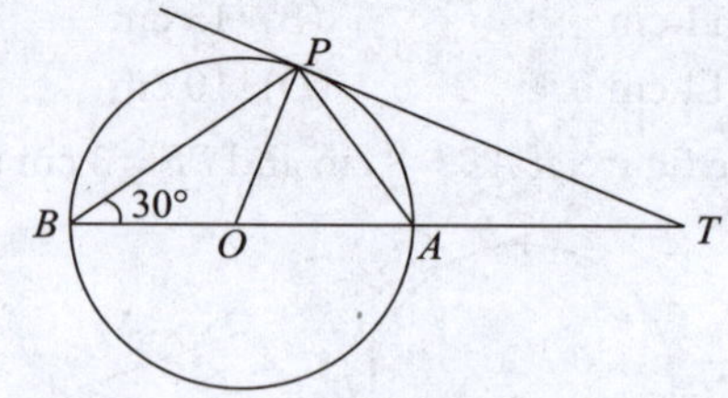

(A) 30° (B) 60°

(C) 90° (D) 45°

40. In the figure, PQ and PR are tangents to the circle with centre O. Also, S is a point on the circle such that $\angle SQC = 50°$ and $\angle SRM = 60°$, then $\angle QSR =$

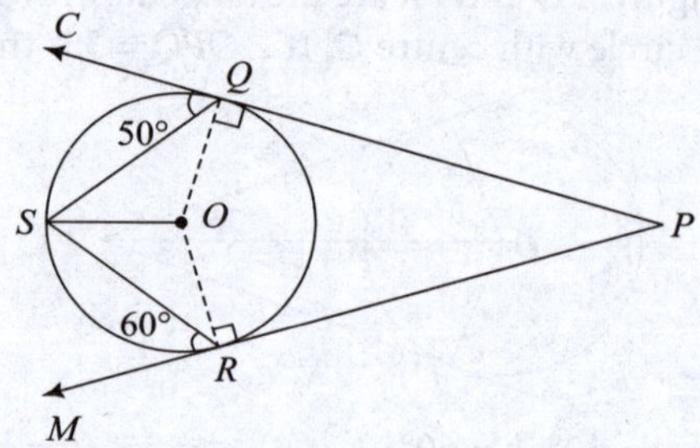

(A) 40° (B) 70°

(C) 80° (D) 45°

41. In the figure, PQ is a chord of a circle and PT is a tangent at P such that $\angle QPT = 60°$. Then $\angle PRQ =$

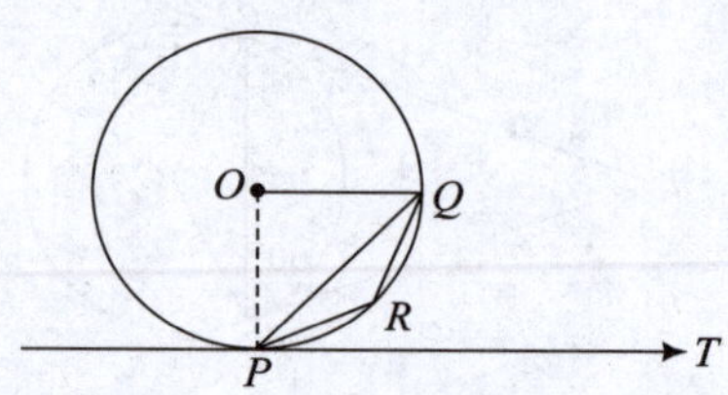

(A) 100° (B) 110°

(C) 120° (D) 150°

42. If the figure, CP and CQ are tangents from an external point C to a circle with centre O. AB is another tangent which touches the circle at R. If $CP = 11$ cm and $BR = 4$ cm, find the length of BC.

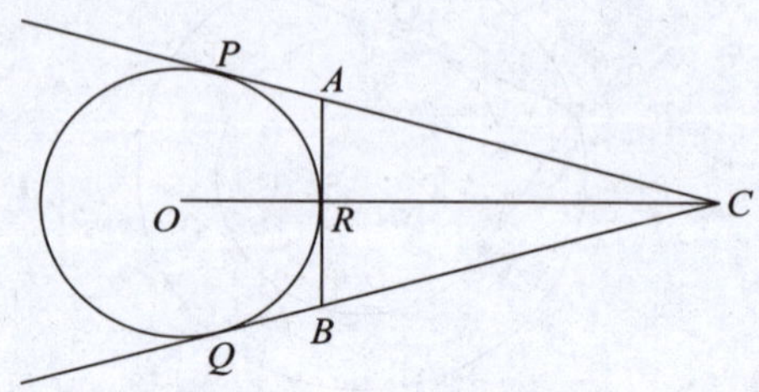

(A) 6 cm (B) 7 cm
(C) 8 cm (D) 9 cm

43. In the figure, CP and CQ are tangents to a circle with centre O. ARB is another tangent touching the circle at R. If $CP = 11$ cm and $BC = 7$ cm then what is the length of BR?

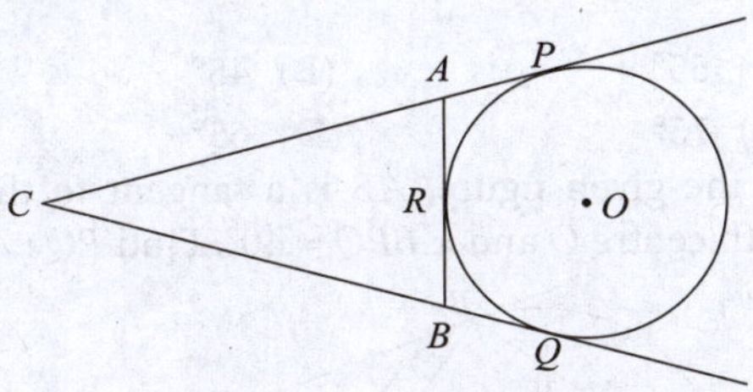

(A) 2 cm (B) 4 cm
(C) 6 cm (D) 8 cm

44. In the figure, O is the centre of the circle and BCD is tangent to it at C. Find the value of $\angle BAC + \angle ACD$.

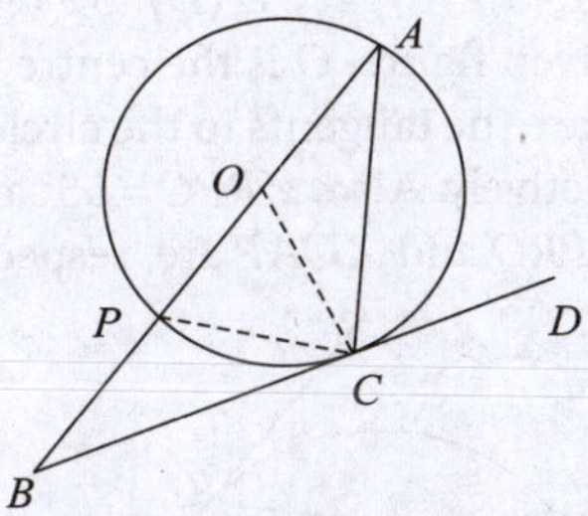

(A) 90° (B) 80°
(C) 45° (D) 60°

45. O is the centre of a circle of radius 5 cm. T is a point such that $OT = 13$ cm and OT intersects the circle at E. If AB is the tangent to the circle at E find the length of AB.

(A) 20 cm (B) $\frac{20}{3}$ cm
(C) 10 cm (D) 25 cm

46. In the figure, triangle ABC is isosceles in which $AB = AC = 6$ cm. If the radius of circle is 9 cm then what is the area of the triangle?

(A) $8\sqrt{2}$ cm^2 (B) 8 cm^2
(C) $2\sqrt{2}$ cm^2 (D) $\frac{8}{\sqrt{2}}$ cm^2

OLYMPIAD AND NTSE LEVEL EXERCISES

47. If the side of the square is 21 cm then the circumference of the circle in the figure is

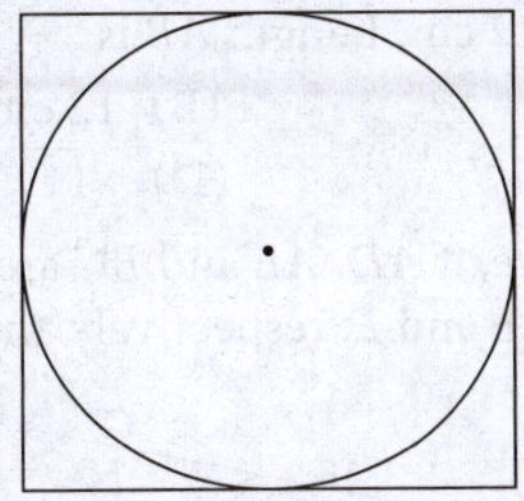

(A) 33 cm (B) 86 cm
(C) 66 cm (D) 132 cm

48. ABC is an equilateral triangle of side x cm. The radius of its incircle is

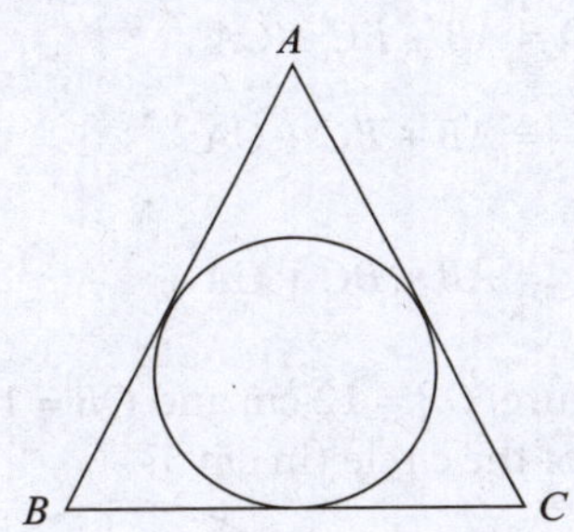

(A) $\frac{x}{\sqrt{3}}$ cm (B) $\sqrt{3}\,x$ cm
(C) $\frac{2x}{\sqrt{3}}$ cm (D) $\frac{x}{2\sqrt{3}}$ cm

49. In the figure, PQ and PT are tangents. AB is a third tangent touching the circle at C. If $PQ = 35$ cm, then the perimeter of ΔPAB is

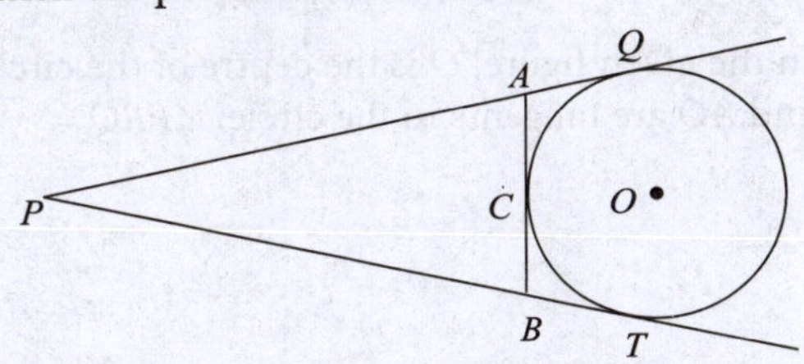

(A) 70 cm (B) $35\sqrt{2}$ cm
(C) $35\sqrt{3}$ cm (D) 35 cm

50. From a point A, the length of the tangent to a circle is 24 cm, and the distance of A from the centre is 25 cm. The radius of the circle is

(A) 7 cm (B) 12 cm
(C) 15 cm (D) 24.5 cm

51. A tangent AB at a point A of a circle of radius 5 cm meets a line through the centre O at a point B so that $OB = 12$ cm, Length AB is

(A) 10 cm (B) 12 cm
(C) 9 cm (D) $\sqrt{119}$ cm

52. In the figure, if AD, AE and BC are tangents to the circle at D, E and F, respectively, then

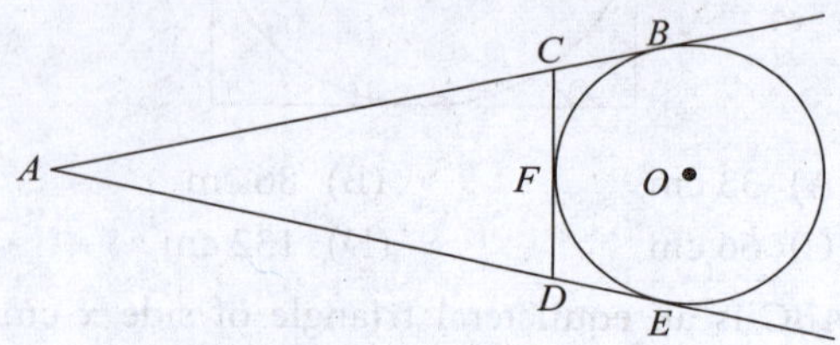

(A) $AD = AB + BC + CA$
(B) $2AD = AB + BC + CA$
(C) $\frac{AD}{2} = AB + BC + CA$
(D) $\frac{AD}{4} = AB + BC + CA$

53. In the figure, $OB = 12$ cm and $OB = 13$ cm. What is the area of the circle (in cm^2)?

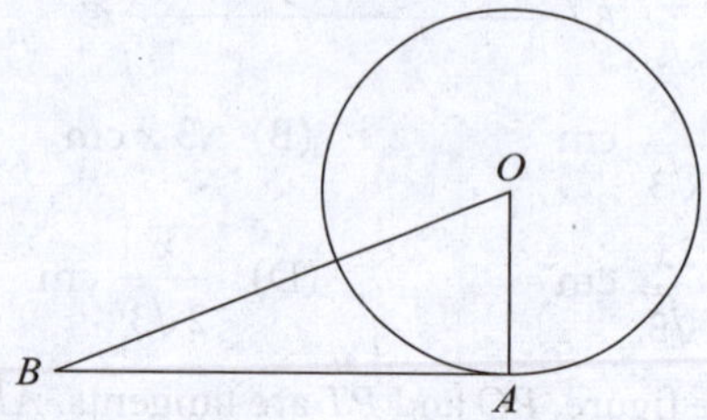

(A) 36π (B) 25π
(C) 5π (D) 50π

54. In the given figure, O is the centre of the circle. AP and AQ are tangents to the circle. $\angle PRQ =$

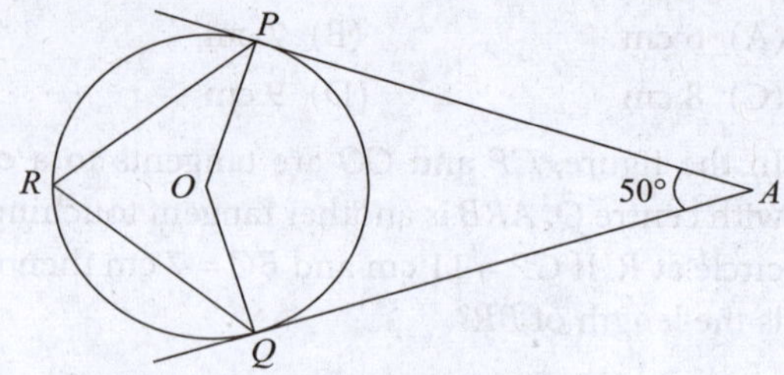

(A) 55° (B) 45°
(C) 75° (D) 65°

55. In the given figure, AB is a tangent to the circle with centre O and $\angle BPQ = 40°$. Find $PQ : AQ$.

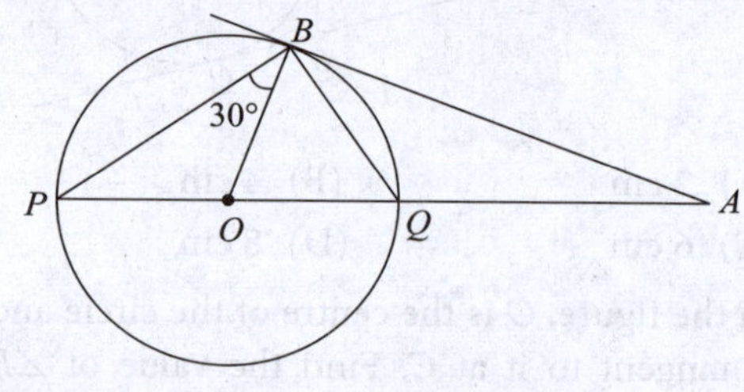

(A) 1 : 2 (B) 1 : 3
(C) 2 : 1 (D) 3 : 1

56. In the given figure, O is the centre of the circle. AC and AB are the tangents to the circle at points P and Q, respectively. Also, $\angle RPC = 55°$ and $\angle RQB = 60°$. Then $\angle PRQ$ and $\angle QAP$ are, respectively,

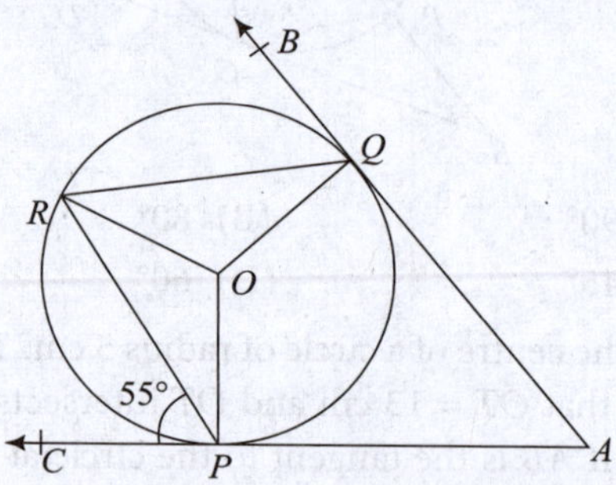

(A) 50° (B) 60°
(C) 30° (D) 40°

15 Coordinate Geometry

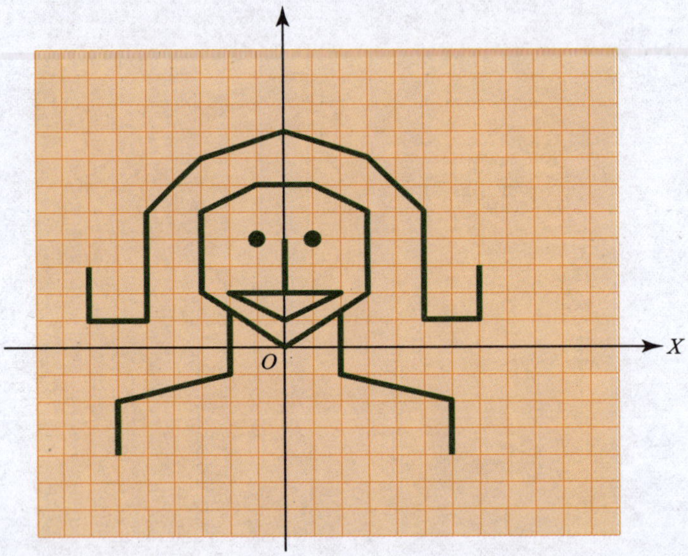

> The essence of mathematics is not to make simple things complicated, but to make complicated things simple
>
> – S. Gudder

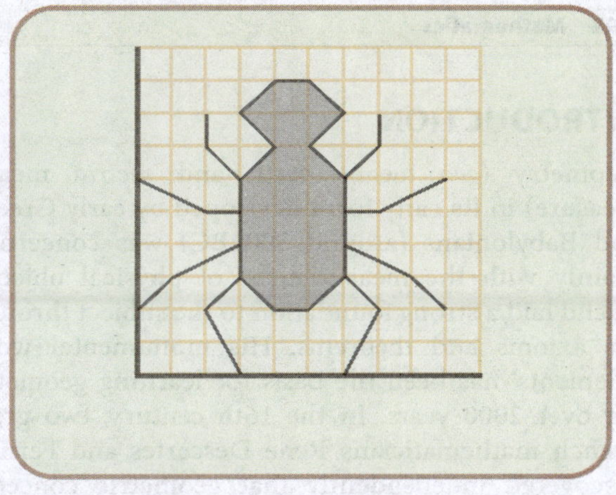

OBJECTIVES

After reading this chapter you should be able to

- know the Cartesian rectangular coordinate system
- learn to represent a point by means of coordinates in a plane
- find the distance between two points in a plane
- find the coordinates of the point dividing the segment internally and externally in a given ratio
- find the centroid of a triangle when the coordinates of the vertices of the triangle are given

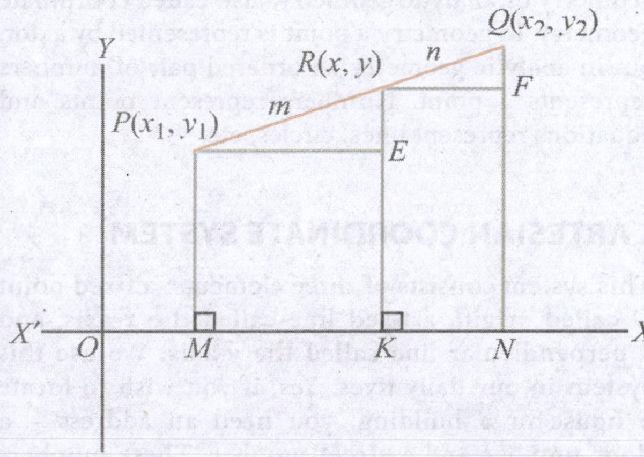

INTRODUCTION

Geometry (*geo* means earth and *metria* means measure) in its early form developed by early Greeks and Babylonians (around 300 BC) was concerned mainly with the measurement of physical objects. Euclid laid a strong foundation to the subject through his axioms and theorems. His monumental work 'Elements' has been the basis for learning geometry for over 2000 years. In the 16th century, two great French mathematicians Rene Descartes and Fermat discovered independently that geometric concepts could be studied very conveniently and often more accurately, by applying algebra. This fusion of geometry and algebra came to be known as algebraic geometry or analytic geometry, also called coordinate geometry. In geometry a point is represented by a dot, but in analytic geometry an ordered pair of numbers represents a point. Numbers represent points and equations represent lines, circles, etc.

CARTESIAN COORDINATE SYSTEM

This system consists of three elements: a fixed point O called origin, a fixed line called the x-axis, and a perpendicular line called the y-axis. We use this system in our daily lives. Yes, if you wish to locate a house or a building, you need an address – a door number and a street number. These numbers are nothing but the coordinates of the building. To locate an address, we have to start from a fixed point which is the origin. Suppose the street number is 6, then you have to walk 6 units in one direction, and if the door number is 20, you have to walk 20 units in a perpendicular direction. Thus, you are using the Cartesian coordinate system without your knowledge.

A horizontal line is drawn and a perpendicular line is drawn cutting the first one at O called origin. The horizontal line is given the name x-axis and represented by XOX' from right to left. The vertical line is given the name y-axis and represented by YOY' from top to bottom. Now, the plane is divided into four parts XOY, YOX', $X'OY'$, and $Y'OX$, respectively, called the first, second, third, and fourth quadrants. It is convention to take lengths measured in the direction of OX and OY as positive and the length measured in OX' and OY' as negative.

Now, have a look at the figure and try to understand the method of representing points on the plane.

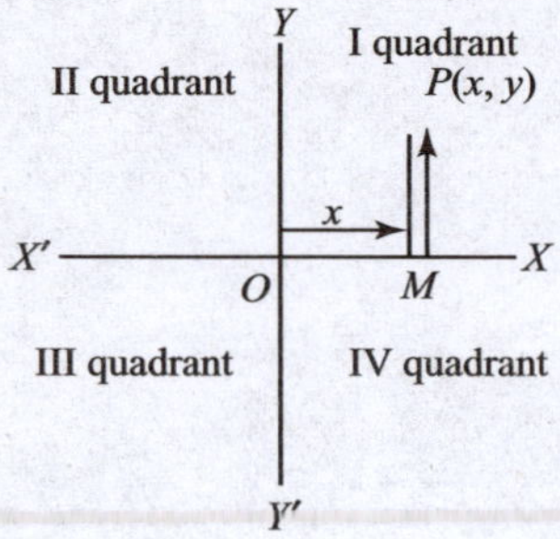

Let P be any point on the plane. Draw PM perpendicular to XOX'. The lengths OM and MP are called the coordinates of the point P, OM being the x-coordinate or abscissa and MP the y-coordinate or the ordinate of the point P. It is customary to denote OM by x and MP by y and the point P by (x, y).

The following table gives the sign of x and y, the coordinates of a point in the four quadrants.

Quadrants	x-coordinate (abscissa)	y-coordinate (ordinate)
First	+	+
Second	–	–
Third	–	–
Fourth	+	–

The system of representing points in a plane is called the Cartesian coordinate system named after the inventor Rene Descartes. When the axes are at right angles as described above, it is called a system of rectangular Cartesian coordinates. Otherwise, it is called oblique coordinate system.

DISTANCE BETWEEN TWO POINTS

Having known how to represent a point in a plane, we are in a position to find the distance between two points represented on a plane.

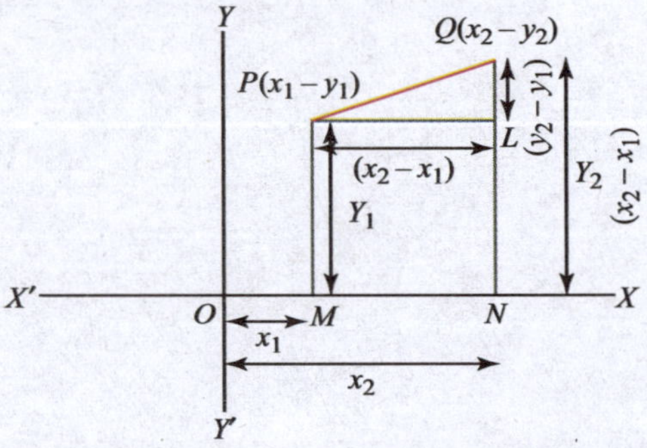

Descartes (1596–1650)

Rene Descartes was a French scientist and philosopher. He gave the first formulation of what is now known as Snell's law of refraction. He stated that the fundamental force of motion was mass times velocity (momentum). He was the first to make a graph, allowing a geometric interpretation of a mathematical function and bearing his name. He advocated systematic doubting of knowledge, believing as Plato did that sense perception and reason deceive us, and that human cannot have real knowledge of nature. The only thing that he believed he could be certain of was his doubting nature, leading to his famous phrase 'Cogito ergo sum' (I think, therefore I am). From this one phrase, he derived the rest of philosophy, including the existence of God.

Let $P(x_1, y_1)$ and $Q(x_2, y_2)$ be two points. To find the distance between them, draw PM and QN perpendicular to the x-axis and draw PL perpendicular to QM to meet QN at L.

Now, $OM = x_1$, $PM = y_1$, $ON = x_2$, $QN = y_2$

$PL = (x_2 - x_1)$ and $QL = (y_2 - y_1)$

Using Pythagoras theorem for this triangle PQL, we get $PQ^2 = PL^2 + LQ^2$.

$$PQ^2 = (x_2 - x_1)^2 + (y_2 - y_1)^2$$

$$\therefore PQ = \sqrt{(x_2 - x_1)^2 + (y_2 - y_1)^2}$$

Corollary

The distance of a point $P(x_1, y_1)$ from the origin is given by $OP = \sqrt{x_1^2 + y_1^2}$.

PQ is found by extracting a square root so that either sign might be given to it. Except in the case of lines parallel to either axis, there is no convention about the sign to be considered positive or negative, and the distance between two points is measured as a positive number. But when we have more than two points, say A, B, C, and D in a straight line, it is necessary to adhere to the rules.

$AB = -BA$

$AB + BC + CD = AD$

Coordinates of the Point Dividing the Segment Joining Two Given Points

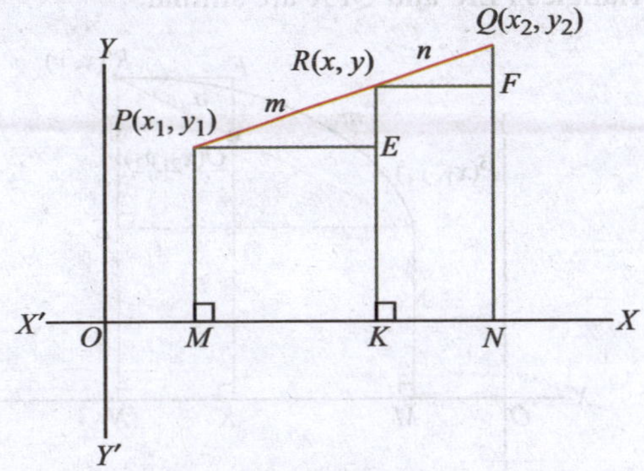

Let $P(x_1, y_1)$ and $Q(x_2, y_2)$ be the two given points and let R divide PQ internally in the ratio $m : n$. Let (x, y) be the coordinates of R.

Draw PM, QN, and RK perpendicular to the x-axis, also PE and RF perpendicular to RK and QN.

Here, $\dfrac{PR}{RQ} = \dfrac{m}{n}$

Triangles PRE and RQF are similar.

$$\therefore \frac{PR}{RQ} = \frac{PE}{RF} = \frac{m}{n}$$

But $PE = x - x_1$ and $RF = x_2 - x$.

$$\therefore \frac{x - x_1}{x_2 - x} = \frac{m}{n}$$

$$\therefore x = \frac{mx_2 + nx_1}{m + n}$$

Similarly, we can write

$$\frac{ER}{FQ} = \frac{m}{n}$$

$$\frac{y - y_1}{y_2 - y} = \frac{m}{n}$$

It gives the following:

$$y = \frac{my_2 + ny_1}{m + n}$$

Therefore, the coordinates of R are as follows:

$$\left(\frac{mx_2 + nx_1}{m + n}, \frac{my_2 + ny_1}{m + n}\right)$$

Let R' divide PQ externally, so that $\frac{PR'}{QR'} = \frac{m}{n}$.

Triangles PER' and QFR' are similar.

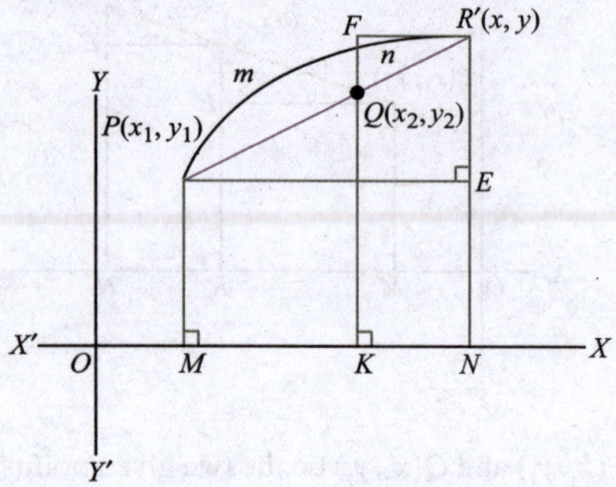

$$\frac{PR'}{R'Q} = \frac{PE}{R'F}$$

i.e., $\frac{m}{n} = \frac{PE}{R'F}$

But $PE = x - x_1$, and $R'F = x - x_2$.

$$\frac{x - x_1}{x - x_2} = \frac{m}{n}$$

$$x = \frac{mx_2 - nx_1}{m - n}$$

Similarly, $y = \frac{my_2 - ny_1}{m - n}$

Therefore, the coordinates of R' are as follows:

$$\left(\frac{mx_2 - nx_1}{m-n}, \frac{my_2 - ny_1}{m-n}\right)$$

Aliter

$$\frac{PR'}{R'Q} = -\frac{m}{n} = \frac{m}{-n}$$

If we change $n = -n$ in the coordinates of R obtained above, then the coordinates of R' are as follows:

$$\left(\frac{mx_2 - nx_1}{m-n}, \frac{my_2 - ny_1}{m-n}\right)$$

Note

The middle point of the line joining points (x_1, y_1) and (x_2, y_2) is $\left(\frac{x_1+x_2}{2}, \frac{y_1+y_2}{2}\right)$.

Centroid of a Triangle when Coordinates of the Vertices of the Triangle are Known

Let the vertices be as shown in the figure.

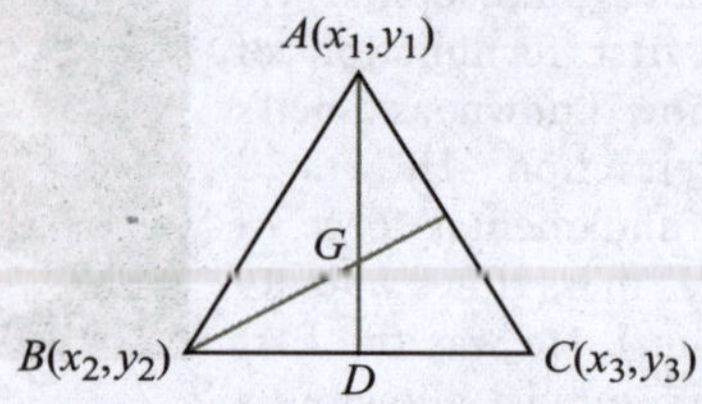

The middle point of BC is $\left(\frac{x_2+x_3}{2}, \frac{y_2+y_3}{2}\right)$.

We know G the centroid divides AD in the ratio 2 : 1 internally, and hence the coordinates of G will be the following:

$$\left(\frac{2\times\frac{x_2+x_3}{2}+1\times x_1}{2+1}, \frac{2\times\frac{y_2+y_3}{2}+1\times y_1}{2+1}\right)$$

$$= \left(\frac{x_1+x_2+x_3}{3}, \frac{y_1+y_2+y_3}{3}\right)$$

Worked Examples

1. Show that the triangle whose vertices are (–3, –4), (2, 6), (–6, 10) is a right triangle.

   Solution

   Let the points be A, B, and C, respectively.

   $AB^2 = (-3-2)^2 + (-4-6)^2 = 125$

   $BC^2 = (2+6)^2 + (6-10)^2 = 80$

   $CA^2 = (-6+3)^2 + (10+4)^2 = 205$

   Since $CA^2 = AB^2 + BC^2$, we conclude that ABC is a right-angled triangle and $\hat{B} = 90°$.

2. Prove that (–7, –3), (5, 10), (15, 8), and (3, –5) are the vertices of a parallelogram.

   Solution

   Name the points A, B, C, and D, respectively. We see that the mid-point of $AC = \left(\frac{-7+15}{2}, \frac{-3+8}{2}\right) = \left(4, \frac{5}{2}\right)$ and the mid-point of $BD = \left(\frac{5+3}{2}, \frac{10-5}{2}\right) = \left(4, \frac{5}{2}\right)$.

   In the quadrilateral $ABCD$, the diagonals bisect each other and hence it is a parallelogram.

3. Show that (4, 2), (7, 5), and (9, 7) lie on a straight line.

Solution

Name the points A, B, and C, respectively.

$\because AB^2 = (4-7)^2 + (2-5)^2 = 18$

$\therefore\ AB = 3\sqrt{2}$

$BC^2 = (7-9)^2 + (5-7)^2 = 8$

$\therefore\ BC = 2\sqrt{2}$

$AC^2 = (4-9)^2 + (2-7)^2 = 50$

$\therefore\ AC = 5\sqrt{2}$

Since $AC = AB + BC$

A, B, and C lie on a straight line.

4. OBC is a triangle where O is the origin. P and Q are points on OB and OC so that $OP : PB = 1 : 2$ and $OQ : QC = 1 : 3$. Find the coordinates of B and C.

Solution

Let (x_1, y_1) and (x_2, y_2) be the coordinates of B and C, respectively.

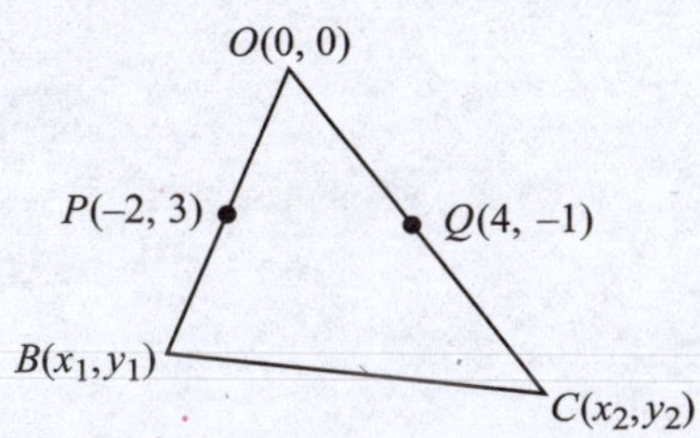

$$\therefore\ \frac{1\times x_1 + 0\times 2}{1+2} = -2$$

i.e., $x_1 = -6$

$$\frac{1\times y_1 + 0\times 2}{3} = 3$$

$y_1 = 9$

$\therefore\ B = (-6, 9)$

$$\text{Again, } C = \frac{1\times x_2 + 0\times 3}{1+3} = 4 \Rightarrow x_2 = 16$$

$$\frac{1\times y_2 + 0\times 3}{1+3} = -1 \Rightarrow y_2 = -4$$

$\therefore\ C = (16, -4)$

5. If $(3, p-5)$ is a point of trisection of the line segment joining $(4, -2)$ and $(1, 4)$, find the value of p.

Solution

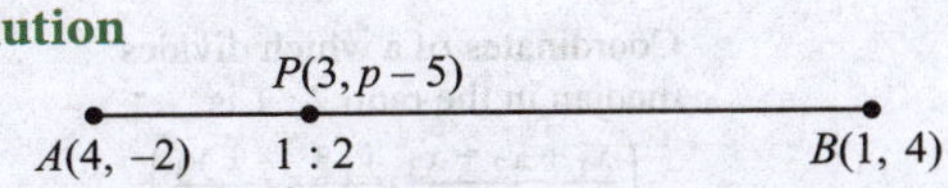

Let $A(4, -2)$, $B(1, 4)$, and $P(3, p-5)$ be as shown in the figure.

To find P, equate the y-coordinate of P:

(a) $\dfrac{1\times 4 + 2(-2)}{1+2} = p-5$

$4-4 \Rightarrow 3p - 15 \Rightarrow 3p = 15 \Rightarrow p = 5$

(b) $\dfrac{2\times 4 - 2}{1+2} \Rightarrow p-5 \Rightarrow 6 = 3p - 15$

$\Rightarrow 3p = 21 \Rightarrow p = 7$

(a) If the point $P(x, y)$ divides the line segment joining the points $A(x_1, y_1)$ and $B(x_2, y_2)$ internally in the ratio $m : n$,

$$P(x, y) = \left(\frac{mx_2 + nx_1}{m+n}, \frac{my_2 + ny_1}{m+n}\right)$$

(b) If $Q(x, y)$ divides the line segment AB externally in the ratio m n,

$$Q(x, y) = \left(\frac{mx_2 - nx_1}{m-n}, \frac{my_2 - ny_1}{m-n}\right)$$

(c) If $R(x, y)$ is the mid-point of the segment AB,

$$R(x, y) = \left(\frac{x_1 + x_2}{2}, \frac{y_1 + y_2}{2}\right)$$

(d) The points of trisection of a line segment

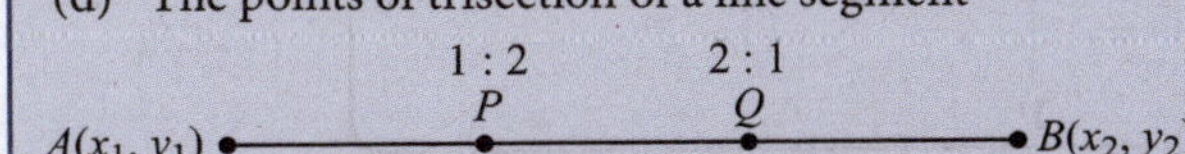

(e) The centroid of the triangle ABC whose vertices are $A(x_1, y_1)$, $B(x_2, y_2)$, and $C(x_3, y_3)$ is given by

$$G(x, y) = \left(\frac{x_1 + x_2 + x_3}{3}, \frac{y_1 + y_2 + y_3}{3}\right)$$

(f) To find the fourth vertex of a parallelogram

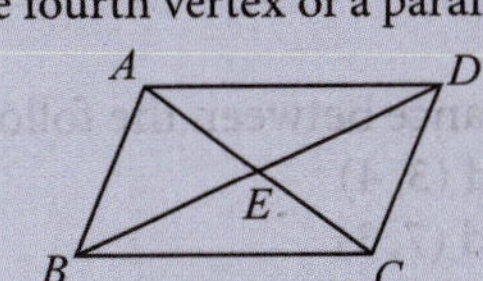

In case of parallelogram, the diagonals bisect each other. E is the middle point of AC and BD.

If AB and C are given, we can find the coordinates of the fourth vertex using the mid-point formula.

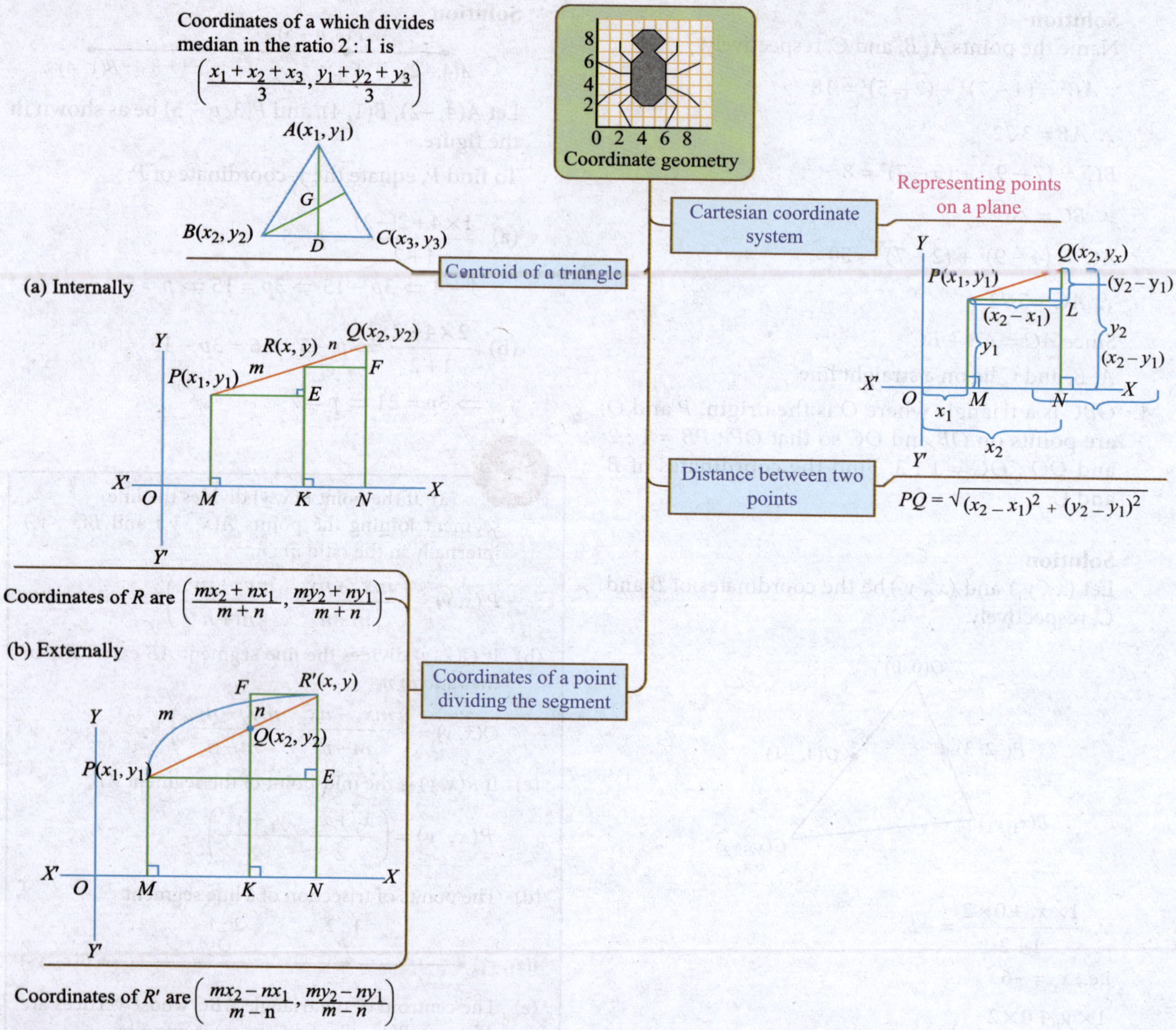

TEST YOURSELF

Level 1

1. Find the distance between the following points:
   (a) (6, 8) and (3, 4)
   (b) (5, 6) and (7, 3)
   (c) (*a*, *b*) and (*c*, *d*)
2. Show that (2, –2), (–1, 2), and (3, 5) are the vertices of an isosceles triangle.
3. Show that the points (3, 7), (6, 5), and (15, –1) lie on a straight line.
4. Find the length of the medians of the triangle formed by (1, 2), (0, 3), and (–1, –2).
5. The points $A(-1, 2)$, $B(x, y)$, and $C(4, 5)$ are such that $AB = BC$. Find a linear relation between x and y.

Level 2

6. Show that the quadrilateral with vertices (3, 2), (0, 5), (–3, 2), and (0, –1) is a square.
7. What points on the x-axis are equidistant from (7, 6) and (–3, 4)?

8. Find the ratio in which the line segment joining (2, –3) and (5, 6) is divided by the x-axis.
9. The distance between (6, 2) and (0, x) is 10. Find the value of x.
10. The centre of a circle is $(2a - 1, 3a + 1)$. Find a if the circle passes through (–3, –1) and the length of its diameter is 20 units.
11. Prove that the point $\left(-\frac{1}{14}, \frac{39}{14}\right)$ is the centre of the circle circumscribing the triangle whose vertices are (1, 1), (2, 3), and (–2, 2).
12. The segment joining (–6, 8) and (8, –6) is divided into four equal parts. Find the coordinates of the points of the intersections.

Level 3

13. The end points of a diagonal of a square are (0, –1) and (0, 5). Find the end points of the other diagonal.
14. Find a point equidistant from the points (8, 6), (8, –2), and (2, –2).
15. Find the vertices of a triangle if the mid-points of the sides of the triangle are (2, 3), (4, 1), and (3, 0). Also, find the lengths of the sides of the triangle.
16. If $x + 3y = 100$, where x and y are positive integers, the number of pairs satisfying the equation is
(A) 100
(B) 97
(C) 34
(D) none of these
17. The values of x, y, and z, respectively on solving the equations $2x + 3y = 0$, $3y + 4z = 14$, and $2x + 4z = 26$ are
(A) 3, –2, 5
(B) –3, 2, 5
(C) –3, 2, –5
(D) 3, 2, 5
18. The equations of the lines on the given graph are

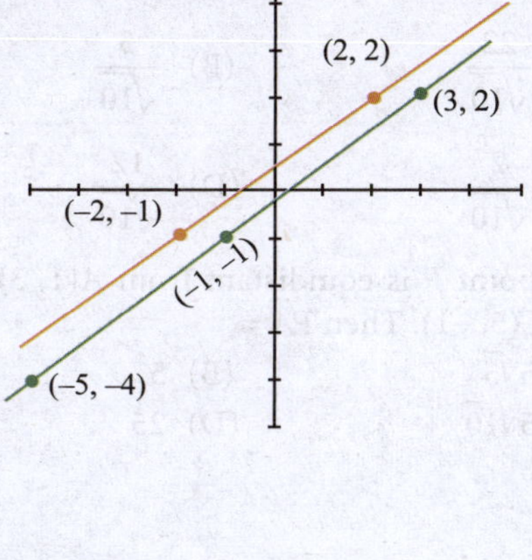

(A) $7x + 2y = 11$, $y - 2x = 3$
(B) $3x - 4y + 2 = 0$, $3x - 4y - 1 = 0$
(C) $3x - 7y = 10$, $3x - 7y = 4$
(D) $3x - 4y = 1$, $6x - 8y = 3$
19. The points (1, 5), (–7, 9), and (–10, –17) form
(A) an acute-angled triangle
(B) a right-angled triangle
(C) an obtuse-angled triangle
(D) a straight line
20. If the point (1, 1) is equidistant from the points $(a + b, b - a)$ and $(a - b, a + b)$, then
(A) $a + b = 0$
(B) $a + b = 1$
(C) $a = b$
(D) $b - a = 1$
21. If the centroid of a triangle is (3, 1) and one of the vertices is (0, –3), then the length of the median is
(A) 5
(B) $\frac{15}{2}$
(C) 15
(D) $\frac{10}{3}$
22. The distance between the points which divide the line joining (8, –5) and (3, 5) internally and externally in the ratio 3 : 2 is
(A) $12\sqrt{5}$
(B) $2\sqrt{5}$
(C) $\sqrt{160}$
(D) none of these
23. If the points (7, K), (–2, 3), and (–1, –5) are collinear, then K is
(A) 15
(B) $\frac{1}{7}$
(C) 69
(D) –69
24. The points (6, 8), (–2, 2), and (–6, –1) when joined form
(A) a straight line
(B) an equilateral triangle
(C) a right-angled triangle
(D) an isosceles triangle

Multiple Choice Questions

25. The distance between the points $P(\sqrt{5}+1, \sqrt{3}-1)$ and $Q(\sqrt{5}-2, \sqrt{3}+2)$ is
    (A) $3\sqrt{2}$ units (B) $4\sqrt{2}$ units
    (C) $3\sqrt{5}$ units (D) $2\sqrt{6}$ units

26. If the distance between the points $A(-3,4)$ and $B(x,7)$ is 5 units, then $x =$
    (A) -1 or 7 (B) 1 or -7
    (C) 5 or -3 (D) -5 or 3

27. If $P(x, y)$ is equidistant from $A(6, -1)$ and $B(2, 3)$, then
    (A) $3x - y = 3$ (B) $x - 3y = 3$
    (C) $x - y = 3$ (D) $x + y = 3$

28. A is a point on the x-axis with abscissa -8 and B is a point on the y-axis with ordinate 15. Distance AB =
    (A) 13 units (B) 15 units
    (C) 17 units (D) 23 units

29. If the points $A(-2,3)$, $B(1,2)$ and $C(k, 0)$ are collinear, then the value of k is
    (A) 5 (B) 6
    (C) 7 (D) 8

30. The ratio in which the point $P(1, 2)$ divides the line segment joining $A(-2, 1)$ and $B(7, 4)$ is
    (A) $1:2$ (B) $3:2$
    (C) $3:4$ (D) $1:3$

31. The ratio in which the line segment joining the points $A(-12, 2)$ and $B(8,3)$ is divided by the y-axis is
    (A) $2:1$ (B) $1:4$
    (C) $1:3$ (D) $3:2$

32. The vertices of ΔABC are $A(0,6)$, $B(8,12)$ and $C(8,0)$. The coordinates of its incentre are
    (A) $(-4, 3)$ (B) $(5, 6)$
    (C) $(8, 11)$ (D) $\left(\frac{16}{3}, 6\right)$

33. A circle passes through the points $A(2,-9)$, $B(5, -8)$ and $C(2,1)$. The centre of the circle is
    (A) $(2, -4)$ (B) $(-3,4)$
    (C) $\left(3, \frac{-16}{3}\right)$ (D) None of these

34. One end of the diameter of a circle is $A(4, 1)$ and its centre is $C(3,3)$. The coordinates of the other end of the diameter are
    (A) $B(5, 2)$ (B) $B(-4, 1)$
    (C) $B(4, -1)$ (D) $B(2, 5)$

OLYMPIAD AND NTSE LEVEL EXERCISES

35. The points (a, b), (c, d) and $\left(\frac{kc+la}{k+l}, \frac{kd+lb}{k+l}\right)$ are
    (A) Vertices of an equilateral triangle
    (B) Vertices of an isosceles triangle
    (C) Vertices of a right angled triangle
    (D) Collinear

36. The points $(0, 0)$, $(a, 0)$ and $\left(\frac{a}{2}, \frac{a\sqrt{3}}{2}\right)$ are vertices of
    (A) Isosceles triangle (B) Equilateral triangle
    (C) Scalene triangle (D) None of these

37. If the vertices of a triangle are $(a, 1)$, $(b, 3)$ and $(4, c)$, then the centroid of the triangle will lie on x-axis, if
    (A) $a + c = -4$ (B) $a + b = -4$
    (C) $c = -4$ (D) $b + c = -4$

38. If the points $(x + 1, 2)$, $(1, x + 2)$ and $\left(\frac{1}{x+1}, \frac{2}{x+1}\right)$ are collinear, then x is
    (A) 4 (B) 0
    (C) -4 (D) None of these

39. If the points $(0, 0)$, $(2, 2\sqrt{3})$ and (p, q) be the vertices of an equilateral triangle, then $(p, q) =$
    (A) $(0, -4)$ (B) $(4, 4)$
    (C) $(4, 0)$ (D) $(5, 0)$

40. The length of altitude through A of the triangle ABC, where $A \equiv (-3, 0)$; $B \equiv (4, -1)$ and $C \equiv (5, 2)$, is
    (A) $\frac{22}{\sqrt{10}}$ (B) $\frac{5}{\sqrt{10}}$
    (C) $\frac{2}{\sqrt{10}}$ (D) $\frac{12}{\sqrt{10}}$

41. The point P is equidistant from $A(1, 3)$, $B(-3, 5)$ and $C(5, -1)$. Then $PA =$
    (A) $5\sqrt{5}$ (B) 5
    (C) $5\sqrt{10}$ (D) 25

42. One vertex of an equilateral triangle is (2, 2) and its centroid is $\left(-\frac{2}{\sqrt{3}}, \frac{2}{\sqrt{3}}\right)$ then length of its side is

(A) $4\sqrt{2}$ (B) $4\sqrt{3}$
(C) $3\sqrt{2}$ (D) $5\sqrt{2}$

43. State T for true and F for false.

(i) The triangle with vertices $(a, 0)$, $(0, b)$ and (0, 0) is a right triangle.

(ii) Centre of the circle passing through (0, 8) and (0, –8) is (1, 0).

(iii) The quadrilateral having vertices (–8, 8), (2, 8), (6, 2) and (–4, 2) is a parallelogram.

(iv) Radius of the circle with centre (1, 1) passing through (–2, 5) is 4 units.

	(i)	(ii)	(iii)	(iv)
(A)	T	T	T	F
(B)	T	F	T	F
(C)	F	T	T	F
(D)	F	F	F	T

44. Fill in the blanks.

(i) Points (–3, 7), (6, 4) and ___*P*___ lie on the same line.

(ii) Points (–1, 3), ___*Q*___ and (7, 3) form a right triangle.

(iii) Points (–1, 4), (2, 4) and ___*R*___ lie on the same circle.

	P	Q	R
(A)	(7, 7)	(2, 7)	(3, 5)
(B)	(2, 7)	(7, 7)	(3, 5)
(C)	(3, 5)	(7, 7)	(2, 7)
(D)	(2, 7)	(3, 5)	(7, 7)

GAME TIME

This activity will help you gain speed in plotting points. You will get practice in familiarising yourself with various positions of points on the Cartesian plane.

Materials Required

1. Graph sheet
2. Pencil and scale

Procedure

Take a graph sheet, and mark the origin *O* in the middle, at the bottom of the page.

Take a suitable scale.

Plot the points (0, –2), (2, –4), (2, –1), (5, –4), (5, –1), (2, 2), (2, 11), and (0, 15) with a scale, and connect the points in the same order.

Next, find the reflections of all the points in the *Y*-axis and connect them in the same order in which you plotted.

What figure do you get?

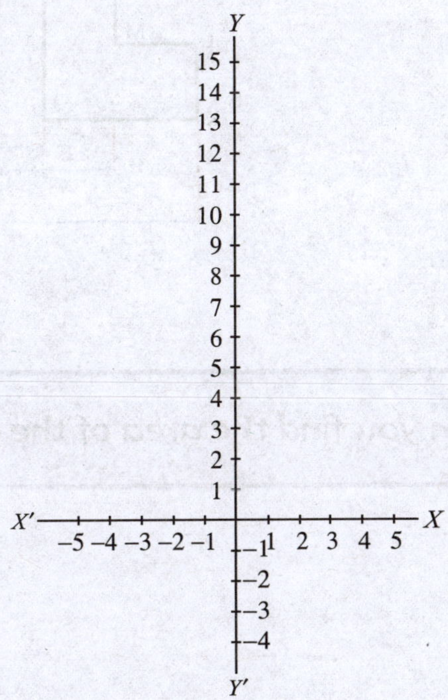

16 Area of a Triangle

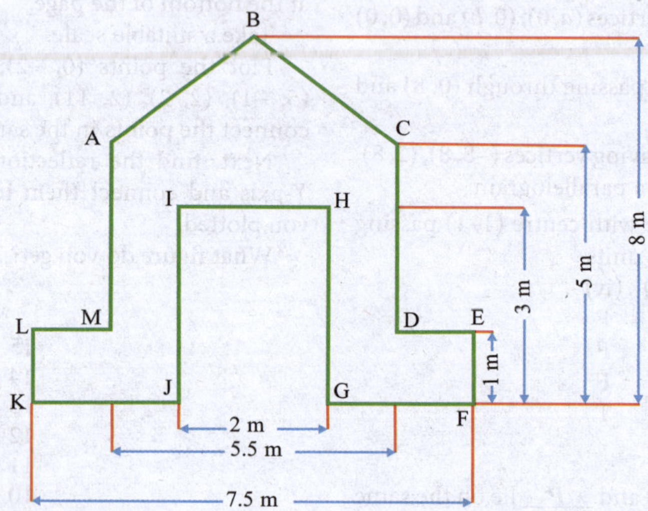

Can you find the area of the polygon ABCDEFGHIJKLMA?

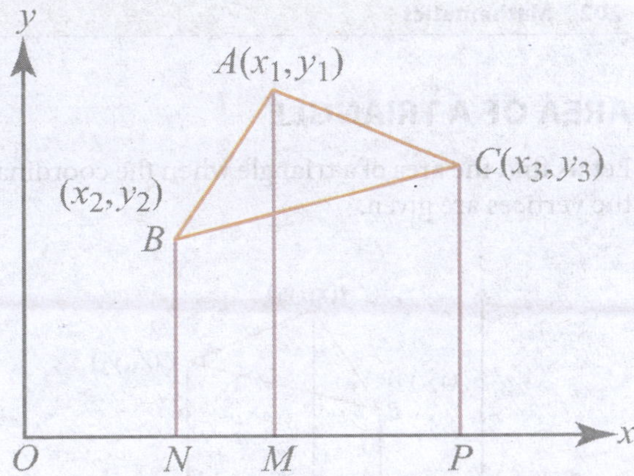

OBJECTIVES

After reading this chapter you should be able to

- find the area of a triangle when coordinates of its vertices are given
- learn about the condition for collinearity of three points
- find the area of a quadrilateral when coordinates of its vertices are given

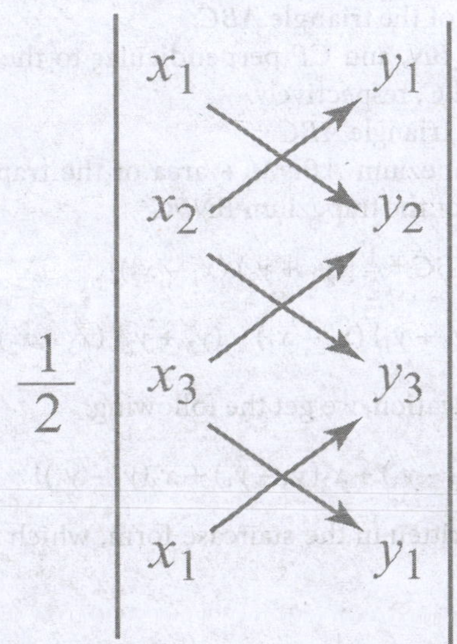

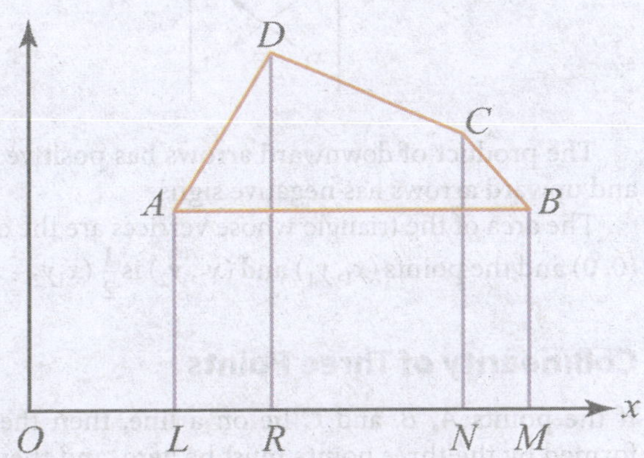

AREA OF A TRIANGLE

Let us find the area of a triangle when the coordinates of the vertices are given.

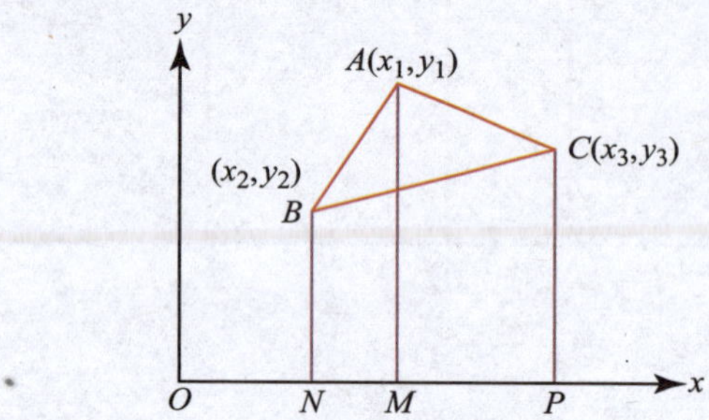

Let $A(x_1, y_1)$, $B(x_2, y_2)$, and $C(x_3, y_3)$ be the coordinates of the vertices of the triangle *ABC*.

Draw *AM*, *BN*, and *CP* perpendicular to the *x*-axis from *A*, *B*, and *C*, respectively.

Area of the triangle *ABC* =
area of the trapezium *ABNM* + area of the trapezium *AMPC* – area of the trapezium *BNPC*

i.e., area of $\Delta ABC = \frac{1}{2}[(y_1 + y_2)(x_1 - x_2)]$

$$+ \frac{1}{2}(y_3 + y_1)(x_3 - x_1) - (y_3 + y_2)(x_3 - x_2)$$

On simplification, we get the following:

$$\text{Area} = \frac{1}{2}[x_1(y_2 - y_3) + x_2(y_3 - y_1) + x_3(y_1 - y_2)]$$

This can be written in the staircase form, which is easy to remember.

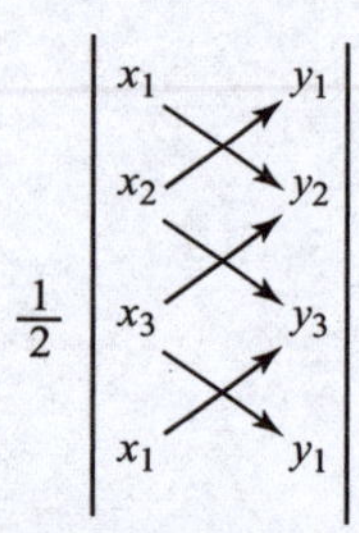

The product of downward arrows has positive signs and upward arrows has negative signs.

The area of the triangle whose vertices are the origin (0, 0) and the points (x_1, y_1) and (x_2, y_2) is $\frac{1}{2}(x_1y_2 - x_2y_1)$.

Collinearity of Three Points

If the points *A*, *B*, and *C* lie on a line, then the area formed by the three points must be zero, and therefore the condition for three points to be collinear is as follows:

$$\frac{1}{2}[x_1(y_2 - y_3) + x_2(y_3 - y_1) + x_3(y_1 - y_2)] = 0$$

i.e., $x_1(y_2 - y_3) + x_2(y_3 - y_1) + x_3(y_1 - y_2) = 0$

i.e., $x_1y_2 + x_2y_3 + x_3y_1 - x_2y_1 - x_3y_2 - x_1y_3 = 0$

Coordinates of the vertices are taken in anticlockwise direction. Otherwise, we will get the area with a negative sign. In this case, only the numerical value has to be taken.

AREA OF A QUADRILATERAL

Let us find the area of a quadrilateral when the coordinates of the vertices are given.

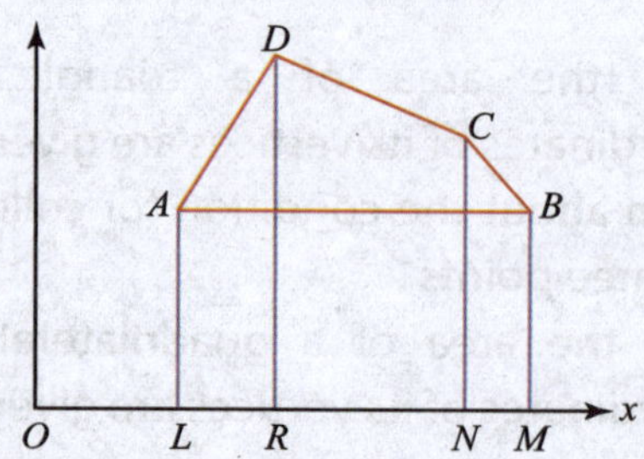

Let the vertices of the quadrilateral, taken in order, be *A*, *B*, *C*, and *D* and let their coordinates be respectively (x_1, y_1), (x_2, y_2), (x_3, y_3), and (x_4, y_4).

Draw *AL*, *BM*, *CN*, and *DR* perpendicular to the *x*-axis.

Now, the area of the quadrilateral *ABCD* = area of the trapezium *ALRD* + area of the trapezium *DRNC* + area of the trapezium *CNMB* area of the trapezium *ALMB*

$$= \frac{1}{2}(x_4 - x_1)(y_1 + y_4) + \frac{1}{2}(x_3 - x_4)(y_3 + y_4)$$
$$+ \frac{1}{2}(x_2 - x_3)(y_2 + y_3) - \frac{1}{2}(x_2 - x_1)(y_1 + y_2)$$

On simplification, we get the following:

$$= \frac{1}{2}\{(x_1y_2 - x_2y_1) + (x_2y_3 - x_3y_2) + (x_3y_4 - x_4y_3)$$
$$+ (x_4y_1 - x_1y_4)\}$$

In a similar manner, it may be shown that the area of a polygon of n sides, the coordinates of whose vertices, taken in order, are $(x_1, y_1), (x_2, y_2), (x_3, y_3), \ldots (x_n, y_n)$, is the following:

$$\frac{1}{2}\{(x_1y_2 - x_2y_1) + (x_2y_3 - x_3y_2) + \cdots + (x_ny_1 - x_1y_n)\}$$

Worked Examples

1. Find the area of a quadrilateral whose vertices are (1, 1), (3, 4), (5, 2), and (4, 7).

Solution

Quadrilateral $ADCB$

= triangle ADC + triangle ACB

$$\Delta ADC = \frac{1}{2}[1(-7+2) + 4(-2-1) + 5(1+7)]$$

$$= \frac{23}{2}$$

$$= [1(-2-4) + 5(4-1) + 3(1+2)] = 9$$

Hence, the area of quadrilateral is 11 + 9 = 20 square units.

2. Find the condition for points (x, y) to lie on the straight line joining the points $(0, b)$ and $(a, 0)$.

Solution

The condition is that the area of the triangle formed by $(a, 0)$, $(0, h)$, and (x, y) is zero.

$$\frac{1}{2}[a(b-y) + 0(y-0) + x(0-b)] = 0$$

$$= ab - ay - bx = 0$$

Therefore, $\frac{x}{a} + \frac{y}{b} = 1$ is the required condition.

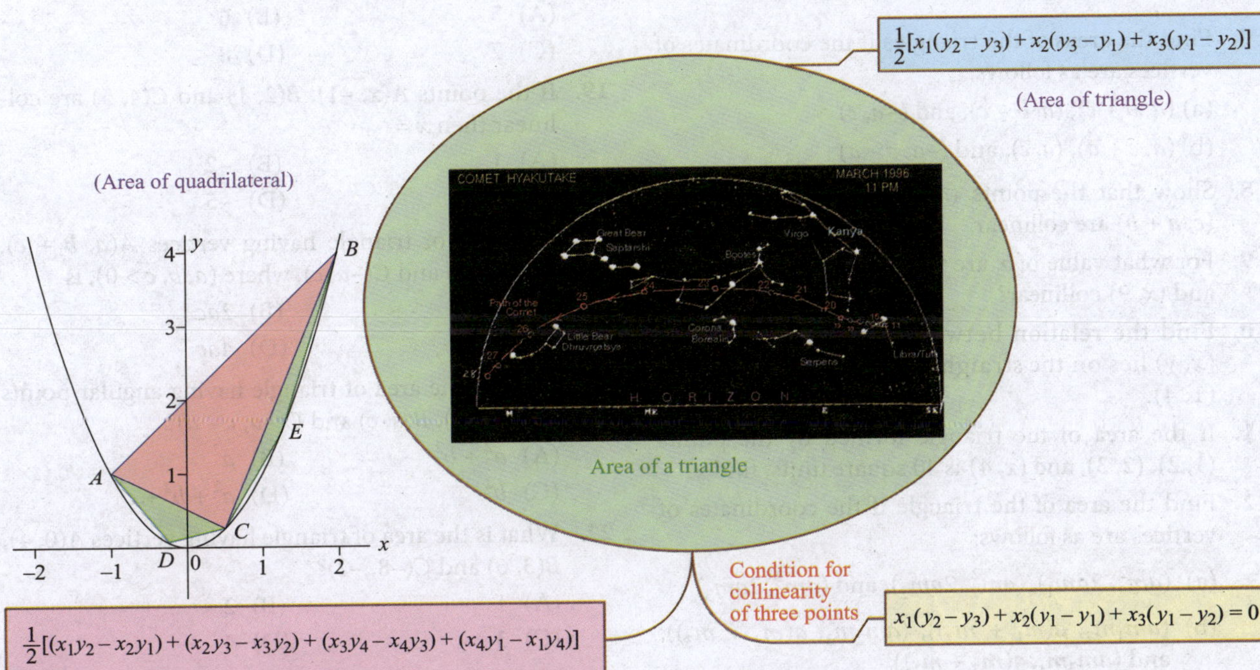

TEST YOURSELF

Level 1

1. Find the area of the triangle formed by the following points:

   (a) (2, 3), (3, –1), (–4, 2)

   (b) (–2, 4), (10, –2), (–2, 3)

2. Find the area of the quadrilateral formed by the following:

   (a) (–1, 6), (–3, –9), (5, –8), (3, 9)

   (b) (6, 2), (10, 16), (–12, 6), (–4, 18)

3. Show that the following points lie on a straight line.

   (a) (–1, 6), (–10, 12), (–16, +16)

   (b) (1, –1), (7, –4), (–5, 2)

Level 2

4. The area of the quadrilateral whose vertices are (1, 1), (3, 4), (5, –2), and (4, –7) is
(A) 20½ $unit^2$ (B) 19½ $unit^2$
(C) 19 $unit^2$ (D) 17½ $unit^2$

5. The vertices of a triangle are $A(0, 0)$, $B(6, 0)$, and $C(0, 8)$. The area of the triangle formed by the mid-points of the sides of the triangle ABC as vertices is
(A) 24 $unit^2$ (B) 12 $unit^2$
(C) 6 $unit^2$ (D) 48 $unit^2$

6. The area of the pentagon whose vertices are (1, 1), (7, 21), (7, –3), (12, 2), and (0, –3) is
(A) 145 (B) 146 (C) 175 (D) 176

Level 3

7. Find the areas of the triangles if the coordinates of vertices are as follows:
(a) $(a, b + c)$, $(a, b - c)$, and $(-a, c)$
(b) $(a, c + a)$, (a, c), and $(-a, c - a)$

8. Show that the points $(a, b + c)$, $(b, c + a)$, and $(c, a + b)$ are collinear.

9. For what value of x are the points (–3, 12), (7, 6), and $(x, 9)$ collinear?

10. Find the relation between x and y if the point (x, y) lies on the straight line joining (2, –3) and (1, 4).

11. If the area of the triangle formed by the points (1, 2), (2, 3), and $(x, 4)$ is 40 square units, find x.

12. Find the area of the triangle if the coordinates of vertices are as follows:
(a) $(am_1^2, 2am_1)$, $(am_2^2, 2am_2)$, and $(am_3^2, 2am_3)$
(b) $(am_1m_2, a(m_1 + m_2))$, $(am_2m_3, a(m_2 + m_3))$, and $(am_3m_1, a(m_3 + m_1))$
(c) $\left(am_1, \frac{a}{m_1}\right)\left(am_2, \frac{a}{m_2}\right)$, and $\left(am_3, \frac{a}{m_3}\right)$

13. For what value of y are the points (1, 4), (3, y), and (–3, 16) collinear?

14. Find the value of k for which the area formed by the triangle with vertices $(k, 2k)$, (–2, 6) and (3, 1) is 5 square units.

15. Find the area of the pentagon whose vertices are $A(4, 3)$, $B(-5, 6)$, $C(-7, -2)$, $D(0, -7)$, and $E(3, -6)$.

16. Find the lengths of the altitudes of the triangle whose vertices are $A(1, 3)$, $B(5, -6)$, and $C(7, -4)$. (*Hint*: Use formula for the area of the triangle as ½ base × height).

Multiple Choice Questions

17. The area of quadrilateral $ABCD$ with $A(1, 1)$, $B(7, -3)$, $C(12, 2)$ and $D(7, 21)$ as its vertices is
(A) 35 sq. units (B) 65 sq. units
(C) 132 sq. units (D) 115 sq. units

18. If the points $A(-2, 3)$, $B(1, 2)$ and $C(K, 0)$ are collinear, then $K =$
(A) 5 (B) 6
(C) 7 (D) 8

19. If the points $A(x, -1)$, $B(2, 1)$ and $C(4, 5)$ are collinear then $x =$
(A) 1 (B) –2
(C) –4 (D) –5

20. The area of triangle having vertices $A(a, b + c)$, $B(a, b - c)$ and $C(-a, c)$, where $(a, b, c > 0)$, is
(A) ac (B) $2ac$
(C) $3ac$ (D) $4ac$

21. What is the area of triangle having angular points $A(a, c + a)$, $B(a, c)$ and $C(-a, c - a)$?
(A) $a^2 + b^2$ (B) a^2
(C) b^2 (D) $a^2 + b^2 + c^2$

22. What is the area of triangle having vertices $A(0, 4)$, $B(3, 6)$ and $C(-8, -2)$?
(A) 1 (B) 2
(C) 3 (D) 4

23. The area of triangle having angular points $P(a \cos \theta_1, b \sin \theta_1)$, $Q(a \cos \theta_2, b \sin \theta_2)$ and $R(a \cos \theta_3, b \sin \theta_3)$ is
(A) $2ab\sin\frac{\theta_1-\theta_2}{2}\sin\frac{\theta_2-\theta_3}{2}\sin\frac{\theta_3-\theta_1}{2}$
(B) $2ab\sin(\theta_1+\theta_2)\sin(\theta_2+\theta_3)\sin(\theta_3+\theta_1)$
(C) $ab\sin\frac{\theta_1-\theta_2}{2}\sin\frac{\theta_2-\theta_3}{2}\sin\frac{\theta_3-\theta_1}{2}$
(D) none of these

24. Area of triangle having angular points $A\{am_1m_2, a(m_1 + m_2)\}$, $B\{am_2m_3, a(m_2 + m_3)\}$ and $C\{am_3m_1, a(m_3 + m_1)\}$ is
(A) $(m_1 - m_2)(m_2 - m_3)(m_3 - m_1)$
(B) $a^2(m_1 - m_2)(m_2 - m_3)(m_3 - m_1)$
(C) $\frac{1}{2}a^2(m_1 - m_2)(m_2 - m_3)(m_3 - m_1)$
(D) none of these

25. The area of quadrilateral coordinates of whose angular points are $A(1, 1)$, $B(3, 4)$, $C(5, -2)$ and $D(4, -7)$ is
(A) 20 (B) $20\frac{1}{2}$
(C) 21 (D) $21\frac{1}{2}$

26. The area of quadrilateral the coordinates of whose angular points, taken in order are $P(-1, 6)$, $Q(-3, -9)$, $R(5, -8)$ and $S(3, 9)$ is
(A) 96 (B) 48
(C) 24 (D) 12

27. If $P\left(\frac{-1}{2}, 3\right), Q(-5, 6)$ and $R(-8, 8)$ are three given points, which of the following is correct?
(A) Points form a triangle having non-zero area
(B) Points lie on a circle
(C) Points lie on a straight line
(D) None of these

OLYMPIAD AND NTSE LEVEL EXERCISES

28. Area of triangle formed by the points $A(a, b + c)$, $B(b, c + a)$, and $C(c, a + b)$ depends upon
(A) a
(B) b
(C) c
(D) independent from a, b and c

29. Let $A(6, 3)$, $B(-3, 5)$, $C(4, -2)$ and $D(x, 3x)$ be four points. If the ratio of area of ΔDBC and ΔABC is 1 : 2, then the value of x is
(A) $\frac{11}{8}$ (B) $\frac{8}{11}$
(C) 3 (D) None of these

30. If the points $(x + 1, 2)$, $(1, x + 2)$ and $\left(\frac{1}{x+1}, \frac{2}{x+1}\right)$ are collinear, then x is
(A) 4 (B) 0
(C) –4 (D) None of these

31. The locus of P such that area of ΔPAB is 12 sq. units, where $A(2, 3)$ and $B(-4, 5)$ is
(A) $(x + 3y - 1)(x + 3y - 23) = 0$
(B) $(x + 3y + 1)(x + 3y - 23) = 0$
(C) $(3x + y - 1)(3x + y - 23) = 0$
(D) $(3x + y + 1)(3x + y + 23) = 0$

32. If $A(1, p^2)$; $B(0, 1)$ and $C(p, 0)$ are the coordinates of three points then the value of p for which the area of the triangle ABC is minimum, is
(A) $\frac{1}{\sqrt{3}}$ (B) $-\frac{1}{\sqrt{3}}$
(C) $\frac{1}{\sqrt{3}}$ or $-\frac{1}{\sqrt{3}}$ (D) None of these

33. $OPQR$ is a square with M and N as the middle points of the sides PQ and QR, respectively. The ratio of the areas of the square and the triangle OMN is
(A) 4 : 1 (B) 2 : 1
(C) 8 : 3 (D) 4 : 3

34. Find the area of quadrilateral whose vertices are $(-4, 5)$, $(0, 7)$, $(5, -5)$ and $(-4, -2)$.
(A) $54\frac{1}{2}$ sq. units (B) $56\frac{1}{2}$ sq. units
(C) $58\frac{1}{2}$ sq. units (D) $60\frac{1}{2}$ sq. units

35. Four points $A(6, 3)$, $B(-3, 5)$, $C(4, -2)$ and $D(x, 2x)$ are given in such a way that $\frac{\text{Area of } \Delta DBC}{\text{Area of } \Delta ABC} = \frac{1}{2}$. The value of x is
(A) $\frac{5}{6}$ (B) $\frac{2}{3}$
(C) $\frac{11}{6}$ (D) $\frac{8}{3}$

36. If area of ΔOPB = area of ΔOPA such that O is origin, $A \equiv (6, 0)$, $B \equiv (0, 4)$ and P lies on line $x + y = 1$ then find the possible coordinates of P.
(A) (1, 2) (B) (–2, 3)
(C) (3, –2) (D) (3, –1)

37. The midpoints of the sides of triangle ABC are $(-1, -2)$, $(6, 1)$, and $(3, 5)$. The area of ΔABC is
(A) 72 sq units (B) 73 sq units
(C) 74 sq units (D) 75 sq units

17 Equation of a Straight Line – I

Letters A through Z are written below. But why are some letters written below the dividing line?

A E F H I K L M N T V W X Y Z

B C D G J O P Q R S U

No human investigation can be called real science if it cannot be demonstrated mathematically."

– Leonardo da Vinci

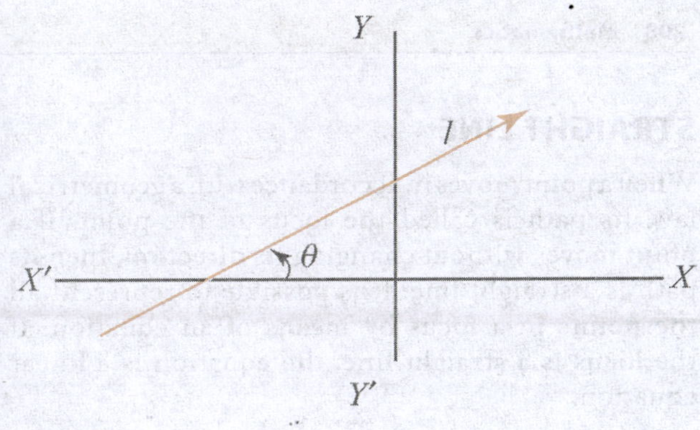

OBJECTIVES

After reading this chapter you should be able to

- define the slope of a line
- find the slope of a line passing through two given points
- find the equation of a line
- find the equation of a straight line in terms of intercepts made by it on the areas

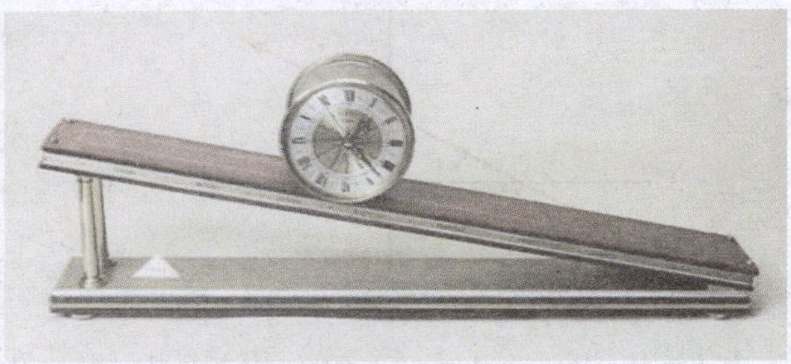

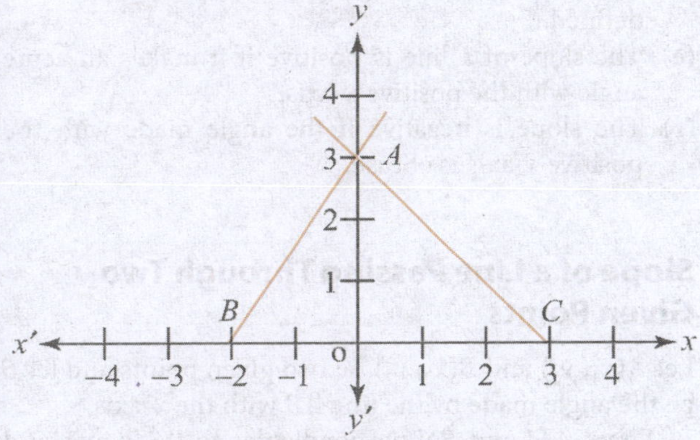

STRAIGHT LINE

When a point moves in accordance with a geometrical law, its path is called the locus of the point. If a point moves without changing its direction, then its path is a straight line. It is possible to represent all the points in a locus by means of an equation. If the locus is a straight line, the equation is a linear equation.

Slope or Gradient of Line

The slope of a line is defined as the tangent of the angle which the straight line makes with the positive direction of the x-axis.

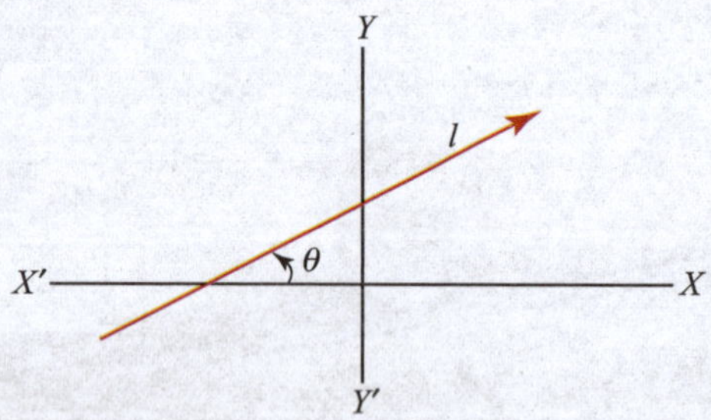

If l is the line making an angle θ with the positive direction of the x-axis, then its slope (gradient) is tan θ which is usually represented by the small letter m.

i.e., $\tan \theta = m$

θ is called inclination of the line.

(a) Slope of the x-axis = tan 0 = 0.
(b) Slope of any line parallel to the x-axis is also 0.
(c) Slope of the y-axis is tan 90° which is not defined.
(d) Slope of any line parallel to the y-axis is also not defined.
(e) The slope of a line is positive if it makes an acute angle with the positive x-axis.
(f) The slope is negative if the angle made with the positive x-axis is obtuse.

Slope of a Line Passing Through Two Given Points

Let $A(x_1, y_1)$ and $B(x_2, y_2)$ be two given points and let θ be the angle made by the line AB with the x-axis.

Draw AM and BN perpendicular to the x-axis and AP perpendicular to BN.

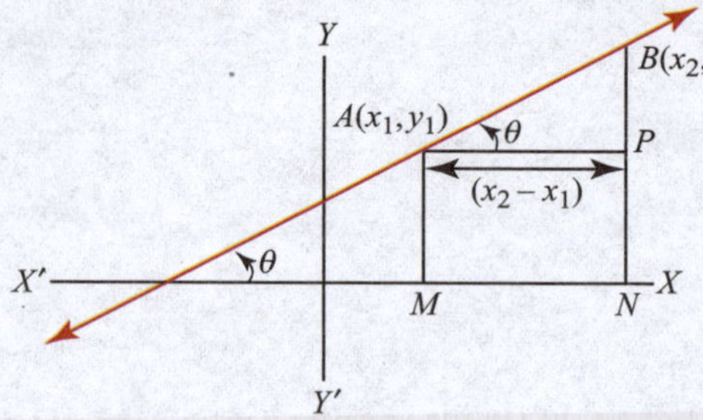

Now $B\hat{A}P = \theta°$

$AP = x_2 - x_1$

$BP = y_2 - y_1$

Therefore, $\tan \theta = \dfrac{y_2 - y_1}{x_2 - x_1}$ is the slope of the line joining A and B.

(a) Slopes of parallel lines are equal.
Let AB and CD be two parallel lines making θ and ϕ with the x-axis.
Since the lines are parallel $\theta = \phi$, and hence $\tan \theta = \tan \phi$.
Therefore, slopes of parallel lines are equal.

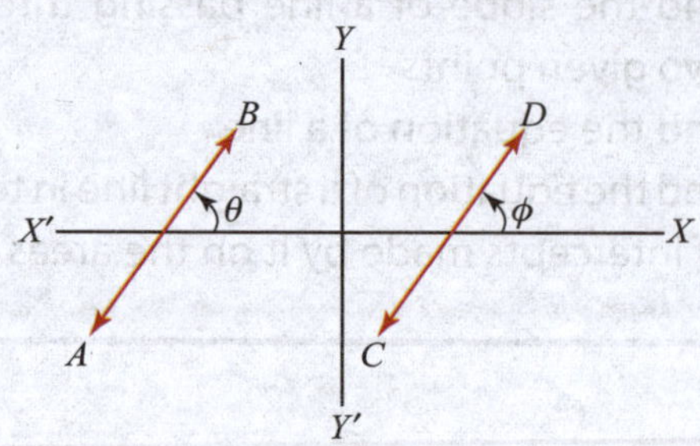

(b) Slopes of perpendicular lines

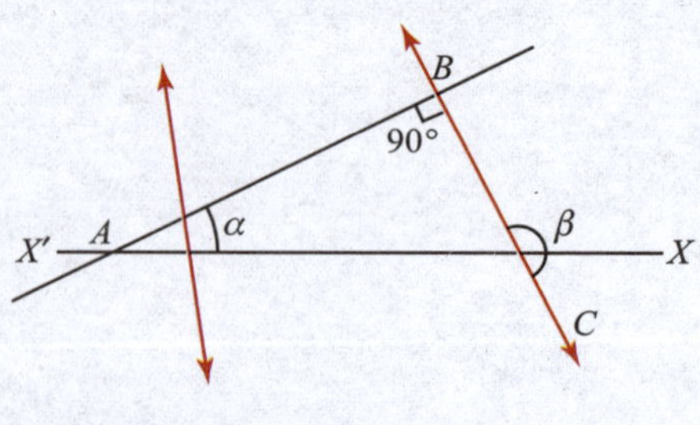

Let AB and CB be perpendicular.
If α and β are the angles made by AB and CB with the x-axis, then we have the following:
$\beta = 90° + \alpha$
$\tan \beta = \tan (90° + \alpha) = -\cot \alpha$

i.e., $\tan\beta = -\dfrac{1}{\tan\alpha}$

i.e., $\tan\alpha\tan\beta = -1$

Hence, the product of slopes of two perpendicular lines is –1.

Conversely, if the product of slopes of two lines is equal to –1, then the lines are perpendicular to each other.

EQUATION OF A LINE

(a) We take a general point (x, y) on the line.
(b) We find the relation between x and y with the given conditions.

Let us find the equation of a line which makes an intercept c on the y-axis and having slope m.

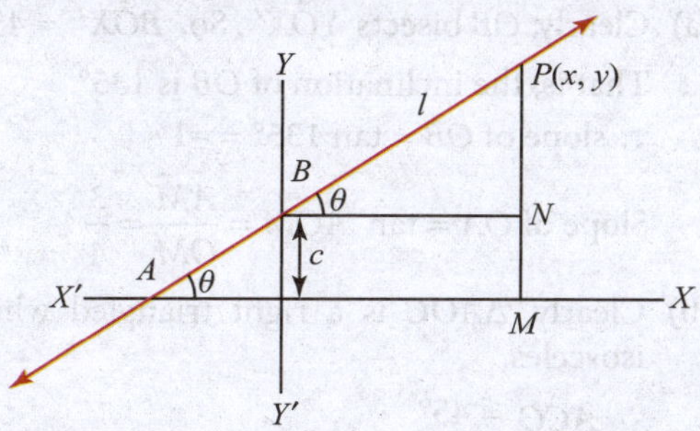

Let l be the line making an angle θ with the positive direction of the x-axis and c the intercept made by it on the y-axis and let A and B be the points of intersection of l with the x and y-axes.

Now, take any point P(x, y) on l.

Draw PM perpendicular to the x-axis and draw BN perpendicular to PM.

In triangle PBN, $P\hat{B}N = \theta$.

$PN = PM - NM$

$= y - c$

$BN = x$

$\therefore \tan\theta = \dfrac{PN}{BN} = \dfrac{y-c}{x} = m$

Therefore, $y - c = mx \Rightarrow y = mx + c$ which is the equation of the line.

(a) If c is zero, then the line passes through the origin.
(b) (i) The equation of a line passing through the origin is $y = mx$.
 (ii) For different values of m, we get different lines passing through the origin.

Equation of a Line in Terms of Intercepts Made by It on the Axes

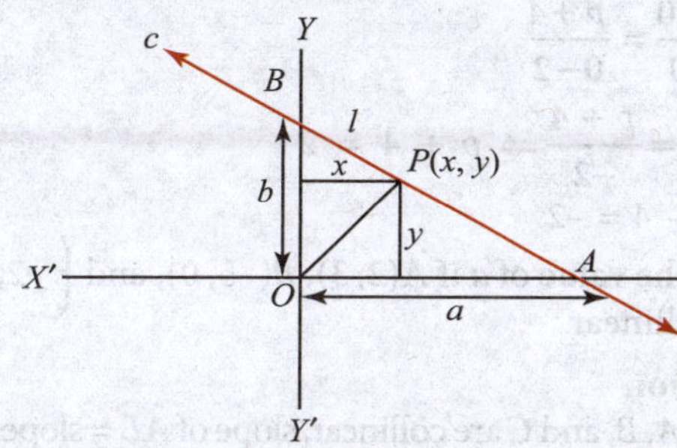

Let l be the line making intercepts a and b on the coordinate axes. Then OA = a, OB = b, and A is (a, 0) and B is (0, b).

Now, take any point P(x, y) on the given line. To find the relation between x, y, a, and b, we proceed as follows:

Area of the triangle OAB

= area of triangle OAP + area of triangle OBP

$\dfrac{1}{2} \times a \times b = \dfrac{1}{2}\, a \times y + \dfrac{1}{2} \times b \times x$

$\Rightarrow bx + ay = ab$

Dividing by ab, we get $\dfrac{x}{a} + \dfrac{y}{b} = 1$, which is the equation of the line in intercept form.

Worked Examples

1. What is the inclination of the line whose slopes are as follows:

(a) 1 (b) $\sqrt{3}$

Solution

(a) Since the slope is 1,

$\tan\theta = 1$

$\therefore \theta = 45°$

(b) Slope is $\sqrt{3}$.

$\tan\theta = \sqrt{3}$

$\therefore \theta = 60°$

2. The line joining A(–6, 8) and B(4, –2) is parallel to the line joining C(2, –4) and D(0, p). Find p.

Solution

Since AB is parallel to CD, slope of AB = slope of CD.

$$\therefore \frac{-2-8}{4-(-6)}=\frac{p-(-4)}{0-2}$$

$$\Rightarrow \frac{-10}{10}=\frac{p+4}{0-2}$$

$$\therefore -1=\frac{p+4}{-2}\Rightarrow p+4=2$$

$p = 2 - 4 = -2$

3. Find the value of a if $A(2, 3)$, $B(-5, 0)$, and $\left(-2, \frac{a}{7}\right)$ are collinear.

Solution

Since A, B, and C are collinear, slope of AB = slope BC.

$$\frac{0-3}{-5-2}=\frac{\frac{1}{7}a-0}{-2+5}\Rightarrow -7\times\frac{1}{7}a=-9$$

$\Rightarrow a = 9$

4. If $A = (5, 2)$, $B = (-9, -3)$, and $C = (-3, -5)$ are the vertices of a triangle ABC, find the slope of the altitude from C to AB.

Solution

$$\text{Slope of } AB=\frac{-3-2}{-9-5}=\frac{-5}{-14}=\frac{5}{14}$$

Therefore, the slope of the altitude from C to AB will be $-\frac{14}{5}$ since $m_1m_2=-1$.

5. A line with slope $-\frac{1}{3}$ passes through $(-1, -2)$ and $(p, 4)$. Find p.

Solution

Let $A = (-1, -2)$ and $B = (p, 4)$.

$$\text{The slope of } AB=\frac{4+2}{p+1}=-\frac{1}{3}$$

$$\frac{6}{p+1}=-\frac{1}{3}$$

$p + 1 = -18$

$\therefore p = -19$

6. (a) In the diagram, find the inclination of the line OB and also find the slope of the straight lines OA and OB.

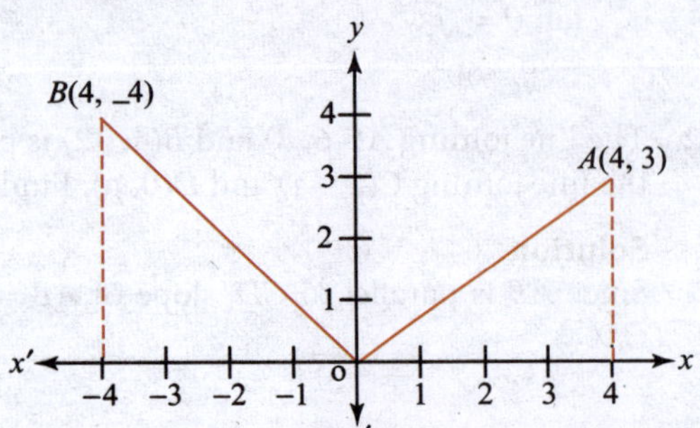

(b) Find the inclination of AC and the slope of AC and intercepts on the x-axis and y-axis by the line AC.

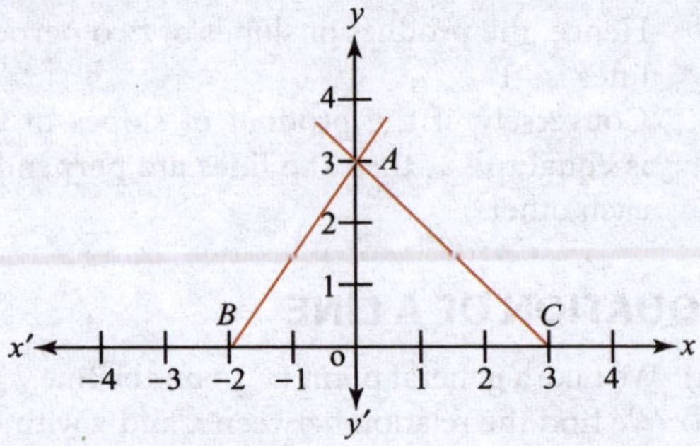

Also, find the slope of AB and the y-intercept made by the line AB.

Solution

(a) Clearly, OB bisects $Y\hat{O}X'$. So, $B\hat{O}X' = 45°$.

That is, the inclination of OB is 135°.

$\therefore$ slope of $OB = \tan 135° = -1$

$$\text{Slope of } OA = \tan A\hat{O}M=\frac{AM}{OM}=\frac{3}{4}$$

(b) Clearly, ΔAOC is a right triangled which is isosceles.

$\therefore A\hat{C}O = 45°$

$\therefore$ inclination of $AC = A\hat{C}X$

$= 180° - 45° = 135°$

Slope of $AC = \tan 135° = -1$

x-intercept = 3

y-intercept = 3

Slope of $AB = \tan A\hat{B}C = \tan A\hat{B}O$

$$=\frac{OA}{BO}=\frac{3}{2}$$

y-intercept = OA = 3

7. Let the equation of a line be $4x - 5y + 2 = 0$.

(a) Does the points (1, 1) lie on the line?

(b) Does the points (2, 2) lie on the line?

(c) If the points $(\alpha, 3)$ lie on the line, find α.

Solution

(a) (1, 1) will be a point on the line if the equation is satisfied by the coordinates.

i.e., $4 \times 1 - 5 \times 1 + 2 = 0$

But this is not true.

Therefore, (1, 1) does not lie on the line.

(b) $4 \times 2 - 5 \times 2 + 2 = 0$ is true.

Therefore, (2, 2) is a point on the line.

(c) $(\alpha, 3)$ will satisfy the equation

$4x - 5y + 2 = 0$

i.e., $4\alpha - 15 + 2 = 0$

$4\alpha - 13 = 0$

$4\alpha = 13$

$\alpha = \dfrac{13}{4}$

8. The equation of a line is $3(x - y) + 4 = 0$. Find its slope, inclination, and intercept on the y-axis.

Solution

The equation of the line is $3(x - y) + 4 = 0$.

$3x - 3y + 4 = 0$

$3y = 3x + 4$

$y = x + \dfrac{4}{3}$

Comparing this with $y = mx + c$

Slope of the line = 1 and its y-intercept = $\dfrac{4}{3}$.

Therefore, inclination = 45° (tan 45° = 1).

9. Find the equation of the line which cuts off intercept –8 and 14 from the x and y-axes.

Solution

Here, $a = -8$ and $b = 14$.

Therefore, the required equation is

$\dfrac{x}{-8} + \dfrac{y}{14} = 1$

$14x - 8y + 112 = 0$

i.e., $7x - 4y + 56 = 0$

10. Find the equation of a line passing through (1, –4) and cutting of intercepts, equal but of opposite sign, from the two axes.

Solution

Let the intercepts cut off from the two axes be m and $-m$.

Then the equation of this line is $\dfrac{x}{m} + \dfrac{y}{-m} = 1$

$x - y = m$ (1)

This passes through (1, –4).

$\therefore 1 - (-4) = m$ i.e., $m = 5$

Therefore, the required equation is $x - y = 5$.

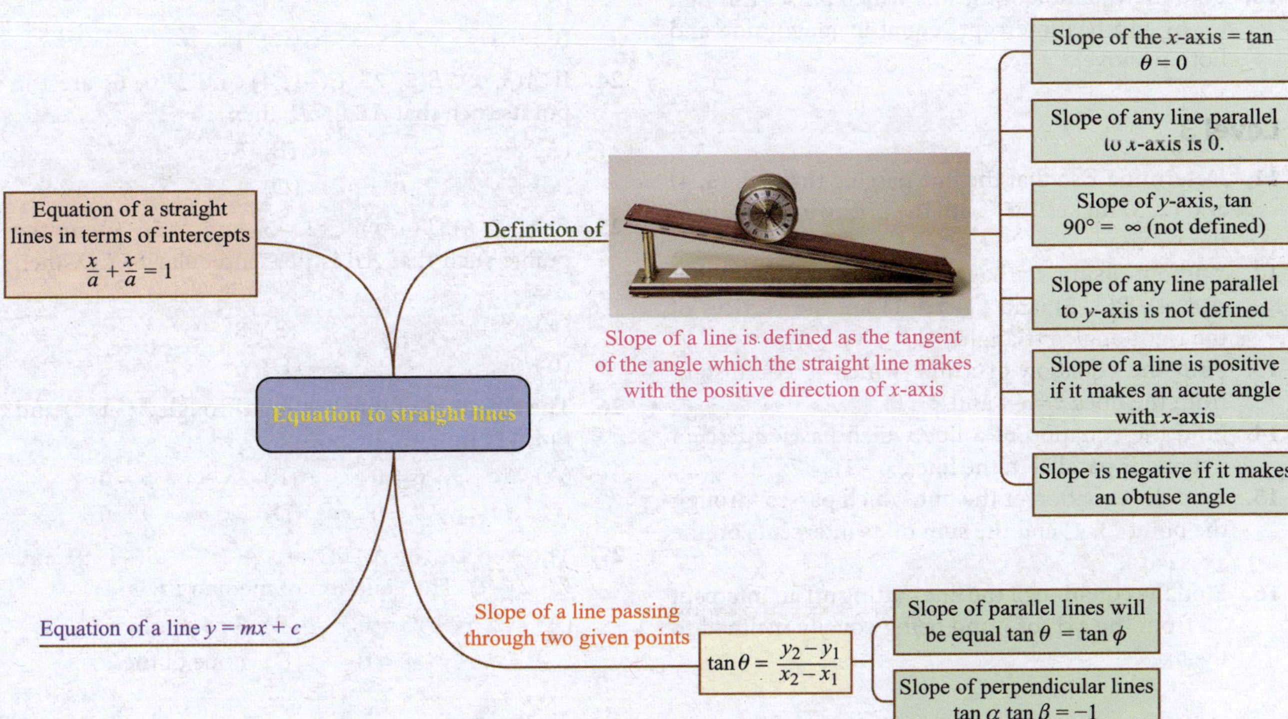

TEST YOURSELF

Level 1

1. What are the x and y-intercepts of the line $3x - 4y + 12 = 0$?
2. If $A = (4, 9)$ and $B = (2, 13)$, what is the slope of a line perpendicular to AB?
3. If $A = (4, 9)$, $B = (2, 6)$, $C = (5, 9)$, and $D = (7, 12)$, prove that $AB \parallel CD$.
4. If $A = (6, 7)$, $B = (9, 9)$, $C = (-1, 4)$, and $D = (1, 1)$, prove that $AB \perp CD$.
5. Find the equation of the line inclined at 135° with the positive side of the x-axis and having the y-intercept 4.

Level 2

6. Find the equation of the straight line cutting off an intercept 2 from the positive direction of the y-axis and inclined at 30° to the axis of x.
7. Using slope, show that the following points are collinear:
   (a) $A(4, 8)$, $B(5, 12)$, $C(9, 28)$
   (b) $A(16, -18)$, $B(3, -6)$, $C(-10, 6)$
8. Find the equation of a line that has y-intercept 4 and is perpendicular to the line joining $(2, -3)$ and $(4, 2)$.
9. Find the equation of the line cutting off intercepts 6 and 4 from the axes.
10. Find the equation of the line which passes through $(5, 6)$ and has intercepts equal in magnitude and both positive.

Level 3

11. Determine x so that the line passing through $(3, 4)$ and $(x, 5)$ makes 135° with the positive direction of the x-axis.
12. Without using Pythagoras' theorem, show that $A(4, 4)$, $B(3, 5)$, and $C(-1, -1)$ are the vertices of the right-angled triangle.
13. Find the equation of a line which is equidistant from the lines $x = -4$ and $x = 8$.
14. Find the equation of a line which has y-intercept $= 4$ and is parallel to the line $2x - 3y = 7$.
15. Find the equation of the line which passes through the point $(3, 4)$ and the sum of its intercepts on the axes is 14.
16. Find the equation of the line cutting off an intercept -5 from the axis of y and being equally inclined to the axes.
17. Find the equation of the line which passes through the point $(-5, 4)$ and is such that the portion intercepted between the axes is divided by the point in the ratio $1 : 2$.
18. The area of the triangle formed by the coordinate axes and a line is 6 square units and the length of the hypotenuse is 5 units. Find the equation of the line.
19. Prove that $A(4, 3)$, $B(6, 4)$, $C(5, 6)$ and $D(3, 5)$ are the angular points of a square.
20. Find the coordinates of the image of the point $A(3, -4)$ when reflected in the line $x + y = 2$.

Multiple Choice Questions

21. If the inclination of the line joining the points $A(x, -3)$ and $B(2, 5)$ is 135°, then the value of x is
    (A) 8 (B) −8
    (C) −10 (D) 10
22. The angle between x-axis and the line joining the points $A(3,-1)$ and $B(4, -2)$ is
    (A) 45° (B) 90°
    (C) 135° (D) 150°
23. If the points $A(x, -1)$, $B(2, 1)$ and $C(4, 5)$ are collinear, then $x =$
    (A) 3 (B) −2
    (C) −4 (D) 1
24. If $A(3, x)$, $B(2, 7)$, $C(-1, 4)$ and $D(0, 6)$ are the points such that $AB \parallel CD$, then $x =$
    (A) 6 (B) 8
    (C) 9 (D) 12
25. If $A(-2, 6)$, $B(4, x)$, $C(3, -3)$ and $D(5, -9)$ are the points such that AB is perpendicular to CD, then $x =$
    (A) 9 (B) 8
    (C) 7 (D) 6
26. The equation of line passing through $A(-1, 8)$ and $B(4, -2)$ is
    (A) $2x - y + 6 = 0$ (B) $2x + y - 6 = 0$
    (C) $3x - y + 9 = 0$ (D) $3x + y - 9 = 0$
27. The vertices of ΔABC are $A(2, 5)$, $B(-4, 9)$ and $C(-2, -1)$. The equation of median BE is
    (A) $x - 5y + 23 = 0$ (B) $8x - y + 15 = 0$
    (C) $7x + 4y - 8 = 0$ (D) none of these

28. The equation of the perpendicular bisector of the line joining the points $A(2, 3)$ and $B(6, -5)$ is

(A) $x + 2y - 6 = 0$ (B) $x - 2y - 6 = 0$
(C) $x + 2y + 6 = 0$ (D) $x - 2y + 6 = 0$

29. If $A(2, 1)$, $B(-2, 3)$ and $C(4, 5)$ are the vertices of ΔABC, then the equation of altitude through B is

(A) $x - 2y + 4 = 0$ (B) $x + 2y - 4 = 0$
(C) $x - 2y - 4 = 0$ (D) $x + 2y + 4 = 0$

30. If $A(1, 4)$, $B(2, -3)$ and $C(-1, -2)$ are the vertices of ΔABC, then the equation of the right bisector of BC is

(A) $3x - y - 4 = 0$ (B) $3x + y - 4 = 0$
(C) $3x - y + 4 = 0$ (D) None of these

31. The equation of a line which is equidistant from the lines $x = -2$ and $x = 6$ is

(A) $x = 4$ (B) $x = 2$
(C) $x = 3$ (D) none of these

32. The equation of a line passing through the point $(3, -4)$ and parallel to the x-axis is

(A) $x - 4 = 0$ (B) $x + 4 = 0$
(C) $y - 4 = 0$ (D) $y + 4 = 0$

33. A line cuts the y-axis at a distance of 3 units above the origin and makes an angle of 30° with the positive direction of the x-axis. The equation of the line is

(A) $y - \sqrt{3}x + 3\sqrt{3} = 0$ (B) $x - \sqrt{3}y + 3\sqrt{3} = 0$
(C) $x + \sqrt{3}y - 3\sqrt{3} = 0$ (D) None of these

34. A line cuts off an intercept of 4 units on negative direction of the y-axis and makes an angle of 120° with the positive direction of the x-axis. The equation of the line is

(A) $\sqrt{3}x - y - 4 = 0$ (B) $\sqrt{3}x + y - 4 = 0$
(C) $\sqrt{3}x + y + 4 = 0$ (D) None of these

35. The equation of a line for which $\tan\theta = \frac{1}{3}$ and x-intercept = 5 (on positive direction) is

(A) $x - 3y - 5 = 0$ (B) $x + 3y - 5 = 0$
(C) $x + 3y + 5 = 0$ (D) None of these

36. The lines $2x + 3y + 7 = 0$ and $27x - 18y + 25 = 0$ are

(A) parallel to each other
(B) coincident
(C) Perpendicular to each other
(D) None of these

37. The angle made by the line $x + \sqrt{3}y - 6 = 0$ with the positive direction of the x-axis is

(A) 45° (B) 60°
(C) 120° (D) 150°

38. A line passes through the point $(2, -5)$ and is parallel to the line $2x - 3y = 7$. The equation of the line is

(A) $2x + 3y - 14 = 0$ (B) $2x - 3y - 21 = 0$
(C) $2x - 3y - 19 = 0$ (D) None of these

39. The y-intercept of a line is -3 and it is perpendicular to the line $3x - 2y + 5 = 0$. The equation of the line is

(A) $2x + 3y + 9 = 0$ (B) $2x - 3y + 9 = 0$
(C) $2x - 3y - 9 = 0$ (D) None of these

40. The equation of line making intercepts 2 and -3 on the x-axis and y-axis, respectively, is

(A) $3x + 2y - 6 = 0$ (B) $3x - 2y + 6 = 0$
(C) $3x - 2y - 6 = 0$ (D) None of these

41. The equation of the line so that the line segment intercepted between the axes is bisected at the point $(2, 3)$ is

(A) $3x + 2y + 12 = 0$ (C) $2x + 3y - 12 = 0$
(C) $2x + 2y - 12 = 0$ (D) $3x + 2y - 12 = 0$

42. If the straight line $\frac{x}{a} + \frac{y}{b} = 1$ passes through the points $A(8,-9)$ and $B(12,-15)$ then

(A) $a = 3, b = 2$ (B) $a = 2, b = 3$
(C) $a = 4, b = 9$ (D) $a = 9, b = 4$

OLYMPIAD AND NTSE LEVEL EXERCISES

43. The equations of the lines which cuts off an intercept -1 from y-axis are equally inclined to the axes are

(A) $x - y + 1 = 0, x + y + 1 = 0$
(B) $x - y - 1 = 0, x + y - 1 = 0$
(C) $x - y - 1 = 0, x + y + 1 = 0$
(D) None of these

44. The equation of the straight line passing through the point $(3, 2)$ and perpendicular to the line $y = x$ is

(A) $x - y = 5$ (B) $x + y = 5$
(C) $x + y = 1$ (D) $x - y = 1$

45. The coordinates of three points A, B and C are $(-1, 5)$, $(0, 0)$ and $(2, 2)$, respectively. If D is the middle point of BC, then the equation of the perpendicular drawn from B to the line AD is

(A) $x + 2y = 0$ (B) $2x + y = 0$
(C) $x - 2y = 0$ (D) $2x - y = 0$

46. If the middle points of the sides BC, CA and AB of the triangle ABC are $(1, 3)$, $(5, 7)$ and $(-5, 7)$, respectively, then the equation of the side AB is

(A) $x - y - 2 = 0$ (B) $x - y + 12 = 0$
(C) $x + y - 12 = 0$ (D) None of these

47. A line meets x-axis and y-axis at the points A and B, respectively. If the middle point of AB is (x_1, y_1), then the equation of the line is

(A) $y_1x + x_1y = 2x_1y_1$ (B) $x_1x + y_1y = 2x_1y_1$
(C) $y_1x + x_1y = x_1y_1$ (D) $x_1x + y_1y = x_1y_1$

48. The equation of the line parallel to the line $2x - 3y = 1$ and passing through the middle point of the line segment joining the points $(1, 3)$ and $(1, -7)$, is

(A) $2x - 3y + 8 = 0$ (B) $2x - 3y = 8$
(C) $2x - 3y + 4 = 0$ (D) $2x - 3y = 4$

49. The equation of a line passing through the point of intersection of the lines $x + 5y + 7 = 0$, $3x + 2y - 5 = 0$, and perpendicular to the line $7x + 2y - 5 = 0$, is given by

(A) $2x - 7y - 20 = 0$ (B) $2x + 7y - 20 = 0$
(C) $-2x + 7y - 20 = 0$ (D) $2x + 7y + 20 = 0$

50. Equations of the two straight lines passing through the point $(3, 2)$ and making an angle of 45° with the line $x - 2y = 3$, are

(A) $3x + y + 7 = 0$ and $x + 3y + 9 = 0$
(B) $3x - y - 7 = 0$ and $x + 3y - 9 = 0$
(C) $x + 3y - 7 = 0$ and $x + 3y - 9 = 0$
(D) None of these

51. Match the following columns:

Column I	Column II
(a) Four lines $x + 3y - 10 = 0$, $x + 3y - 20 = 0$, $3x - y + 5 = 0$ and $3x - y - 5 = 0$ form a figure which is	(p) a quadrilateral which is neither a parallelogram nor a trapezium nor a kite
(b) The points $A(1, 2)$, $B(2, -3)$, $C(-1, -5)$ and $D(-2, 4)$ in order are the vertices of	(q) a parallelogram
(c) The lines $7x + 3y - 33 = 0$, $3x - 7y + 19 = 0$, $3x - 7y - 10$ and $7x + 3y - 4 = 0$ form a figure which is	(r) a rectangle of area 10 sq. units
(d) Four lines $4y - 3x - 7 = 0$, $3y - 4x + 7 = 0$, $4y - 3x - 21 = 0$ and $3y - 4x + 14 = 0$ form a figure which is	(s) a square

52. The coordinates of the foot of the perpendicular drawn from $(2, 4)$ to the line $x + y = 1$ are given by

(A) $\left(\frac{1}{3}, \frac{3}{2}\right)$ (B) $\left(-\frac{1}{2}, \frac{3}{2}\right)$
(C) $\left(\frac{4}{3}, \frac{1}{2}\right)$ (D) $\left(\frac{3}{4}, -\frac{1}{2}\right)$

18 Equation of a Straight Line – II

Can you guess the story behind the 4 straight lines in the box below?

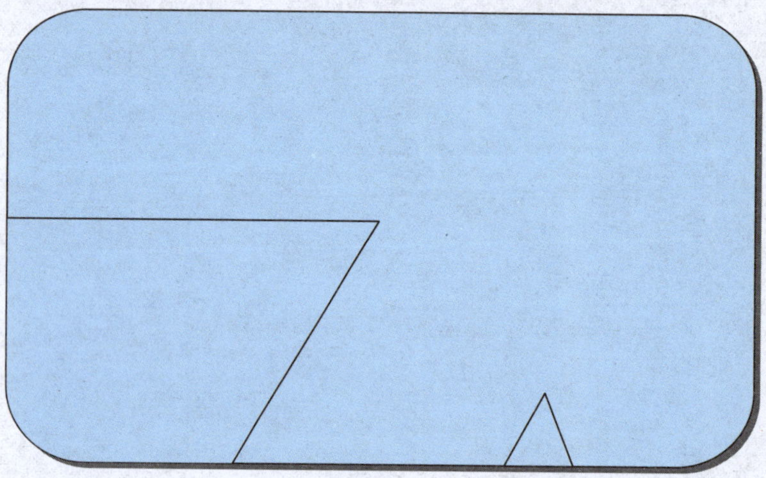

The notion of infinity is our greatest friend; it is also the greatest enemy of our peace of mind.

– James Pierpont

(A ship arriving too late to save a drowning boat!)

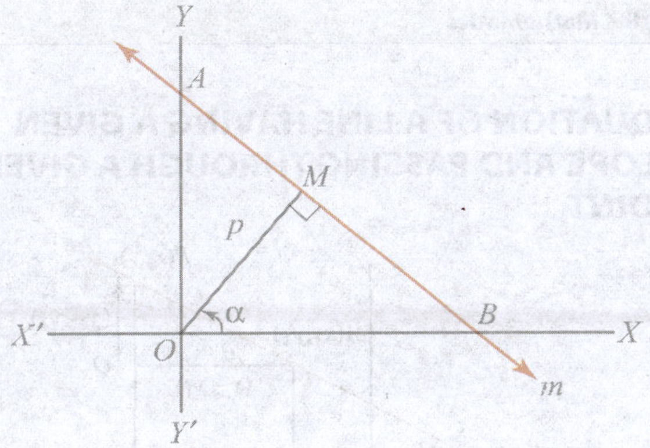

OBJECTIVES

After reading this chapter you should be able to find the equation of a line

- having a given slope and passing through a given point
- passing through two given points
- in terms of the length of the perpendicular distance from the origin
- passing through a given point and making an angle θ

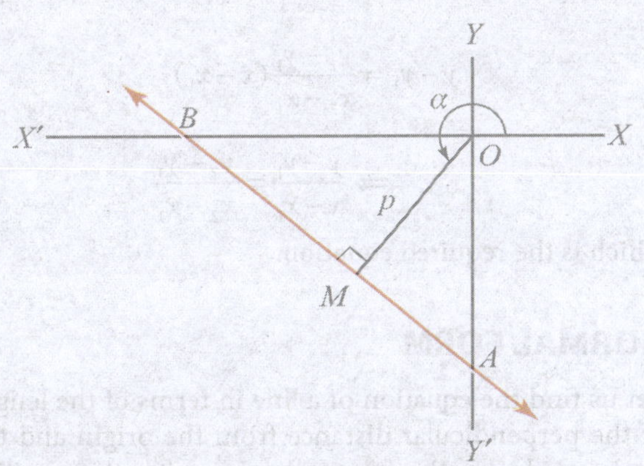

EQUATION OF A LINE HAVING A GIVEN SLOPE AND PASSING THROUGH A GIVEN POINT

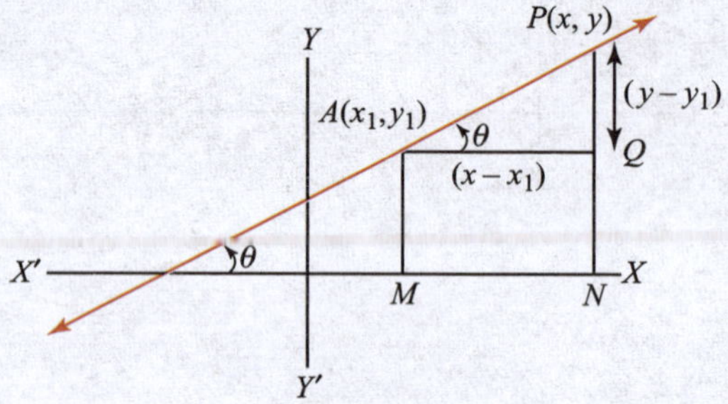

Let $A(x_1, y_1)$ be the given point on the line making an angle θ with the x-axis. Take $P(x, y)$ as any point on the line.

Draw AM, PN perpendicular to the x-axis, and AQ perpendicular to PN.

Now $\quad P\hat{A}Q = \theta$

$$PQ = (y - y_1)$$

$$AQ = (x - x_1)$$

$$\frac{y - y_1}{x - x_1} = \tan\theta = m$$

$\Rightarrow (y - y_1) = m(x - x_1)$ which is the required equation.

EQUATION OF A LINE PASSING THROUGH TWO GIVEN POINTS

Let the given points be (x_1, y_1) and (x_2, y_2).

$$\therefore \qquad m = \frac{y_2 - y_1}{x_2 - x_1}$$

Using the previous equation, $y - y_1 = m(x - x_1)$, we get

$$y - y_1 = \frac{y_2 - y_1}{x_2 - x_1}(x - x_1)$$

$$\Rightarrow \frac{y - y_1}{x - x_1} = \frac{y_2 - y_1}{x_2 - x_1}$$

which is the required equation.

NORMAL FORM

Let us find the equation of a line in terms of the length of the perpendicular distance from the origin and the angle made by the perpendicular with the positive direction of the x-axis.

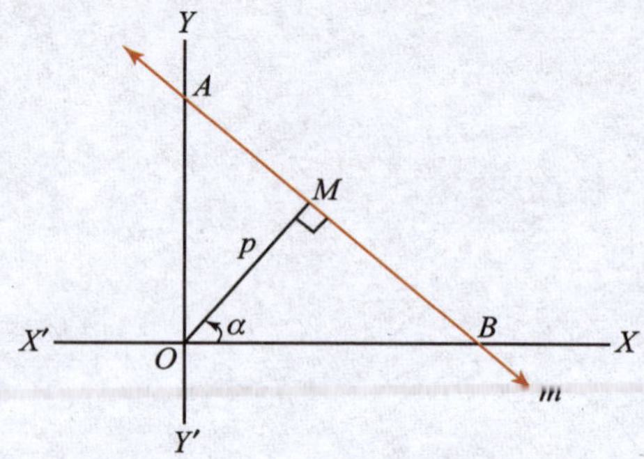

Let m be the line cutting x and y-axes at B and A, OM be perpendicular to the line from the origin, and $OM = p$. Let OM make an angle α with OX.

$$\text{Now, } \frac{OM}{OB} = \cos\alpha \Rightarrow OB = \frac{OM}{\cos\alpha}$$

$$\frac{OM}{OA} = \cos(90° - \alpha) = \sin\theta \Rightarrow OA = \frac{OM}{\sin\alpha}$$

Using the intercept form for the line m, we get

$$\frac{x}{OB} + \frac{y}{OA} = 1$$

$$\frac{x}{\frac{OM}{\cos\alpha}} + \frac{y}{\frac{OM}{\sin\alpha}} = 1$$

$$x\cos\alpha + y\sin\alpha = OM = p$$

which is the required equation, called normal form.

> p is considered positive in all positions of the line, and if $X\hat{O}M$ is always measured from OX in the positive direction, this equation, $x \cos \alpha + y \sin \alpha = p$, holds in every case.

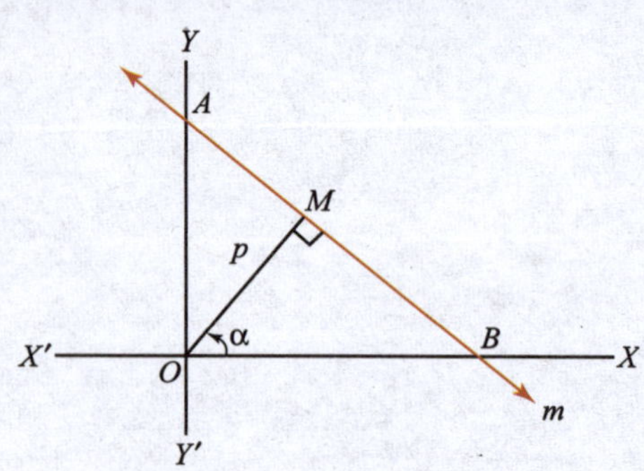

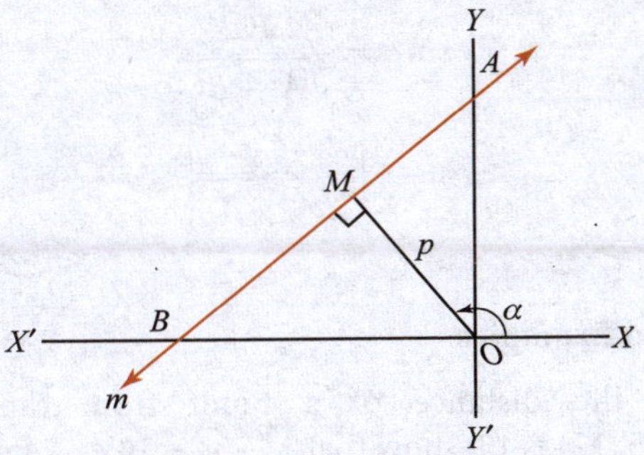

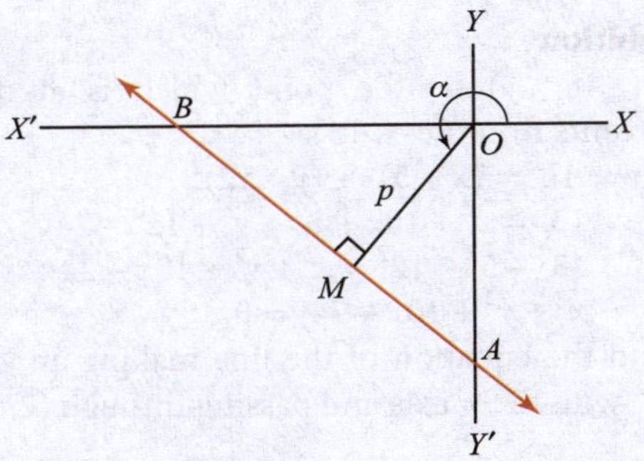

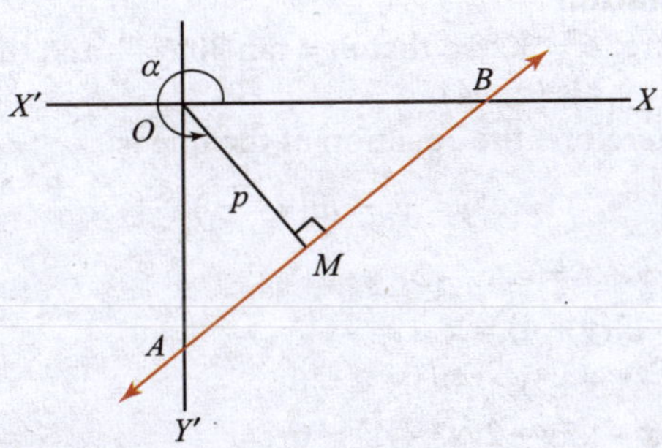

PARAMETRIC FORM

Let is find the equation of a straight line passing through a given point $Q(x_1, y_1)$ and making an angle θ with OX.

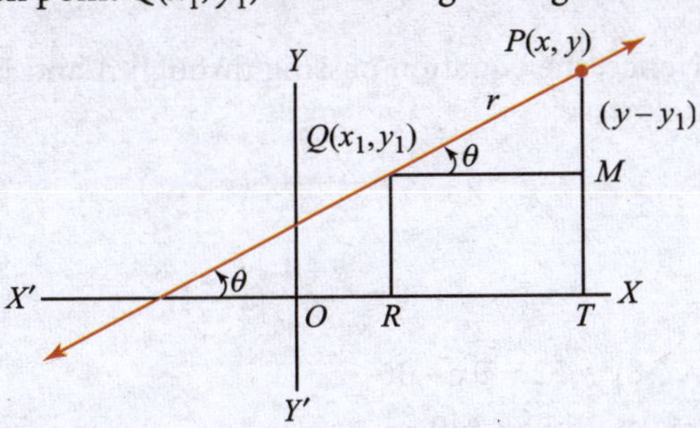

Let $P(x, y)$ be any point in the line at a distance r from the given point $Q(x_1, y_1)$.

Draw QR, PT perpendicular to the x-axis, and QM perpendicular to PT.

Now $QM = QP\cos\theta = r\cos\theta$

$PM = QP\sin\theta = r\sin\theta$

i.e., $x - x_1 = r\cos\theta$

$y - y_1 = r\sin\theta$

Hence, we have

$$\frac{x-x_1}{\cos\theta} = \frac{y-y_1}{\sin\theta} = r$$

This equation is known as *parametric form* of a line, where θ and r are parameters.

This form is very useful when we wish to bring into consideration the distance of a variable point from a fixed point.

The coordinates of any point on the line $\frac{x-x_1}{\cos\theta} = \frac{y-y_1}{\sin\theta} = r$ are $x = x_1 + r\cos\theta$ and $y = y_1 + r\sin\theta$.

Representation of $Ax + By + C = 0$ in Different Forms

$Ax + By + C = 0$ Represents a Straight Line

Let (x_1, y_1), (x_2, y_2), and (x_3, y_3) be three points on the locus represented by

$$Ax + By + C = 0 \qquad \text{(A)}$$

$$\therefore \quad Ax_1 + By_1 + C = 0 \qquad (1)$$

$$Ax_2 + By_2 + C = 0 \qquad (2)$$

$$Ax_3 + By_3 + C = 0 \qquad (3)$$

Subtracting Eq. (1) from Eq. (2), we get the following:

$$A(x_1 - x_2) + B(y_1 - y_2) \qquad (4)$$

Subtracting Eq. (1) from Eq. (3), we get the following:

$$A(x_1 - x_3) + B(y_1 - y_3) \qquad (5)$$

Multiplying Eq. (4) by $(x_1 - x_3)$, we have the following:

$$A(x_1 - x_2)(x_1 - x_3) + B(y_1 - y_2)(x_1 - x_3) = 0 \qquad (6)$$

Multiplying Eq. (5) by $(x_1 - x_2)$, we have the following:

$$A(x_1 - x_2)(x_1 - x_3) + B(y_1 - y_3)(x_1 - x_2) = 0 \qquad (7)$$

Subtracting Eq. (6) from Eq. (7), we get the following:

$$B[(y_1 - y_2)(x_1 - x_3) - (y_1 - y_3)(x_1 - x_2)] = 0$$

As $B \neq 0$, we have the following:

$$x_1y_1 - x_3y_1 - x_1y_2 + y_2x_3 - y_1x_1 + y_1x_2 + x_1y_3 - y_3x_2 = 0$$

$$x_1(y_3 - y_2) + x_2(y_1 - y_3) + x_3(y_3 - y_1) = 0$$

i.e., $$x_1(y_2 - y_3) + x_2(y_3 - y_1) + x_3(y_1 - y_3) = 0$$

This is the condition for (x_1, y_1), (x_2, y_2), and (x_3, y_3) to be on a straight line, and hence, Eq. (A) represents a straight line

$Ax + By + C = 0$ in Slope Intercept Form

$$Ax + By + C = 0 \quad (1)$$

$$\Rightarrow y = -\frac{A}{B}x - \frac{C}{B} \quad (2)$$

which is of the form $y = mx + c$, where m is slope and c is intercept on the y axis.

Therefore, slope of Eq. (1) is

$$-\frac{A}{B} = -\frac{\text{coefficient of } x}{\text{coefficient of } y}$$

Intercept on the y-axis is

$$-\frac{C}{A} = -\frac{\text{constant term}}{\text{coefficient of } y}$$

$Ax + By + C = 0$ in Intercept Form

$$Ax + By + C = 0$$

$$\frac{x}{\left(-\frac{C}{A}\right)} + \frac{y}{\left(-\frac{C}{B}\right)} = 1$$

which is of the form $\frac{x}{a} + \frac{y}{b} = 1$, where a is x-intercept and b is y-intercept, and hence

$$x\text{-intercept} = -\frac{\text{constant term}}{\text{coefficient of } x}$$

$$y\text{-intercept} = -\frac{\text{constant term}}{\text{coefficient of } y}$$

$Ax + By + C = 0$ in the Perpendicular Form

$$Ax + By + C = 0$$

$$Ax + By = -C$$

Dividing throughout by $\sqrt{A^2 + B^2}$, we get

$$\frac{A}{\sqrt{A^2+B^2}}x + \frac{B}{\sqrt{A^2+B^2}}y = \frac{-C}{\sqrt{A^2+B^2}}$$

which is of the form $x \cos\alpha + y \sin\alpha = p$.

$$\therefore \quad \cos\alpha = \frac{A}{\sqrt{A^2+B^2}}$$

$$\sin\alpha = \frac{B}{\sqrt{A^2+B^2}}$$

$$p = -\left|\frac{C}{\sqrt{A^2+B^2}}\right|$$

Worked Examples

1. If the distance of a point from the point (–5, 12) is 13, show that $x^2 + y^2 + 10x - 24y = 0$.

Solution

Let $A(x, y)$ be the point which is at distance 13 units from $B(-5, 13)$.

Now, $AB^2 = (x + 5)^2 + (y - 12)^2$

$$13^2 = x^2 + 10x + 5^2 + y^2 + 12^2 - 24y$$

$$13^2 - 5^2 - 12^2 = x^2 + y^2 + 10x - 24y$$

$$\Rightarrow \quad x^2 + y^2 + 10x - 24y = 0$$

2. Find the equation of the line making an angle of 30° with the x-axis and passing through (2, –4).

Solution

Here, $\alpha = 30°$ so that $m = \tan 30° = \frac{1}{\sqrt{3}}$ and the point (x_1, y_1) is (2, –4).

Therefore, the equation of the line is

$$y - y_1 = m(x - x_1)$$

$\Rightarrow y + 4 = \frac{1}{\sqrt{3}}(x - 2)$

$\Rightarrow \sqrt{3}(y + 4) = x - 2$

$\Rightarrow -x + \sqrt{3}y + 4\sqrt{3} + 2 = 0$

or $x - \sqrt{3}y - 2(\sqrt{3} + 1) = 0$

3. Find the equation of the line passing through the points (4, –4) and (–4, 5).

Solution

The given point $A(4, -4)$ is (x_1, y_1) and $B(-4, 5)$ is (x_2, y_2).

Hence, the equation passing through A and B is as follows:

$$y - y_1 = \frac{y_2 - y_1}{x_2 - x_1}(x - x_1)$$

$$y + 4 = \frac{5+4}{-4-4}(x - 4)$$

$\therefore -8y - 32 = 9x - 36$

$\Rightarrow 9x + 8y - 4 = 0$

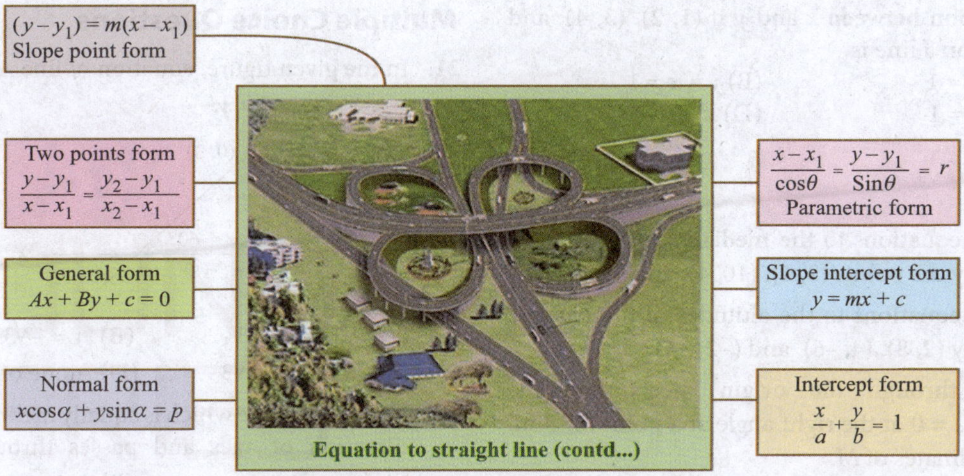

Equation to straight line (contd...)

TEST YOURSELF

Level 1

1. Find the equation to the straight lines passing through the following points:
   (a) (0, 0) and (3, –3)
   (b) (6, 8) and (10, 12)
   (c) (–2, 6) and (12, –14)
   (d) (a, b) and $(c, -d)$
   (e) $(0, a)$ and $(-b, 0)$
2. Find the equations of the sides of the triangles if the coordinates of vertices are as follows:
   (a) (1, 4), (2, –3), and (–1, –2)
   (b) (0, 1), (2, 0), and (–1, –2)
3. Find the equation of the diagonals of the rectangle formed by (2, 2), (8, 2), (8, 8), and (2, 8).
4. Find the equations of the straight lines in the following cases where p is the perpendicular distance of the line from the origin and α the angle made by the perpendicular with the positive x-axis.
   I (a) $p = 5$ units and $\alpha = 30°$
   (b) $p = 7$ units and $\alpha = 45°$
   (c) $p = 8$ units and $\alpha = 60°$
   II (a) $p = a$ units and $\alpha = 120°$
   (b) $p = 9$ units and $\alpha = 135°$
   (c) $p = 5½$ units and $\alpha = 150°$
   III (a) $p = m$ units and $\alpha = 210°$
   (b) $p = 3½$ units and $\alpha = 225°$
   (c) $p = 6$ units and $\alpha = 240°$
   IV (a) $p = q$ units and $\alpha = 300°$
   (b) $p = 7½$ units and $\alpha = 315°$
   (c) $p = 9$ units and $\alpha = 330°$

Level 2

5. The distance of the line $3x + 4y - 5 = 0$ from the origin is
   (A) 1 unit (B) 5 units
   (C) 3 units (D) 4 units
6. The angle made by the perpendicular to the line $x + y = 5$ from the origin is
   (A) 30° (B) 45° (C) 60° (D) 75°
7. The point where the line $3x + 5y = 7$ meets the y-axis is
   (A) $\left(0, \frac{7}{5}\right)$ (B) $\left(0, \frac{5}{7}\right)$
   (C) $\left(0, \frac{7}{3}\right)$ (D) $\left(\frac{3}{7}, 0\right)$
8. The coordinate of a point at a distance 4 units from the point (5, 5) on the line $x = y$ is
   (A) $(5+2\sqrt{2}, 5-2\sqrt{2})$ (B) $(5-2\sqrt{2}, 5+2\sqrt{2})$
   (C) $(5+2\sqrt{2}, 5+2\sqrt{2})$ (D) $(5+\sqrt{2}, 5+\sqrt{2})$
9. The line $x + y - 5 = 0$ is the perpendicular bisector of the line joining AB, where A is (0, 0). The coordinates of B are
   (A) (–5, 5) (B) $\left(\frac{-5}{2}, \frac{-5}{2}\right)$
   (C) (5, 5) (D) $\left(\frac{5}{2}, 5\right)$

10. The relation between x and y if (1, 2), (3, 4), and (x, y) lie on a line is
(A) $x + y = 1$ (B) $y - x = 1$
(C) $x - y = 1$ (D) $x + y = 2$

Level 3

11. Find the equations to the medians of the triangle formed by the points (2, 4), (10, 4), and (6, 6).

12. Find the equations to the altitudes of the triangle formed by (2, 8), (4, –6), and (–2, –4).

13. A line through the origin meets the line $3y - x + 2 = 0$ at the right angle at a point M. Find the coordinates of M.

14. A is a point whose coordinates are (3, 5). A line through A makes an angle θ with the positive x-axis. Find the coordinates of a point at a distance of 6 units from A (use parametric form) when
(a) $\theta = 45°$ (b) $\theta = 60°$ (c) $\theta = 30°$

15. The line $x - 3y - 5 = 0$ is the perpendicular bisector of the line joining AB, where A is (1, –3). Find the coordinates of B.

16. ABC is a triangle, whose vertices are $A(5, 3)$, $B(1, 5)$, and $C(3, -1)$. Find equations of the perpendicular bisectors of BC and AB and hence find the coordinates of circumcentre.

17. A point $P(-3, 7)$ has a line passing through it making an angle α with the positive x-axis. Find the coordinates of a point at a distance of 6 units from P, when
(a) $\alpha = 120°$ (b) $\alpha = 135°$ (c) $\alpha = 150°$

18. The relation between x and y if (x, y) is equidistant from (3, 4) and (5, –2) is
(A) $x + 3y = 1$ (B) $x - 3y = 1$
(C) $3y - x = 1$ (D) $x + 2y = 1$

19. The vertices of a triangle are A(0, 0), $B(6, 0)$, and $C(0, 8)$. The area of the triangle formed by the mid-points of the sides of the triangle ABC as vertices is
(A) 24 units square (B) 12 units square
(C) 6 units square (D) 48 units square

20. The points $A(-4, -1)$, $B(-2, -4)$, $C(4, 0)$, and $D(2, 3)$ are vertices of
(A) square (B) rhombus
(C) parallelogram (D) rectangle

Multiple Choice Questions

21. In the given figure, equation of line AB is

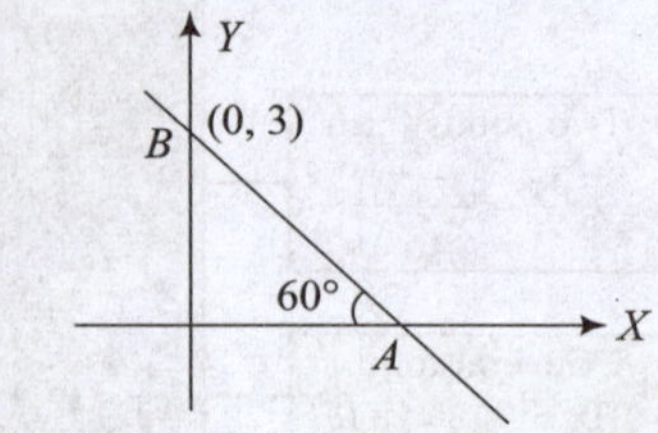

(A) $y = \sqrt{3}x - 3$ (B) $y = -\sqrt{3}x + 3$
(C) $\sqrt{3}y = x + \sqrt{3}$ (D) none of these

22. Equation of line which is equally inclined to positive direction of axes and passes through origin is
(A) $y = x$ (B) $y + x = 0$
(C) $y = \frac{x}{2}$ (D) none of these

23. Equation of line passing through the point (2, –4) and making an angle of 30° with the x-axis is
(A) $\sqrt{3}y - x + 4\sqrt{3} = 0$
(B) $\sqrt{3}y - x + 2 = 0$
(C) $\sqrt{3}y - x + 4\sqrt{3} + 2 = 0$
(D) none of these

24. Equation of the line passing through the points (3, 2) and (3, 4) is
(A) $y = 4$ (B) $x + y = 2$
(C) $x = 3$ (D) $y = 0$

25. Equation of line passing through the points (2, 1) and (3, 1) is
(A) $x = 1$ (B) $x = 3$
(C) $y = 1$ (D) $y = 3$

26. Equation of line which makes intercepts 2 and –3 on the x-axis and y-axis, respectively, is
(A) $2x - 3y - 6 = 0$ (B) $3x - 2y - 6 = 0$
(C) $3x + 3y + 7 = 0$ (D) $3x - 2y + 6 = 0$

27. Which of the following is NOT an equation of the line which passes through the point (3, 4) and the sum of whose intercepts on the axes is 14?
(A) $2x + 3y - 18 = 0$ (B) $4x + 3y - 24 = 0$
(C) $x + y - 7 = 0$ (D) $8x + 6y = 48$

28. What is the equation of the line if $M(a, b)$ is the mid-point of a line segment intercepted between the axes?

(A) $bx + ay = ab$ (B) $ax + by = ab$
(C) $bx + ay = 2ab$ (D) none of these

29. What is the equation of line which passes through the point (–3, 7) and makes intercepts on the axes equal in magnitude but opposite in sign?
(A) $x - y + 7 = 0$ (B) $x - y + 10 = 0$
(C) $x - y + 6 = 0$ (D) $x + y + 10 = 0$

30. Equation of line which is bisected at (2, 3) between the axes is
(A) $3x + 2y - 12 = 0$ (B) $2x + 3y - 12 = 0$
(C) $3x + 3y - 11 = 0$ (D) $2x + 2y - 7 = 0$

31. Equation of line so that segment intercepted between the axes is divided by the point $P(-5, 4)$ in the ratio 1 : 2 is
(A) $8x - 5y - 60 = 0$ (B) $8x - 5y + 60 = 0$
(C) $7x - 4y + 27 = 0$ (D) $x + 2y + 7 = 0$

32. A line passes through the point of intersection of the lines $4x + 3y - 3 = 0$ and $2x - 3y + 1 = 0$. Also, it has equal intercepts on the axes. The equation of line is
(A) $12x + 12y - 7 = 0$ (B) $9x + 9y - 8 = 0$
(C) $12x + 13y - 7 = 0$ (D) $8x + 8y - 9 = 0$

33. What is the area of the triangle formed by the co-ordinate axes and the line $ax + by = 2ab$?
(A) ab sq. units (B) $4ab$ sq. units
(C) $2ab$ sq. units (D) $\frac{ab}{2}$ sq. units

34. What is the equation of a line whose perpendicular distance from the origin is 5 units and the angle between the positive direction of the x-axis and perpendicular is 30°?
(A) $\sqrt{3}x - y - 10 = 0$
(B) $\sqrt{3}x + y - 10 = 0$
(C) $x + y + 10\sqrt{3} = 0$
(D) $\sqrt{3}x + 10y + 3 = 0$

35. What is the equation of a line whose perpendicular distance from the origin is $\sqrt{8}$ units and the angle between the positive direction of the x-axis and perpendicular is 135°?
(A) $x - y + 4 = 0$ (B) $2x + 3y + 7 = 0$
(C) $3x - 2y + 8 = 0$ (D) none of these

36. What is the equation of a line whose perpendicular distance from the origin is 2 units and the angle between perpendicular segment and the positive direction of the x-axis is 240°?
(A) $\sqrt{3}x + y + 4 = 0$ (B) $x - \sqrt{3}y + 4 = 0$
(C) $2x + y + 5 = 0$ (D) $x + \sqrt{3}y + 4 = 0$

37. If the straight line $\frac{x}{a} + \frac{y}{b} = 1$ passes through the points (8, –9) and (12, –15), then
(A) $a = 2, b = 3$ (B) $a = 4, b = 3$
(C) $a = 5, b = 2$ (D) $a = 7, b = 7$

38. For the equation $\sqrt{3}x + y + 2 = 0$,
(A) slope $= \frac{1}{\sqrt{3}}$, y-intercept = 2
(B) slope $= \sqrt{3}$, y-intercept = 2
(C) slope $= -\sqrt{3}$, y-intercept = -2
(D) slope $= -\frac{1}{\sqrt{3}}$, y-intercept = 2

39. If the normal form of equation $\sqrt{3}x + y + 2 = 0$ is $x \cos \alpha + y \sin \alpha = P$, then
(A) $\alpha = 210°, P = 1$ (B) $\alpha = 150°, P = 1$
(C) $\alpha = 120°, P = \sqrt{3}$ (D) $\alpha = 60°, P = 1$

40. What is the angle between positive x-axis and the perpendicular from the origin to the line $x + \sqrt{3}y + 5 = 0$?
(A) 120° (B) 180°
(C) 90° (D) 240°

41. Equation of straight line which forms a triangle with coordinate axes having area 6 sq. units and the length of hypotenuse is 5 units is
(A) $3x + 4y - 12 = 0$ (B) $12x + 5y - 13 = 0$
(C) $2x + y - 7 = 0$ (D) $x + 2y - 7 = 0$

42. If the normal form of line $y + 4 = 0$ is $x \cos \alpha + y \sin \alpha = P$, then
(A) $\alpha = 90°, P = 4$ (B) $\alpha = 240°, P = 4$
(C) $\alpha = 270°, P = 4$ (D) $\alpha = 120°, P = 4$

43. Equation of line parallel to the y-axis and drawn through the point of intersection of the lines $x - 7y + 15 = 0$ and $2x + y = 0$ is

(A) $x = 1$ (B) $x + 1 = 0$

(C) $x + \frac{3}{2} = 0$ (D) $x + 2 = 0$

44. Find the value of k for which the lines $3x + y = 2$, $kx + 2y = 3$ and $2x - y = 3$ may intersect at a point.

(A) 2 (B) 5

(C) −2 (D) −4

45. The coordinates of a point at a distance 2 units from the point (2, 2) on the line $y = x$ is

(A) $(2 + \sqrt{2}, 2 - \sqrt{2})$ (B) $(2 - \sqrt{2}, 2 + \sqrt{2})$

(C) $(2 + \sqrt{2}, 2 + \sqrt{2})$ (D) none of these

46. The area of triangle which is formed by the points $A(2, 3)$, $B(1, 1)$ and $C(4, 1)$ is

(A) 2 sq. units (B) 3 sq. units

(C) 4 sq. units (D) 5 sq. units

47. The equation of median of ΔABC whose vertices are $A(2, 5)$, $B(-4, 9)$ and $C(-2, -1)$ passing through the point $A(2, 5)$ is

(A) $5x + y - 23 = 0$ (B) $5x + 2y - 10 = 0$

(C) $x - 5y + 23 = 0$ (D) $7x + 4y - 8 = 0$

48. If A(2, 1), B(−2, 3) and C(4, 5) are the vertices of a ΔABC then the equation of the altitude through point B is

(A) $2x + y + 4 = 0$ (B) $x + 2y - 4 = 0$

(C) $2x - y - 8 = 0$ (D) $x + 2y - 8 = 0$

49. If $A(2, 1)$, $B(-2, 3)$ and $C(4, 5)$ are the vertices of ΔABC then the equation of the right bisector of side BC is

(A) $x + 3y + 7 = 0$ (B) $3x + y - 7 = 0$

(C) $x + 3y - 7 = 0$ (D) $x + y - 7 = 0$

OLYMPIAD AND NTSE LEVEL EXERCISES

50. The line joining the points (x_1, y_1) and (x_2, y_2) subtends a right angle at the point (1, 1) then $x_1x_2 + y_1y_2 =$

(A) 0

(B) $x_1 + x_2 + y_1 + y_2 - 2$

(C) $3(x_1 + x_2 + y_1 + y_2)$

(D) $x_1x_2y_1y_2$

51. In the xy-plane, how many straight lines whose x-intercept is a prime number and whose y-intercept is a positive integer pass through the point (4, 3)?

(A) 1 (B) 2

(C) 3 (D) 4

52. The coordinates of the points A and B are $(a, 0)$ and $(-a, 0)$, respectively. If a point P moves so that $PA^2 - PB^2 = 2k^2$, where k is constant, then the equation to the line of the point P, is

(A) $2ax - k^2 = 0$ (B) $2ax + k^2 = 0$

(C) $2ay - k^2 = 0$ (D) $2ay + k^2 = 0$

53. The number of rational values of m for which the y-coordinate of the point of intersection of the lines $3x + 2y = 10$ and $x = my + 2$ is an integer is

(A) 2 (B) 4

(C) 6 (D) 8

54. The equation of the line perpendicular to the line $\frac{x}{a} - \frac{y}{b} = 1$ and passing through a point at which it cuts the x-axis, is

(A) $\frac{x}{a} + \frac{y}{b} + \frac{a}{b} = 0$ (B) $\frac{x}{b} + \frac{y}{a} = \frac{b}{a}$

(C) $\frac{x}{b} + \frac{y}{a} = 0$ (D) $\frac{x}{b} + \frac{y}{a} = \frac{a}{b}$

55. The equations of the lines which pass through the origin and are inclined at an angle $\tan^{-1} m$ to the line $y = mx + c$, are given by

(A) $x = 0, 2mx + (m^2 - 1)y = 0$

(B) $y = 0, 2mx + (m^2 - 1)y = 0$

(C) $y = 0, 2mx + (1 - m^2)y = 0$

(D) None of these

56. The equations $y - y_1 = m(x - x_1)$ for different m and fixed (x_1, y_1) represent

(A) a family of parallel lines

(B) a single line

(C) a family of lines which are not concurrent

(D) a family of concurrent lines

57. The locus of the mid-point of the portion of intercept between the axes by the line $x \cos \alpha + y \sin \alpha = P$ where P is a constant is

(A) $x^2 + y^2 = 4P^2$ (B) $\frac{1}{x^2} + \frac{1}{y^2} = \frac{4}{P^2}$

(C) $x^2 + y^2 = \frac{4}{P^2}$ (D) $\frac{1}{x^2} + \frac{1}{y^2} = \frac{2}{P^2}$

58. The straight lines $4ax + 3by + c = 0$ where $a + b + c = 0$, are concurrent at the point

(A) (4, 3) (B) (1/4, 1/3)

(C) (1/2, 1/3) (D) None of these

59. Let O be the origin and B be a point on the x-axis at a distance of 2 units from the origin.

Now, match the following columns.

Column I	Column II
(A) If ΔAOB is equilateral then the coordinates of A can be	(p) $(-1, \sqrt{3})$
(B) If ΔAOB is isosceles such that $\angle OAB$ is 30° then coordinates of A can be	(q) $(-1, 2 - \sqrt{3})$
(C) If OB is one side of rhombus of area $\sqrt{3}$ units then other vertices of rhombus can be	(r) $(-3, -\sqrt{3})$
(D) If OB is a chord of circle with radius equal to OB then coordinates of point A on the circumference of the circle such that ΔOAB is isosceles can be	(s) $(1, 2 + \sqrt{3})$

19 Measurement of Angles

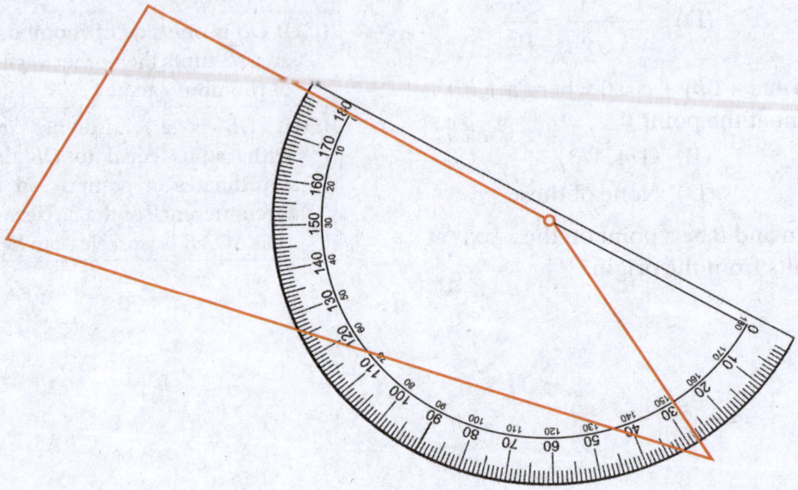

Where there is matter, there is geometry.

– Johannes Kepler

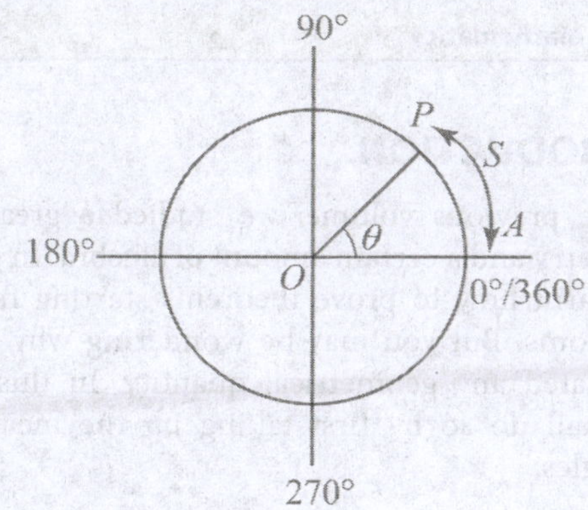

OBJECTIVES

After reading this chapter you should be able to

- measure an angle using a protractor
- use the sexagesimal system (British system) and the centesimal system (French system) to measure an angle
- use radian measure/circular measure/absolute measure

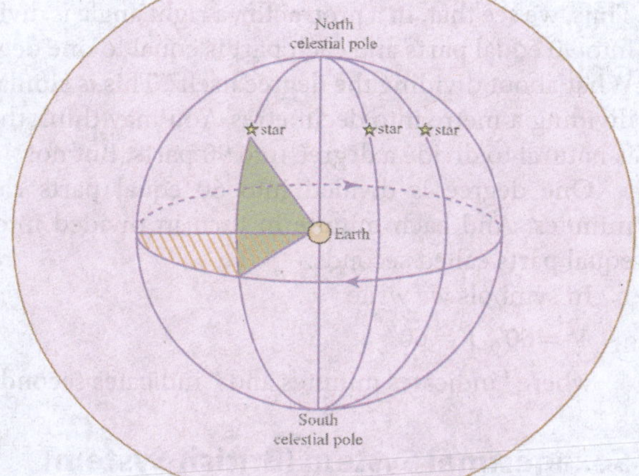

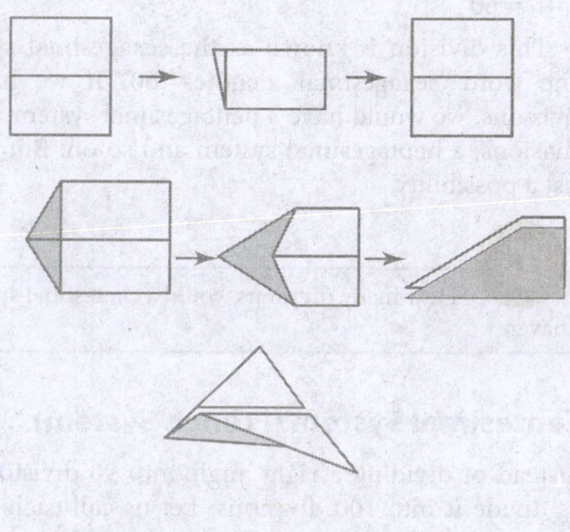

INTRODUCTION

In the previous volume, we studied a great deal of geometry and a certain amount of algebra. In geometry we learnt how to prove theorems starting from a set of axioms. But you may be wondering why we never calculated any geometrical quantity. In this volume, we shall do so by first taking up the measurement of angles.

UNITS OF MEASURE OF ANGLES

You know how to use a protractor for measuring angles. Using a protractor, we measure angles in 'degrees'. Recall that a protractor reads 0°, 90°, and 180°. With the circular protractor we can measure 270° and 360°. Thus, we see that, in a protractor, a right angle is divided into 90 equal parts and each part is equal to one degree. What about dividing the degree itself? This is similar to dividing a metre into decimetres. You may think that it is natural to divide a degree into 90 parts. But no!

One degree is divided into 60 equal parts called minutes. And each minute in turn in divided into 60 equal parts called seconds.

In symbols we write

$1° = 60'$, $1' = 60''$

where $'$ indicates minutes and $''$ indicates seconds.

Sexagesimal System (British System)

The above discussion of a protractor may be summarised as follows:

1 right angle is divided into 90 degrees

$1° = 60'$

$1' = 60''$

This division is known as the sexagesimal system. The word 'sexagesimal' denotes '60'. If we had 50 divisions, we would have a pentagesimal system, for 70 divisions, a heptagesimal system and so on. But this is just a possibility.

> How many divisions would a centesimal system have?

Centesimal System (French System)

Instead of dividing a right angle into 90 divisions, let us divide it into 100 divisions. Let us call each part a 'grade'. Then divide each grade into 100 equal divisions. Now each part is called a minute. A minute is again divided into 100 equal divisions. Each part is called a second.

We summarise this as follows:

1 right angle is divided into 100^g

[g for grade]

$1^g = 100^{\backprime}$ [$^{\backprime}$ for grade-minute]

$1^{\backprime} = 100^{\backprime\backprime}$ [$^{\backprime\backprime}$ for grade-seconds]

We should not confuse between the minutes and seconds of degrees and the minutes and seconds of grades.

> (a) Relate $1'$ to $1'$ and then $1''$ to $1''$.
> (b) Take a protractor and mark grades along with degrees. What do you infer?

> It is not known who exactly introduced 360° for the division of 4 right angles and 90 divisions for 1 right angle. The selection of this number seems to have been made in ancient times in Babylonia. It is thought that the number 360 arises because it corresponds roughly to the 365 days in a year. The division of 90° into 60 was perhaps because 60 divides 360 into 6 parts, an integral number.
>
> The sexagesimal system was replaced by the centesimal system by the French who adopted the metric system. Since they wanted the measurement of angles to be consistent with other metric systems, it was decided to divide a right angle into 100 equal parts and then each part into yet another 100, and that, in turn, into another 100 equal parts.

RADIAN MEASURE OR CIRCULAR MEASURE

In studying temperature in physics, recall that instead of using the centigrade or Fahrenheit units, we adopt the kelvin or absolute scale. This is because whereas the former scales are arbitrary the latter is unique and thereby serves to determine the temperature in an absolute sense.

> Can you introduce an absolute scale for measuring angles? If so, how would you do it?

Let us consider a protractor with the entire 360 divisions marked on it.

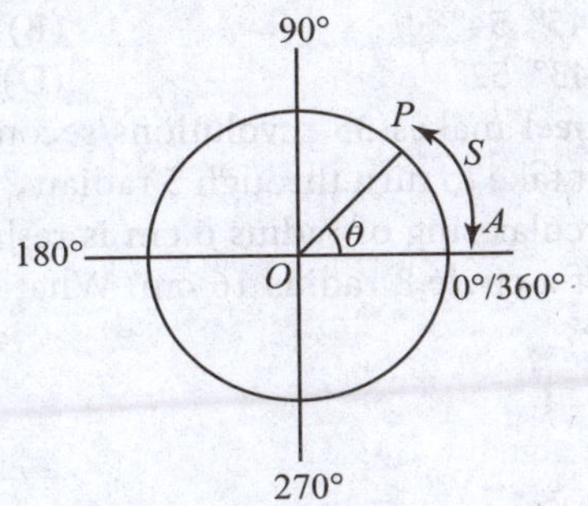

Suppose we wish to choose a scale which is unique. We need units which do not depend on any arbitrary division. How to do this? We look for something in a circle which is a constant. The obvious choice is the radius. Let us use the length of the radius as a reference. But we need to measure angles not lengths. So, we have to relate an angle to the radius, we observe that given an angle θ the arc length AP depends on θ.

$$s \propto \theta \quad \text{or} \quad s = K\theta \tag{1}$$

where K is a constant.

When s becomes the full circle, θ is 360°.

$$\therefore \quad 2\pi r = K \times 360° \tag{2}$$

Now, the only possible constant in a circle is the radius. Therefore, Eq.(1) becomes

$$s = r\theta \tag{3}$$

If we wish to have a unit angle, we should choose s to be equal to the radius.

Then, the unit angle θ is defined as 1 radian.

Substituting Eq.(3) in Eq.(2), we get

$2\pi r = r \times 360°$ or 2π radians $= 360°$

$2\pi^c = 360°$

where ^{c} stands for radians.

Thus, the radian is defined to be a unit of the angle such that the arc length equals to radius of the circle. In short, a radian is an angle subtended by the arc of a circle whose length is equal to the radius.

$2\pi^c = 360°$ or $\pi^c = 180°$

$$\therefore 1^c = \left(\frac{180}{\pi}\right)^o$$

Relation among the three systems of measurement of angles is as follows:

$$\frac{D}{90} = \frac{G}{100} = \frac{C}{\pi/2}.$$

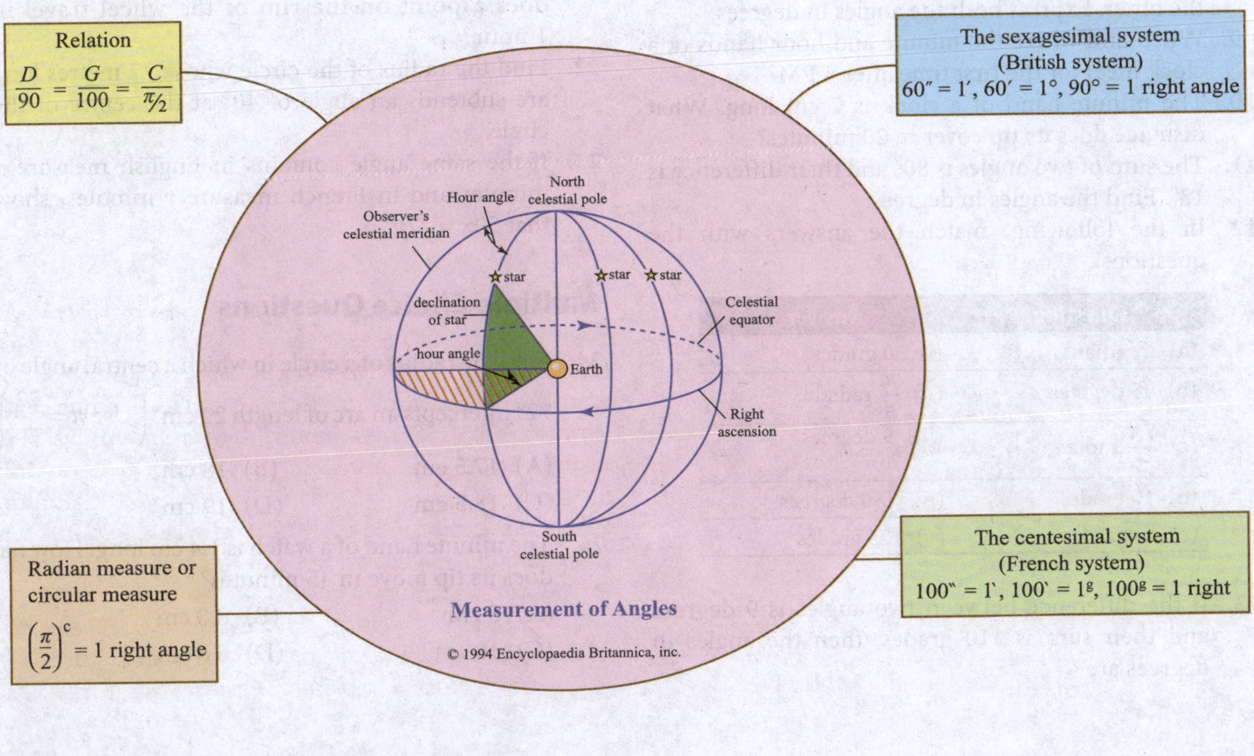

TEST YOURSELF

Level 1

1. Express the following angles in grades, minutes, and seconds:
(a) 30° (b) 81° (c) 138°30′
2. How many degrees, minutes, and seconds are, respectively, passed by in $11\frac{1}{9}$ minutes by the hour and minute hands of a watch?
3. Express the following in terms of a right angle:
(a) 66° (b) 75°15′
(c) 130°30′ (d) 60°30′30″
4. Express the following in terms of a right angle:
(a) 140^g (b) 150^g
(c) 50^g 50` 50″ (d) $220^g$ 15` 15″
5. How many degrees, minutes, and seconds are, respectively, passed by in $22\frac{2}{9}$ min by the hour and minute hands of a wall clock?
6. Express the following in degrees:
(a) $\frac{\pi^c}{3}$ (b) $\frac{\pi^c}{6}$ (c) $10\pi^C$

Level 2

7. The angles of a triangle are in the ratio 2 : 3 : 4. Find the angles in sexagesimal and centesimal systems.
8. The number of degrees in one acute angle of a right-angled triangle is equal to the number of grades in the other. Express both the angles in degrees.
9. When and where the minute and hour hands of a clock meet for the first time after 4 PM?
10. The minute hand of a clock is 7 cm long. What distance does its tip cover in 20 minutes?
11. The sum of two angles is 80^g and their difference is 18°. Find the angles in degrees.
12. In the following, match the answers with the questions.

Question	Answer
(a) π radian	(i) 50 grades
(b) 45 degrees	(ii) $\frac{\pi}{5}$ radian
(c) $\frac{7\pi}{8}$ radian	(iii) 9 degrees
(d) 10 grades	(iv) 180 degrees
(e) 40 grades	(v) 175 grades

13. If the difference between two angles is 9 degrees and their sum is 110 grades, then the angles in degrees are
(A) 45°, 54° (B) 44°, 53°
(C) 43°, 52° (D) 40°, 49°
14. A wheel makes 35 revolutions/second. How long will it take to turn through 5 radians?
15. A circular ring of radius 6 cm is reshaped into an arc of a circle 8 radius 16 cm. What is the central angle?

Level 3

16. Divide 44°8′ into two parts such that the number of sexagesimal seconds in one part may be equal to the number of centesimal seconds in the other part.
17. Find the value of one radian in sexagesimal measure up to approximate value of seconds.
18. What is the distance between two places on the equator making an angle 60° at the centre of the earth? Take the radius of the earth to be 6400 km.
19. Express an angle of a regular polygon of 40 sides in circular and in degrees.
20. If the radius of the earth is 6400 km, what is its circumference?
21. The radius of a carriage wheel is 50 cm, and in 1/9th of a second, it turns through 80° about its centre, which is fixed. How many kilometres does a point on the rim of the wheel travel in 1 hour?
22. Find the radius of the circle whose 77 metres long arc subtends an angle of 40° at the centre of the circle.
23. If the same angle contains in English measure x minutes and in French measure y minutes, show that $50x = 27y$.

Multiple Choice Questions

24. Find the radius of a circle in which a central angle of 72° intercepts an arc of length 22 cm. $\left(\text{Use } \pi = \frac{22}{7}\right)$
(A) 17.5 cm (B) 18 cm
(C) 18.5 cm (D) 19 cm
25. The minute hand of a watch is 1.4 cm long. How far does its tip move in 45 minutes?
(A) 5 cm (B) 8.8 cm
(C) 10 cm (D) 6.6 cm

26. The angle of 36° is equivalent to

(A) $\left(\frac{\pi}{4}\right)^c$ (B) $\left(\frac{\pi}{5}\right)^c$

(C) $\left(\frac{\pi}{7}\right)^c$ (D) π^c

27. Angle $\left(\frac{5\pi}{12}\right)^c$ can be expressed in degrees as

(A) 75° (B) 80°
(C) 85° (D) 90°

28. 45° 20′ 10″ is equivalent to

(A) 50° 30′ 30″ (B) 50° 45′ 45″
(C) 50° 37′ 34″ (D) None of these

29. The angles of a triangle are in the ratio 3 : 4 : 5. The smallest angle in degrees is

(A) 15° (B) 30°
(C) 45° (D) 60°

30. In the circle of radius 5 cm, what is the length of arc which subtends an angle of 33°5′ at the centre?

(A) 3 cm (B) $3\frac{65}{72}$ cm

(C) 2 cm (D) $2\frac{65}{72}$ cm

31. The wheel of a railway carriage is 4 feet in diameter and makes 6 revolutions in a second. What is the speed of the train?

(A) 20 π ft/sec (B) 22 π ft/sec
(C) 24 π ft/sec (D) None of these

32. What is the measure of each interior angle of a regular pentagon in radian system?

(A) $\frac{\pi}{4}$ (B) $\frac{3\pi}{5}$

(C) $\frac{\pi}{3}$ (D) $\frac{\pi}{6}$

33. If the arcs of the same length in two circles subtend angles of 60° and 75° at their respective centres, then what is the ratio of their radii?

(A) $\frac{3}{2}$ (B) $\frac{4}{5}$

(C) $\frac{5}{4}$ (D) $\frac{7}{8}$

34. In a right-angled triangle, the difference between the two acute angles is $\left(\frac{\pi}{15}\right)^c$. The angles in degrees are

(A) 60°, 30°, 90° (B) 51°, 39°, 90°
(C) 45°, 45°, 90° (D) None of these

35. A horse is tied to a post by a rope. The horse moves along a circular path, always keeping the rope tight and describes 66 meters when it traces 72° at the centre. Length of the rope is (take $\pi = \frac{22}{7}$)

(A) 70 m (B) 52.50 m
(C) 55 m (D) 60 m

36. Find the length of an arc of a circle of radius 14 cm which subtends an angle of 36° at the centre.

(A) 7 cm (B) 8.8 cm
(C) 8 cm (D) 10 cm

37. Interior angles of a quadrilateral are in AP and the greatest angle is double the least. Find the least angle in radian.

(A) $\frac{\pi}{4}$ (B) $\frac{\pi}{6}$

(C) $\frac{\pi}{2}$ (D) $\frac{\pi}{3}$

38. A wire of length 121 cm is bent so as to lie along the arc of a circle of radius 180 cm. What is the angle subtended by the wire at the centre?

(A) 37° (B) 38° 30′
(C) 40° (D) 45°

39. A train is moving in a circular curve of radius 1500 m at the rate of 66 km per hour. Through what angle does it turn in 10 seconds?

(A) 10° (B) 15°
(C) 7° (D) 17°

40. A wheel makes 180 revolutions in 1 minute. Through how many radians does it turn in 1 second?

(A) 4π (B) 6π
(C) 2π (D) π

41. The interior angles of a polygon are in AP. The smallest angle is 120° and the common difference is 5. The number of sides of the polygon is

(A) 12 (B) 10
(C) 9 (D) 8

OLYMPIAD AND NTSE LEVEL EXERCISES

42. The sum of the angles of a pentagon is

(A) 360° (B) 540°
(C) 180° (D) 720°

43. The sum of the acute angles of an obtuse triangle is 70° and their difference is 10°. The bigger acute angle is

(A) 40° (B) 70°
(C) 110° (D) 30°

44. If one angle of a triangle is equal to the sum of the other two, the triangle is always

(A) isosceles (B) equilateral
(C) scalene (D) right-angled

45. In ΔABC, $2\angle A = 3\angle B$, $5\angle B = 2\angle C$. Then the triangle is

(A) isosceles
(B) right-angled
(C) equilateral
(D) right-angled and isosceles

46. In ΔPQR, if $\angle R = 90°$, then angles P and Q are

(A) equal (B) supplementary
(C) complementary (D) 45° each

47. If one angle of a triangle is equal to half the sum of the other two equal angles, then the triangle is

(A) isosceles (B) scalene
(C) equilateral (D) right-angled

48. From 5 am to 5 pm, the number of times the angle between the hands of a clock is zero degrees is

(A) 2 (B) 12
(C) 13 (D) 11

49. The complement angle of the supplement angle of 150° is

(A) 90° (B) 70°
(C) 60° (D) 75°

50. The number of isosceles triangles with 90° as one of the angles is

(A) 0 (B) 1
(C) 2 (D) 3

51. In ΔABC, angle B is 25° more than angle A and angle C is 5° more than four times angle A. The biggest angle is

(A) 50° (B) 25°
(C) 105° (D) 115°

GAME TIME

To see and understand different types of angles, we have the following paper-folding activity.

Start with an A4 sheet and fold as shown in the diagrams.

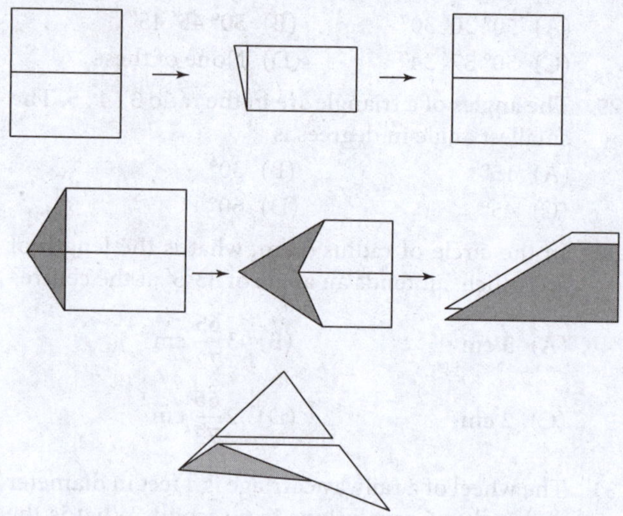

You have just made an aeroplane which flies. Now, when you unfold completely, you will see the following creases.

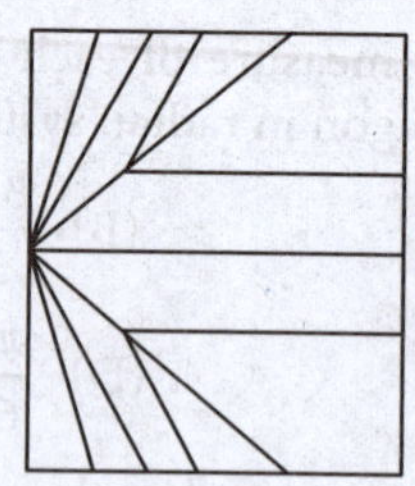

Measure all the angles and express them in degrees and radians.

20 Trigonometry

> After exponential quantities the circular functions, sine and cosine, should be considered because they arise when imaginary quantities are involved in the exponential.
>
> – Leonhard Euler

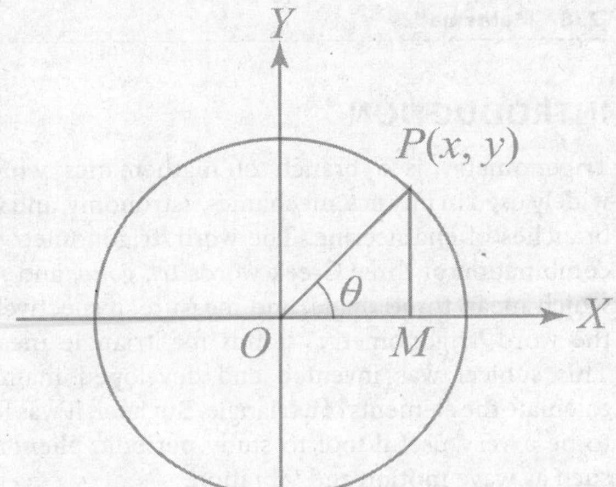

OBJECTIVES

After reading this chapter you should be able to

- define trigonometric ratios of an angle
- understand the relation between t-ratios and solve simple problems
- understand the signs of t-ratios in different quadrants

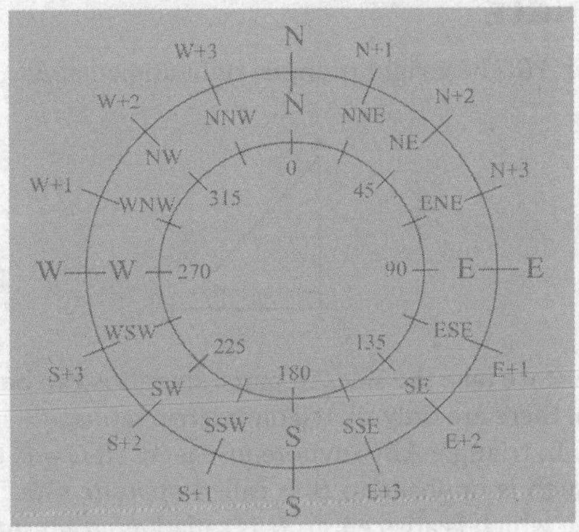

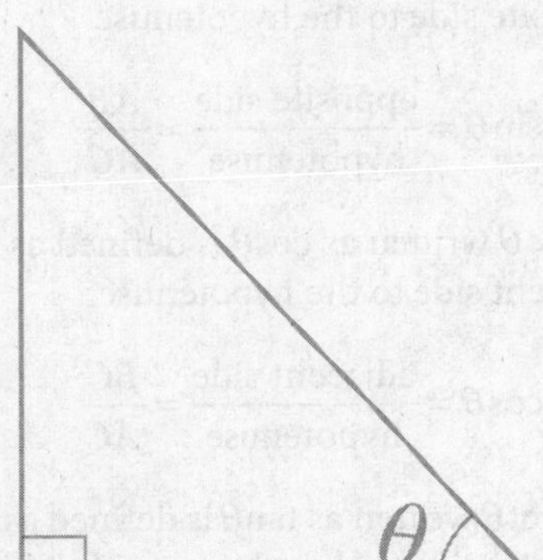

INTRODUCTION

Trigonometry is a branch of mathematics which is widely used in physics, mechanics, astronomy, and many branches of engineering. The word 'trigonometry' is a combination of three Greek words *tri, gono,* and *metry* which mean three, angle, and measure, respectively. So the word 'trigonometry' stands for 'triangle measure'. This subject was invented and developed mainly to calculate the elements of a triangle. But later, it was found to be a very useful tool to study periodic phenomena such as wave motion and vibration.

The subject was developed by many mathematicians such as Johann Bernoulli, Euler, Fourier, Gauss, and Aryabhata.

TRIGONOMETRIC RATIOS OF AN ACUTE ANGLE

Let ABC be a right triangle, right angled at B.

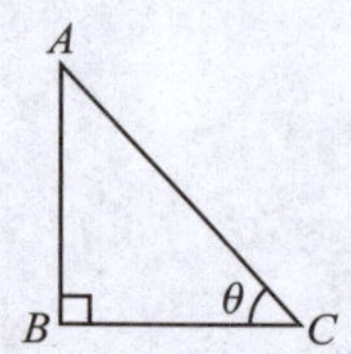

If we take the sides in pairs, six ratios can be formed. So, there are only six trigonometric ratios.

In triangle ABC, hypotenuse is AC. If $C = \theta$, then AB, which is opposite to θ, is called *opposite side*. The side BC, which is adjacent to θ, is called *adjacent side*.

Definition

(a) sine θ written as $\sin\theta$ is defined as the ratio of the opposite side to the hypotenuse.

$$\text{i.e., } \sin\theta = \frac{\text{opposite side}}{\text{hypotenuse}} = \frac{AB}{AC}$$

(b) cosine θ written as $\cos\theta$ is defined as the ratio of the adjacent side to the hypotenuse.

$$\text{i.e., } \cos\theta = \frac{\text{adjacent side}}{\text{hypotenuse}} = \frac{BC}{AC}$$

(c) tangent θ written as $\tan\theta$ is defined as the ratio of the opposite side to the adjacent side.

$$\text{i.e., } \tan\theta = \frac{\text{opposite side}}{\text{adjacent side}} = \frac{AB}{BC}$$

(d) cotangent θ written as $\cot\theta$ is defined as the ratio of the adjacent side to the opposite side.

$$\text{i.e., } \cot\theta = \frac{\text{adjacent side}}{\text{opposite side}} = \frac{BC}{AB}$$

(e) secant θ written as $\sec\theta$ is defined as the ratio of the hypotenuse to the adjacent side.

$$\text{i.e., } \sec\theta = \frac{\text{hypotenuse}}{\text{adjacent side}} = \frac{AC}{CB}$$

(f) cosecant θ written as $\text{cosec}\,\theta$ is defined as the ratio of the hypotenuse to the opposite side.

$$\text{i.e., } \text{cosec}\,\theta = \frac{\text{hypotenuse}}{\text{opposite side}} = \frac{AC}{AB}$$

Trigonometric Ratios of an Angle Θ Starting with a Circle

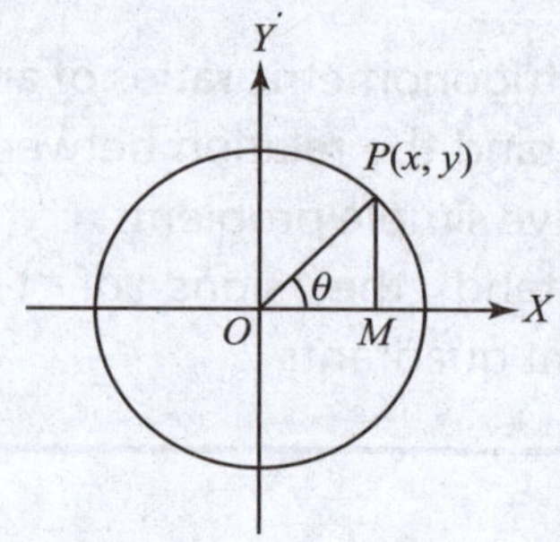

Take a unit circle (radius = 1) with center at O. Let P be any point (x, y) on the circle and let XOP be θ. Then $\sin\theta = y$ and $\cos\theta = x$. If the radius is r, then $\sin\theta = \frac{y}{r}$ and $\cos\theta = \frac{x}{r}$.

Note that this definition is more general because it holds good for all positions of P on the circle.

Furthermore, if P is in the II quadrant and θ is obtuse, then x is negative and hence $\cos\theta$ is negative. If P is in the III quadrant, θ lies between 180° and 270° and both sin and cos are negative. If P is in the IV quadrant, x is + and y is – and hence sin is negative. From this definition, we can write down the values of sine and cosine of quadrantal angles (0°, 90°, 180°, 270°, and 360°).

To find the trigonometric ratios of 0° (or 360°), we have to take P as (1, 0) on the x-axis. We get sin 0° = y coordinate of P = 0 and cos 0° = x coordinate of P = 1.

To find sin 90° and cos 90°, we have to take P at (0, 1) on the y-axis. We have sin 90° = 1 and cos 90° = 0.

To find sin 180° and cos 180°, we have to take P as (–1, 0) on the x-axis to the left of origin. We get sin 180° = 0 and cos 180° = –1.

For 270°, take P at (0, –1) on the negative side of the y-axis. We get sin 270° = –1 and cos 270° = 0. Once sine and cosine are known, we can easily write down the values of the other ratios.

We can also prove some important identities which connect trigonometric ratios.

(a) $\sin^2\theta + \cos^2\theta = 1$

In the given diagram, $PM^2 + OM^2 = OP^2$ (Pythagoras' theorem).

That is, $y^2 + x^2 = 1^2$ from which $\sin^2\theta + \cos^2\theta = 1$, since $\sin\theta = y$ and $\cos\theta = x$.

(b) $1 + \tan^2\theta = \sec^2\theta$

Dividing both sides of the previous identity by $\cos^2\theta$, we get the following:

$$\frac{\sin^2\theta}{\cos^2\theta} + \frac{\cos^2\theta}{\cos^2\theta} = \frac{1}{\cos^2\theta}$$

From this we get

$\tan^2\theta + 1 = \sec^2\theta$

(c) $1 + \cot^2\theta = \operatorname{cosec}^2\theta$

Dividing both sides of the first identity by $\sin^2\theta$, we get

$$\frac{\sin^2\theta}{\sin^2\theta} + \frac{\cos^2\theta}{\sin^2\theta} = \frac{1}{\sin^2\theta}$$

$\therefore 1 + \cot^2\theta = \operatorname{cosec}^2\theta$

Worked Examples

1. Prove that $\sin^4\theta - \cos^4\theta = \sin^2\theta - \cos^2\theta$.

Solution

$$\begin{aligned}\text{LHS} &= \sin^4\theta - \cos^4\theta \\ &= (\sin^2\theta + \cos^2\theta)(\sin^2\theta - \cos^2\theta) \\ &= \sin^2\theta - \cos^2\theta\end{aligned}$$

2. Show that $\sqrt{\dfrac{1-\sin\theta}{1+\sin\theta}} = \sec\theta - \tan\theta$.

Solution

$$\text{LHS} = \sqrt{\frac{(1-\sin\theta)\,(1-\sin\theta)}{(1+\sin\theta)\,(1-\sin\theta)}}$$

$$= \frac{1-\sin\theta}{\sqrt{1-\sin^2\theta}} = \frac{1-\sin\theta}{\cos\theta}$$

$$= \frac{1}{\cos\theta} - \frac{\sin\theta}{\cos\theta}$$

$$= \sec\theta - \tan\theta$$

3. Prove that $\dfrac{\cot A + \tan B}{\cot B + \tan A} = \cot A \tan B$.

Solution

$$LHS = \frac{\cot A + \tan B}{\cot B + \tan A} = \frac{\dfrac{1}{\tan A} + \tan B}{\dfrac{1}{\tan B} + \tan A}$$

$$= \frac{\dfrac{1+\tan A\tan B}{\tan A}}{\dfrac{1+\tan A\tan B}{\tan B}} = \frac{\tan B}{\tan A}$$

$$= \cot A \times \tan B$$

4. Prove the following:

$$\frac{\tan A + \sec A - 1}{\tan A - \sec A + 1} = \frac{1+\sin A}{\cos A} = \frac{\cos A}{1-\sin A}$$

Solution

$$LHS = \frac{\tan A + \sec A - (\sec^2 A - \tan^2 A)}{\tan A - \sec A + 1}$$

$(\because 1 + \tan^2 A = \sec^2 A)$

$$= \frac{(\tan A + \sec A) - (\sec A - \tan A)\,(\sec A + \tan A)}{(\tan A - \sec A + 1)}$$

$$= \frac{(\tan A + \sec A)[1 - \sec A + \tan A]}{(\tan A - \sec A + 1)}$$

$$= \tan A + \sec A = \frac{\sin A}{\cos A} + \frac{1}{\cos A} = \frac{1+\sin A}{\cos A} \qquad (1)$$

LHS of Eq. (1) = $(\sec A + \tan A) = \dfrac{1}{\sec A - \tan A}$

$(\because \sec^2 A - \tan^2 A = 1)$

$$= \frac{1}{\dfrac{1}{\cos A} - \dfrac{\sin A}{\cos A}} = \frac{\cos A}{1-\sin A}$$

Note

The signs of $\sin\theta$, $\cos\theta$, and $\tan\theta$ are shown in the diagram. Their reciprocals $\operatorname{cosec}\theta$, $\sec\theta$, and $\cot\theta$ will have same signs, respectively.

$\sin\theta$ +ve $\cos\theta$ −ve $\tan\theta$ −ve	$\sin\theta$ +ve $\cos\theta$ +ve $\tan\theta$ +ve
$\sin\theta$ −ve $\cos\theta$ −ve $\tan\theta$ +ve	$\sin\theta$ −ve $\cos\theta$ +ve $\tan\theta$ −ve

S	A
T	C

sine positive in the second quadrant	all positive in the first quadrant
tangent is positive in the third quadrant	cosine is positive in the fourth quadrant

A: All +
S: sin +
T: tan +
C: cos +

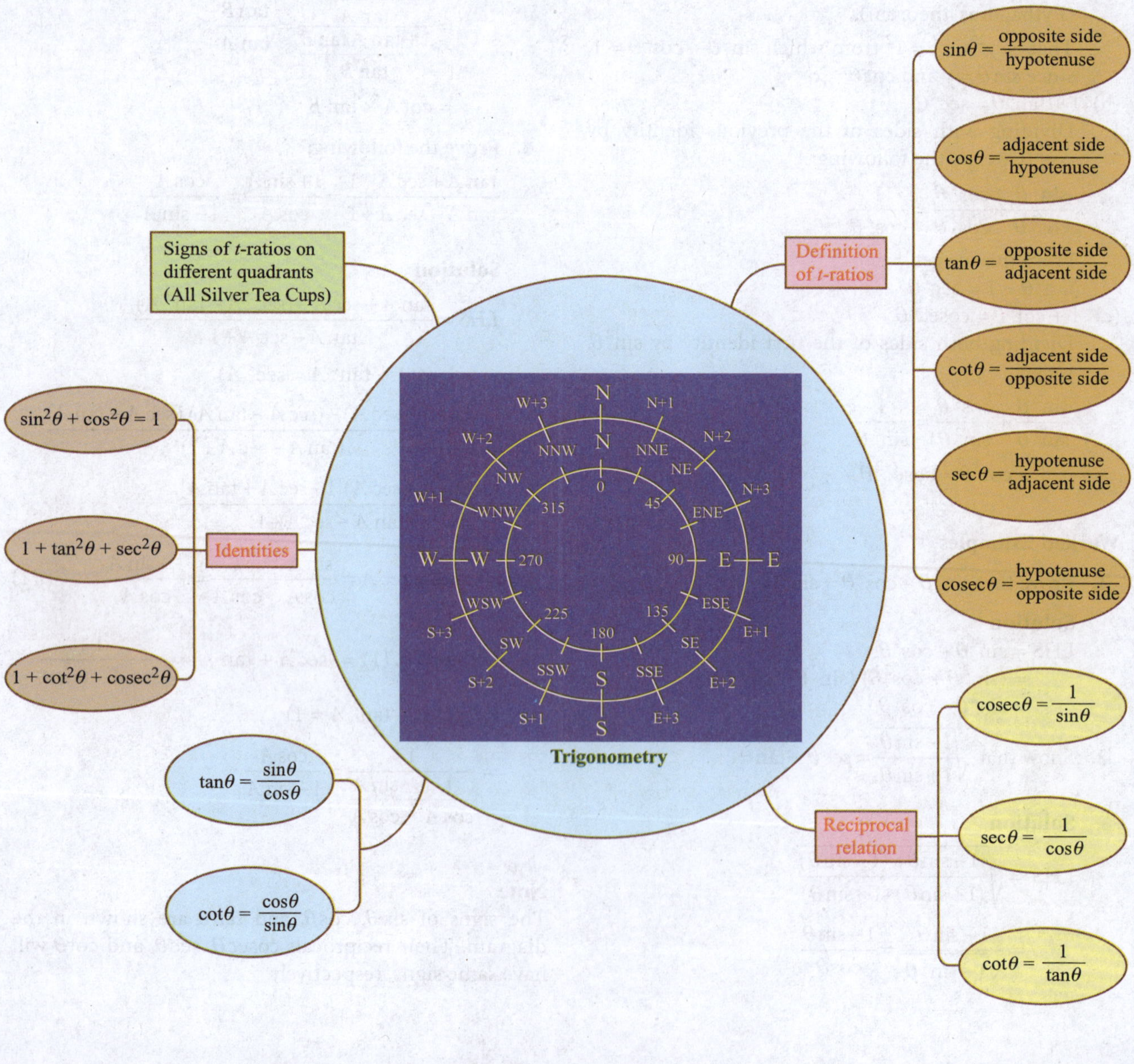

TEST YOURSELF

Level 1

Prove the following identities:

1. $\cos\theta\tan\theta = \sin\theta$
2. $(1 - \cos^2\theta)\text{cosec}^2\theta = 1$
3. $(1 - \cos^2\theta)\sec^2\theta = \tan^2\theta$
4. $\cos\beta\ \text{cosec}\,\beta\sqrt{\sec^2\beta - 1} = 1$
5. $\sin A \sec A\sqrt{\text{cosec}^2 A - 1} = 1$
6. $\sin^2 A \cot^2 A + \sin^2 A = 1$
7. $\sin^2 A \sec^2 A = \sec^2 A - 1$
8. $\dfrac{1}{\sec^2 A} + \dfrac{1}{\text{cosec}^2 A} = 1$
9. $\dfrac{\sin A}{\text{cosec}\,A} + \dfrac{\cos A}{\sec A} = 1$
10. $(\sec A \cot A)^2 - (\cos A\ \text{cosec}\,A)^2 = 1$

Level 2

11. If $\sin\theta = \frac{3}{5}$, find the value of $\tan\theta$.
12. If $\cot\theta = \frac{\sqrt{5}}{2}$, find the value of $\cos\theta$.
13. Express $\sin\theta$ in terms of $\tan\theta$.
14. Express $\cos\theta$ in terms of $\text{cosec}\,\theta$.
15. Prove that $\dfrac{x^2}{a^2} + \dfrac{y^2}{b^2} = 1$ if $x = a\cos\theta$ and $y = b\sin\theta$.
16. Prove that $(\sin A + \cos A)^2 + (\sin A - \cos A)^2 = 2$.

Level 3

Prove the following statements:

17. $\dfrac{1}{\cot A + \tan A} = \sin A \cos A$
18. $\dfrac{1 - \tan A}{1 + \tan A} = \dfrac{\cot A - 1}{\cot A + 1}$
19. $\dfrac{\cos A}{1 - \tan A} + \dfrac{\sin A}{1 - \cot A} = \sin A + \cos A$
20. $\sec^4 A - \sec^2 A = \tan^4 A + \tan^2 A$
21. $(1 + \cot A - \text{cosec}\,A)(1 + \tan A + \sec A) = 2$

Multiple Choice Questions

22. If $\sec\theta + \tan\theta = a$, then the value of $\sec\theta - \tan\theta$ is
    (A) a^2 (B) $\dfrac{1}{a}$
    (C) $\dfrac{1}{a^2}$ (D) a
23. What is the value of $(1 + \tan A + \sec A)(1 + \cot A - \text{cosec}\,A)$?
    (A) 1 (B) 0
    (C) 2 (D) –1
24. If $\sin\theta = \dfrac{5}{7}, 0 < \theta < 90^\circ$, what is the value of $\tan\theta$?
    (A) $\dfrac{5}{2}$ (B) $\dfrac{5}{\sqrt{6}}$
    (C) $\dfrac{5}{2\sqrt{6}}$ (D) none of these
25. The value of $(\cos\theta + \sin\theta)^2 + (\cos\theta - \sin\theta)^2$ is
    (A) 1 (B) 0
    (C) 3 (D) 2
26. What is the value of $\dfrac{1}{1 - \sin\alpha} + \dfrac{1}{1 + \sin\alpha}$?
    (A) $2\tan^2\alpha$ (B) $2\cos^2\alpha$
    (C) $\sec^2\alpha$ (D) $2\sec^2\alpha$
27. What is the value of $\dfrac{\sin\theta}{1 + \cos\theta} + \dfrac{1 + \cos\theta}{\sin\theta}$?
    (A) 2 (B) $2\,\text{cosec}\,\theta$
    (C) $2\sec\theta$ (D) $\tan\theta$
28. The value of $\sqrt{\dfrac{1 + \cos\theta}{1 - \cos\theta}}$ is
    (A) $\dfrac{1 + \sin\theta}{\cos\theta}$ (B) $\dfrac{1 + \cos\theta}{\cos\theta}$
    (C) $\dfrac{1 - \cos\theta}{\sin\theta}$ (D) $\dfrac{1 + \cos\theta}{\sin\theta}$
29. The value of $\dfrac{\sec\theta - 1}{\sec\theta + 1}$ is
    (A) $(\cot\theta - \text{cosec}\,\theta)^2$ (B) $(\cot\theta + \text{cosec}\,\theta)^2$
    (C) $(\sin\theta - \cos\theta)^2$ (D) $(\sin\theta + \cos\theta)^2$
30. The value of $1 + \dfrac{\cot^2\theta}{1 + \text{cosec}\,\theta}$ is
    (A) $\sin\theta$ (B) $\cos\theta$
    (C) $\text{cosec}\,\theta$ (D) $\cot\theta$

31. The value of $\dfrac{\cos\theta}{\text{cosec}\,\theta-1}+\dfrac{\cos\theta}{\text{cosec}\,\theta+1}$ is
(A) $2\sin\theta$ (B) $2\cot\theta$
(C) $2\tan\theta$ (D) $\tan\theta$

32. If $x\cos A = 1$ and $\tan A = y$, then the value of $x^2 - y^2$ is
(A) 0 (B) −1
(C) 2 (D) 1

33. If $\cot\theta=\dfrac{7}{8}$ then the value of $\dfrac{(1+\sin\theta)(1-\sin\theta)}{(1+\cos\theta)(1-\cos\theta)}$ is
(A) $\dfrac{49}{64}$ (B) $\sqrt{\dfrac{7}{8}}$
(C) $\dfrac{7}{8}$ (D) $\dfrac{14}{21}$

34. The value of θ for $\cos^4\theta-\sin^4\theta=\dfrac{1}{2}$ $(0<\theta<90°)$ is
(A) $\dfrac{\pi}{2}$ (B) $\dfrac{\pi}{3}$
(C) $\dfrac{\pi}{4}$ (D) $\dfrac{\pi}{6}$

35. The value of $(a\sin\theta - b\cos\theta)^2\,(a\sin\theta + b\cos\theta)^2$ is
(A) 0 (B) $a^2 - b^2$
(C) $a^2 + b^2$ (D) none of these

36. The value of θ $(0<\theta<90°)$ for which $\dfrac{\sin\theta}{1+\cos\theta}+\dfrac{\sin\theta}{1-\cos\theta}=2$ is
(A) $\dfrac{\pi}{3}$ (B) $\dfrac{\pi}{6}$
(C) $\dfrac{\pi}{4}$ (D) $\dfrac{\pi}{2}$

37. The value of θ $(0<\theta<90°)$ for which $\dfrac{\cos\theta}{1+\sin\theta}+\dfrac{1+\sin\theta}{\cos\theta}=4$ is
(A) $\dfrac{\pi}{3}$ (B) $\dfrac{\pi}{6}$
(C) $\dfrac{\pi}{2}$ (D) $\dfrac{\pi}{4}$

38. The value of $(1 + \cos A)(1 - \cos A)(1 + \cot^2 A)$ is
(A) 0 (B) 1
(C) 2 (D) 3

39. The expression $\dfrac{\text{cosec}\,A}{1-\text{cosec}\,A}+\dfrac{\text{cosec}\,A}{\text{cosec}\,A+1}$ is equivalent to
(A) $2\sin A\cos A$ (B) $2\tan A\,\text{cosec}\,A$
(C) $-2\tan A\sec A$ (D) none of these

40. The value of $\sqrt{\dfrac{1+\sin\theta}{1-\sin\theta}}+\sqrt{\dfrac{1-\sin\theta}{1+\sin\theta}}$ is
(A) 1 (B) $2\sec\theta$
(C) $3\tan\theta$ (D) $2\sin\theta$

41. The value of A for which $(1 + \cot A - \text{cosec}\,A)(1 + \tan A + \sec A) = 2\sec A$ is
(A) 0° (B) 60°
(C) 45° (D) 90°

42. The value of $(\sec\theta\cot\theta)^2 - (\cos\theta\,\text{cosec}\,\theta)^2$ is
(A) 0 (B) 1
(C) 2 (D) 3

43. If $x = a\cos\theta$ and $y = a\sin\theta$, then which of the following is correct?
(A) $x^2 - y^2 = a^2$ (B) $x^2 + y^2 = 0$
(C) $x^2 + y^2 = a^2$ (D) $x^2 - y^2 = 0$

44. If $x = A\sec\alpha$ and $y = B\tan\alpha$, then which of the following is correct?
(A) $\dfrac{x^2}{A^2}+\dfrac{y^2}{B^2}=1$ (B) $\dfrac{x^2}{A^2}+\dfrac{y^2}{B^2}=-1$
(C) $\dfrac{x^2}{A^2}-\dfrac{y^2}{B^2}=1$ (D) none of these

45. If $x = a\cos\theta$ and $y = b\sin\theta$, then the value of $\dfrac{x^2}{a^2}+\dfrac{y^2}{b^2}$ is
(A) 0 (B) 1/2
(C) −1 (D) 1

46. The value of $(\sin\theta + \cos\theta)^2 + (\sin\theta - \cos\theta)^2$ is
(A) 1 (B) 2
(C) 3 (D) 0

47. If $\sec^2\theta\,(1 + \sin\theta)(1 - \sin\theta) = k$ then the value of k is
(A) 0 (B) $\dfrac{1}{2}$
(C) 1 (D) 2

48. If $x = r \sin\theta \cos\phi$, $y = r \sin\theta \sin\phi$ and $z = r \cos\theta$ then the value of $x^2 + y^2 + z^2$ is

(A) $r^2 + 2$ (B) $r^2 - 2$

(C) r (D) r^2

49. The value of $\dfrac{\cos A}{1-\tan A} + \dfrac{\sin A}{1-\cot A}$ is

(A) $\sin A + \cos A$ (B) $\sin A - \cos A$

(C) $2\sin A + \cos A$ (D) $\sin A + 2\cos A$

50. If $a \sec\alpha - c \tan\alpha = d$ and $b \sec\alpha + d \tan\alpha = c$ then which of the following is correct?

(A) $a^2 + b^2 = 2c^2$ (B) $a^2 + b^2 = c^2 - d^2$

(C) $a^2 + b^2 = c^2 + d^2$ (D) $a^2 + b^2 = c^2 - d^2$

51. If $3 \sin\theta + 4 \cos\theta = 5$, then the value of $4 \sin\theta - 3 \cos\theta$ is

(A) 1 (B) 0

(C) 16 (D) 9

OLYMPIAD AND NTSE LEVEL EXERCISES

52. $(m + 2) \sin\theta + (2m - 1) \cos\theta = 2m + 1$, if

(A) $\tan\theta = \dfrac{3}{4}$ (B) $\tan\theta = \dfrac{4}{3}$

(C) $\tan\theta = \dfrac{2m}{m^2+1}$ (D) None of these

53. The value of the expression

$1 - \dfrac{\sin^2 y}{1+\cos y} + \dfrac{1+\cos y}{\sin y} - \dfrac{\sin y}{1-\cos y}$ is equal to

(A) 0 (B) 1

(C) $\sin y$ (D) $\cos y$

54. If $p = \dfrac{2\sin\theta}{1+\cos\theta+\sin\theta}$, and $q = \dfrac{\cos\theta}{1+\sin\theta}$, then

(A) $pq = 1$ (B) $\dfrac{q}{p} = 1$

(C) $q - p = 1$ (D) $q + p = 1$

55. If $\tan\theta - \cot\theta = a$ and $\sin\theta + \cos\theta = b$, then $(b^2 - 1)^2 (a^2 + 4)$ is equal to

(A) 2 (B) -4

(C) ± 4 (D) 4

56. If $\tan A = \dfrac{1}{2}, \tan B = \dfrac{1}{3}$, then $\cos 2A =$

(A) $\sin B$ (B) $\sin 2B$

(C) $\sin 3B$ (D) None of these

57. If $\sin A + \sin B = C$, $\cos A + \cos B = D$, then the value of $\sin(A + B) =$

(A) CD (B) $\dfrac{CD}{C^2+D^2}$

(C) $\dfrac{C^2+D^2}{2CD}$ (D) $\dfrac{2CD}{C^2+D^2}$

58. If θ lies in the second quadrant, then the value of $\sqrt{\left(\dfrac{1-\sin\theta}{1+\sin\theta}\right)} + \sqrt{\left(\dfrac{1+\sin\theta}{1-\sin\theta}\right)}$ is

(A) $2\sec\theta$ (B) $-2\sec\theta$

(C) $2\cos\theta$ (D) None of these

59. If $2y \cos\theta = x \sin\theta$ and $2x \sec\theta - y \operatorname{cosec}\theta = 3$, then $x^2 + 4y^2 =$

(A) 4 (B) -4

(C) ± 4 (D) None of these

60. Match the functions in Column I with their ranges in Column II.

Column I	Column II
(A) $A = \sin^2\theta + \cos^4\theta$	(p) $A \in [-1, 1]$
(B) $A = 3\cos^2\theta + \sin^4\theta$	(q) $A \in [3/4, 1]$
(C) $A = \sin^2\theta - \cos^4\theta$	(r) $A \in [2\sqrt{2}, \infty)$
(D) $A = \tan^2\theta + 2\cot^2\theta$	(s) $A \in [1, 3]$

61. $\dfrac{\sin 2A}{1+\cos 2A} \times \dfrac{\cos A}{1+\cos A} =$

(A) $\tan\dfrac{A}{2}$ (B) $\cot\dfrac{A}{2}$

(C) $\sec\dfrac{A}{2}$ (D) $\operatorname{cosec}\dfrac{A}{2}$

21 Standard and Allied Angles

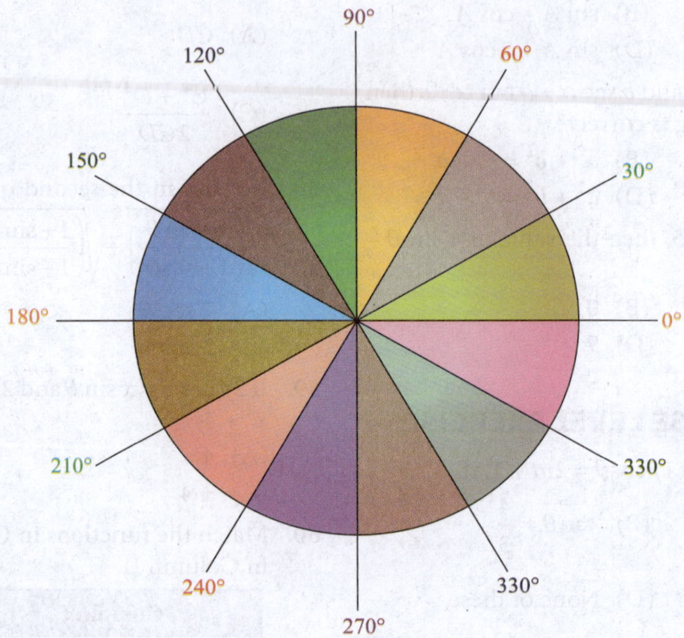

> One person's constant is another person's variable.
>
> – Susan Gerhart

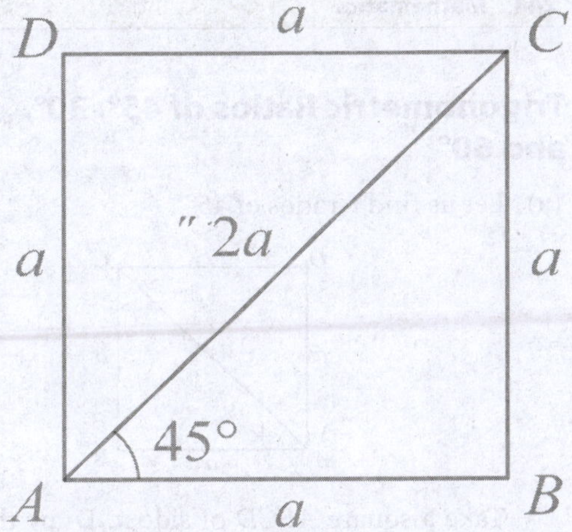

OBJECTIVES

After reading this chapter you should be able to

- find the *t*-ratios of standard angles and solve simple problems

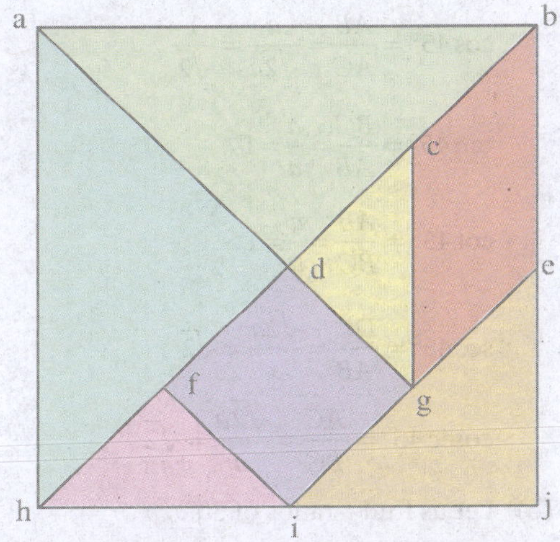

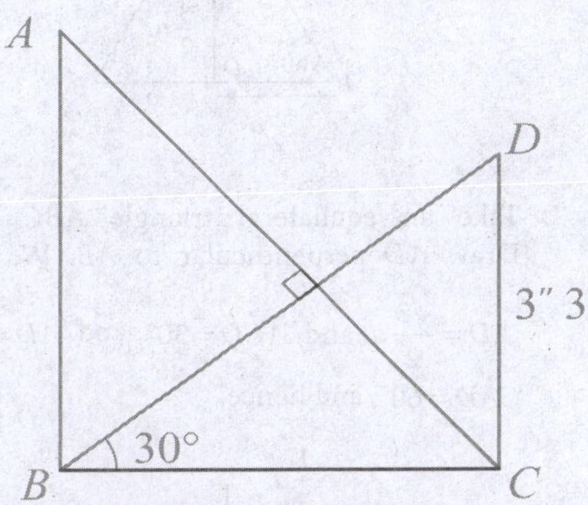

Trigonometric Ratios of 45°, 30°, and 60°

(a) Let us find t-ratios of 45°.

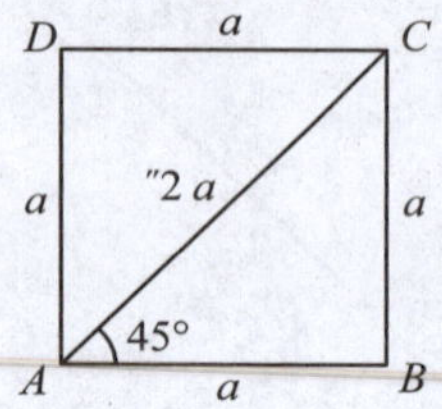

Take a square $ABCD$ of side a. Draw the diagonal AC. We know that $AC = \sqrt{2}a$ and $C\hat{A}B = 45°$.

$$\sin 45° = \frac{BC}{AC} = \frac{a}{\sqrt{2}a} = \frac{1}{\sqrt{2}}$$

$$\cos 45° = \frac{AB}{AC} = \frac{a}{\sqrt{2}a} = \frac{1}{\sqrt{2}}$$

$$\tan 45° = \frac{BC}{AB} = \frac{a}{a} = 1$$

$$\cot 45° = \frac{AB}{BC} = \frac{a}{a} = 1$$

$$\sec 45° = \frac{DC}{AB} = \frac{\sqrt{2}a}{a} = \sqrt{2}$$

$$\text{cosec}\, 45° = \frac{AC}{BC} = \frac{\sqrt{2}a}{a} = \sqrt{2}$$

(b) Let us find t-ratios of 30°.

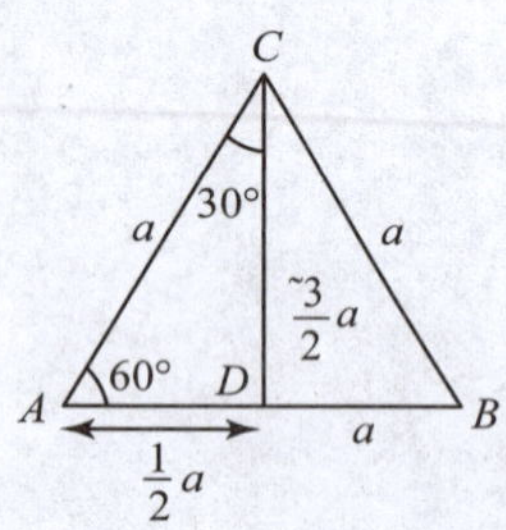

Take an equilateral triangle ABC of side a. Draw AD perpendicular to AB. We know that $AD = \frac{\sqrt{3}}{2}a$ and $A\hat{C}D = 30°$ and $AD = \frac{1}{2} = a$ and $C\hat{A}D = 60°$, and hence

$$\sin 30° = \frac{AD}{AC} = \frac{\frac{1}{2}a}{a} = \frac{1}{2}$$

$$\cos 30° = \frac{CD}{AC} = \frac{\frac{\sqrt{3}}{2}a}{a} = \frac{\sqrt{3}}{2}$$

$$\tan 30° = \frac{AD}{DC} = \frac{\frac{1}{2}a}{\frac{\sqrt{3}}{2}a} = \frac{1}{\sqrt{3}}$$

$$\cot 30° = \frac{DC}{AD} = \frac{\frac{\sqrt{3}}{2}a}{\frac{1}{2}a} = \sqrt{3}$$

$$\sec 30° = \frac{AC}{CD} = \frac{a}{\frac{\sqrt{3}}{2}a} = \frac{2}{\sqrt{3}}$$

$$\text{cosec}\, 30° = \frac{AC}{AD} = \frac{a}{\frac{1}{2}a} = 2$$

(c) We can find the t-ratios of 60° using the same figure.

$$\sin 60° = \frac{CD}{AC} = \frac{\frac{\sqrt{3}}{2}a}{a} = \frac{\sqrt{3}}{2}$$

$$\cos 60° = \frac{AD}{AC} = \frac{\frac{1}{2}a}{a} = \frac{1}{2}$$

$$\tan 60° = \frac{CD}{AD} = \frac{\frac{\sqrt{3}}{2}a}{\frac{1}{2}a} = \sqrt{3}$$

$$\cot 60° = \frac{AC}{AD} = \frac{\frac{1}{2}a}{\frac{\sqrt{3}}{2}a} = \frac{1}{\sqrt{3}}$$

$$\sec 60° = \frac{AC}{AD} = \frac{a}{\frac{1}{2}a} = 2$$

$$\text{cosec}\, 60° = \frac{AC}{CD} = \frac{a}{\frac{\sqrt{3}}{2}a} = \frac{2}{\sqrt{3}}$$

Angle	0°	30°	45°	60°	90°
Sine	0	$\frac{1}{2}$	$\frac{1}{\sqrt{2}}$	$\frac{\sqrt{3}}{2}$	1
Cosine	1	$\frac{\sqrt{3}}{2}$	$\frac{1}{\sqrt{2}}$	$\frac{1}{2}$	0
Tangent	0	$\frac{1}{\sqrt{3}}$	1	$\sqrt{3}$	Not defined

Worked Examples

1. Find the value of the following:
$4\cos^2 60° + 4\tan^2 45° - \text{cosec}^2 30°$

Solution

$$4\left(\frac{1}{2}\right)^2 + 4(1)^2 - (2)^2$$

$$= 4 \times \frac{1}{4} + 4 - 4 = 1$$

2. $\sin^2 30° + \tan^2 45° + \tan^2 60°$

Solution

$$\left(\frac{1}{2}\right)^2 + (1)^2 + (\sqrt{3})^2$$

$$= \frac{1}{4} + 1 + 3 = 4\frac{1}{4}$$

Trigonometric Functions of (90° – θ) in Terms of Those of θ

The two triangles *MOP* and *M′P′O* are congruent.

$OP = OP' = r,\ OM' = PM = y = x'$

$OM = M'P' = y' = x$

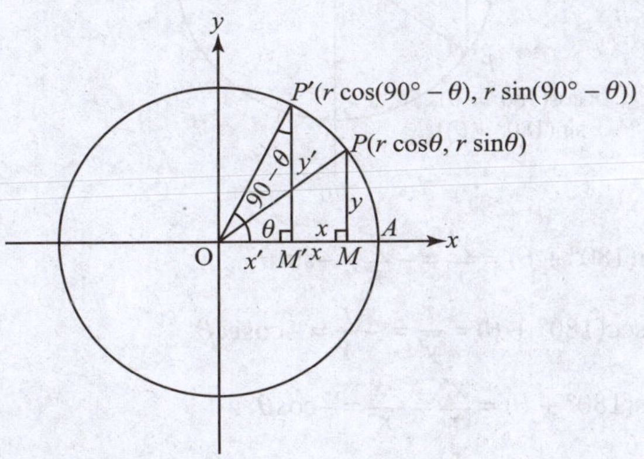

From ΔMOP

$\sin\theta = \frac{y}{r}$ $\quad\cos\theta = \frac{x}{r}$ $\quad\tan\theta = \frac{y}{x}$

$\text{cosec}\,\theta = \frac{r}{y}$ $\quad\sec\theta = \frac{r}{x}$ $\quad\cot\theta = \frac{x}{y}$

From $\Delta M'P'O$

$$\sin(90° - \theta) = \frac{y'}{r} = \frac{x}{r} = \cos\theta$$

$$\cos(90° - \theta) = \frac{x'}{r} = \frac{y}{r} = \sin\theta$$

$$\tan(90° - \theta) = \frac{y'}{x'} = \frac{x}{y} = \cot\theta$$

$$\text{cosec}(90° - \theta) = \frac{x}{y'} = \frac{r}{x} = \sec\theta$$

$$\sec(90° - \theta) = \frac{x'}{r} = \frac{y}{r} = \text{cosec}\,\theta$$

$$\cot(90° - \theta) = \frac{x'}{y'} = \frac{y}{x} = \tan\theta$$

Trigonometric Functions of (90° + θ) in Terms of Those of θ

The two triangles *MOP* and *M′P′O* are congruent.

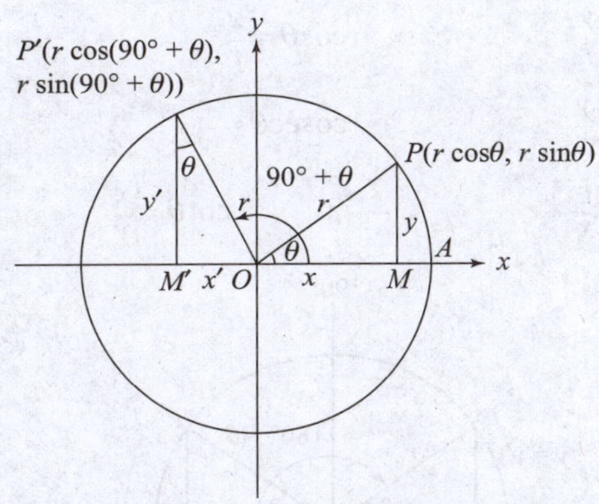

$M\hat{O}P = M'\hat{P}'O = \theta$ $\qquad OP = OP' = r$

$MO = x = M'P = y'$

$MP = y = OM' = -x'$

From ΔMOP

$\sin\theta = \frac{y}{x}$ $\qquad\cos\theta = \frac{x}{r}$

$\tan\theta = \frac{y}{x}$ $\qquad\text{cosec}\,\theta = \frac{r}{y}$

$\sec\theta = \frac{r}{x}$ $\qquad\cot\theta = \frac{x}{y}$

From the $\Delta M'P'O$

$$\sin(90° + \theta) = \frac{y'}{x} = \frac{x}{r} = \cos\theta$$

$$\cos(90° + \theta) = \frac{x'}{r} = \frac{-y}{r} = -\sin\theta$$

$$\tan(90° + \theta) = \frac{y'}{x'} = \frac{x}{-y} = -\cot\theta$$

$$\text{cosec}(90^\circ + \theta) = \frac{r}{y'} = \frac{r}{x} = \sec\theta$$

$$\sec(90^\circ + \theta) = \frac{r}{x'} = \frac{r}{-y} = -\text{cosec}\theta$$

$$\cot(90^\circ + \theta) = \frac{x'}{y'} = \frac{y}{x} = -\tan\theta$$

Trigonometric Function of (180° – θ) in Terms of Those of θ

The two triangles OMP and $OM'P'$ are congruent.

$M\hat{O}P = M'\hat{O}P' = \theta$, $y' = y,\ x' = -x$

$OP = OP' = r$

From ΔOMP

$\sin\theta = \frac{y}{r}$ $\cos\theta = \frac{x}{r}$

$\tan\theta = \frac{y}{x}$ $\text{cosec}\theta = \frac{r}{y}$

$\sec\theta = \frac{r}{x}$ $\cot\theta = \frac{x}{y}$

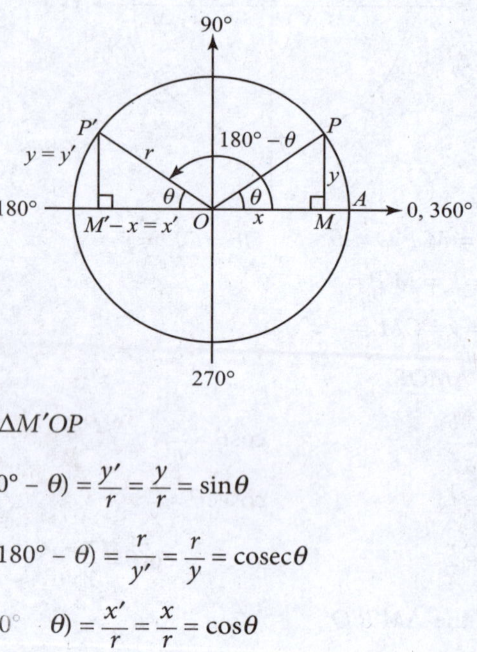

From $\Delta M'OP$

$$\sin(180^\circ - \theta) = \frac{y'}{r} = \frac{y}{r} = \sin\theta$$

$$\text{cosec}(180^\circ - \theta) = \frac{r}{y'} = \frac{r}{y} = \text{cosec}\theta$$

$$\cos(180^\circ - \theta) = \frac{x'}{r} = \frac{x}{r} = \cos\theta$$

$$\sec(180^\circ - \theta) = \frac{r}{x'} = \frac{-r}{x} = \sec\theta$$

$$\tan(180^\circ - \theta) = \frac{y'}{x'} = \frac{y}{-x} = -\tan\theta$$

$$\cot(180^\circ - \theta) = \frac{x'}{y} = \frac{x}{y} = \cot\theta$$

Trigonometric Functions of (180° + θ) in Term of Those of θ

Two triangles MOP and $M'OP'$ are congruent.

$M'\hat{O}P' = M\hat{O}P = \theta$, $OP' = OP = r$,

$OM = x$

$OM' = -OM = x' = -x$

$M'P' = y' = -MP = -y$

From ΔOMP

$\sin\theta = \frac{y}{r}$ $\text{cosec}\theta = \frac{r}{y}$

$\cos\theta = \frac{x}{r}$ $\sec\theta = \frac{r}{x}$

$\tan\theta = \frac{y}{x}$ $\cot\theta = \frac{x}{y}$

From $\Delta M'OP'$

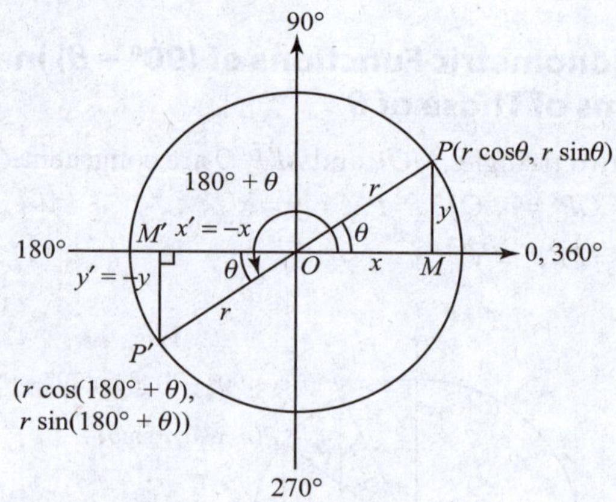

$$\sin(180^\circ + \theta) = \frac{y'}{r} = -\frac{y}{r} = -\sin\theta$$

$$\text{cosec}(180^\circ + \theta) = \frac{r}{y'} = -\frac{r}{y} = -\text{cosec}\theta$$

$$\cos(180^\circ + \theta) = \frac{x'}{r} = \frac{-x}{r} = -\cos\theta$$

$$\sec(180^\circ + \theta) = \frac{r}{x'} = -\frac{r}{x} = -\sec\theta$$

$$\tan(180^\circ + \theta) = \frac{x'}{y'} = \frac{-x}{-y} = \tan\theta$$

$$\cot(180^\circ + \theta) = \frac{y'}{x'} = \frac{-y}{-x} = \cot\theta$$

Trigonometric Functions of (270° – θ) in Terms of Those of θ

The two triangles OMP and $OP'M'$ are congruent.

$M\hat{O}P = M'\hat{P}'O = \theta$ $PM = OM' = x' = -y$

$OM = M'P' = y' = -y$ $OP' = OP = r$

From ΔOMP

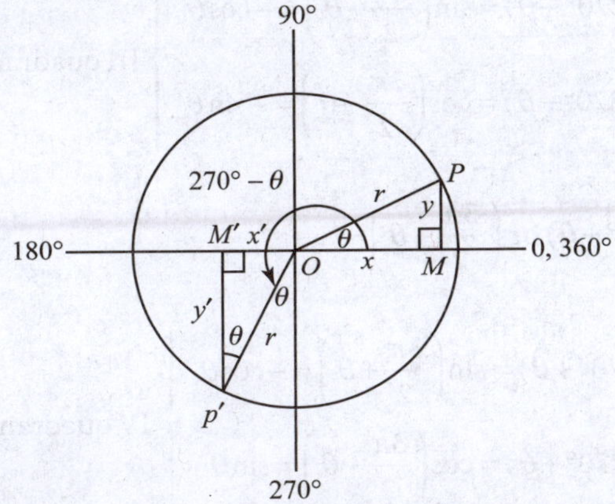

$\sin\theta = \frac{y}{r}$ $\quad\quad$ $\text{cosec}\,\theta = \frac{r}{y}$

$\cos\theta = \frac{x}{r}$ $\quad\quad$ $\sec\theta = \frac{r}{x}$

$\tan\theta = \frac{y}{x}$ $\quad\quad$ $\cot\theta = \frac{x}{y}$

From $\Delta OP'M'$

$$\sin(270° - \theta) = \frac{y'}{r} = \frac{-x}{r} = -\cos\theta$$

$$\text{cosec}(270° - \theta) = \frac{r}{y'} = -\frac{r}{x} = -\sec\theta$$

$$\cos(270° - \theta) = \frac{y'}{r} = -\frac{y}{r} = -\sin\theta$$

$$\sec(270° - \theta) = \frac{r}{x'} = -\frac{r}{y} = \text{cosec}\,\theta$$

$$\tan(270° - \theta) = \frac{y'}{x'} = \frac{-x}{-y} = \cot\theta$$

$$\cot(270° - \theta) = \frac{x'}{y'} = \frac{-y'}{-x'} = \tan\theta$$

Trigonometric Function of (360° − θ) and (− θ) in Terms of Those of θ

Two triangles MOP and MOP' are congruent.

From ΔMOP

$\sin\theta = \frac{y}{r}$ $\quad\quad$ $\text{cosec}\,\theta = \frac{r}{y}$

$\cos\theta = \frac{x}{r}$ $\quad\quad$ $\sec\theta = \frac{r}{x}$

$\tan\theta = \frac{y}{x}$ $\quad\quad$ $\cot\theta = \frac{x}{y}$

From $\Delta MOP'$

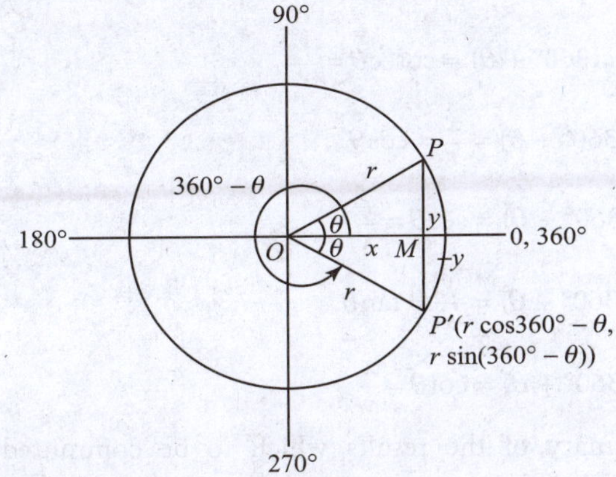

$$\sin(360° - \theta) = \frac{-y}{r} = -\sin\theta$$

$$\text{cosec}(360° - \theta) = -\frac{r}{y} = -\text{cosec}\,\theta$$

$$\cos(360° - \theta) = \frac{x}{r} = \cos\theta$$

$$\sec(360° - \theta) = \frac{r}{x} = \sec\theta$$

$$\tan(360° - \theta) = -\frac{y}{x} = -\tan\theta$$

$$\cot(360° - \theta) = -\frac{x}{y} = -\cot\theta$$

Trigonometric Functions of (360° + θ) in Terms of Those of θ

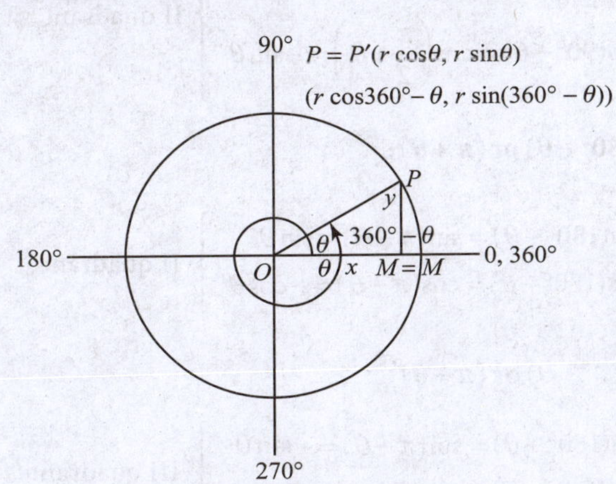

The two congruent triangles coincide.

From MOP and $M'OP'$

$OP = OP' = r,\ OM = OM' = x$,

$P'M' = PM = y,\quad P'\hat{O}M' = P\hat{O}M = \theta$

$\sin(360° + \theta) = \dfrac{y}{r} = \sin\theta$

$\text{cosec}(360° + \theta) = \text{cosec}\theta = \dfrac{r}{y}$

$\cos(360° + \theta) = \dfrac{x}{r} = \cos\theta$

$\sec(360° + \theta) = \sec\theta = \dfrac{r}{x}$

$\tan(360° + \theta) = \dfrac{y}{x} = \tan\theta$

$\cot(360° + \theta) = \cot\theta = \dfrac{x}{y}$

Summary of the results which to be committed to memory.

$(90° - \theta)$ or $\left(\dfrac{\pi}{2} - \theta\right)$:

$$\left.\begin{aligned}\sin(90° - \theta) &= \sin\left(\frac{\pi}{2} - \theta\right) = \cos\theta \\ \cos(90° - \theta) &= \cos\left(\frac{\pi}{2} - \theta\right) = \sin\theta\end{aligned}\right\}\text{ I quadrant}$$

$(90° + \theta)$ or $\left(\dfrac{\pi}{2} + \theta\right)$:

$$\left.\begin{aligned}\sin(90° + \theta) &= \sin\left(\frac{\pi}{2} + \theta\right) = \cos\theta \\ \cos(90° + \theta) &= \cos\left(\frac{\pi}{2} + \theta\right) = -\sin\theta\end{aligned}\right\}\text{ II quadrant}$$

$(180° - \theta)$ or $(\pi + \theta)$:

$$\left.\begin{aligned}\sin(180° - \theta) &= \sin(\pi - \theta) = \sin\theta \\ \cos(180° - \theta) &= \cos(\pi - \theta) = -\cos\theta\end{aligned}\right\}\text{ II quadrant}$$

$(180° + \theta)$ or $(\pi + \theta)$:

$$\left.\begin{aligned}\sin(180° + \theta) &= \sin(\pi + \theta) = -\sin\theta \\ \cos(180° + \theta) &= \cos(\pi + \theta) = -\cos\theta\end{aligned}\right\}\text{ III quadrant}$$

$(270° - \theta)$ or $\left(\dfrac{3\pi}{2} - \theta\right)$

$$\left.\begin{aligned}\sin(270° - \theta) &= \sin\left(\frac{3\pi}{2} - \theta\right) = -\cos\theta \\ \cos(270° - \theta) &= \cos\left(\frac{3\pi}{2} - \theta\right) = -\sin\theta\end{aligned}\right\}\text{ III quadrant}$$

$(270° + \theta)$ or $\left(\dfrac{3\pi}{2} + \theta\right)$

$$\left.\begin{aligned}\sin(270° + \theta) &= \sin\left(\frac{3\pi}{2} + \theta\right) = -\cos\theta \\ \cos(270° + \theta) &= \cos\left(\frac{3\pi}{2} + \theta\right) = \sin\theta\end{aligned}\right\}\text{ IV quadrant}$$

$(360° - \theta)$ or $(2\pi - \theta)$

$$\left.\begin{aligned}\sin(360° - \theta) &= \sin(2\pi - \theta) = -\sin\theta \\ \cos(360° - \theta) &= \cos(2\pi - \theta) = \cos\theta\end{aligned}\right\}\text{ IV quadrant}$$

$(360° + \theta)$ or $(2\pi + \theta)$

$$\left.\begin{aligned}\sin(360° + \theta) &= \sin(2\pi + \theta) = -\sin\theta \\ \cos(360° + \theta) &= \cos(2\pi + \theta) = \cos\theta\end{aligned}\right\}\text{ I quadrant}$$

Worked Examples

Find the numerical values of the following:

1. $\cos 135° = \cos(90° + 45°) = \sin 45° = -\dfrac{1}{\sqrt{2}}$
2. $\sin 150° = \sin(180° - 30°) = \sin 30° = -\dfrac{1}{2}$
3. $\tan 240° = \tan(180° + 60°) = \tan 60° = \sqrt{3}$
4. $\tan(-120°) = -\tan 120°$
   $= -\tan(180° - 60°) = -(\tan 60°)$
   $= \tan 60° = \sqrt{3}$
5. $\text{cosec}\left(\dfrac{3\pi}{4}\right) = \text{cosec}\left(\pi - \dfrac{\pi}{4}\right) = \text{cosec}\,\dfrac{\pi}{4} = \sqrt{2}$
6. $\cot\left(-\dfrac{5\pi}{6}\right) = \cot\left(\dfrac{5\pi}{6}\right)$,
   $\cot\left(\pi - \dfrac{\pi}{6}\right) = -\left(-\cot\dfrac{\pi}{6}\right) = \cot\dfrac{\pi}{6} = \sqrt{3}$

Find the values of the following:

1. $\cos(270° + A) = -\sin A$ (IV quadrant)
2. $\cot(270° - A) = \tan A$ (III quadrant)
3. $\sin(90° - A) = \cos A$ (I quadrant)
4. $\sec(A - 180°) = \sec(180° - A) = -\sec A$
   $= \sec[-(180° - A)]$
5. $\sin(270° - A) = -\cos A$
6. $\cot(A - 90°) = -\cot(90° - A) = -\tan A$

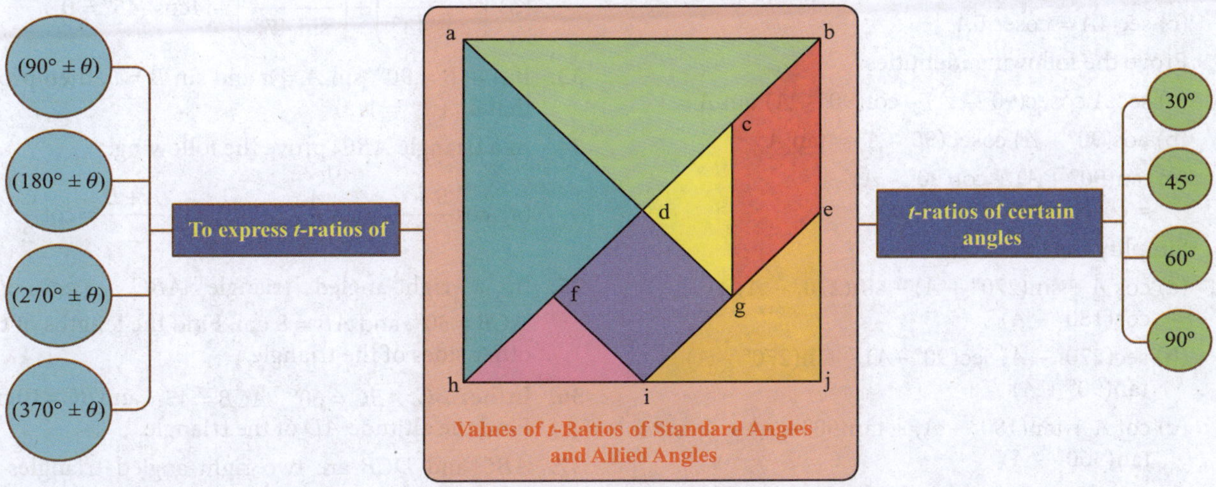

TEST YOURSELF

Level 1

Evaluate the following:

1. $\sin^2 60° + \cos^2 30° + \cot^2 45° + \sec^2 60° - \text{cosec}^2 30°$
2. $\sin 60° \cos 30° + \cos 60° \sin 30°$
3. $\sin 45° \cos 30° - \cos 45° \sin 30°$
4. Verify that $\sin^2 30° + \sin^2 45° + \sin^2 60° = \frac{3}{2}$.
5. Verify that $4 \cot^2 45° - \sec^2 60° + \sin^3 30° = \frac{1}{8}$.

Verify that if A = 30°.

6. $\cos 2A = \cos^2 A - \sin^2 A = 2\cos^2 A - 1$
7. $\sin 3A = 3 \sin A - 4 \sin^3 A$
8. $\tan 2A = \dfrac{2\tan A}{1 - \tan^2 A}$

If A = 45°, verify the following:

9. $\sin 2A = 2 \sin A \cos A$
10. $\cos 2A = 1 - 2\sin^2 A$
11. Using the formula $\sin(A + B) = \sin A \cos B + \cos A \sin B$, find the value of sin 75°
12. Using the formula $\cos(A - B) = \cos A \cos B + \sin A \sin B$, find the value of cos 15°.

Express as a function of A:

13. $\cos(270° - A)$
14. $\sin(360° - A)$
15. $\tan(A - 360°)$

Level 2

Find the numerical value of the following:

16. $\cot^2 30° + 4\sin^2 45° + 3\,\text{cosec}^2 60°$
17. $\frac{1}{3}\cos^2 30° - \frac{1}{2}\text{cosec } 30° \cot^2 60° + \frac{4}{3}\sin^2 45° \cot^2 30°$
18. If $\tan^2 45° - \sin^2 30° = x \cos 45° \sin 45° \cot 30°$, find x.
19. Find x from the following equation:

$$x \cos 60° \sin^2 45° = \frac{\tan^2 60° \,\text{cosec } 30° \cot 45°}{\sec^2 45° \sec 60°}$$

20. If $7\sin^2\theta + 3\cos^2\theta = 4$, find $\tan\theta$.

Find the numerical values of the following:

21. sin 135°
22. cos(–120°)
23. $\tan\left(\dfrac{3\pi}{4}\right)$

Level 3

24. Find the value of A for which each of the following equation is true:

(a) $\sin 3A = \cos 7A$

(b) $\cot 2A = \tan 2A$

(c) $\sec 4A = \text{cosec } 5A$

25. Prove the following identities:

(a) $\sec A \text{ cosec}(90° - A) - \cot(90° - A) \tan A = 1$

(b) $\cos(90° - A) \text{ cosec}(90 - A) = \tan A$

(c) $\tan(90° - A) + \cot(90° - A)$
$= \text{cosec } A \text{ cosec}(90° - A)$

26. Simplify the following:

(a) $\cos A + \sin(270° + A) - \sin(270° - A) + \cos(180° - A)$

(b) $\sec(270° - A) \sec(90° - A) - \tan(270° - A) \tan(90° + A)$

(c) $\cot A + \tan(180° + A) + \tan(90° + A) - \tan(360° - A)$

27. Express the following in the simplest form:

(a) $\dfrac{\sin(-A)}{\sec(180° + A)} - \dfrac{\tan(90° + A)}{\cot A} + \dfrac{\cos A}{\sin(90° + A)}$

(b) $\dfrac{\cos(90° + A)\sec(-A)\tan(180° - A)}{\sec(180° + A)\sin(180° + A)\cot(90° - A)}$

(c) $\dfrac{\sin(180° - \theta)}{\tan(180° + \theta)} \times \dfrac{\cot(90° - \theta)}{\tan(90° + \theta)} \times \dfrac{\cos(360° - \theta)}{\sin(-\theta)}$

28. Express the following in terms of ratios of a positive angle, which is less than 45°:

(a) $\cos(-83°)$ (b) $\tan 140°$

(c) $\tan 1145°$ (d) $\cot(-1056°)$

(e) $\sec 1326°$

29. State the quadrant in which the following radius angles lie.

(a) 140° (b) −150° (c) $\dfrac{5\pi}{6}$

(d) $\dfrac{11\pi}{6}$ (e) $\dfrac{-12\pi}{6}$

30. For each of the following angles, state which of $\sin\theta$, $\cos\theta$, and $\tan\theta$ are positive.

(a) 475° (b) 575° (c) −235° (d) $17\dfrac{\pi}{6}$

(e) 1200 (f) $-\dfrac{11\pi}{6}$ (g) $-10\dfrac{\pi}{3}$

31. Evaluate (without using tables) the following:

(a) $\sin 10° \cos 80° + \cos 10° \sin 80°$

(b) $3 \cos 80° \text{ cosec } 10° + 2 \cos 59° \text{ cosec } 31°$

32. Prove the following:

(a) $\cos 1° - \cos 89° = \sin 89° - \sin 1°$

(b) $\tan 15° \tan 30° \tan 45° \tan 60° \tan 75° = 1$

(c) $\sin^2 85° + \sin^2 80° + \sin^2 10° + \sin^2 5° = 2$

(d) $\left(\dfrac{\sin 47°}{\cos 43°}\right) + \left(\dfrac{\cos 43°}{\sin 47°}\right) - 4\cos^2 45° = 0$

33. If $A + B = 90°$, $\sin A = a$ and $\sin B = b$, then prove that $a^2 + b^2 = 1$.

34. In a triangle ABC, prove the following:

(a) $\cos\dfrac{B+C}{2} = \sin\dfrac{A}{2}$ (b) $\tan\dfrac{A+B}{2} = \cot\dfrac{C}{2}$

35. In a right-angled triangle ABC, $A\hat{B}C = 90°$, $A\hat{C}B = 30°$, and $AB = 8$ cm. Find the lengths of the other sides of the triangle.

36. In the ΔBC, $A\hat{B}C = 60°$, $A\hat{C}B = 45°$, and $BC = 10$ cm. Find the altitude AD of the triangle.

37. ABC and DCB are two right-angled triangles in which $A\hat{B}C = 90°$, $D\hat{C}B = 90°$, $D\hat{B}C = 30°$, and $AC \perp BD$. If $DC = 3\sqrt{3}$ cm, find the length of AB.

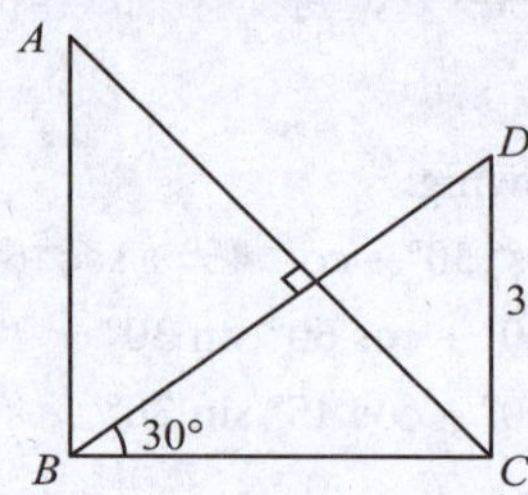

Multiple Choice Questions

38. The value of cot (−135°) is

(A) 1 (B) $\sqrt{3}$

(C) $\dfrac{1}{\sqrt{3}}$ (D) −1

39. What is the value of cosec (−1410°)?

(A) $\dfrac{1}{2}$ (B) 2

(C) $\sqrt{3}$ (D) 1

40. The value of $\cos\theta + \sin(270° + \theta) - \sin(270° - \theta) + \cos(180° + \theta)$ is

(A) 1 (B) 0

(C) 2 (D) −1

41. The value of

$$\frac{\cos(90^\circ+\theta)\sec(270^\circ+\theta)\sin(180^\circ+\theta)}{\text{cosec}(-\theta)\cos(270^\circ-\theta)\tan(180^\circ+\theta)} \text{ is}$$

(A) $\cos\theta$ (B) $\sin\theta$
(C) $\tan\theta$ (D) $\cot\theta$

42. What is the value of $\cos\left(\frac{3\pi}{2}+\theta\right)\cos(2\pi+\theta)$

$\left[\cot\left(\frac{3\pi}{2}-\theta\right)+\cot(2\pi+\theta)\right]$?

(A) –1 (B) 2
(C) 1 (D) –2

43. What is the value of $\sin^2 60° + \tan 45° - \cos 60° + \text{cosec}^2 45°$?

(A) $\frac{5}{2}$ (B) $\frac{13}{4}$
(C) $\frac{7}{2}$ (D) $\frac{13}{2}$

44. The value of $\sin^2 30° + \sin^2 45° + \sin^2 60° + \tan 45°$ is

(A) $\frac{3}{2}$ (B) $\frac{7}{2}$
(C) $\frac{5}{2}$ (D) 2

45. If $A+B+C=180°$, then the value of $\tan\left(\frac{A+B}{2}\right)=$

(A) $\sin\frac{C}{2}$ (B) $\cos\frac{C}{2}$
(C) $\cot\frac{C}{2}$ (D) none of these

46. The value of $\frac{\cos(\pi+\theta)\cos(-\theta)}{\cos(\pi-\theta)\cos\left(\frac{\pi}{2}+\theta\right)}$ is

(A) $\tan\theta$ (B) $-\tan\theta$
(C) $\cot\theta$ (D) $-\cot\theta$

47. The value of

$$\frac{\dfrac{\sin(180^\circ+\theta)\cos(90^\circ+\theta)\tan(270^\circ-\theta)}{\cot(360^\circ-\theta)}}{\dfrac{\sin(360^\circ-\theta)\cos(360^\circ+\theta)\text{cosec}(-\theta)}{\sin(270^\circ+\theta)}} \text{ is}$$

(A) 1 (B) 2
(C) 3 (D) 4

48. The value of $\frac{\cos 135^\circ - \cos 120^\circ}{\cos 135^\circ + \cos 120^\circ}$ is

(A) $2+2\sqrt{2}$ (B) $3-2\sqrt{2}$
(C) $3+2\sqrt{2}$ (D) $2-2\sqrt{2}$

49. The value of

$3\tan^2 45^\circ - \sin^2 60^\circ - \frac{1}{2}\cot^2 30^\circ + \frac{1}{8}\sec^2 45^\circ$ is

(A) 1 (B) $\frac{1}{2}$
(C) $\sqrt{3}$ (D) $\frac{1}{\sqrt{3}}$

50. The value of tan 1° tan 2° tan 3° ... tan 87° tan 88° tan 89° is

(A) 0 (B) 2
(C) 1 (D) –1

51. The value of $\sin^2 5° + \sin^2 10° + \sin^2 15° + ... + \sin^2 90°$ is

(A) $8\frac{1}{2}$ (B) $9\frac{1}{2}$
(C) $10\frac{1}{2}$ (D) $11\frac{1}{2}$

52. The value of

$\cos^2\frac{\pi}{16}+\cos^2\frac{3\pi}{16}+\cos^2\frac{5\pi}{16}+\cos^2\frac{7\pi}{16}$ is

(A) 1 (B) 0
(C) –1 (D) 2

53. The value of

$\cot^2 30^\circ - 2\cos^2 60^\circ - \frac{3}{4}\sin^2 45^\circ - 4\sin^2 30^\circ$ is

(A) $\frac{1}{8}$ (B) $\frac{7}{8}$
(C) $\frac{5}{8}$ (D) $\frac{9}{8}$

54. The value of $\frac{\sec 480^\circ \cdot \text{cosec}\, 570^\circ \cdot \tan 330^\circ}{\sin 600^\circ \cdot \cos 660^\circ \cdot \cot 405^\circ}$ is

(A) $\frac{14}{3}$ (B) $\frac{16}{3}$
(C) $\frac{13}{3}$ (D) $\frac{17}{3}$

55. The value of $\sin^2 6° + \sin^2 12° + \sin^2 18° + ... + \sin^2 84° + \sin^2 90°$ is

(A) 1 (B) 2
(C) 8 (D) 4

56. The value of $\tan 9° \cdot \tan 27° \cdot \tan 45° \cdot \tan 63° \cdot \tan 81°$ is

(A) 1 (B) 2
(C) 3 (D) 4

57. If $4n\alpha = \pi$, then find the value of $\tan\alpha \cdot \tan 2\alpha \cdot \tan 3\alpha \cdot ... \cdot \tan(2n-2)\alpha \cdot \tan(2n-1)\alpha$.

(A) 3 (B) 1
(C) 13 (D) 7

58. The value of $2(\sin^6\theta + \cos^6\theta) - 3(\sin^4\theta + \cos^4\theta) + 1$ is

(A) 0 (B) 1
(C) $\frac{3}{2}$ (D) 2

59. The value of

$$\tan^2\frac{\pi}{3} + 2\cos^2\frac{\pi}{4} + 3\sec^2\frac{\pi}{6} + 4\cos^2\frac{\pi}{2} \text{ is}$$

(A) 2 (B) 4
(C) 6 (D) 8

60. The value of

$$\sin\frac{\pi}{6}\cos 0 + \sin\frac{\pi}{4}\cos\frac{\pi}{4} + \sin\frac{\pi}{3}\cos\frac{\pi}{6} \text{ is}$$

(A) $\frac{3}{4}$ (B) $\frac{7}{4}$
(C) $\frac{5}{4}$ (D) $\frac{9}{4}$

61. The value of

$$\frac{\sin(-A)}{\sec(180^\circ + A)} - \frac{\tan(90^\circ + A)}{\cot A} + \frac{\cos(360^\circ - A)}{\sin A} + \frac{\cos A}{\sin(90^\circ + A)} \text{ is}$$

(A) $\sin A \cos A - \cot A + 2$
(B) $\sin A \cos A + \cot A + 2$
(C) $\sin A \cos A + 2$
(D) $\sin A \cos A - \cot A$

62. The value of

$$\frac{\cos(90^\circ + A)\sec(-A)\tan(180^\circ - A)}{\sec(180^\circ + A)\sin(180^\circ + A)\cot(90^\circ - A)} \text{ is}$$

(A) 1 (B) 0
(C) –1 (D) cos A

63. The value of $\sin 10° \cos 80° + \cos 10° \sin 80°$ is

(A) 0 (B) 1
(C) –1 (D) 2

64. The value of x if $\tan^2 45° - \sin^2 30° = x \sin^2 45° \cdot \tan 60°$ is

(A) sin 60° (B) sin 30°
(C) tan 30° (D) sin 45°

65. The value of x which satisfies the equation

$$x \sin 60^\circ \cos^2 45^\circ = \frac{\tan^2 45^\circ \operatorname{cosec} 30^\circ \cot 45^\circ}{\sec 45^\circ \sec^2 60^\circ} \text{ is}$$

(A) $\frac{2}{\sqrt{3}}$ (B) $\frac{\sqrt{3}}{2}$
(C) $\sqrt{\frac{2}{3}}$ (D) $\sqrt{\frac{3}{2}}$

66. In ΔABC, $\angle ABC = 60°$, $\angle ACB = 45°$ and $BC = 10$ cm. Find the altitude AD of the triangle.

(A) $\frac{10\sqrt{3}}{7}$ (B) $\frac{10\sqrt{3}}{\sqrt{3}+1}$
(C) $\frac{7\sqrt{3}}{5}$ (D) none of these

OLYMPIAD AND NTSE LEVEL EXERCISES

67. The value $\cos 105° + \sin 105°$ is

(A) $\frac{1}{2}$ (B) 1
(C) $\sqrt{2}$ (D) $\frac{1}{\sqrt{2}}$

68. $\tan\left(\frac{\pi}{4} + \theta\right) - \tan\left(\frac{\pi}{4} - \theta\right) =$

(A) $2\tan 2\theta$ (B) $2\cot 2\theta$
(C) $\tan 2\theta$ (D) $\cot 2\theta$

69. $\sin 50° - \sin 70° + \sin 10° =$

(A) 1 (B) 0
(C) 1/2 (D) 2

70. $\tan 20° + \tan 40° + \sqrt{3}\tan 20° \tan 40° =$

(A) $\frac{1}{\sqrt{3}}$ (B) $\sqrt{3}$
(C) $-\frac{1}{\sqrt{3}}$ (D) $-\sqrt{3}$

71. If $\tan x+\tan\left(\frac{\pi}{3}+x\right)+\tan\left(\frac{2\pi}{3}+x\right)=3$, then

(A) $\tan x = 1$ (B) $\tan 2x = 1$
(C) $\tan 3x = 1$ (D) None of these

72. The expression $2\cos\frac{\pi}{13}\cos\frac{9\pi}{13}+\cos\frac{3\pi}{13}+\cos\frac{5\pi}{13}$ is equal to

(A) – 1 (B) 0
(C) 1 (D) None of these

73. The value of 2 cos 10° + sin 100° + sin 1000° + sin 10000° is

(A) 0 (B) sin 10°
(C) cos 10° (D) –1

74. The value of sec 11° · sec 19° – 2 cot 71° is equal to

(A) 2 cot 11° (B) tan 19°
(C) 2 tan 11° (D) $\frac{1}{9}$ tan 19°

75. The value of $\frac{\cot 84^\circ \cot 48^\circ}{\cot 66^\circ \cot 78^\circ}$ is equal to

(A) 1 (B) 0
(C) $\sqrt{3}$ (D) none of these

76. $\sin\frac{9\pi}{14}\sin\frac{11\pi}{14}\sin\frac{13\pi}{14}$ is equal to

(A) $\frac{1}{64}$ (B) $-\frac{1}{64}$
(C) $\frac{1}{8}$ (D) $-\frac{1}{8}$

22 Heights and Distances

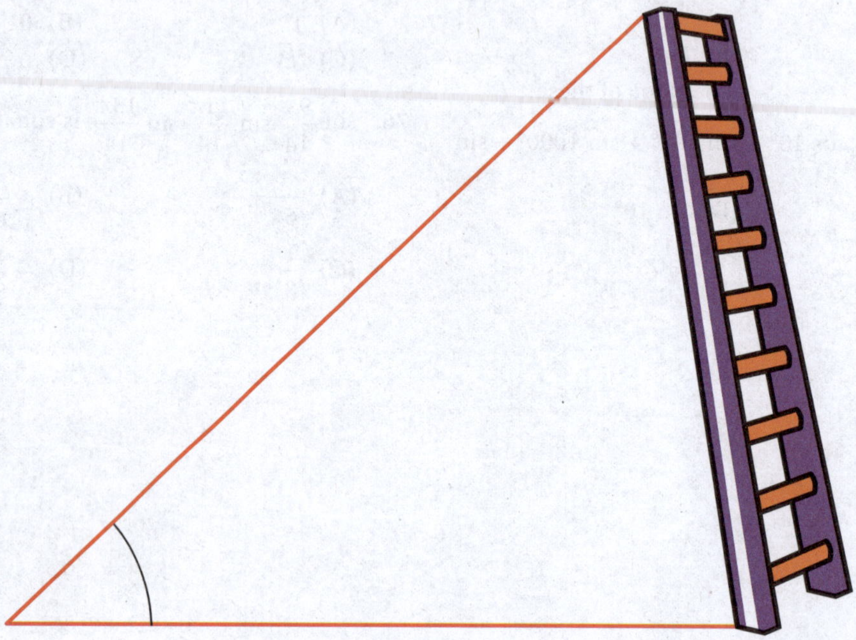

> Mathematics, even in its present and most abstract state, is not detached from life. It is just the ideal handling of the problems of life.
>
> – Cassius Jackson Keyser

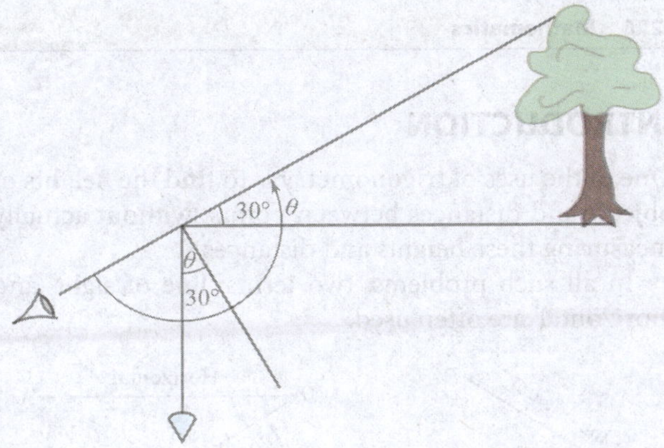

OBJECTIVES

After reading this chapter you should be able to

- understand the meaning of line of sight, horizontal, angle of elevation, and angle of depression
- understand the construction and working of clinometer to measure the angle of elevation and depression

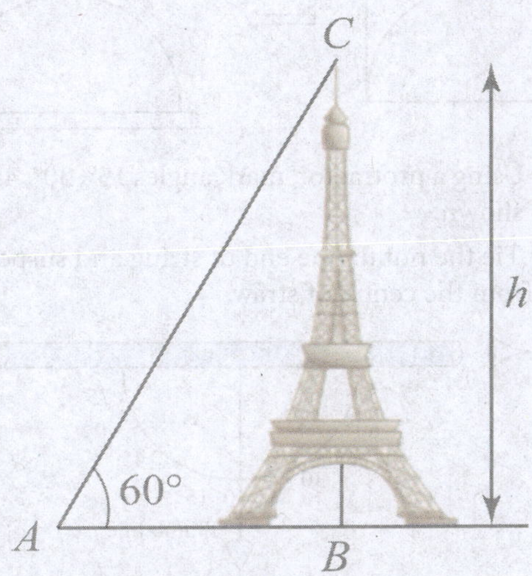

INTRODUCTION

One of the uses of trigonometry is to find the heights of objects and distances between points, without actually measuring these heights and distances.

In all such problems, two terms 'line of sight' and 'horizontal' are often used.

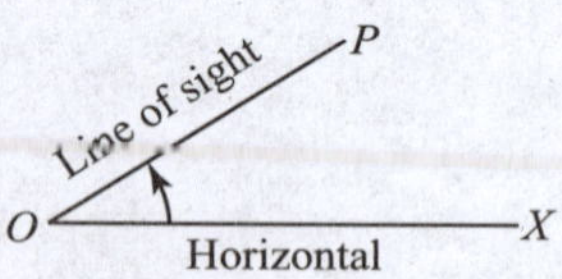

Fig. 22.1.

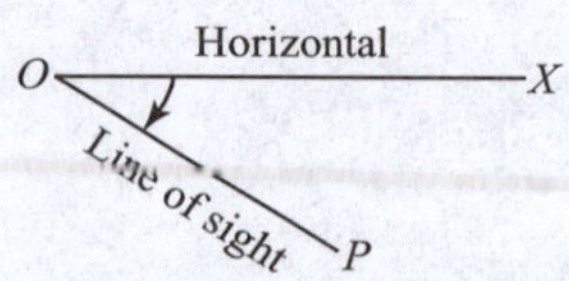

Fig. 22.2.

If through O, the eye of the observer, a horizontal line OX is drawn and the object P is in a vertical plane through OX, then $X\hat{O}P$ is called the angle of

(a) elevation of P as seen from O, if P is above OX, as in Fig. 22.1.

(b) depression of P as seen from O, if P is below OX, as in Fig. 22.2.

The instrument used to measure these angles is called *clinometer*.

Clinometer can be made easily. You will need a cardboard (14×9), a thick straw of length 15 cm, thread, a small steel nut, glue, and a protractor.

Procedure

Step 1: Draw a semicircle on the cardboard and cut it out.

Step 2: Stick a straw along the diameter.

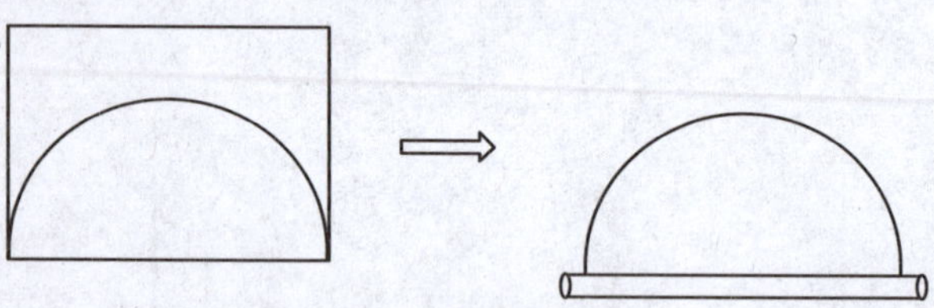

Step 3: Using a protractor, mark angles 15°, 30°, 45°, 60°, etc., as shown.

Step 4: Tie the nut to one end of string and suspend the string from the centre of straw.

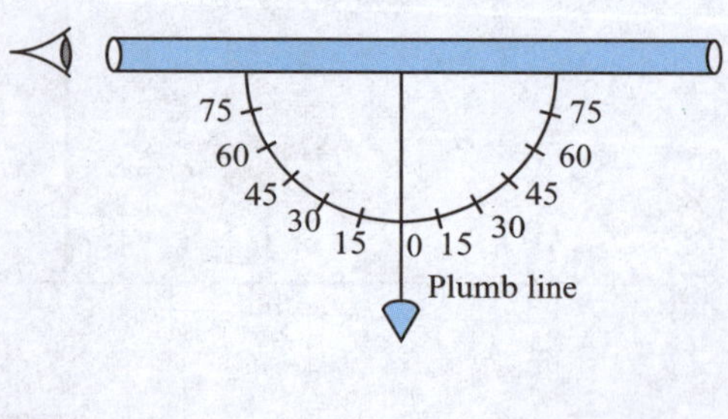

Thus, a clinometer is made up of a semicircular disc having a small tube fixed along the diameter with a plumb line fixed at the centre as shown in the figure.

Angle of Elevation

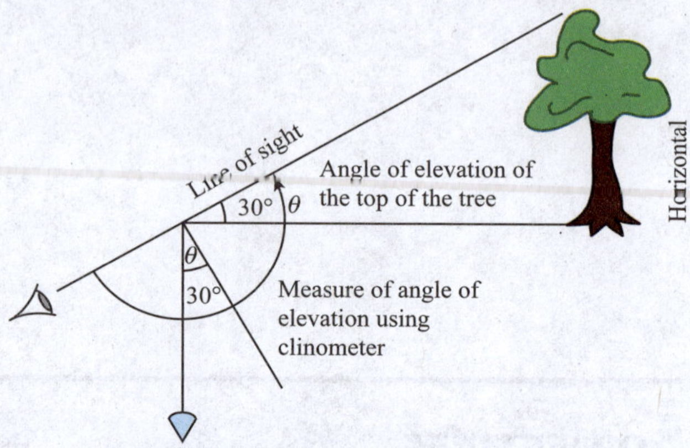

Angle of Depression

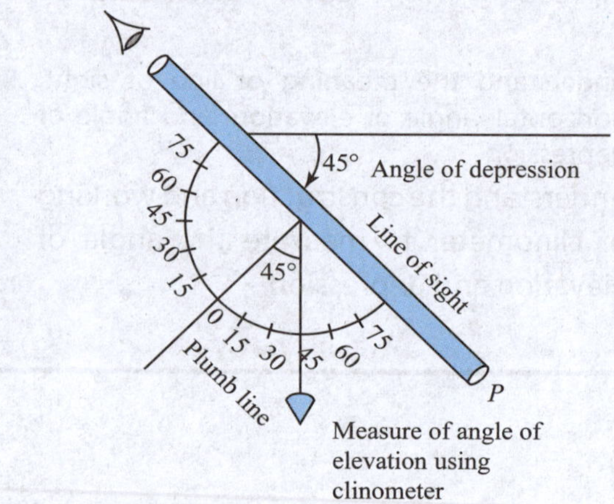

Worked Examples

1. The angle of elevation of the top of a tower from a place 360 m from the base of the tower is 60°. Find the height of the tower.

Solution

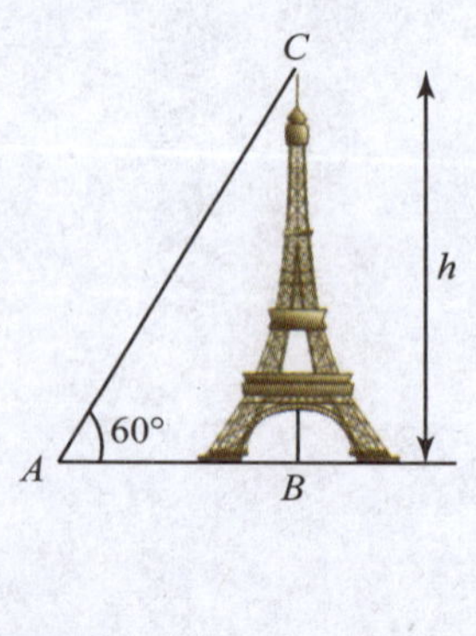

Using the rough figure as shown above,

$$\tan 60° = \frac{BC}{AB} \Rightarrow BC = AB \tan 60°$$

i.e. $360 \times \sqrt{3} = 360 \times 1.732 = 623.52$

The height of the tower is 623.52 m.

2. The height of a hill is 150 m. From the top of the hill, the angles of depression of two objects lying towards east of the hill are 45° and 30°. Find the distance between the objects.

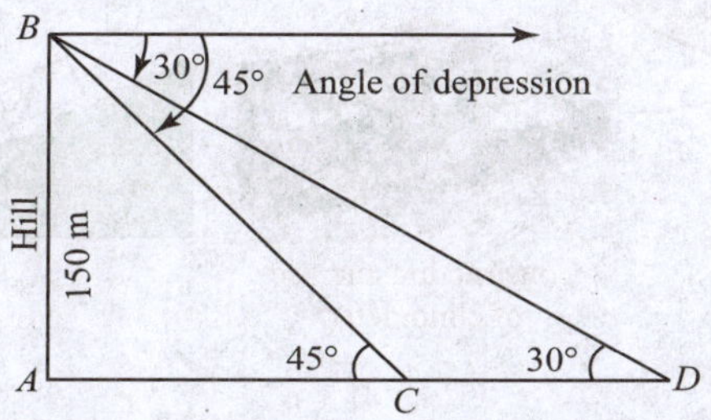

Solution

Let C and D be the objects.

Let CD be the distance between the objects.

Using the rough figure, we get

$$\frac{AB}{AC} = \tan 45° = 1$$

$AB = AC = 150$ m

$$\frac{AB}{AD} = \tan 30°$$

$AD \tan 30° = 150$

$$AD \frac{1}{\sqrt{3}} = 150$$

$AD = 150 \times \sqrt{3} = 259.8$ m

$CD = AD - AC = 259.8 - 150 = 109.8$ m

3. The shadow of a tower standing in a level plane is found to be 18 m shorter when the sun's altitude changes from 30° to 45°. Show that the height of the tower is $9(1+\sqrt{3})$ m.

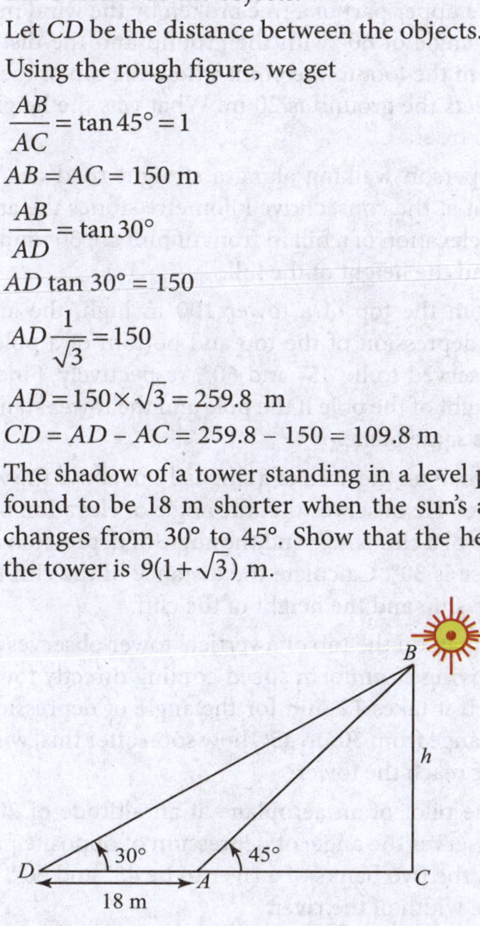

Solution

AC is the shadow of the tower when the altitude is 45°.

CD is the shadow when the altitude is 30°.

$DA = 18$ m

If h is the height of the tower,

$AC = CB = h$ (from ΔBAC)

Also $\frac{DC}{BC} = \cot 30°$ (from ΔBDC)

$\therefore DC = h\sqrt{3}$

But $DC - AC = DA = 18$ m

$\therefore h = (\sqrt{3} - 1) = 18$

$$\text{or } h = \frac{18}{\sqrt{3}-1} = \frac{18(\sqrt{3}+1)}{(\sqrt{3}-1)(\sqrt{3}+1)}$$

$$= \frac{18}{2}(\sqrt{3}+1) = 9(\sqrt{3}+1)$$

Hence, this is the result.

4. The angles of elevation of the top of a tower from two points 8 m and 32 m from the base and in the same straight line with it are complementary. Find the height of the tower.

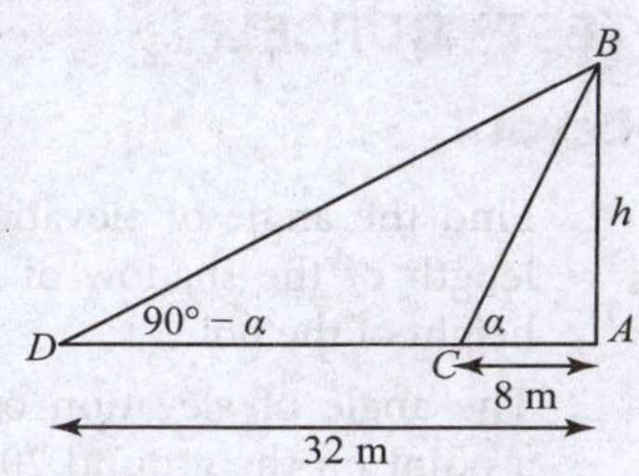

Let h be the height of the tower AB.

$AC = 8$ m and $AD = 32$ m, where C and D are two points of observation.

Let $A\hat{C}B = \alpha$ and $A\hat{D}B = 90° - \alpha$.

Now from ΔABC,

$$\frac{AB}{AC} = \tan \alpha$$

$AB = AC \tan \alpha$

$h = 8 \tan \alpha$ (1)

From ΔABD

$$\frac{AB}{AD} = \tan A\hat{D}B = \tan(90° - \alpha) = \cot \alpha$$

$AB = AD \cot \alpha$

$h = 32 \cot \alpha$ (2)

Multiplying Eqs. (1) and (2), we have the following:

$h^2 = 8 \times 32 \tan \alpha \cot \alpha$

$h^2 = 8 \times 32$

$h^2 = 16 \times 16$

$h = 16$ m

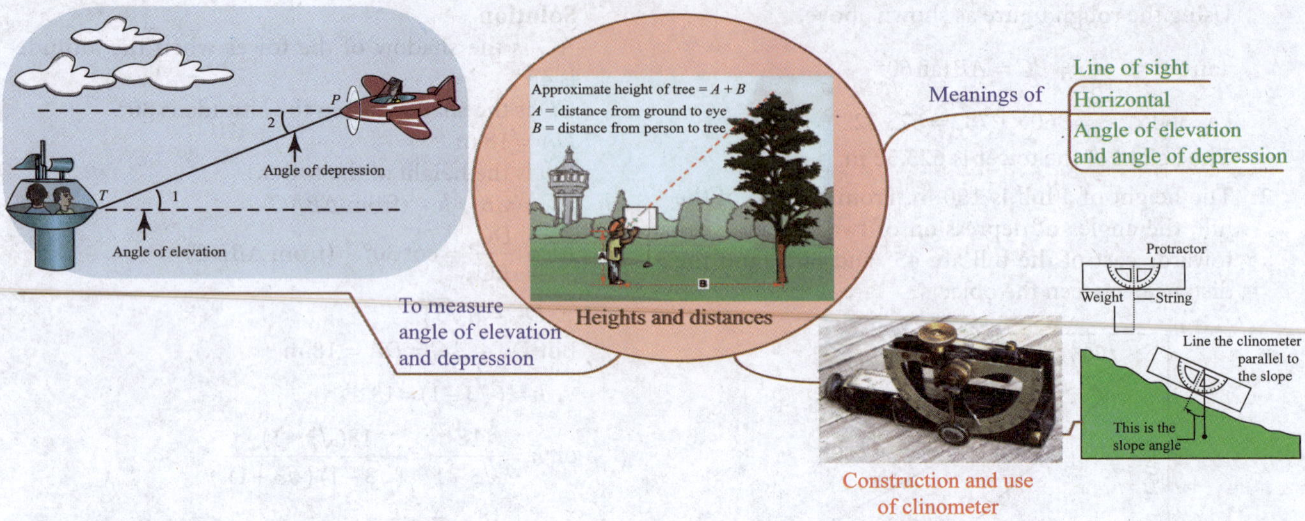

TEST YOURSELF

Level 1

1. Find the angle of elevation of the sun when the length of the shadow of a pole is $\sqrt{3}$ times the height of the pole.
2. The angle of elevation of the top of a tower at a point on the ground 20 m from the foot of the tower is 30°. What is the height of the tower?
3. From the top of a building 30 m tall, the angle of depression of an object on the ground is 60°. How far is the object from the building?

Level 2

4. The angles of depression of two ships from the top of a lighthouse are 45° and 30° towards east. If the ships are 100 m apart, find the height of the lighthouse.
5. Two pillars of equal height are on either side of a roadway which is 30 m wide. At a point on the roadway between the pillars, the elevations of the top of the pillars are 60° and 30°. Find the height of the pillars and the position of the point.
6. A person standing on the bank of a river observes that the angle subtended by a tree on the opposite bank is 60°; when he retires 14 m from the bank, he finds the angle to be 30°. Find the height of the tree and the breadth of the river.

Level 3

7. The upper part of a tree broken by the wind makes an angle of 60° with the ground and the distance from the foot to the point where the top of the tree meets the ground is 20 m. What was the height of the tree?
8. A person walking along a straight road observes that at the consecutive kilometre stones the angles of elevation of a hill in front of him are 30° and 45°. Find the height of the hill.
9. From the top of a tower 100 m high, the angles of depression of the top and bottom of a pole are observed to be 45° and 60°, respectively. Find the height of the pole if the pole and the tower stand on the same plane.
10. A man on the deck of a ship is 12 m above the water level. He observes that the angle of elevation of the top of a cliff is 45° and the angle of depression of its base is 30°. Calculate the distance of the cliff from the ship and the height of the cliff.
11. A man on the top of a vertical tower observes a car moving at uniform speed coming directly towards it. If it takes 12 min for the angle of depression to change from 30° to 45°, how soon after this, will the car reach the tower?
12. The pilot of an aeroplane at an altitude of 200 m observes the angle of depression of opposite points on the two banks of a river to be 45° and 60°. Find the width of the river.

13. From the top of a lighthouse, the angles of depression of two stations on opposite sides of it at a distance a apart are α and β. Find the height of the lighthouse.
14. The angle of elevation of an aeroplane from a point on the ground is 45°. After 15 s the angle changes to 30°. If the plane is flying at a height of 2500 m, find the speed of the plane.
15. The horizontal distance between two towers is 60 m and the angular depression of the top of the second tower which is 150 m high is 30°. The height of the first is
    (A) 120 m (B) $10(15+2\sqrt{3})$m
    (C) $10(15-2\sqrt{3})$m (D) $10(15+\sqrt{3})$m

Multiple Choice Questions

16. The angle of elevation of a tower at a point d metres away from its base is 30°. If the tower is 20 metres high, then d is equal to
    (A) $10\sqrt{3}$ m (B) $20\sqrt{3}$ m
    (C) $\frac{20}{\sqrt{3}}$ m (D) 10 m
17. A person standing on the bank of a river observes that the angle subtended by a tree on the opposite bank is 60°. When he retreats 40 m from the bank, he finds the angle to be 30°. The breadth of the river is
    (A) 20 m (B) 40 m
    (C) 30 m (D) 60 m
18. From a 60 m high tower, angle of depression of the top and bottom of a house are α and β, respectively. If the height of the house is $\frac{60\sin(\beta-\alpha)}{x}$, then the value of x is
    (A) $\sin\alpha\sin\beta$ (B) $\cos\alpha\cos\beta$
    (C) $\sin\alpha\cos\beta$ (D) $\cos\alpha\sin\beta$
19. A tree of height 100 feet subtends a right angle at the top of another tree. If the height of the other tree is 64 metres then the distance between the two trees is
    (A) 48 m (B) 36 m
    (C) 54 m (D) 72 m
20. An observer in a boat finds the angle of elevation of a tower standing on the top of a cliff as 60° and that of the top of cliff as 30°. If the height of the tower is 60 m then the height of the cliff is
    (A) $60\sqrt{3}$ m (B) 30 m
    (C) $20\sqrt{3}$ m (D) None of these
21. A tower subtends an angle α at a point A in the plane of its base and the angle of depression of the foot of the tower at a point l m just above A is β. The height of the tower is
    (A) $l\tan\beta\cot\alpha$ (B) $l\cot\alpha\cot\beta$
    (C) $l\tan\alpha\tan\beta$ (D) $l\tan\alpha\cot\beta$
22. The angle of elevation of a tower from a point A due south of it is 30° and from a point B due west of it is 45°. If the height of the tower is 100 m, then AB =
    (A) 150 m (B) 200 m
    (C) 173.2 m (D) 141.4 m
23. The angle of elevation of the Sun, when the shadow of a pole is $\sqrt{3}$ times its height is
    (A) 60° (B) 30°
    (C) 45° (D) 15°
24. A ladder rests against a wall so that its top touches the roof of the house. If the ladder makes an angle of 60° with the horizontal, and height of the house be $6\sqrt{3}$ m, then the length of the ladder is
    (A) $12\sqrt{3}$ m (B) $3\sqrt{3}$ m
    (C) $\frac{12}{\sqrt{3}}$ m (D) 12 m
25. If the angles of elevation of two towers from the middle point of the line joining their feet are 60° and 30° then the ratio of their heights is
    (A) 2 : 1 (B) $1:\sqrt{2}$
    (C) 3 : 1 (D) $1:\sqrt{3}$
26. The base of a cliff is circular. From the extremities of a diameter of the base, the angle of elevation of the top of the cliff is 30° and 60°. If the height of the cliff is 500 metres, then the diameter of the base of the cliff is
    (A) $1000\sqrt{3}$ m (B) $2000/\sqrt{3}$ m
    (C) $1000/\sqrt{3}$ m (D) $2000\sqrt{3}$ m
27. The angle of elevation of the top of a tower from top of a house is 60° and the angle of depression of its base is 30°. If the horizontal distance between

the house and the tower is 12 m, then the height of the tower is

(A) $48\sqrt{3}$ m (B) $\frac{16}{\sqrt{3}}$ m

(C) $24\sqrt{3}$ m (D) $16\sqrt{3}$ m

28. The angle of depression of a ship from the top of a 30 m high tower is 60°. The distance of ship from the base of the tower is

(A) 30 m (B) $30\sqrt{3}$ m

(C) $10\sqrt{3}$ m (D) 10 m

29. A 6 metres high flagstaff placed on the top of a tower throws a shadow of $2\sqrt{3}$ m on the ground. The angle (in degrees) that the sun makes with the ground is

(A) 60° (B) 80°

(C) 75° (D) none of these

30. The angles of elevation of a cliff from a point A on the ground and a point B, 100 m vertically above A are α and β, respectively. The height of the cliff is

(A) $\frac{100\cot\alpha}{\cot\alpha-\cot\beta}$ (B) $\frac{100\cot\beta}{\cot\alpha-\cot\beta}$

(C) $\frac{100\cot\beta}{\cot\beta-\cot\alpha}$ (D) $\frac{100\cot\beta}{\cot\beta+\cot\alpha}$

31. Two men are on the opposite sides of tower. They measure the angles of elevation of the top of the tower as 45° and 30°. If the height of the tower is 40 m then the distance between the men is

(A) 40 m (B) $40\sqrt{3}$ m

(C) 68.280 m (D) 109.28 m

32. The angle of elevation of the top of a pole from any point A on the ground is 15°. On walking 40 metres towards the pole, the angle becomes 30°. The height of the pole is

(A) 40 m (B) 20 m

(C) $20\sqrt{3}$ m (D) $\frac{40}{\sqrt{3}}$ m

33. The shadow of a tower standing on a level ground is x metres long when the sun's altitude is 30°, while it is y metres long when the sun's altitude is 60°. If the height of the tower is $45\frac{\sqrt{3}}{2}$ m then the value of $x - y$ is

(A) 45 m (B) $45\sqrt{3}$ m

(C) $\frac{45}{\sqrt{3}}$ m (D) $45\frac{\sqrt{3}}{2}$ m

34. For a man, the angle of elevation of the highest point of a temple due east of him is 60°. On walking 240 metres towards north, the angle of elevation is reduced to 30°. The height of the temple is

(A) $60\sqrt{6}$ m (B) 60 m

(C) $50\sqrt{3}$ m (D) $30\sqrt{3}$ m

OLYMPIAD AND NTSE LEVEL EXERCISES

35. The angle of elevation of the top of a tower at point on the ground is 30°. If on walking 20 metres towards the tower, the angle of elevation become 60°, then the height of the tower is

(A) 10 metres (B) $\frac{10}{\sqrt{3}}$ metres

(C) $10\sqrt{3}$ metres (D) None of these

36. An observer on the top of a tree, finds the angle of depression of a car moving towards the tree to be 30°. After 3 minutes, this angle becomes 60°. After how much more time, the car will reach the tree

(A) 4 minutes (B) $4\frac{1}{2}$ minutes

(C) $1\frac{1}{2}$ minutes (D) 2 minutes

37. The length of the shadow of a pole inclined at 10° to the vertical towards the sun is 2.05 metres. When the elevation of the sun is 38°, the length of the pole is

(A) $\frac{2.05\sin 38°}{\sin 42°}$ (B) $\frac{2.05\sin 42°}{\sin 38°}$

(C) $\frac{2.05\cos 38°}{\cos 42°}$ (D) None of these

38. An aeroplane flying horizontally 1 km above the ground is observed at an elevation of 60° and after 10 seconds the elevation, it is observed to be 30°. The uniform speed of the aeroplane (in km/h) is

(A) 240 (B) $240\sqrt{3}$

(C) $60\sqrt{3}$ (D) None of these

39. The base of a cliff is circular. From the extremities of a diameter of the base the angles of elevation of the top of the cliff are 30° and 60°. If the height of the cliff is 500 metres, then the diameter of the base of the cliff is

(A) $1000\sqrt{3}$ m (B) $2000/\sqrt{3}$ m
(C) $1000/\sqrt{3}$ m (D) $2000\sqrt{3}$ m

40. For a man, the angle of elevation of the highest point of the temple situated east of him is 60°. On walking 240 metres to north, the angle of elevation is reduced to 30°. The height of the temple is

(A) $60\sqrt{6}$ m (B) 650 m
(C) $50\sqrt{3}$ m (D) $30\sqrt{6}$ m

41. A vertical tower stands on a declivity which is inclined at 15° to the horizon. From the foot of the tower, a man ascends the declivity from 870 feet and then finds that the tower subtends an angle of 30°. Then the height of the tower is

(A) $40(\sqrt{6} - \sqrt{2})$ (B) $20(\sqrt{6} - \sqrt{2})$
(C) $40(\sqrt{3} - \sqrt{2})$ (D) None of these

42. A flagstaff stands in the centre of a rectangular field whose diagonal is 1200 m, and subtends angles 15° and 45° at the mid points of the sides of the field. The height of the flagstaff is

(a) 200 m (b) $300\sqrt{2+\sqrt{3}}$ m
(c) $300\sqrt{2-\sqrt{3}}$ m (d) 400 m

43. A vertical pole consists of two parts, the lower part being one third of the whole. At a point in the horizontal plane through the base of the pole and at a distance 20 metres from it, the upper part of the pole subtends an angle whose tangent is 1/2. Find the possible height of the pole.

(A) 20 m (B) 45 m
(C) 60 m (D) *A* and *C*

44. The angle of elevation of a stationary cloud from a point 2500 m above a lake is 15° and the angle of depression of its reflection in the lake is 45°. Then find the height of cloud above the lake level.

(A) $1000\sqrt{3}$ m (B) $1500\sqrt{3}$ m
(C) $2500\sqrt{3}$ m (D) $3000\sqrt{3}$ m

Hints and Answers to Test Yourself

Permutations and Combinations

1. (a) 720 (b) 68140
   (c) 5550 (d) 15120
2. (a) 288 (b) 1728
3. 20160
4. 3628800
5. (a) 56 (b) 1365
6. (a) $n = 8$ (b) $n = 7$ (c) $n = 2$
7. 1320 ways
8. 5040; 720
9. (a) 190 (b) 276 (c) 4914
10. 120
11. 56
12. 45
13. 21
14. 364
15. (a) 9! (b) 720 (c) 8!
16. 119
17. (a) 7 (b) 7
18. 3
19. 3
20. (A)

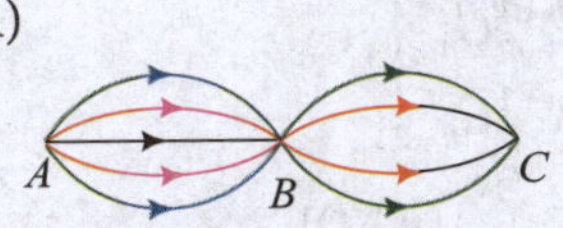

From fundamental principal of multiplication,
Total number of ways = 5 × 4 = 20

21. (A) By fundamental principal of addition,
    Total number of ways = 15 + 10 = 25
22. (B) $(n + 2)! = 60 \times (n - 1)!$

$(n + 2)(n + 1)n\,(n - 1)! = 60 \times (n - 1)!$

$(n + 2)(n + 1)n = 60$

$(n + 2)(n + 1)n = 5 \times 4 \times 3$

$n = 3$

23. (B) $\dfrac{1}{a!}+\dfrac{1}{b!}=\dfrac{v}{c!}$

$$\frac{1}{a!}+\frac{1}{(a+1)!}=\frac{v}{(a+2)!}$$

$$\frac{1}{a!}+\frac{1}{(a+1)a!}=\frac{v}{(a+2)(a+1)a!}$$

$$1+\frac{1}{(a+1)}=\frac{v}{(a+2)(a+1)}$$

$$\frac{a+2}{a+1}=\frac{v}{(a+2)(a+1)}$$

$$v=(a+2)^2=c^2$$

$$\sqrt{v}=c$$

24. (A) $n!$, $3 \times n!$ and $(n + 1)!$ are in GP

$(3 \times n!)^2 = n! \times (n + 1)!$

$9 \times n! \times n! = n! \times (n + 1) \times n!$

$9 = n + 1$

$n = 8$

Now,

$n!$, $5 \times n!$ $(n + 1)!$

$8!$, $5 \times 8!$, $9!$

$$\frac{8!+9!}{2}=\frac{8!+9\times 8!}{2}=\frac{10}{2}\times 8!=5\times 8!$$

Hence, these are in AP.

25. (C) $\dfrac{{}^{56}P_{r+6}}{{}^{54}P_{r+3}}=\dfrac{30800}{1}$

$$\frac{56\times 55\times {}^{54}P_{r+4}}{{}^{54}P_{r+3}}=\frac{30800}{1}$$

$$\frac{{}^{54}P_{r+4}}{{}^{54}P_{r+3}}=10$$

$54 - (r + 4) + 1 = 10$

$\Rightarrow \quad r = 41$

${}^{r}P_2 = {}^{41}P_2 = 41 \times 40 = 1640$

26. (A) ${}^{n+5}P_{n+1} = \frac{11(n-1)}{2}\,{}^{n+3}P_n$

$\Rightarrow \quad (n+5)(n+4)\,{}^{n+3}P_{n-1} = \frac{11(n-1)}{2}\,{}^{n+3}P_n$

$\Rightarrow \quad \frac{(n+5)(n+4)}{(n+3-n+1)} = \frac{11(n-1)}{2}$

$\Rightarrow \quad n^2 - 13n + 42 = 0$

$\Rightarrow \quad n = 6, 7$

27. (A) ${}^{m+n}P_2 = 90 = 10 \times 9 = {}^{10}P_2$

$\Rightarrow \quad m + n = 10$

${}^{m-n}P_2 = 30 = 6 \times 5 = {}^{6}P_2$

$\Rightarrow \quad m - n = 6$

28. (C) ${}^{8}P_5 + 5 \times {}^{8}P_4 = {}^{9}P_r$

$\Rightarrow \quad {}^{8}P_4\left(\frac{{}^{8}P_5}{{}^{8}P_4} + 5\right) = {}^{9}P_r$

$\Rightarrow \quad {}^{8}P_4(4+5) = {}^{9}P_r$

$\Rightarrow \quad 9 \times {}^{8}P_4 = {}^{9}P_r$

$\Rightarrow \quad {}^{9}P_5 = {}^{9}P_r$

$\Rightarrow \quad r = 5$

29. (B) There are 9 letters and all are different.

$\therefore$ Total permutations = ${}^{9}P_9 = 9! = 362880$

30. (D) The signals can be made by using one or more flags at a time.

$\therefore$ Total number of signals

$= {}^{4}P_1 + {}^{4}P_2 + {}^{4}P_3 + {}^{4}P_4$

$= 4 + 12 + 24 + 24$

$= 64$

31. (C) ${}^{m+n}P_2 = 56 = 8 \times 7 = {}^{8}P_2$

$\Rightarrow \quad m + n = 8$

$\Rightarrow \quad {}^{m+n}P_3 = 24 = 4 \times 3 \times 2 = {}^{4}P_3$

$\Rightarrow \quad m - n = 4$

$\therefore \quad m = 6$ and $n = 2$

$\frac{{}^{6}P_3}{{}^{2}P_2} = \frac{6 \times 5 \times 4}{1 \times 2} = 60$

32. (C) ${}^{9}P_5 + 5 \times {}^{9}P_4 = 5 \times {}^{9}P_4 + 5 \times {}^{9}P_4$

$= 10 \times {}^{9}P_4 = {}^{10}P_5$

33. (C) ${}^{15}C_{3r} = 15C_{r+3}$

$3r = r + 3$

$2r = 3$

$r = \frac{3}{2}$ (not possible because r should be an integer)

${}^{15}C_{3r} = {}^{15}C_{15-(r+3)}$

$\Rightarrow \quad {}^{15}C_{3r} = {}^{15}C_{12-r}$

$\Rightarrow \quad 3r = 12 - r$

$\Rightarrow \quad 4r = 12$

$\Rightarrow \quad r = 3$

$\Rightarrow \quad {}^{r}C_2 = {}^{3}C_2 = {}^{3}C_1 = 3$

34. (B) ${}^{n}C_r + {}^{n}C_{r-1} + {}^{n}C_{r-1} + {}^{n}C_{r-2}$

$= \left({}^{n}C_{r-1} + {}^{n}C_r\right) + \left({}^{n}C_{r-2} + {}^{n}C_{r-1}\right)$

$= {}^{n+1}C_r + {}^{n+1}C_{r-1} = {}^{n+2}C_r$

35. (C) Let the number of teams be n.

${}^{n}C_2 = 36$

$\Rightarrow \quad \frac{n(n-1)}{1 \times 2} = 36$

$\Rightarrow \quad n(n-1) = 72$

$\Rightarrow \quad n(n-1) = 9 \times 8$

$\Rightarrow \quad n = 9$

36. (A) ${}^{n}C_3 + {}^{n}C_4 > {}^{n+1}C_3$

$\Rightarrow \quad {}^{n+1}C_4 > {}^{n+1}C_3$

$\Rightarrow \quad {}^{n+1}C_4 > {}^{n+1}C_{n+1-3}$

$\Rightarrow \quad {}^{n+1}C_4 > {}^{n+1}C_{n-2}$

$\Rightarrow \quad \frac{(n+1)!}{4!(n-3)!} > \frac{(n+1)!}{(n-2)!(3)!}$

$\Rightarrow \quad \frac{1}{4(n-3)!} > \frac{1}{(n-2)!}$

$\Rightarrow \quad \frac{1}{4(n-3)!} > \frac{1}{(n-2)(n-3)!}$

$\Rightarrow \quad n - 2 > 4$

$\Rightarrow \quad n > 6$

37. (B) ${}^nC_r = {}^nC_{r-1}$

$$\frac{n!}{r!(n-r)!} = \frac{n!}{(r-1)!(n-r+1)!}$$

$$\frac{1}{r} = \frac{1}{n-r+1}$$

$\Rightarrow \quad r = n-r+1$

$\Rightarrow \quad n-2r+1 = 0$

$${}^nP_r = {}^nP_{r+1}$$

$$\Rightarrow \quad \frac{n!}{(n-r)!} = \frac{n!}{(n-r-1)!}$$

$$\Rightarrow \quad \frac{1}{n-r} = 1$$

$\Rightarrow \quad n-r = 1$

$r = n-1$

Now, $n - 2(n-1) + 1 = 0$

$\Rightarrow \quad n - 2n + 3 = 0$

$\Rightarrow \quad n = 3$

38. (B) $\dfrac{{}^nC_r}{{}^nC_{r-1}} = \dfrac{84}{36}$

$$\Rightarrow \quad \frac{n-r+1}{r} = \frac{7}{3}$$

$\Rightarrow \quad 3n - 3r + 3 = 7r$

$\Rightarrow \quad 3n - 10n + 3 = 0$

$$\frac{{}^nC_{r+1}}{{}^nC_r} = \frac{n-(r+1)+1}{r+1} = \frac{126}{84}$$

$\Rightarrow \quad 2n - 2r = 3r + 3$

$\Rightarrow \quad 5r - 2n = -3$

$\Rightarrow \quad 10r - 4n = -6$

$\therefore \quad r = 3$

Hints and answers to Olympiad and NTSE level exercises

39. (C) To be divisible by 5, the digit 5 must be at unit place, so now to be between 3000 and 4000, 3 must be at thousand place. Hence the required number of ways is ${}^4P_2$.

40. (C) Required number of ways are n^{mn}, since each letter may be posted in n ways.

41. (D) The letters can be select in ${}^5C_3 \times {}^4C_2$ ways. Therefore the number of arrangements are $({}^5C_3 \times {}^4C_2)5!$.

42. (B) Since 5 are always to be excluded and 6 always to be included, therefore 5 players to be chosen from 14. Hence required number of ways are ${}^{14}C_5 = 2002$.

43. (A) 5 persons are to be seated on 8 chairs i.e. ${}^8C_3 \times 5!$ or 6720.
{Since 5 chairs can be select in ${}^8C_5$ and then 5 persons can be arranged in 5! ways}.

44. (C) The required number of ways = ${}^8C_3 - {}^5C_3 - {}^3C_3$.
{Since total points are 8, but 5 are collinear and other three are also collinear}.

45. (A)

(P) The number of permutations which contain the word 'INDIA'
(INDIA), N, O, I, L
Number of ways = 5! = 120

(Q) The number of permutations which contain the word 'OIL'
(OIL), I, N, D, I, A, N
Number of ways = 7!/2!2! = 1260

(R) The number of permutations which contains neither word 'OIL' nor 'INDIA'
Total number of permutation of all letters (III NN D A O L) = 9!/3!2! = 30240
Number of words which contains both words 'INDIA' and 'OIL' = 3!
So required number of ways
= 9!/3!2! – 5! – 7!/2!2! + 3! = 28866

46. (A) The batting order of 11 players can be decided in 11! ways. Now, Yuvraj, Dhoni and Pathan can be arranged in batting order in 3! ways. But the order of these 3 players is fixed, i.e., Yuvraj-Dhoni-Pathan. Hence, the number of ways in which the team can bat is 11!/3!.

47. (D) (P) ${}^{10}C_2 - {}^4C_2 + 1 = 45 - 6 + 1 = 40$

(Q) $1 \times {}^{10}C_2 = 45$

(R) $2 \times {}^6C_2 = 30$

(S) $4 \times {}^6C_2 = 60$.

48. (B) $\dfrac{(2n)!}{(2n-3)!\,.\,3!} \times \dfrac{2!\times(n-2)!}{n!} = \dfrac{44}{3}$

$\Rightarrow \quad n$

$\Rightarrow \quad 4(2n - 1) = 44$

$\Rightarrow \quad 2n = 12$

$\Rightarrow \quad n = 6$

Now ${}^6C_r = 15$

$\Rightarrow \quad {}^6C_r = {}^6C_2$

or $\quad {}^6C_4$

$\Rightarrow \quad r = 2, 4.$

Binomial Theorem

1. $a^6 + 6a^5b + 15a^4b^2 + 20a^3b^3 + 15a^2b^4 + 6ab^5 + b^6$
2. $1 + 5x + 10x^2 + 10x^3 + 5x^4 + x^5$
3. $a^8 + 8a^6 + 28a^4$
4. $-\frac{20}{x^8} + \frac{1}{x^{10}}$ **5.** 1.2664 **6.** 1.22987
7. 0.95099 **8.** $\frac{13440}{x^2}$
9. ${}^{20}C_9 \times 3^9 \times 2^{11} \times x^2$
10. $126x$ and $\frac{126}{x}$
11. $-308x^7$ and $693x^4$
12. 724
13. $140\sqrt{2}$
14. $3^6 \times {}^{13}C_6 x^{-17/2}$ and $-3^7 \times {}^{13}C_7 \times x^{-11}$
15. $2688x^6$ and $-3024x^3$
16. 462
17. T_5
18. 512
19. $-17C_4 \times 2^{13}$
20. $\frac{2}{3}$ and 9
21. (C) Number of terms in the expansion of $(x + y)^n$ is $n + 1$.
22. (C)

$$\left(\sqrt{x}+\sqrt{y}\right)^8 = {}^8C_0\left(\sqrt{x}\right)^8\left(\sqrt{y}\right)^0 + {}^8C_1\left(\sqrt{x}\right)^7\left(\sqrt{y}\right)^1 + {}^8C_2\left(\sqrt{x}\right)^6\left(\sqrt{y}\right)^2 + {}^8C_3\left(\sqrt{x}\right)^5\left(\sqrt{y}\right)^3 + {}^8C_4\left(\sqrt{x}\right)^4\left(\sqrt{y}\right)^4 + {}^8C_5\left(\sqrt{x}\right)^3\left(\sqrt{y}\right)^5 + {}^8C_6\left(\sqrt{x}\right)^2\left(\sqrt{y}\right)^6 + {}^8C_7\left(\sqrt{x}\right)\left(\sqrt{y}\right)^7 + {}^8C_8\left(\sqrt{x}\right)^0\left(\sqrt{y}\right)^8$$

$$\left(5x-\sqrt{y}\right)^8 = {}^8C_0\left(\sqrt{x}\right)^8\left(\sqrt{y}\right)^0 - {}^8C_1\left(\sqrt{x}\right)^7\left(\sqrt{y}\right)^1 + {}^8C_2\left(\sqrt{x}\right)^6\left(\sqrt{y}\right)^2 - {}^8C_3\left(\sqrt{x}\right)^5\left(\sqrt{y}\right)^3 + {}^8C_4\left(\sqrt{x}\right)^4\left(\sqrt{y}\right)^4 - {}^8C_5\left(\sqrt{x}\right)^3\left(\sqrt{y}\right)^5 + {}^8C_6\left(\sqrt{x}\right)^2\left(\sqrt{y}\right)^6 - {}^8C_7\left(\sqrt{x}\right)\left(\sqrt{y}\right)^7 + {}^8C_8\left(\sqrt{x}\right)^0\left(\sqrt{y}\right)^8$$

Then

$$\left(\sqrt{x}+\sqrt{y}\right)^8 + \left(\sqrt{x}-\sqrt{y}\right)^8 = 2\left\{{}^8C_0(\sqrt{x})^8(\sqrt{y})^0 + {}^8C_2(\sqrt{x})^6(\sqrt{y})^2 + {}^8C_4(\sqrt{x})^4(\sqrt{y})^4 + {}^8C_6(\sqrt{x})^2(\sqrt{y})^6 + {}^8C_8(\sqrt{x})^0(\sqrt{y})^8\right\}$$

23. (A) $(5x + 2y)^7 - (5x - 2y)^7$

$$= 2\left\{{}^7C_1(5x)^6(2y)^1 + {}^7C_3(5x)^4(2y)^3 + {}^7C_5(5x)^2(2y)^5 + {}^7C_7(5x)^0(2y)^7\right\}$$

24. (D) $(\sqrt{2}x + \sqrt{3}y)^{10} - (\sqrt{2}x - \sqrt{3}y)^{10}$

$$= 2\left\{{}^{10}C_1\left(\sqrt{2}x\right)^9\left(\sqrt{3}y\right)^1 + {}^{10}C_3\left(\sqrt{2}x\right)^7\left(\sqrt{3}y\right)^3 + {}^{10}C_5\left(\sqrt{2}x\right)^5\left(\sqrt{3}y\right)^5 + {}^{10}C_7\left(\sqrt{2}x\right)^3\left(\sqrt{3}y\right)^7 + {}^{10}C_9\left(\sqrt{2}x\right)^1\left(\sqrt{3}y\right)^9\right\}$$

25. (C) Out of 34 terms, 16 terms will be cancelled and remaining 18 terms will be reduced to sum of 9 terms.
26. (D) $T_{9+1} = {}^{12}C_9\left(2x^2\right)^3\left(\frac{1}{x^2}\right)^9$
27. (C) $T_{3+1} = {}^{12}C_3(x)^9(-2y)^3$
28. (B) $T_{13} = T_{12+1} = {}^{18}C_{12}(9x)^6\left(-\frac{1}{3\sqrt{x}}\right)^{12}$
29. (A) $T_3' = T_{6-3+2} = T_5 = T_{4+1} = {}^6C_4\left(x\right)^2\left(\frac{1}{x}\right)^4$
30. (C) $T_{14}' = T_{17-14+2} = T_5 = T_{4+1} = {}^{17}C_4\left(\sqrt{x}\right)^{13}\left(-\sqrt{y}\right)^4$
31. (A) Total number of terms in expansion = 10 + 1 = 11

Middle term $= \frac{11+1}{2} = 6$

$$T_6 = T_{5+1} = {}^{10}C_5(x)^5\left(\frac{-1}{2y}\right)^5$$

32. (B) Total number of terms in expansion = 10 + 1 = 11

Middle term $=\left(\frac{11+1}{2}\right)^{th}$ term $= 6^{th}$ term

$$T_6 = T_{5+1} = {}^{10}C_5 (x^2)^5 \left(\frac{-2}{x}\right)^5$$

33. (B) $T_{r+1} = {}^6C_r (3x)^{6-r}\left(\frac{-1}{x}\right)^r = (-1)^r\ {}^6C_r (3)^{6-r} x^{6-2r}$

Here, $6 - 2r = 2$

$\Rightarrow\ 2r = 4$

$\Rightarrow\ r = 2$

Coefficient of $x^2 = (-1)^2\,{}^6C_2(3)^4$

34. (D) $T_{r+1} = {}^9C_r (3x^2)^{9-r}\left(\frac{-1}{3x}\right)^r$

$= (-1)^r\ {}^9C_r 3^{9-r}\left(\frac{1}{3}\right)^r x^{18-3r}$

Here, $18-3r=6 \Rightarrow 3r = 12 \Rightarrow r = 4$

Coefficient of $x^6 = (-1)^4\ {}^9C_4 3^5\left(\frac{1}{3}\right)^4$

35. (D) $T_{r+1} = {}^{10}C_r\left(\sqrt{x}\right)^{10-r}\left(\frac{1}{3x^2}\right)^r$

$= {}^{10}C_r\left(\frac{1}{3}\right)^r x^{\frac{10-r}{2}-2r}$

$= {}^{10}C_r\left(\frac{1}{3}\right)^r x^{\frac{10-5r}{2}}$

Here, $\frac{10-5r}{2} = 0 \Rightarrow r = 2$

36. (A) $T_{r+1} = {}^{12}C_r x^{12-r}\left(\frac{-1}{x}\right)^r = (-1)^r\ {}^{12}C_r x^{12-2r}$

Here, $12 - 2r = 0 \quad\Rightarrow\quad r = 6$

37. (C) $T_{r+1} = {}^{15}C_r\left(x^4\right)^{15-r}\left(\frac{-1}{x^3}\right)^r = (-1)^r\ {}^{15}C_r\ x^{60-7r}$

Here, $60 - 7r = 32$

$\Rightarrow\ 7r = 28$

$\Rightarrow\ r = 4$

38. (B) $T_{r+1} = {}^9C_r\left(\frac{3}{2}x^2\right)^{9-r}\left(\frac{-1}{3x}\right)^r$

$= {}^9C_r\left(\frac{3}{2}\right)^{9-r}\left(\frac{-1}{3}\right)^r x^{18-2r-r}$

Here, $18 - 3r = 0$

$\Rightarrow\ r = 6$

39. (D) $T_{r+1} = {}^8C_r (3x)^{8-r}\left(\frac{-1}{2x}\right)^r$

$= {}^8C_r 3^{8-r}\left(\frac{-1}{2}\right)^r x^{8-2r}$

Here, $8 - 2r = 3$

$\Rightarrow\ 2r = 5 \quad\Rightarrow\quad x = 5/2$

There is no term having x^3.

40. (C) $T_{r+1} = {}^8C_r\left(\frac{4x}{5}\right)^{8-r}\left(\frac{5}{2x}\right)^r$

$= {}^8C_r\left(\frac{4}{5}\right)^{8-r}\left(\frac{5}{2}\right)^r x^{8-2r}$

Here, $8 - 2r = -4$

$\Rightarrow\ 12 = 2r \quad\Rightarrow\quad r = 6$

41. (B) $(x+k)^{100} + (x-k)^{100} =$

$2\left\{{}^{100}C_0 x^{100}k^0 + {}^{100}C_2 x^{98}k^2 + {}^{100}C_4 x^{96}K^4 + \ldots + {}^{100}C_{100}k^{100}\right\}$

42. (C) $T_2 = {}^nC_1x^1$; $T_3 = {}^nC_2x^2$; $T_4 = {}^nC_3x^3$

Given that ${}^nC_1$, ${}^nC_2$ and ${}^nC_3$ are in A.P..

$\therefore\ 2({}^nC_2) = {}^nC_1 + {}^nC_3$

43. (A) $T_{p+1} = {}^{p+q}C_p x^p$ and $T_{q+1} = {}^{p+q}C_q x^q$

${}^{p+q}C_p = {}^{p+q}C_{p+q-p} = {}^{p+q}C_q$

44. (A) In the expansion of $(1+x)^{2n}$, we have

$T_{r+1} = {}^{2n}C_r x^{2n-r}$

$\Rightarrow\ T_{n+1} = {}^{2n}C_n x^n$

$\Rightarrow\ {}^{2n}C_n = P$

In the expansion of $(1+x)^{2n-1}$, we have

$T_{r+1} = {}^{2n-1}C_r x^{2n-1-r}$

Putting $2n - 1 - r = n$, we get $r = n - 1$

$T_n = T_{(n-1)+1} = {}^{2n-1}C_{n-1}x^n \Rightarrow {}^{2n-1}C_{n-1} = q$

$$\frac{P}{q} = \frac{{}^{2n}C_n}{{}^{2n-1}C_{n-1}} = 2 \quad\Rightarrow P = 2q$$

45. (C) Let the three consecutive terms be T_r, T_{r+1}, T_{r+2}.

Then $^nC_{r-1} = 220, ^nC_r = 495$ and $^nC_{r+1} = 792$

Therefore, $n = 12$ and $r = 4$.

46. (C) $T_{r+1} = {}^nC_r\ 2^{n-r}\left(\frac{x}{3}\right)^r = {}^nC_r \frac{2^{n-r}}{3^r} x^r$

$r = 7 \Rightarrow T_8 = {}^nC_7\ 2^{n-7}\ 3^{-7}\ x^7$

$r = 8 \Rightarrow T_9 = {}^nC_8\ 2^{n-8}\ 3^{-8}\ x^8$

Here,

$^nC_7\ 2^{n-7}\ 3^{-7} = {}^nC_8\ 2^{n-8}\ 3^{-8}$

$n = 55$

Hints and answers to Olympiad and NTSE level exercises

47. (C) $T_r = {}^nC_{r-1}\, p^{n-r+1} . q^{r-1}$

$\Rightarrow\ T_{r+1} = {}^nC_r\, p^{n-r} . q^r$ (given)

$$\Rightarrow\ p^{n-r+1} . q^{r-1} = \frac{n-r+1}{r} p^{n-r} . q^r$$

$$\Rightarrow\ \frac{p}{(n-r+1)} = \frac{q}{r}$$

$$\Rightarrow\ \frac{(n+1)\,q}{(p+q)\,r} = 1$$

48. (D) $T_r = {}^{14}C_{r-1}\, x^{r-1}; T_{r+1} = {}^{14}C_{r-1}\, x^r;$

$T_{r+2} = {}^{14}C_{r+1}\, x^{r+1}$

By the given condition

$$2.\ {}^{14}C_r = {}^{14}C_{r-1} + {}^{14}C_{r+1} \quad \ldots\text{(i)}$$

$$\Rightarrow 2.\frac{14!}{r!\,(14-r)!} = \frac{14!}{(r-1)!\,(15-r)!} + \frac{14!}{(r+1)!(13-r)!}$$

$$\Rightarrow\ \frac{2}{r.(r-1)!.(14-r).(13-r)!}$$

$$= \frac{1}{(r-1)!.(15-r).(14-r).(13-r)!} + \frac{1}{(r+1)r(r-1)!(13-r)!}$$

$$\Rightarrow\ \frac{2}{r(14-r)} = \frac{1}{(15-r)(14-r)} + \frac{1}{(r+1)r}$$

$$\Rightarrow\ \frac{1}{r(14-r)} - \frac{1}{(15-r)(14-r)} = \frac{1}{(r+1)r} - \frac{1}{r(14-r)}$$

$$\Rightarrow\ \frac{(15-r)-r}{r(15-r)(14-r)} = \frac{(14-r)-(r+1)}{(r+1)r(14-r)}$$

$$\Rightarrow\ \frac{15-2r}{15-r} = \frac{13-2r}{r+1}$$

$$\Rightarrow\ 15r+15-2r^2-2r = 195-30r-13r+2r^2$$

$$\Rightarrow\ 4r^2-56r+180 = 0 \Rightarrow r^2-14r+45 = 0$$

$$\Rightarrow\ (r-5)(r-9) = 0 \Rightarrow r = 5, 9$$

But 5 is not given. Hence $r = 9$.

Trick : Put the value of r from options in equation (i), only (d) satisfy it.

49. (A) $T_{r+1} = {}^{15}C_r\, (x^2)^{15-r}\left(\frac{2}{x}\right)^r = {}^{15}C_r\, 2^r\, x^{30-3r}$

For coefficient of x^{15}, set $r = 5$ for independent term, put $r = 10$.

50. (A) $(1.0002)^{3000} = (1 + 0.0002)^{3000}$

$$= 1 + (3000)(0.0002) + \frac{(3000)(2999)}{1.2}(0.0002)^2 + \frac{(3000)(2999)(2998)}{1.2.3}(0.0002)^3 +$$

$= 1 + (3000)(0.0002) = 1.6$

We want to get answer correct to only one decimal places and as such we have left further expansion.

51. (C) $10^n + 3(4^{n+2}) + 5$

Taking $n = 2$; $10^2 + 3 \times 4^4 + 5$

$= 100 + 768 + 5 = 873$

Therefore this is divisible by 9.

52. (C) $99^n + 1 = (100 - 1)^n + 1$

$= 100^n - {}^nC_1\, 100^{n-1} + ... + (-1)^{n-1}\, {}^nC_{n-1}\, 100$

$= 100$ m

where $m = 100^{n-1} - {}^nC_1\, 100^{n-2} + ... + {}^nC_{n-1}$

As the unit's digit of m is unit's digit of n and n is odd, there are exactly two zeros at the end of $99^n + 1$.

53. (C) Let the three consecutive coefficients be $^nC_{r-1} = 28, {}^nC_r = 56$ and $^nC_{r+1} = 70$, so that

$$\frac{^nC_r}{^nC_{r-1}} = \frac{n-r+1}{r} = \frac{56}{28} = 2$$

and $$\frac{^nC_{r+1}}{^nC_r} = \frac{n-r}{r+1} = \frac{70}{56} = \frac{5}{4}$$

This gives $n + 1 = 3r$ and $4n - 5 = 9r$

$$\therefore\ \frac{4n-5}{n+1} = 3 \Rightarrow n = 8$$

54. (B) $T_{r+1} = {}^{642}C_r (5^{1/2})^{642-r} . (7^{1/6})^r$

Obviously, r should be a multiple of 6.

Total numbers $= \frac{642}{6} = 107$; But first term for $r = 0$ is also integral. Hence total terms are $107 + 1 = 108$.

55. (a) → (q), (r), (s);

(b) → (p), (q), (r), (s);

(c) → (p), (q), (r);

(d) → (p), (q)

(a) ${}^{(n+1)}C_4 + {}^{(n+1)}C_3 + {}^{(n+2)}C_3 = {}^{(n+3)}C_4$

$\Rightarrow \quad {}^{(n+3)}C_4 > {}^{(n+3)}C_3 \quad \Rightarrow \quad \frac{{}^{n+3}C_4}{{}^{n+3}C_3} > 1$

$\Rightarrow \quad n > 4 \quad$ or $\quad n \geq 5$.

(b) As $(3053)^{456} - (2417)^{333}$

$= (339 \times 9 + 2)^{456} - (269 \times 9 - 4)^{333}$

Remainder of given number is same as remainder of $2^{456} + 4^{333}$

and $2^{456} + 4^{333} = (64)^{76} + (64)^{111}$

$= (1 + 63)^{76} + (1 + 63)^{111}$

$= (1 + 9 \times 7)^{76} + (1 + 9 \times 7)^{111}$

Hence the required remainder is 2.

(c) We know that $n!$ terminates in 0 for $n \geq 5$ and 3^{4n} terminates in 1 [$\because 3^4 = 81$]

$\therefore \quad 3^{180} = (3^4)^{45}$ terminates in 1

Also $3^3 = 27$ terminates in 7

Hence $183 ! + 3^{183}$ terminates in 7

i.e. the digit in the unit place = 7

(d) We are given ${}^mC_0 + {}^mC_1 + {}^mC_2 = 46$

$\Rightarrow \quad 2m + m(m - 1) = 90$

$\Rightarrow \quad m^2 + m - 90 = 0$

$\Rightarrow \quad m = 9$ as $m > 0$

Now, $(r + 1)$th term of $\left(x^2 + \frac{1}{x}\right)^m$ is

$${}^mC_r (x^2)^{m-r} \left(\frac{1}{x}\right)^r = {}^mC_r \, x^{2m-3r}$$

For this to be independent of x, $2m - 3r = 0$

$\Rightarrow \quad r = 6$

56. (a) → (p), (r);

(b) → (p), (q), (r), (s) ;

(c) → (p), (q) ;

(d) → (p), (q), (r), (s)

(a) Let consecutive coefficients ${}^nC_r$ and ${}^nC_{r+1}$.

$\Rightarrow \quad \frac{n!}{(n-r)!\, r!} = \frac{n!}{(n-r-1)!(r+1)!}$

$\Rightarrow \quad \frac{1}{(n-r)(n-r-1)!r!} = \frac{1}{(n-r-1)!\,(r+1)r!}$

$\Rightarrow \quad r + 1 = n - r$

$\Rightarrow \quad n = 2r + 1$. Hence n is odd.

(b) $E = (19 - 4)^{23} + (19 + 4)^{23}$

$= 2\,[19^{23} + {}^{23}C_2 \cdot 19^{21} \cdot 4^2 + ... + {}^{23}C_{22} \cdot 19 \cdot 4^{22}]$

$= 2 \cdot 19\,[19^{22} + {}^{23}C_2 \cdot 19^{20} \cdot 4^2 + ... + {}^{23}C_{22} \cdot 4^{22}]$

$\Rightarrow \quad E$ is divisible by 19

$\Rightarrow \quad$ Remainder = 0

(c) ${}^{10}C_0\, {}^{20}C_{10} - {}^{10}C_1\, {}^{18}C_{10} + {}^{10}C_2\, {}^{16}C_{10} - ....$

= coefficient of x^{10} in $[{}^{10}C_0 (1 + x)^{20} - {}^{10}C_1 (1 + x)^{18} + {}^{10}C_2 (1 + x)^{16} - ....]$

= coefficient of x^{10} in $[{}^{10}C_0 ((1 + x)^2)^{10} - {}^{10}C_1 ((1 + x)^2)^9 + {}^{10}C_2 ((1 + x)^2)^8 - ....]$

= coefficient of x^{10} in $[(1 + x)^2 - 1]^{10}$

= coefficient of x^{10} in $[2x + x^2]^{10}$

$= 2^{10}$

(d) $T_r = {}^{14}C_{r-1}\, x^{r-1}$; $T_{r+1} = {}^{14}C_r\, x^r$;

$T_{r+2} = {}^{14}C_{r+1} x^{r+1}$

By the given condition

$2.{}^{14}C_r = {}^{14}C_{r-1} + {}^{14}C_{r+1}$ ... (i)

$\Rightarrow \quad 2 = \frac{{}^{14}C_{r-1}}{{}^{14}C_r} + \frac{{}^{14}C_{r+1}}{{}^{14}C_r}$

$\Rightarrow \quad 2 = \frac{r}{14-r+1} + \frac{14-(r+1)+1}{r+1}$

$\Rightarrow \quad 2 = \frac{r}{15-r} + \frac{14-r}{r+1}$

$\Rightarrow \quad r = 5 \quad$ or $\quad r = 9$

Remainder Theorem

1. Remainder: +39, quotient: $x^2 + 8x + 19$

2. Remainder: +4, quotient: $2x^2 - 2x - 3$

3. Remainder: +7, quotients: $x^2 - 3x - 2$

4. Remainder: +1, quotient: $x^2 + 3$

5. Remainder: 0,
quotient: $x^5 + x^4y + x^3y^2 + x^2y^3 + xy^4 + y^5$

6. Remainder: 0,
quotient: $x^4 - x^3y + x^2y^2 - xy^3 + y^4$

7. Remainder: 0,
quotient: $x^4 + 2x^3 + 4x^2 + 8x + 16$

8. Remainder: 0, quotient: $2x^2 + 2x - 11$

9. Remainder: 0, quotient: $2x^2 + 3$

10. Remainder: 0, quotient: $x^4 + x^2 + 1$

11. $(2x + 7)(x + 2)$
$2 \times 3 \times 3 \times 3 \times 17 \times 23$

12. 4

13. $a = -2, b = +6$

14. $k = 5$

15. $p = 2$

16. $k = -11$

17. -20

18. $l = 2$ and $m = -18$

19. (i) $R = 34$
(ii) $R = 60$

20. $p = \dfrac{399}{406}$ and $q = 77$

21. $l = -1, m = -3$

22. $R = 7$

23. (a) $-10x + 15$
(b) $39x + 27$
(c) $46x - 93$
(d) $85x + 128$
(e) $25x - 77$

24. (C)

25. (C)

26. (B) $P(x) = x^3 - 4x^2 + 3x - 11$
Remainder is
$$P(4) = 4^3 - 4 \times 4^2 + 3 \times 4 - 11$$
$$= 64 - 64 + 12 - 11$$
$$= 1$$

27. (A) $x - 1 = 0 \Rightarrow x = 1$
$$P(x) = 2x^3 - 13x + 11$$
$$P(1) = 2 \times 1^3 - 13 \times 1 + 11$$
$$= 2 - 13 + 11$$
$$= 0$$

28. (D) $P(x) = x^3 - 6x + k$
$$P(2) = 2^3 - 6 \times 2 + k$$
$$\Rightarrow 8 - 12 + k = 0$$
$$\Rightarrow k = 4$$

29. (C) $P_1(x) = ax^3 + 4x^2 + 3x - 4$
$$P_1(3) = a \times 3^3 + 4 \times 3^2 + 3 \times 3 - 4$$
$$= 27a + 36 + 9 - 4$$
$$= 27a + 41$$
And,
$$P_2(x) = x^3 - 4x + a$$
$$P_2(3) = 3^3 - 4 \times 3^2 + a$$
$$= 27 - 12 + a$$
$$= 15 + a$$
Since,
$$27a + 41 = 15 + a$$
$$26a = -26$$
$$a = -1$$

30. (C) $P(x) = x^4 - 2x^3 + 3x^2 - ax + 3a - 7$
$$P(-1) = (-1)^4 - 2(-1)^3 + 3(-1)^2 - a(-1) + 3a - 7$$
Also, $P(-1) = 19$
$$\Rightarrow 1 + 2 + 3 + a + 3a - 7 = 19$$
$$\Rightarrow 4a = 20$$
$$\therefore a = 5$$
Then,
$$P(-1) = x^4 - 2x^3 + 3x^2 - 5x + 15 - 7$$
$$= x^4 - 2x^3 + 3x^2 - 5x + 8$$
Then,
$$P(-2) = (-2)^4 - 2(-2)^3 + 3(-2)^2 - 5(-2) + 8$$
$$= 16 + 16 + 12 + 10 + 8$$
$$= 62$$

31. (A) $P(x) = px^2 + 5x + r$
$$P(2) = p \times 2^2 + 5 \times 2 + r = 0$$
$$\Rightarrow 4p + 10 + r = 0 \quad \text{...(i)}$$
And,
$$P\left(\frac{1}{2}\right) = p.\left(\frac{1}{2}\right)^2 + 5 \times \frac{1}{2} + r = 0$$
$$\Rightarrow \frac{p}{4} + \frac{5}{2} + r = 0$$
$$\Rightarrow p + 10 + 4r = 0 \quad \text{...(ii)}$$
Now, (i) – (ii)
$$3p - 3r = 0$$
$$\Rightarrow p = r$$

32. (B) $x^{50} + 1 = Q(x)(x^2 - 4) + ax + b$

Put $x = 2$

$2^{50} + 1 = 2a + b$ ...(i)

Put $x = -2$

$2^{50} + 1 = -2a + b$ ...(ii)

Now, from (i) and (ii), we get

$2b = 2(2^{50} + 1)$

$\Rightarrow \; b = 2^{50} + 1$

33. (A) $P(x) = x^{100} - 1$

$P(1) = 1^{100} - 1 = 0$

$P(-1) = (-1)^{100} - 1 = 0$

$(x - 1)$ and $(x + 1)$ both are factors, so $x^2 - 1$ is also factor of $x^{100} - 1$.

34. (A) If $(x^2 - 3x + 2)$ is a factor, then $(x - 1)$ and $(x - 2)$ both are factors.

$P(x) = x^4 + ax^3 + 2x^2 + 5x + b$

$P(1) = 1^4 + a \times 1^3 + 2 \times 1^2 + 5 \times 1 + b = 0$

$\Rightarrow \; a + b = -8$ ...(i)

And,

$P(2) = 2^4 + a \times 2^3 + 2 \times 2^2 + 5 \times 2 + b = 0$

$= 16 + 8a + 8 + 10 + b = 0$

$\Rightarrow \; 8a + b = -34$ ...(ii)

Now, solving (i) and (ii), we get

$$\begin{array}{r} a + b = -8 \\ \underline{8a + b = -34} \\ -7a = 26 \end{array}$$

$$a = -\frac{26}{7}$$

$a + b = -8$

$$\Rightarrow \left(-\frac{26}{7}\right) + b = -8$$

$$\Rightarrow b = \frac{26}{7} - 8$$

$$\Rightarrow b = -\frac{30}{7}$$

35. (B) $f(x) = Q(x)\,(x - 2)\,(x - 1) + (3ax + 2b)$

$f(2) = 6a + 2b = 8$

$\Rightarrow \; 3a + b = 4$ ...(i)

And,

$f(1) = 3a + 2b = 9$ ...(ii)

Now, solving (i) and (ii), we get

$$\begin{array}{r} 3a + 2b = 9 \\ \underline{3a + b = 4} \\ b = 5 \end{array}$$

$$\therefore \; a = -\frac{1}{3},\; b = 5$$

36. (B) $P(x) = x^4 + ax^3 - bx^2 + 18$

$P(2) = 2^4 + a \times 2^3 - b \times 2^2 + 18$

$\Rightarrow \; 16 + 8a - 4b + 18 = 0$

$\Rightarrow \; 8a - 4b = -34$

$\Rightarrow \; 4a - 2b = -17$

$\Rightarrow \; 4a - 2b + 17 = 0$

37. (B) $P(x) = x^3 + x^2 + x + 1$

$P(-1) = (-1)^3 + (-1)^2 + (-1) + 1$

$= -1 + 1 - 1 + 1$

$= 0$

38. (D) $(25x^2 - 1) + (1 + 5x)^2$

$= (5x + 1)(5x - 1) + (1 + 5x)^2$

$= (5x + 1)\{(5x - 1) + (5x + 1)\}$

$= (5x + 1)(10x)$

39. (A) $P(x) = 2x^3 + 6px^2 - 2qx - q - 2p + 1$

$P(1) = 2 \times 1^3 + 6p \times 1^2 - 2q \times 1 - q - 2p + 1 = 0$

$\Rightarrow \; 2 + 6p - 2q - q - 2p + 1 = 0$

$\Rightarrow \; 4p - 3q = -3$

$$\Rightarrow \; p = \frac{3q - 3}{4} \quad \text{...(i)}$$

And,

$P(2) = 2 \times 2^3 + 6p \times 2^2 - 2q \times 2 - q - 2p + 1 = 0$

$\Rightarrow \; 16 + 24p - 4q - q - 2p + 1 = 0$

$\Rightarrow \; 22p - 5q = -17$

$$\Rightarrow \; p = \frac{5q - 17}{22} \quad \text{...(ii)}$$

On comparing values from (i) and (ii), we get

$$\frac{3q - 3}{4} = \frac{5q - 17}{22}$$

$\Rightarrow \; 66q - 66 = 20q - 68$

$$\Rightarrow \; q = \frac{-2}{46} = -\frac{1}{23}$$

$$p = \frac{\frac{-3}{23} - 3}{4} = -\frac{18}{23}$$

40. (A) $P(x) = 5x^4 - 4x^3 + ax^2 + b$

$P(-1) = 5(-1)^4 - 4(-1)^3 + a(-1)^2 + b$

$\Rightarrow 5 + 4 + a + b = 10$

$\Rightarrow a + b = 1$

41. (B) $P(x) = 4x^3 - 5x^2 - kx + 32$

$$P\left(\frac{1}{4}\right) = 4\left(\frac{1}{4}\right)^3 - 5\left(\frac{1}{4}\right)^2 - k\left(\frac{1}{4}\right) + 32$$

$$\Rightarrow \frac{4}{64} - \frac{5}{16} - \frac{k}{4} + 32 = 0$$

$$\Rightarrow \frac{k}{4} = 32 + \frac{1}{16} - \frac{5}{16}$$

$$\Rightarrow \frac{k}{4} = 32 - \frac{4}{16}$$

$\Rightarrow k = 128 - 1$

$\therefore k = 127$

42. (A) $P(x) = ax^3 + 8x^2 + 7$

$P(1) = a + 8 + 7 = a + 15$

$Q(x) = 9x^2 + 7x + b$

$Q(1) = 9 + 7 + b = 16 + b$

Now,

$a + 15 = 16 + b$

$a - b = 1$

43. (A) $P(x) = ax^3 + bx^2 + x - 6$

$P(-2) = a(-2)^3 + b(-2)^2 + (-2) - 6$

$\Rightarrow -8a + 4b - 8 = 0$

$\Rightarrow -2a + b = 2$

And,

$P(-2) = a(2)^3 + b(2)^2 + (2) - 6$

$\Rightarrow 8a + 4b - 4 = 4$

$\Rightarrow 2a + b = 2$

Now,

$$\begin{array}{r} -2a + b = 2 \\ 2a + b = 2 \\ \hline 2b = 4 \end{array}$$

$b = 2$

$a = 0$

44. (C) $R_1 = 5a - 6$

$R_2 = 4a - 10$

Now,

$2R_1 + R_2 = 6$

$\Rightarrow 2(5a - 6) + (4a - 10) = 6$

$\Rightarrow 10a - 12 + 4a - 10 = 6$

$\Rightarrow 14a = 28$

$\Rightarrow a = 2$

45. (D) $P(x) = ax^2 - 3x + 5$

$P(1) = a - 3 + 5 = a + 2$

$Q(x) = 3x^2 + 2a + 5$

$Q(3) = 27 + 2a + 5 = 2a + 32$

Now,

$a + 2 = 2a + 32$

$a = -30$

Hints and answers to Olympiad and NTSE level exercises

46. (B) Let $q(x)$ be the quotient and $ax + b$ be the remainder.

Therefore, $x^{37} = (x^2 - 4x + 3)\,q(x) + ax + b$

When $x = 1$,

$1^{37} = (1 - 4 + 3)\,q(x) + a + b$

$\Rightarrow 1 = a + b$ ...(1)

When $x = 3$,

$3^{37} = (3^2 - 4 \times 3 + 3)\,q(x) + 3a + b$

$\Rightarrow 3^{37} = 3a + b$ ...(2)

Solving (1) and (2), we get

$2a = 3^{37} - 1$

$$\therefore a = \frac{3^{37} - 1}{2}$$

Hence $b = 1 - \dfrac{3^{37} - 1}{2} = \dfrac{3 - 3^{37}}{2}$

Thus, the required remainder is

$$\left(\frac{3^{37} - 1}{2}\right)x + \left(\frac{3 - 3^{37}}{2}\right)$$

47. (C) $(5x^2 - 2x + 1)(px + q) = ax^3 - 19x^2 + bx - 3$

$\Rightarrow 5px^3 + 5qx^2 - 2px^2 - 2qx + px + q$

$= ax^3 - 19x^2 + bx - 3$

$\Rightarrow 5px^3 + (5q - 2p)x^2 + (-2q + p)x + q$

$= ax^3 - 19x^2 + bx - 3$

Comparing the coefficients on both the sides, we get

$5p = a$ ...(1)

$(5q - 2p) = -19$ ...(2)

$(-2q + p) = b$ ...(3)

$q = -3$ ...(4)

Solving these, we get $a = 10$ and $b = 8$.

48. (D) Let $P(x) = ax^2 + bx + c$

Then

$a + b + c = 2$ ...(1)

$4a - 2b + c = 5$ ...(2)

$9a + 3b + c = 1$ ...(3)

Solving these, we get

$$a = \frac{1}{10}, b = -\frac{9}{10}, c = \frac{28}{10}$$

$$\therefore \quad P(x) = \frac{1}{10}\left(x^2 - 9x + 28\right)$$

49. (A) Let $(x - p)$ be the common factor.

Then

$p^2 - p + a = 0$ ...(1)

$2p^2 + 5p + b = 0$ ...(2)

By 2 × (1), we get

$2p^2 - 2p + 2a = 0$ ...(3)

Solving (2) and (3), we get

$7p + b - 2a = 0$

$$p = \frac{2a - b}{7}$$

Putting this value of p into (1), we get

$$\left(\frac{2a-b}{7}\right)^2 - \left(\frac{2a-b}{7}\right) + a = 0$$

$$\Rightarrow \frac{4a^2 + b^2 - 4ab}{49} - \frac{2a - b}{7} + a = 0$$

$$\Rightarrow \frac{4a^2 + b^2 - 4ab}{49} - \frac{14a - 7b}{49} + \frac{49a}{49} = 0$$

$\Rightarrow 4a^2 + b^2 - 4ab - 14a + 7b + 49a = 0$

$\Rightarrow 4a^2 + b^2 - 4ab + 35a + 7b = 0$

50. (C) According to the question,

$f(y) = P(y)(y + 2)$ ...(1)

Also,

$f(y) = [P(y) + y - 1]\,(y - 3) + Q(y)$ ...(2)

$\therefore \quad P(y)(y + 2) = [P(y) + y - 1]\,(y - 3) + Q(y)$

$\Rightarrow y\,P(y) + 2P(y) = [yP(y) + y^2 - y] - [3P(y) + 3y - 3] + Q(y)$

$\Rightarrow y\,P(y) + 2P(y) = yP(y) + y^2 - y - 3P(y) - 3y + 3 + Q(y)$

$\Rightarrow 5P(y) = y^2 - 4y + 3 + Q(y)$

$\Rightarrow Q(y) = 5P(y) - y^2 + 4y - 3$

51. (B) According to the question,

$a^2 - 3a - 3 = 1$

$\Rightarrow a^2 - 3a - 4 = 0$

$\Rightarrow (a - 4)\,(a + 1) = 0$

$\therefore \quad a = 4, -1$

52. (A) $x^2 + 3x + 2 = (x + 1)\,(x + 2)$

So, $ax^3 + 3x^2 - 5x + b$ is divisible by both $(x + 1)$ and $(x + 2)$.

$\therefore \quad a\,(-1)^3 + 3(-1)^2 - 5(-1) + b = 0$

$\Rightarrow -a + b + 8 = 0$ ...(1)

And $a(-2)^3 + 3(-2)^2 - 5(-2) + b = 0$

$\Rightarrow -8a + 12 + 10 + b = 0$

$\Rightarrow -8a + b + 22 = 0$ ...(2)

Solving (1) and (2), we get

$-7a + 14 = 0$

$\Rightarrow a = 2$

$\therefore \quad b = -6$

53. (B)

54. (D)

55. (A) $p - q + r = 1$ ...(1)

$4p + 2q + r = 0$ ...(2)

$9p + 3q + r = -1$ ...(3)

Solving these, we get

$$p = -\frac{1}{6}, q = -\frac{1}{6}, r = 1$$

Algebraic Expressions

1. (i) 10
(ii) $3a^2$

2. (i) –1
(ii) –3
(iii) –5
(iv) $3a^2 + a - 5$

3. –7
–2, 9
0

4. $h + 2x + 2$

5. 0

6. (i) $a^2b + b^2c + c^2a$
(ii) $bc(b - c) + ca(c - a) + ab(a - b)$

(iii) $a(b-c)^2 + b(c-a)^2 + c(a-b)^2$

(iv) $(b-c) + (c-a) + (a-b)$

(v) $a(b-c) + b(c-a) + c(a-b)$

7. (i) 0

(ii) 0

(iii) $2x + 2y + 2z$

(iv) 0

(v) 0

8. (i) Σa^2bc (ii) $\Sigma(x-y)^2$

9. (i) $xy + 2x^2$ (ii) $x^2(x+y)$

10. $-a^2(b^2-c^2) - b^2(c^2-a^2) - c^2(a^2-b^2)$

11. $a = 2, b = 3$

12. $-3x + 9$

13. $\dfrac{4x}{3x+1}$

14. x

15. $4x + 9$

16. $(x-y)(y-z)(z-x)$

17. $(x-y)(y-z)(z-x)$

18. $(a+b)(b+c)(c+a)$

19. $-(x-y)(y-z)(z-x)$

20. $(a+b)(b+c)(c+a)$

21. $(a-b)(b-c)(c-a)(a+b+c)$

22. x

23. $\dfrac{19x+3}{x+12}$

24. $\dfrac{3a+2}{2}$

25. 2

26. $(t^2 - 2t - t + 2 - 1)(t^2 - 2t - t + 2 - 2)$

27. $(a+b+c)(ab+bc+ca)$

28. $-(a-b)(b-c)(c-a)(a+b+c)$

29. $(x+y+z)(x+y-z)(x-y+z)(-x+y+z)$

30. $(a-b)(b-c)(c-a)(a+b+c)$

31. $(a+b+c)(ab+bc+ca)$

32. $-(a-b)(b-c)(c-a)(a^2+b^2+c^2+ab+bc+ca)$

33. $(a^2+b^2+c^2+ab+bc+ca)$

34. $-(x-y)(y-z)(z-x)(xy+yz+zx)$

35. $(x-y)(y-z)(z-x)(xy+yz+zx)$

36. $(a+b)(a-b)(b+c)(b-c)(c+a)(c-a)$

37. $5(a+b)(b+c)(c+a)$

$(a^2+b^2+c^2+ab+bc+ca)$

38. $(a-b)(b-c)(c-a)$

39. $24abc$

40. $2abc(a+b+c)$

41. (B) $a^3+b^3+c^3-3abc$

$= (a+b+c)(a^2+b^2+c^2-ab-bc-ca)$

If $(a+b+c) = 0$

$a^3+b^3+c^3-3abc = 0$

$a^3+b^3+c^3 = 3abc$

42. (C) $8a^3 - 27b^3 - c^3 - 18abc$

$= (2a)^3 + (-3b)^3 + (-c)^3 - 3(2a)(-3b)(-c)$

$= (2a-3b-c)(4a^2+9b^2+c^2+6ab-3bc+2ac)$

Therefore, $(2a-3b-c)$ is a factor.

43. (A) $8p^3 - 1 - m^3 - 6pm$

$= (2p)^3 + (-1)^3 + (-m)^3 - 3(2p)(-1)(-m)$

$= (2p-1-m)(4p^2+1+m^2+2p-m+2pm)$

If $8p^3 - 1 - m^3 - 6pm = 0$

$(2p-1-m) = 0$

$\Rightarrow\ 2p = 1 + m$

44. (D) $x^2y^2(x-y) + y^2z^2(y-z) + z^2x^2(z-x)$

$= x^3y^2 - x^2y^3 + y^3z^2 - y^2z^3 + z^3x^2 - z^2x^3$

$= y^2(x^3 - z^3) - y^3(x^2 - z^2) + z^2x^2(z-x)$

$= (x-z)[x^2y^2 + xy^2z + y^2z^2 - y^3x - y^3z - z^2x^2]$

$= (x-z)[xy^2(x-y) + z^2(y-x)(y+x) + y^2z(x-y)]$

$= (x-z)(x-y)[xy^2 - z^2x + y^2z - yz^2]$

$= (x-z)(x-y)(y-z)(xy+yz+zx)$

45. (A) $a^3(b-c) + b^3(c-a) + c^3(a-b) = 0$

$\Rightarrow\ (a-b)(b-c)(c-a)(a+b+c) = 0$

$a+b+c = 0$

46. (A) $f(a, 3) = 5a^2 - 10(a)(3) + 12 \times 9$

$= 5a^2 - 30(a) + 108$

$f(2, a) = 5(2)^2 - 10(2)(a) + 12a^2$

$= 20 - 20a + 12a^2$

Since

$f(a, 3) = f(2, a)$

$\Rightarrow\ 5a^2 - 30a + 108 = 20 - 20a + 12a^2$

$\Rightarrow\ 7a^2 + 10a - 88 = 0$

47. (B) $f(1) = 4 + 1 + 5b = 5 + 5b$

$f(-1) = 4 + 1 + 5b = 5 + 5b$

Since $f(1) = f(-1)$

For all real values of b.

48. (A) $f(x) = 2x^2 - 3x + 2$

$f(0) = 2$

$f(1) = 2 - 3 + 2 = 1$

$f(2) = 8 - 6 + 2 = 4$

$\therefore \quad f(0) + f(1) + f(2) = 2 + 1 + 4 = 7$

49. (D) $f(x) = ax^2 + bx + c$

$f(-1) = a - b + c$

$f(0) = c$

$f(1) = a + b + c$

Given that $f(-1)$, $f(0)$ and $f(1)$ are in A.P..

$\therefore \quad 2f(0) = f(-1) + f(1)$

$2c = a - b + c + a + b + c$

$\Rightarrow \quad 2c = 2a + 2c$

$\Rightarrow \quad 2a = 0$

$\therefore \quad a = 0$

50. (A) Since p, q and r are in A.P.

$2q = p + r$

Now,

$f(x) = px^2 + qx + r$

$f(1) = p + q + r$

$f(1) = 3$

$\Rightarrow \quad p + q + r = 3$

$\Rightarrow \quad (p + r) + q = 3$

$\Rightarrow \quad 2q + q = 3$

$\Rightarrow \quad 3q = 3$

$\therefore \quad q = 1$

51. (B) homogenous equation contains only same degree terms.

52. (A) $f(x) = \dfrac{x+1}{x-1}$

$$f(f(x)) = \frac{f(x)+1}{f(x)-1} = \frac{\frac{x+1}{x-1}+1}{\frac{x+1}{x-1}-1}$$

$$= \frac{x+1+x-1}{x+1-x+1} = \frac{2x}{2} = x$$

53. (C) $f(x) - x^2 - x + 1$

$$f(x+1) = (x+1)^2 - (x+1) + 1$$
$$= x^2 + 1 + 2x - x - 1 + 1$$
$$= x^2 + x + 1$$

$$(x+1) f(x) - (x-1) f(x+1)$$
$$= (x+1)(x^2 - x + 1) - (x-1)(x^2 + x + 1)$$
$$= (x^3 - x^2 + x + x^2 - x + 1)$$
$$- (x^3 + x^2 + x - x^2 - x - 1)$$
$$= 2$$

54. (D) $f(x) = 3x^2 + 4x + 1$

$f(-x) = 3x^2 - 4x + 1$

$f(x) - f(-x) = 2 + 6x^2$

55. (A) $f(x) = \dfrac{x+1}{x-1}$

$$f(f(x)) = \frac{f(x)+1}{f(x)-1}$$
$$= \frac{\frac{x+1}{x-1}+1}{\frac{x+1}{x-1}-1}$$
$$= \frac{x+1+x-1}{x+1-x+1}$$
$$= \frac{2x}{2}$$
$$= x$$

$\therefore \quad f(f(5)) = 5$

56. (D) $f(x) = x^2 + 4x + 4$

$f(x + a) = (x + a)^2 + 4(x + a) + 4$

$= x^2 + a^2 + 2ax + 4x + 4a + 4$

Now, $f(x + a) - f(x)$

$= x^2 + a^2 + 2ax + 4x + 4a + 4 - (x^2 + 4x + 4)$

$a^2 + 2ax + 4a$

57. (A) $P = (a + b - c)$

$Q = (a - b + c)$

$R = (-a + b + c)$

$P + Q + R = a + b + c = 0$

58. (B) $a(b - c) + b(c - a) + c(a - b) = 0$

59. (A) $2b^2c^2 + 2c^2a^2 + 2a^2b^2 - a^4 - b^4 - c^4$

$= 4b^2c^2 + 2c^2a^2 + 2a^2b^2 - 2b^2c^2 - a^4 - b^4 - c^4$

$= (2bc)^2 - (a^4 + b^4 + c^4 - 2a^2b^2 + 2b^2c^2 - 2c^2a^2)$

$= (2bc)^2 - (a^2 - b^2 - c^2)^2$

$= (2bc + a^2 - b^2 - c^2)(2bc - a^2 + b^2 + c^2)$

$= \left[a^2 - (b-c)^2\right]\left[(b+c)^2 - a^2\right]$

$= (a+b+c)(a-b+c)(a+b-c)(-a+b+c)$

60. (D) $f(x) = 2x^2 - x + a + b$

$f(0) = a + b = 1$

$f(1) = 2 - 1 + a + b = -2$

$a + b = -3$

Hints and answers to Olympiad and NTSE level exercises

61. (B) $a^2 + b^2 + c^2 - ab - bc - ca = 0$

or $\frac{1}{2}[2a^2 + 2b^2 + 2c^2 - 2ab - 2bc - 2ca] = 0$

or $\frac{1}{2}[(a-b)^2 + (b-c)^2 + (c-a)^2] = 0$

Since the sum of the three squares is zero, each must be zero. Therefore,

$a - b = 0, b - c = 0, c - a = 0$ or $a = b = c$

62. (A) Given $a^2 - (b-c)^2$

This is of the form $a^2 - b^2 = (a+b)(a-b)$. Therefore,

$a^2 - (b-c)^2 = (a+b-c)[a-(b-c)]$

$= (a+b-c)(a-(b+c)$

63. (A) $(x-1)^2 + 20(x-1) + 36$

Let $(x-1)$ be y.

Then the given equation becomes

$y^2 + 20y + 36 = y^2 + 2y + 18y + 36$

$= y(y-2) + 18(y+2)$

$= (y+2) + (y+18)$

$= (x-1+2)(x-1+18)$

$= (x+1)(x+17)$

64. (C) $x^5 + x^4 - x - 1 = x^4(x+1) - 1(x+1)$

$= (x+1)(x^4 - 1)$

$= (x+1)(x^2+1)(x^2-1)$

$= (x+1)(x^2+1)(x+1)(x-1)$

$= (x^2+1)(x+1)^2(x-1)$

65. (C) $x^6 + 27 = (x^2)^3 + 3^3$

This is of the form $a^3 + b^3 = (a+b)(a^2 - ab + b^2)$. Therefore,

$(x^2)^3 + 3^3 = (x^2+3)(x^4 - 3x^2 + 9)$

66. (D) $\frac{a}{x+a} + \frac{b}{x-a} - \frac{c}{x^2-a^2} = \frac{(x-a)a + (x+a)b - c}{x^2 - a^2}$

67. (D) $\frac{1}{(x+1)(x+2)} + \frac{1}{(x+2)(x+3)} + \frac{1}{(x+3)(x+1)}$

$= \frac{x+3+x+1+x+2}{(x+1)(x+2)(x+3)}$

$= \frac{3x+6}{(x+1)(x+2)(x+3)}$

$= \frac{3(x+2)}{(x+1)(x+2)(x+3)}$

$= \frac{3}{(x+1)(x+3)}$

68. (B) $4(a-b)^2 - 9(a+b)^2 = [2(a-b)]^2 - [3(a+b)]^2$

This is of the form $a^2 - b^2 = (a+b)(a-b)$. Therefore,

$[2(a-b)]^2 - [3(a+b)]^2$

$= [2(a-b) + 3(a+b)][2(a-b) - 3(a+b)]$

$= (2a - 2b + 3a + 3b)(2a - 2b - 3a + 3b)$

$= (5a + b)(-a - 5b)$

$= -(a + 5b)(5a + b)$

69. (A) $x^2 - x + \frac{1}{4} = (x)^2 - 2\cdot\frac{1}{2}x + \left(\frac{1}{2}\right)^2 = \left(x - \frac{1}{2}\right)^2$

70. (C) $a^2 + ac + bc - b^2$

$= a^2 - b^2 + c(a+b)$

$= (a+b)(a-b+c)$

Game Time

1. Distances covered by the ships are equal. If s is the distance covered, we can write for ship 1

   $4.5 \times X = s$

   For ship 2

   $3.5 \times Y = s$

   From the above equations, we get

   $\frac{9}{2}X = \frac{7}{2}Y \Rightarrow Y = \frac{9}{7}X$

2. Let t_1, t_2, and t_3 be the time duration to cover the first, second and third 1/3rd distances, respectively.

Given that

$$t_2 = \left(1+\frac{1}{2}\right)t_1 = \frac{3}{2}t_1$$

$$t_3 = \left(1+\frac{1}{4}\right)t_2 = \frac{5}{4}\times\frac{3}{2}t_1$$

Total time $t_1 + t_2 + t_3 = 5$ h 50 min

$$\therefore t_1 + \frac{3}{2}t_1 + \frac{15}{8}t_1 = 350 \text{ min}$$

$$\Rightarrow \left(\frac{35}{8}\right)t_1 = 350 \text{ min} \Rightarrow t_1 = 80 \text{ min}$$

3. Let x be the number.

25% of it is $\frac{x}{4}$

20% of $\frac{x}{4}$ is $\left(\frac{x}{4}\right)\times\frac{1}{5} = \frac{x}{20}$

Given that $\frac{x}{20} = 100$

$x = 2000$

4. Let the share of the property to be given to the girl child be $\frac{1}{3}x$. The mother's share is double the girl child's share $= \frac{2}{3}x$.

The son's share is double the mother's share $= \frac{4}{3}x$, all put together is 1.

$$\left(\frac{1}{3}+\frac{2}{3}+\frac{4}{3}\right)x = 1$$

$$\frac{7}{3}x = 1, \; x = \frac{3}{7}$$

$\therefore \frac{1}{3}x$ is $\frac{1}{7}$, $\frac{2}{3}x$ is $\frac{2}{7}$, and $\frac{4}{3}x$ is $\frac{4}{7}$

5. Let the hare cover x m before the hound overtakes it. Let t be the time taken by the hare to cover x m. In this time interval, the hound will cover a distance $(x + 50 \times 1.75)$ m. In this time, the hare takes $4n$ leaps while hound takes $3n$ leaps.

Therefore, we can write

$x = 4n \times 1.75$

$(x + 52 \times 1.75) = 3n \times 2.75$

$\Rightarrow 4n \times 1.75 + 87.5 = 3n \times 2.75$

$\Rightarrow 1.25n = 87.5$

$n = 70$

The number of leaps taken by the hound $= 3n = 210$.

Simple and Simultaneous Equations

1. $5x = 6y$
2. $3x = 2y - 2$
3. $x + 10 = 2(x - 5)$
4. $3x - 7 = 8$
5. $2x + 6 = 3x$
6. $x + y = 40$
   $x - y = 10$
7. $x = 3, y = 4$
8. $x = 2, y = 3$
9. $x = 7, y = 2$
10. $x = 2, y = 3$
11. $x = 2, y = 1$
12. $x = 3, y = 12$
13. $x = 0, y = -3$
14. $x = 7, y = 6$
15. $x = 7, y = 9$
16. $x = 27, y = -4$
17. $x = \dfrac{a^2 - b^2}{ma - nb}, \; y = \dfrac{a^2 - b^2}{na - mb}$
18. $x = \dfrac{-be + cd}{ad - bc}, \; y = \dfrac{ae - c^2}{ad - bc}$
19. $x = \dfrac{ac + b^2}{b + a^2}, \; y = \dfrac{c - ab}{-b - a^2}$
20. $x = \dfrac{al - am}{bl - m^2}, \; y = \dfrac{ab - am}{bl - m^2}$
21. $x = \dfrac{16}{7}, \; y = \dfrac{31}{-14}$
22. $x = \dfrac{68}{17}, \; y = \dfrac{119}{17}$
23. $x = 2, y = 3$
24. $x = \dfrac{a^2 + b^2}{2ab}, \; y = \dfrac{b^2 - a^2 + 2ab}{2ab}$
25. $x = \dfrac{125}{41}, \; y = \dfrac{-89}{41}$
26. $x = \dfrac{1}{3}, \; y = \dfrac{9}{5}$
27. $x = 3, y = 5$
28. 12
29. 42 years and 18 years
30. Length of rectangle = 24 m
    Breadth of rectangle = 13 m

31. $x = 14, y = 21$

32. $x = 1$ and $y = -\frac{1}{2}$

33. The graphs of the equation coincide with the given system and has an infinite number of solutions.

34. $x = -1,\ y = \frac{3}{2}$

35. (A) Let a number be x and another number be y.

Thrice of $x = 3x$

Twice of $y = 2y$

$$3x = 2y - 10$$

36. (A) Ratio of coefficients of $x \neq$ Ratio of coefficients of y

$$\frac{2}{3} \neq \frac{-3}{2}$$

Thus, only one solution is possible.

37. (B) $\frac{1}{3} = \frac{2}{6} = \frac{3}{9}$

Ratio of coefficients of x = ratio of coefficients of y = ratio of absolute terms.

38. (C) $\frac{2}{6} = \frac{-3}{-9} \neq \frac{2}{5}$

39. (B) Since $x = 1$ and $y = 3$ satisfy both the equations,

$$2(1) + k(3) = 3$$
$$\Rightarrow\ 3k = 1$$
$$\Rightarrow\ k = \frac{1}{3}$$
$$1 - n(3) = 2$$
$$\Rightarrow\ 3n = -1$$
$$\Rightarrow\ n = \frac{-1}{3}$$

40. (C) Element of solution set should belong to set A. So, (1, 5) can't be a solution.

41. $0.2u + 0.3v = 1.3$

$$\Rightarrow\ \frac{2u}{10} + \frac{3v}{10} = \frac{13}{10}$$
$$\Rightarrow\ 2u + 3v = 13$$
$$\Rightarrow\ v = \frac{13 - 2u}{3}$$

$$0.4u + 0.5v = 2.3$$
$$\Rightarrow\ \frac{4u}{10} + \frac{5v}{10} = \frac{23}{10}$$
$$\Rightarrow\ 4u + 5v = 23$$
$$\Rightarrow\ v = \frac{23 - 4u}{5}$$

then,

$$\frac{13 - 2u}{3} = \frac{23 - 4u}{5}$$
$$\Rightarrow\ 65 - 10u = 69 - 12u$$
$$\Rightarrow\ 2u = 4$$
$$\Rightarrow\ u = 2$$
$$v = \frac{23 - 4(2)}{5} = \frac{15}{5} = 3$$

42. (B) $\sqrt{2}r + \sqrt{2}s = 0$

$$\Rightarrow\ r = \frac{-\sqrt{3}}{\sqrt{2}}s$$
$$\sqrt{3}\left(\frac{-\sqrt{3}}{\sqrt{2}}\right)s - \sqrt{8}s = 0$$
$$\Rightarrow\ s\left(-\frac{3}{\sqrt{2}} - \sqrt{8}\right) = 0$$
$$\Rightarrow\ s = 0$$

43. (D)

$$\frac{x}{2} + \frac{2y}{3} + 1 = 0 \qquad \text{...(i)}$$
$$x - \frac{y}{3} - 3 = 0$$
$$\text{or } 2x - \frac{2y}{3} - 6 = 0 \qquad \text{...(ii)}$$

Adding (i) and (ii), we get

$$\frac{5x}{2} - 5 = 0$$
$$\Rightarrow \frac{5x}{2} = 5$$
$$\Rightarrow x = 2$$
$$\frac{y}{3} = x - 3$$
$$\Rightarrow y = 3(2 - 3) = -3$$

44. (A) $2x + 3y = 7$

$$(p - q)\,x + (p + q)\,y = 3p + q - 2$$

$$\frac{2}{p-q}=\frac{3}{p+q}=\frac{7}{3p+q-2}$$
$$2(p+q)=3(p-q)$$
$$\Rightarrow \quad 2p+2q=3p-3q$$
$$\Rightarrow \quad 5q=p$$
$$\frac{3}{p+q}=\frac{7}{3p+q-2}$$
$$\Rightarrow \quad 9p+3q-6=7p+7q$$
$$\Rightarrow \quad 2p-4q=6$$

now

$$2(5q)-4q=6$$
$$\Rightarrow \quad 6q=6$$
$$\therefore \quad q=1$$
$$p=5$$

45. (A) $6u+3v=6uv$

$$\Rightarrow \quad \frac{6}{v}+\frac{3}{u}=6 \quad \text{...(i)}$$
$$2u+4v=5uv$$
$$\Rightarrow \quad \frac{2}{v}+\frac{4}{u}=5 \quad \text{...(ii)}$$

Now, (i) − (ii) × 3

$$\frac{1}{u}(3-12)=-9$$
$$u=1 \text{ and } v=2$$

46. (B) Let $x+y=u$ and $x-y=v$.

$$\frac{2}{u}+\frac{3}{v}=2 \quad \text{...(i)}$$
$$\frac{5}{u}+\frac{10}{v}=\frac{35}{6} \quad \text{...(ii)}$$

Now, operating (i) × 5 − (ii) × 2, we get

$$\frac{1}{v}(15-20)=10-\frac{70}{6}$$
$$\Rightarrow \quad \frac{1}{v}(-5)=\frac{60-70}{6}$$
$$\Rightarrow \quad \frac{-5}{v}=\frac{-10}{6}$$

So, $v=3$ and $u=2$

Therefore, $x+y=2, x-y=3$

Then $x=\frac{5}{2}$ and $y=\frac{-1}{2}$

47. (C) Assume

$$x=\frac{1}{3u+v}, v=\frac{1}{3u-v}$$

Now, equations will reduce to

$$x+y=\frac{3}{4},$$

and $\frac{x}{2}-\frac{y}{2}=\frac{-1}{8}$

$$\Rightarrow \quad x-y=\frac{-1}{4}$$

Solving these two equations, we get

$$x=\frac{1}{4}, y=\frac{1}{2}$$

now,

$$x=\frac{1}{3u+v}$$
$$\Rightarrow \quad \frac{1}{4}=\frac{1}{3u+v}$$
$$\Rightarrow \quad 3u+v=4 \quad \text{...(i)}$$
$$y=\frac{1}{3u-v}$$
$$\Rightarrow \quad \frac{1}{2}=\frac{1}{3u-v}$$
$$\Rightarrow \quad 3u-v=2 \quad \text{...(ii)}$$

(i) + (ii)

$$6u=6$$
$$\Rightarrow \quad u=1, v=1$$

48. (A) $ax+by=c$ ...(i)

$bx+ay=1+c$ ...(ii)

$b\times$(i) $-\ a\times$(ii)

$$abx+b^2y=bc$$
$$abx+a^2y=a+ac$$
$$(b^2-a^2)\,y=bc-a-ac$$
$$y=\frac{bc-a-ac}{b^2-a^2}$$
$$x=\frac{bc-ac+b}{b^2-a^2}$$

49. (C) $ax+by=a-b$ ...(i)

$bx-ay=a+b$ ...(ii)

(i) $\times a$ + (ii) $\times b=0$

$$a^2x + aby = a^2 - ab$$
$$b^2x - aby = ab + b^2$$
$$x(a^2 + b^2) = a^2 + b^2$$
$$x = 1, y = -1$$

50. (B) $x - y = 0.9$

$$\frac{11}{2(x+y)} = 1$$

$$2x = \frac{9}{10} + \frac{11}{2} = \frac{64}{10}$$
$$x = \frac{32}{10} = \frac{16}{5}$$
$$y = \frac{23}{10}$$

51. (D) $x + y = a + b$ ...(i)

$ax - by = a^2 - b^2$ ...(ii)

(i) × b + (ii)

$$bx + by = ba + b^2$$
$$ax - by = a^2 - b^2$$
$$x(a + b) = a(a + b)$$
$$x = a, y = b$$

52. (D)

$$\frac{x}{a} + \frac{y}{b} = a + b \quad \text{...(i)}$$
$$\frac{x}{a^2} + \frac{y}{b^2} = 2 \quad \text{...(ii)}$$
$$\left(\frac{1}{b}\right) \times \text{(i)} - \text{(ii)}$$
$$\frac{x}{ab} + \frac{y}{b^2} = \frac{a}{b} + 1$$
$$\frac{x}{a^2} + \frac{y}{b^2} = 2$$

$$x\left(\frac{1}{ab} - \frac{1}{a^2}\right) = \frac{a}{b} - 1$$
$$\Rightarrow x\left(\frac{a-b}{a^2b}\right) = \left(\frac{a-b}{b}\right)$$
$$x = a^2 \text{ and } y = b^2$$

53. (C) $\frac{p}{3} \neq \frac{2}{1}$

$p \neq 6$

54. (A) $\frac{3}{2t-1} = \frac{1}{t-1} \neq \frac{1}{2t+1}$

$$\Rightarrow 3(t-1) = 2t - 1$$
$$\Rightarrow 3t - 3 = 2t - 1$$
$$\Rightarrow t = 2$$

Also,

$$\frac{1}{t-1} \neq \frac{1}{2t+1}$$
$$\frac{1}{1} \neq \frac{1}{5} \text{ (true for } t = 2)$$

55. (A) $\frac{k}{12} = \frac{3}{k} = \frac{k-3}{k}$

$$k^2 = 36$$
$$\Rightarrow k = \pm 6$$
$$3k = k^2 - 3k$$
$$\Rightarrow k^2 - 6k = 0$$
$$\Rightarrow k(k-6) = 0$$
$$\Rightarrow k = 0, 6$$

So, only $k = 6$ is true.

56. (B) $\frac{2}{k+2} = \frac{3}{6} = \frac{4}{3k+2}$

$$3k + 6 = 12$$
$$\Rightarrow k = 2$$
$$9k + 6 = 24$$
$$\Rightarrow 9k = 18$$
$$\Rightarrow k = 2$$

57. (D) $\frac{2}{2\alpha} = \frac{3}{\alpha+\beta} = \frac{7}{28}$

$$\frac{2}{2\alpha} = \frac{7}{28}$$
$$\Rightarrow \frac{1}{\alpha} = \frac{1}{4}$$
$$\Rightarrow \alpha = 4$$
$$\frac{3}{\alpha+\beta} = \frac{1}{4}$$
$$\Rightarrow \alpha + \beta = 12$$
$$\Rightarrow 4 + \beta = 12$$
$$\Rightarrow \beta = 8$$

58. (C) $\frac{3}{2m-1}=\frac{1}{m-1}\neq\frac{1}{2m+1}$

Now,

$$\frac{3}{2m-1}=\frac{1}{m-1}$$

$\Rightarrow$ $3m-3=2m-1$

$\Rightarrow$ $m=2$

Also,

$$\frac{1}{m-1}\neq\frac{1}{2m+1}$$

$\Rightarrow$ $\frac{1}{2-1}\neq\frac{1}{5}$

$\Rightarrow$ $1\neq\frac{1}{5}$

59. (D) $p(x+y)+q(x-y)=p^2-pq+q^2$ ...(i)

$p(x+y)-q(x-y)=p^2+pq+q^2$ ...(ii)

Adding (i) and (ii),

$$2p(x+y)=2(p^2+q^2)$$

$$x+y=\frac{p^2+q^2}{p} \quad ...(iii)$$

Now,

(i) − (ii)

$$2q(x-y)=-2pq$$

$$x-y=-p \quad (iv)$$

From (iii) and (iv)

$$2x=\frac{p^2+q^2}{p}-p$$

$$2x=\frac{q^2}{p}$$

$\Rightarrow$ $x=\frac{q^2}{2p}, y=p+\frac{q^2}{2p}$

60. (B) Let x and y be the digits at unit and ten places, respectively. Then the number is $10x + y$.

When the digits are reversed, the number becomes $10y + x$.

According to the question,

$$x+y=11 \quad ...(i)$$

and $10x + y + 45 = 10y + x$

$\Rightarrow$ $9x - 9y = -45$

$\Rightarrow$ $x - y = -5$ ...(ii)

Solving (i) and (ii), we get

$x+y=11$

$x-y=-5$

$2x=6$

$\Rightarrow$ $x=3, y=8$

61. (A) Sum of opposite angles of a cyclic quadrilateral is 180°.

$\angle A+\angle C=180°$

$x+y+10+x+y-30°=180°$

$\Rightarrow$ $2x+2y=200°$

$\Rightarrow$ $x+y=100°$

$\angle B+\angle D=180°$

$y+20+x+y=180°$

$\Rightarrow$ $x+2y=160°$

$x+y=100$ ...(i)

$x+2y=160$ ...(ii)

(i) − (ii)

$-y=-60$

$\Rightarrow$ $y=60, x=40$

62. (A) Let the present ages of father and his son be x and y, respectively.

$$\frac{x}{y}=\frac{3}{1}$$

$\Rightarrow$ $x=3y$

$$\frac{x+14}{y+14}=\frac{2}{1}$$

$\Rightarrow$ $x+14=2y+28$

$\Rightarrow$ $x-2y=14$

Now,

$3y-2y=14$

$\Rightarrow$ $y=14, x=42$

63. (B) Let us assume that length of the rectangle is x metres and breadth is y metres. Then area is xy sq. metres.

$(x + 3)(y + 2) = xy + 67$

$xy + 2x + 3y + 6 = xy + 67$

$2x + 3y = 61$ ...(i)

$(x - 5)(y + 3) = xy - 9$

$xy + 3x - 5y - 15 = xy - 9$

$3x - 5y = 6$ ...(ii)

$3 \times$ (i) $- 2 \times$ (ii)

$$\begin{aligned} 6x + 9y &= 183 \\ 6x - 10y &= 12 \\ \hline 19y &= 171 \end{aligned}$$

$\Rightarrow \quad y = 9,$

$$x = \frac{6+5(9)}{3} = \frac{51}{3} = 17$$

Hence, area = 17 × 9 = 153 sq. units

61. (C) $\angle R = \angle Q + 9$

$\beta = 3\alpha - 2 + 9$

$\beta = 3\alpha + 7$ ...(i)

$\angle P + \angle Q + \angle R = 180°$

(sum of angles of a triangle is 180°)

$\alpha + 3\alpha - 2 + \beta = 180°$

$\Rightarrow \quad 4\alpha + \beta = 182$

$\Rightarrow \quad \beta = 182 - 4\alpha$ ...(ii)

Now,

From (i) and (ii)

$3\alpha + 7 = 182 - 4\alpha$

$\Rightarrow \quad 7\alpha = 175$

$\therefore \quad \alpha = 25, \beta = 82$

$\angle P = \alpha = 25°$

$\angle Q = 3\alpha - 2 = 3(25) - 2 = 73°$

$\angle R = 82°$

Hints and answers to Olympiad and NTSE level exercises

65. (A)

(i) $\frac{3}{6} = \frac{-5}{-10} = \frac{20}{40}$; Infinitely many solutions

(ii) $\frac{1}{3} = \frac{-3}{-9} \neq \frac{3}{2}$; No solution

(iii) $\frac{3}{7} \neq \frac{-5}{2} \neq \frac{25}{15}$; Exactly one solution

66. (B) x men can do the work in y days.

1 men can do the work in xy days.

z men can do the work in $\frac{xy}{z}$ days

67. (A) $-\frac{2}{x} + \frac{5}{y} = \frac{7}{6}$ (1)

or $\frac{36}{x} - \frac{24}{y} = 1$ (2)

Equation (1) × 18 : $-\frac{36}{x} + \frac{90}{y} = 1$ (3)

Eq. (2) + (3): $\frac{66}{y} = 22$ or $y = 3$

Using $y = 3$ in (2), we get $x = 4$. Therefore, $x = 4$, $y = 3$

68. (C) Let the fraction be x/y. Then

$\frac{x+9}{y} = 2$ or $x - 2y + 9 = 0$ (1)

$\frac{x}{y-2} = 1$ or $x - y + 2 = 0$ (2)

Solving (1) and (2), we get $x = 5, y = 7$.

Therefore, answer is 5/7.

69. (B) $\frac{a_1}{a_2} = \frac{b_1}{b_2} = \frac{c_1}{c_2}$ should be true to have infinitely many solutions.

$3x + 2y - 8 = 0$

$6x + 4y - p = 0$

$\frac{3}{6} = \frac{2}{4} = \frac{8}{p}$

$\therefore \quad p = 16$

70. (C) $4 \times 4 = 3xm$ or $m \neq \frac{16}{3}$

71. (B) Let the numbers be x and y. Then

$x + y = 640$ (1)

$x - y = 280$ (2)

Equation (1) + equation (2): $2x = 920$

$x = 460$

Using in (1), we get $x + 460 = 640$

$x = 640 - 460 = 180$

Therefore, the numbers are 180 and 460.

72. (C) Let the tenths digit be y and units digit be x.

The number will be $10y + x$

As per the problems, $10y + x = 4(x + y)$ (1)

$10x + y = (10y + x)4$ (2)

Solving (1) and (2), we get $x = 2, y = 1$.

Therefore, the number is 12.

73. (C) Let the present ages of father and son be x and y, respectively.

As per the problem, six years ago,

$(x - 6) = 3(y - 6)$

Six years later, $(x + 6) = 2(y + 6)$

Solving these two equations, we get that father's age is 42 years, and son's age is 18 years.

74. (C) Let the man's salary be x and the salary of woman be y.

As per the problem,

$$15x + 8y = 3050 \quad (1)$$
$$3x - 5y = -50 \quad (2)$$

Equation (1) × 1: $15x + 8y = 3050$ (3)

Equation (2) × 5: $15x - 25y = -250$ (4)

Equation (3) – equation (4):

$$33y = 3300$$
$$\therefore \quad y = 100$$

Using this in (2). we get

$$3x - 500 = -50$$
$$\Rightarrow \quad 3x = 450$$
$$\therefore \quad x = 150$$

Total salary = $3 \times 150 + 5 \times 100$

= 450 + 500 = ₹ 950

Quadratic Equations

1. $x = \pm 3$
2. $x = \pm 2$
3. $x = \pm 4$
4. $x = \pm\sqrt{30}$
5. $x = \pm\sqrt{11}$
6. $x = \pm 5$
7. $x = \frac{2}{3}$ or $-\frac{7}{3}$
8. $x = -5$ or $\frac{1}{3}$
9. $x = 5$ or $-\frac{1}{5}$
10. $x = \frac{a}{b}$ or $\frac{b}{a}$
11. $x = b \pm 1$
12. $x = 0$ or $(a + b)$
13. $x = 4$ or 8
14. $x = 5$ or -3
15. $x = -12$ or 2
16. $x = 6$ or -3
17. $x = \frac{1}{2}$ or 6
18. $2 \pm \sqrt{-2}$
19. $x = a$ or $-\frac{1}{a}$
20. $x = a \pm 1$
21. $x = -4$ or 1
22. $x = 2$ (repeated root)
23. $x = \frac{1}{4}$ or $-\frac{5}{8}$
24. $x = \frac{9}{4}$ or -3
25. $x = \frac{a}{(a-b)}$ or $-\frac{b}{(a-b)}$
26. $x = \pm 4$ or $\pm\frac{2}{3}$
27. $x = 1$ or $\frac{1}{5}$
28. $x = \frac{1}{9}$ or $-\frac{1}{7}$
29. $x = 125$ or -27
30. $x = 1$ or -1
31. $x = 1$ or 2 or $\frac{1}{2}$
32. $x = \frac{1}{2}$ or 2 or $\frac{1}{3}$ or -3
33. $x = 0$ or -5
34. $x = 1$ or -6
35. $x = 2$ or -4
36. $x = 3$ or $-\frac{3}{2}$
37. $x = 0$ or 4
38. $x = \frac{2}{3}$ or $\frac{1}{2}$
39. $x = 0$ or 3
40. $x^2 + 5x + 6 = 0$
41. $x^2 - a^2 = 0$
42. $4x^2 - 36x + 81 = 0$
43. $x^2 - 2x - 1 = 0$
44. $x^2 - 4x + 1 = 0$
45. (a) $\frac{9}{4}$ (b) $\frac{3}{512}$
46. (a) $x^2 - 9x - 5 = 0$
    (b) $8x^2 - 40x + 5 = 0$
47. (a) $x^2 - 13x + 36 = 0$
    (b) $6x^2 - 5x + 1 = 0$
    (c) $6x^2 - 13x + 6 = 0$
    (d) $x^2 - 35x + 216 = 0$
    (e) $144x^2 + 468x + 16 = 0$
    (f) $6x^2 - 35x + 30 = 0$

48. $x = 3$ or -3

49. 0 or -2

50. $x = 2$ or -1

51. $x = 9$ and -18 is extraneous root

52. $x = 4$, $x = -21$ is extraneous root

53. $x = \dfrac{-7 \pm \sqrt{33}}{4}$

54. $x = 5$ or 6

55. Roots are distinct and rational.

56. Roots are distinct and rational.

57. Roots are distinct and irrational.

58. Roots are not real.

59. Roots are not real.

60. $k = \pm 12$

62. $\dfrac{a^3 - 3ab}{b^3}$

63. $x^2 - 2px + p^2 - q = 0$

65. $m = 8, 9$

66. Length = 12 cm

Breadth = 5 cm

67. $x = 35$ breadth-wise

$x = 40$ length-wise

68. $n = 2$

5 and 7

69. 50 km/h

70. Time taken by the smaller pipe = 75 min

Time taken by the larger pipe = 60 min

71. 36

72. 5 km/h

73. 30 km/h

74. 3 h 30 min

75. $x = 2, -6$

76. $x = 0.73, -2.73$

77. $x = 1, 3$

78. $x = 2, -3$

79. $x = 3, -0.5$

80. 21 is the minimum value when $x = -1$

81. $-\dfrac{88}{9}$ is the minimum value when $x = \dfrac{+4}{9}$

82. -35 is the minimum value when $x = \dfrac{3}{4}$

83. (B) $4^x - 3(2^{x+3}) + 128 = 0$

$\Rightarrow 2^{2x} - 3 \times 2^x \times 2^3 + 128 = 0$

$\Rightarrow 2^{2x} - 24 \times 2^x + 128 = 0$

$\Rightarrow 2^x = y$

$\Rightarrow y^2 - 24y + 128 = 0$

$\Rightarrow (y - 16)(y - 8) = 0$

$\Rightarrow y = 16, 8$

$\therefore x = 4, 3$

84. (D) $f(x) = x^2 - 5x + k$

Since α and β are the roots, we have

$f(\alpha) = \alpha^2 - 5\alpha + k = 0$ ...(i)

$f(\beta) = \beta^2 - 5\beta + k = 0$ ...(ii)

From (i) – (ii), we get

$\alpha^2 - \beta^2 - 5\alpha + 5\beta = 0$

$\Rightarrow (\alpha - \beta)(\alpha + \beta) - 5(\alpha - \beta) = 0$

$\Rightarrow (\alpha - \beta)(\alpha + \beta - 5) = 0$

$\Rightarrow 1 \times (\alpha + \beta - 5) = 0$ (Given that $\alpha - \beta = 1$)

$\Rightarrow \alpha + \beta = 5$

$\therefore \alpha = 3$ and $\beta = 2$

85. (B) $P(x) = ax^2 + bx + c$

$D_1 = b^2 - 4ac$

$Q(x) = -ax^2 + dx + c$

$D_2 = d^2 + 4ac$

Since $ac \neq 0$, ac is either positive or negative. Then at least one of D_1 and D_2 is positive.

86. (A) $(q - r)x^2 + (r - p)x + (p - q) = 0$

$D = (r - p)^2 - 4(q - r)(p - q) = 0$

$= (r^2 + p^2 - 2pr) - 4(pq - q^2 - rp + qr)$

$= r^2 + p^2 - 2pr - 4pq + 4q^2 + 4pr - 4qr$

Since the roots are equal,

$D = 0$

$\Rightarrow r^2 + p^2 + 2pr + 4q^2 - 4pq - 4qr = 0$

$\Rightarrow r^2 + p^2 + 4q^2 + 2pr - 4pq - 4qr = 0$

$\Rightarrow (r)^2 + (p)^2 + (-2q)^2 + 2pr + 2(p)(-2q) + 2(-2q)(r) = 0$

$\Rightarrow (r + p - 2q)^2 = 0$

$\Rightarrow r + p = 2q$

$\Rightarrow q = \dfrac{r + p}{2}$

87. (B) $3p^2 - 5p - 2 = 0$ and $3q^2 - 5q - 2 = 0$

From the above equations, it is clear that p and q are the roots of the equation $3x^2 - 5x - 2 = 0$.

Product of roots $= pq = -2/3$

88. (C) $x^2 + bx - c = 0$

$\alpha + \beta = -b$

$\alpha\beta = -c$

Now, the equation having b and c as its roots will be

$x^2 - (b + c)x + bc = 0$

$x^2 + \{(\alpha + \beta) + \alpha \cdot \beta\} + (\alpha + \beta)\alpha\beta = 0$

89. (D) $px^2 + qx + 1 = 0$

For real roots

$q^2 - 4p \geq 0$

90. (B) $(x + 2)(x + 3)(x + 8)(x + 12) = 4x^2$

$(x + 2)(x + 12)(x + 3)(x + 8) = 4x^2$

$(x^2 + 14x + 24)(x^2 + 11x + 24) = 4x^2$

$$\left(x+14+\frac{24}{x}\right)\left(x+11+\frac{24}{x}\right)=4$$

Put $x+\frac{24}{x}=y$

Then $y=-15,-10$

$\therefore \quad x=\frac{-15-\sqrt{129}}{2},\frac{-15+\sqrt{129}}{2},-6,-4$

91. (A) $(6 - x)^4 + (8 - x)^4 = 16$

Put $y=\frac{6-x+8-x}{2}=7-x$

$y = 7 - x$

$\Rightarrow x = 7 - y$

Then, $y^4 + 6y^2 - 7 = 0$

$\Rightarrow \quad (y^2 - 1)(y^2 + 7) = 0$

$y = \pm 1$

$\therefore \quad x = 6, 8$

92. (B) $\sqrt{2x^2+5x-2}-\sqrt{2x^2+5x-9}=1$

$u^2 = 2x^2 + 5x - 2$

$v^2 = 2x^2 + 5x - 9$

then $u - v = 1$

$\Rightarrow \quad u^2 - v^2 = (2x^2 + 5x - 2) - (2x^2 + 5x - 9) = 7$

$\Rightarrow \quad (u - v)(u + v) = 7$

$\Rightarrow \quad u + v = 7$

then, $u = 4, v = 3$

then, $u^2 = 2x^2 + 5x - 2$

$\Rightarrow \quad 16 = 2x^2 + 5x - 2$

$\Rightarrow \quad 2x^2 + 5x - 18 = 0$

$\Rightarrow \quad 2x^2 - 4x + 9x - 18 = 0$

$\Rightarrow \quad 2x(x - 2) + 9(x - 2) = 0$

$\therefore \quad x = -9/2, 2$

93. (D) $64 \times 9^x - 84 \times 12^x + 27 \times 16^x = 0$

$$\Rightarrow \quad 64\left(\frac{9}{12}\right)^x-84+27\left(\frac{16}{12}\right)^x=0$$

$$\Rightarrow \quad 64\left(\frac{3}{4}\right)^x-84+27\left(\frac{4}{3}\right)^x=0$$

Let $\left(\frac{3}{4}\right)^x=y.$

Then $64y-84+\frac{27}{y}=0$

$\Rightarrow \quad y=\frac{3}{4},\frac{9}{16}$

94. (A) Let r be the common ratio of the G.P.. Then $\beta = \alpha r$, $\gamma = \alpha r^2$ and $\delta = \alpha r^3$

$\alpha + \beta = 1$

$\Rightarrow \alpha(1 + r) = 1$

$\alpha\beta = p$

$\Rightarrow \quad \alpha(\alpha r) = p$

$\Rightarrow \quad \alpha^2 r = p$

$\gamma + \delta = 4$

$\Rightarrow \quad \alpha r^2 + \alpha r^3 = 4$

$\Rightarrow \quad \alpha r^2(1 + r) = 4$

$\gamma\delta = q$

$\Rightarrow \quad (\alpha r^2)(\alpha r^3) = q$

$\Rightarrow \quad \alpha^2 r^5 = q$

$r^2 = 4$

$\therefore \quad r = 2, -2$

By substituting value of r, we get $p = -2$, $q = -32$.

95. (D) $x^2 - 3x + 5 = 0$

$\alpha + \beta = 3, \alpha \cdot \beta = 5$

$x^2 + 5x - 3 = 0$

$\gamma + \delta = -5, \gamma \cdot \delta = -3$

Equation having $(\alpha\gamma + \beta\delta)$ and $(\alpha\gamma + \beta\delta)$ as the roots is

$$x^2 - \{(\alpha\gamma + \beta\delta) + (\alpha\delta + \beta\gamma)\}x + (\alpha\gamma + \beta\delta)(\alpha\delta + \beta\gamma) = 0$$

96. (D) $\cos^4\theta - \sin^4\theta = (\cos^2\theta)^2 - (\sin^2\theta)^2$

$= (\cos^2\theta - \sin^2\theta)(\cos^2\theta + \sin^2\theta)$

$= \cos 2\theta$

$(\cos^4\theta + p) - (\sin^4\theta + p) = (\cos^2\theta + q) - (\sin^2\theta + q)$

$\Rightarrow \sqrt{4a^2 - 4a} = \sqrt{16-8}$

$\Rightarrow 4a^2 - 4a = 8$

$\Rightarrow a^2 - a - 2 = 0$

$\Rightarrow (a-2)(a+1) = 0$

$\therefore \quad a = -1, 2$

97. (A) $x^2 - px - p - c = 0$

$\Rightarrow \alpha + \beta = p$ and $\alpha\beta = -p - c$

$(\alpha+1)(\beta+1) = \alpha\beta + \alpha + \beta + 1$

$= -p - c + p + 1$

$= 1 - c$

$$\frac{\alpha^2+2\alpha+1}{\alpha^2+2\alpha+c} + \frac{\beta^2+2\beta+1}{\beta^2+2\beta+c}$$

$$= \frac{(\alpha+1)^2}{(\alpha+1)^2 - (1-c)} + \frac{(\beta+1)^2}{(\beta+1)^2 - (1-c)}$$

$$= \frac{\alpha+1}{\alpha-\beta} + \frac{\beta+1}{\beta-\alpha}$$

$$= \frac{\alpha-\beta}{\alpha-\beta}$$

$= 1$

98. (C) $\dfrac{x_1}{x_1-1} + \dfrac{x_2}{x_2-1} = 2$

$\Rightarrow x_1(x_2-1) + x_2(x_1-1) = 2(x_1-1)(x_2-1)$

$\Rightarrow x_1x_2 - x_1 + x_1x_2 - x_2 = 2(x_1x_2 - x_2 - x_1 + 1)$

$\Rightarrow 2x_1x_2 - x_1 - x_2 = 2x_1x_2 - 2x_2 - 2x_1 + 2$

$\Rightarrow x_1 + x_2 = 2$

The required equation is $x^2 - 2x + 4 = 0$.

99. (D) Even prime = 2

$f(x) = x^2 - dx + 12 = 0$

$f(2) = 4 - 2d + 12 = 0$

$\Rightarrow 2d = 16$

$\Rightarrow d = 8$

$x^2 + dx + \mu = 0$

$\Rightarrow x^2 + 8x + \mu = 0$

For equal roots

$64 - 4(1)(\mu) = 0$

$\Rightarrow \mu = 16$

100. (B) Since parabola is downward, $a < 0$. Also, y intercept of the equation is negative i.e., $c < 0$.

Therefore $\dfrac{c}{a} > 0$.

101. (B)

$x + \dfrac{1}{x} = 5$

$$x^3 + \frac{1}{x^3} - 5\left(x^2 + \frac{1}{x^2}\right) + \left(x + \frac{1}{x}\right)$$

$$\Rightarrow \left[\left(x+\frac{1}{x}\right)^3 - 3\left(x+\frac{1}{x}\right)\right] - 5\left[\left(x+\frac{1}{x}\right)^2 - 2.x.\frac{1}{x}\right] + \left(x+\frac{1}{x}\right)$$

$$\Rightarrow \left[\left(x+\frac{1}{x}\right)^3 - 3\left(x+\frac{1}{x}\right)\right] - 5\left[\left(x+\frac{1}{x}\right)^2 - 2\right] + \left(x+\frac{1}{x}\right)$$

$\Rightarrow \left[(5)^3 - 3(5)\right] - 5\left[(5)^2 - 2\right] + 5$

$\Rightarrow 125 - 15 - 5(25-2) + 5$

$\Rightarrow 110 - 5\times 23 + 5$

$\Rightarrow 110 + 5 - 115 = 0$

102. (C) $(a^2-1)x^2 + (a-1)x + (a^2-4a+3) = 0$ is an identity in x only when

$a^2 - 1 = 0$ and $a - 1 = 0$ and $(a-1)(a-3) = 0$

$a = 1, -1$ and $a = 1$ and $a = 1, 3$

Hence, $a = 1$

103. (A) $ax^2 - bx(x-1) + c(x-1)^2 = 0$

$$a\left(\frac{x}{x-1}\right)^2 - b\left(\frac{x}{x-1}\right) + c = 0$$

$$a\left(\frac{x}{1-x}\right)^2 + b\left(\frac{x}{1-x}\right) + c = 0$$

$\alpha = \dfrac{x}{1-x}$ and $\beta = \dfrac{x}{1-x}$

$\alpha - \alpha x = x$ and $\beta - \beta x = x$

$\alpha = x + \alpha x$ and $\beta = x + \beta x$

$x = \dfrac{\alpha}{1+\alpha}$ and $x = \dfrac{\beta}{1+\beta}$

104. (C) $(2x-1)^{1/3}+(x-1)^{1/3}=1$

Cubing both sides, we get

$$(2x-1)+(x-1)+3(2x-1)^{1/3}(x-1)^{1/3}\left\{(2x-1)^{1/3}+(x-1)^{1/3}\right\}=1$$

$$(3x-2)+3(2x-1)^{1/3}(x-1)^{1/3}=1$$

$$3\{(2x-1)(x-1)\}^{1/3}=1-(3x-2)$$

$$3\{2x^2-3x+1\}^{1/3}=1-3x+2$$

$$3(2x^2-3x+1)^{1/3}=3(1-x)$$

$$(2x^2-3x+1)^{1/3}=(1-x)$$

Cubing both sides, we get

$$(2x^2-3x+1)=(1-x)^3$$

$$2x^2-3x+1=1-x^3-3\times1\times x(1-x)$$

$$2x^2-3x+1=1-x^3-3x+3x^2$$

$$x^3-x^2=0$$

$$\Rightarrow\quad x^2(x-1)=0$$

$$\therefore\quad x=0,1$$

But $x = 0$ does not satisfy the original equation.

105. (A) $\sqrt{x+1}-\sqrt{x-1}=\sqrt{4x+1}$

Squaring both sides of the equation, we get

$$(x+1)+(x-1)-2\sqrt{x^2-1}=4x-1$$

$$\Rightarrow\quad 2x-2\sqrt{x^2-1}=4x-1$$

$$\Rightarrow\quad -2\sqrt{x^2-1}=2x-1$$

Again squaring, we get

$$4(x^2-1)=4x^2+1-4x$$

$$\Rightarrow\quad 4x^2-4=4x^2+1-4x$$

$$\Rightarrow\quad 4x=5$$

$$\therefore\quad x=\frac{5}{4}$$

But this does not satisfy the original equation.

Hints and answers to Olympiad and NTSE level exercises

106. (C) $(b+c-2a)x^2+(c+a-2b)x+(a+b-2c)=0$

Disc. $=(c+a-2b)^2-4(b+c-2a)(a+b-2c)$

$$=c^2+a^2+4b^2+2ac-4ab-4bc-4(ab+b^2-2bc+ac+bc-2c^2-2a^2-2ab+4ac)$$

$$=c^2+a^2+4b^2+2ac-4ab-4bc-4ab-4b^2+8bc-4ac-4bc+8c^2+8a^2+8ab-16ac$$

$$=9a^2+9c^2-18ac$$

$$=9(a-c)^2$$

$$=\{3(a-c)\}^2$$

Here, $\{3(a-c)\}^2$ is a perfect square.

107. (A) $(12x-1)(6x-1)(4x-1)(3x-1)=5$

$$\Rightarrow\quad [(12x-1)(3x-1)][(6x-1)(4x-1)]=5$$

$$\Rightarrow\quad (36x^2-15x+1)(24x^2-10x+1)=5$$

$$\Rightarrow\quad \left(12x^2-5x+\frac{1}{3}\right)\left(12x^2-5x+\frac{1}{2}\right)=\frac{5}{6}$$

Putting $y=12x^2-5x$, we get

$$\left(y+\frac{1}{3}\right)\left(y+\frac{1}{2}\right)=\frac{5}{6}$$

$$\Rightarrow\quad y=\frac{1}{2},-\frac{4}{3}$$

Now, $12x^2-5x=\frac{1}{2}$

$$\Rightarrow\quad x=\frac{1}{2},-\frac{1}{12}$$

$12x^2-5x=-\frac{4}{3}$ has no real roots

108. (D) Let

$$\frac{1-\alpha}{1+\alpha}=x$$

$$\Rightarrow\quad \alpha=\frac{1-x}{1+x}$$

then, $3\left(\frac{1-x}{1+x}\right)^2+2\left(\frac{1-x}{1+x}\right)+1=0$

$$\Rightarrow\quad x^2-2x+3=0$$

So, $\frac{1-\alpha}{1+\alpha}+\frac{1-\beta}{1+\beta}=2$

And $\left(\frac{1-\alpha}{1+\alpha}\right)\left(\frac{1-\beta}{1+\beta}\right)=3$

$$\left(\frac{1-\alpha}{1+\alpha}\right)^3+\left(\frac{1-\beta}{1+\beta}\right)^3$$

$$= \left[\left(\frac{1-\alpha}{1+\alpha}\right)+\left(\frac{1-\beta}{1+\beta}\right)\right]^3 - 3\left(\frac{1-\alpha}{1+\alpha}\right)\left(\frac{1-\beta}{1+\beta}\right)\left[\frac{1-\alpha}{1+\alpha}+\frac{1-\beta}{1+\beta}\right]$$

$$= (2)^3 - 3(3)(2)$$

$$= 8 - 18$$

$$= -10$$

109. (A) $x^2 + px + 45 = 0$

Let α and β be the roots of the equation.

$\alpha + \beta = -p$ and $\alpha\beta = 45$

$(\alpha - \beta)^2 = 144$

$\Rightarrow \alpha^2 + \beta^2 - 2\alpha\beta = 144$

$\Rightarrow (\alpha + \beta)^2 - 4\alpha\beta = 144$

$\Rightarrow p^2 - 4 \times 45 = 144$

$\Rightarrow p^2 = 144 + 180 = 324$

$\therefore p = \pm 18$

110. (B)

Statement 1:

We have $x^{1/8} = (3x^4 + 4)^{1/64}$

$\Rightarrow x^8 = 3x^4 + 4$

$\Rightarrow x = \sqrt{2}$

Statement 2:

$a^2 \geq 8b$ and $4b^2 \geq 4a$

Now $b^2 \geq a$

$\Rightarrow b^4 \geq a^2 \geq 8b$ $\quad (a > 0,\ b > 0)$

$\therefore b^3 \geq 8$

$\Rightarrow b \geq 2$ ....(1)

Again $a^2 \geq 8b$ and $b \geq 2$

$a^2 \geq 16$

$\Rightarrow a \geq 4$ ....(2)

From (1) and (2) $(a + b)_{least} = 6$

111. (D) $f(x, y) = (x - 2)^2 + (y - 1)^2 = 0$

$\Rightarrow$ x = 2 and y = 1

$$\therefore E = \frac{(\sqrt{2}-1)^2 + 4\sqrt{2}}{2+\sqrt{2}} = \frac{(\sqrt{2}+1)^2}{\sqrt{2}(\sqrt{2}+1)} = \frac{\sqrt{2}+1}{\sqrt{2}}$$

112. (B) Since $x = 4$ is a root of the equation

$x^2 + px + 12 = 0$,

$\therefore 16 + 4p + 12 = 0$

$\Rightarrow p = -7$

The equation $x^2 + px + q = 0$ has equal roots.

Therefore $p^2 = 4q$

$\Rightarrow 49 = 4q$

$\Rightarrow q = 49/4$

113. (A) Here $\alpha^2 + \beta^2 = (\alpha + \beta)^2 - 2\alpha\beta = (2)^2 - 2(4) = -4$

$\alpha^3 + \beta^3 = (\alpha + \beta)^3 - 3\alpha\beta(\alpha + \beta)$

$= (2)^3 - 3(4)(2) = -16$

Now, $(\alpha^2 + \beta^2)(\alpha^3 + \beta^3) = \alpha^5 + \beta^5 + \alpha^2\beta^2(\alpha + \beta)$

$\Rightarrow (-4)(-16) = \alpha^5 + b^5 + (16)(2)$

$\Rightarrow \alpha^5 + \beta^5 = 32$

114. a. → q., r., s.; b → r.; c. → p.; d. → q.

(a) $d + a - b = 0$ and $d + b - c = 0$

$\therefore b - a = c - b$

$\Rightarrow 2b = a, b, c$ are in A.P.

also $x = 1$ satisfies the 2nd equation

Therefore, other root is also 1.

$\therefore$ Product of roots = 1

$\therefore c(a - b) = a(b - c)$

$\Rightarrow b = \dfrac{2ac}{a + c}$

$\Rightarrow a, b, c$ are in H.P.

$\therefore a, b, c$ are in A.P. and a, b, c in H.P.

$\Rightarrow a, b, c$ are in G.P.

(b) $2\log b = \log a + \log c$

$\Rightarrow b^2 = ac$

$\Rightarrow a, b, c$ are in G.P.

(c) $(x - 1)^3 = 0$

$\Rightarrow x = 1$ is the common root

hence, $a + b + c = 0$

(d) $D = (a + c)^2 + 4b^2 - 4b(a + c) \leq 0$

$\Rightarrow ((a + c) - 2b)^2 \leq 0$

$\Rightarrow a + c = 2b$

$\Rightarrow a, b, c$ in A.P.

115. a. → p., b. → p., q., r., s., c. → p., q., s., d. → r., s.

(a) $y = \dfrac{x^2 - 2x + 4}{x^2 + 2x + 4}$

$\Rightarrow x^2y + 2xy + 4y = x^2 - 2x + 4$

$\Rightarrow \quad (y-1)\,x^2 + 2(y+1)\,x + 4\,(y-1) = 0$

As $\quad x$ is real $D \geq 0$

$\Rightarrow \quad 4(y+1)^2 - 16(y-1)^2 \geq 0$

$\Rightarrow \quad (y+1)^2 - (2y-2)^2 \geq 0$

$\Rightarrow \quad (3y-1)\,(3-y) \geq 0$

$\Rightarrow \quad (3y-1)\,(y-3) \leq 0$

$\Rightarrow \quad y \in [1/3, 3]$

(b) by similar method for $y = \dfrac{x^2 - 3x - 2}{2x - 3}$,

we get $y \in R$

(c) For $y = \dfrac{2x^2 - 2x + 4}{x^2 - 4x + 3}$,

we get $y \geq 1$ or $y \leq -7$

(d) $f(x) = x^2 - (a-3)x + 2 < 0, \forall x \in [-2, -1]$

$\Rightarrow \quad f(-2) < 0$ and $f(-1) < 0$

$\Rightarrow \quad 4 + 2(a-3) + 2 < 0$ and $1 + (a-2) + 2 < 0$

$\Rightarrow \quad a < 0$ and $a < -1$

$\Rightarrow \quad a < -1$

$\Rightarrow \quad a \in \{-10, -3\}$

Relations and Functions

1. $\{(5, 1), (3, 2), (1, 3)\}$
2. True
3. $\{(3, 0), (3, 1), (3, 3), (5, 0), (5, 1), (5, 3)\}$
4. $\{(-2, 3), (-2, 4), (-2, 5), (0, 3), (0, 4), (0, 5), (2, 3), (2, 4), (2, 5), (4, 3), (4, 4), (4, 5)\}$
5. $\{(1, 1), (1, 2), (2, 1), (3, 1)\}$
6. f is not a function from A to B.
7. (a) $f = \{(1, 2), (3, 6), (5, 10)\}$

   (b) Range = $\{2, 6, 10\}$, 10

   (c) It is one-to-one relation
8. (a) $\dfrac{2}{3}$ (b) $\dfrac{2}{3}$
9. One-to-one
10. Many-to-one
11. Many-to-one
12. $(2, c)$

    $A = \{2, 3\}$, $B = \{a, b, c\}$
13. 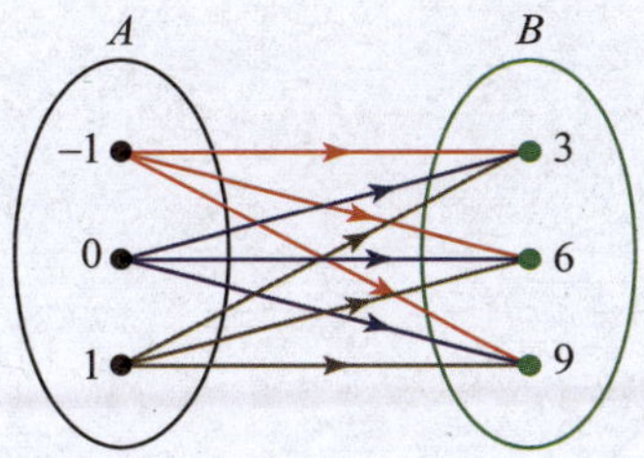

14. $\{(3, 2), (5, 2), (5, 4)\}$

    Domain = $\{3, 5\}$

    Range = $\{2, 4\}$
15. (a) -11 (b) $\dfrac{5}{2}, \dfrac{-5}{2}$
16. $S = \{(-2, 5), (-1, -1), (0, -3), (1, -1), (2, 5)\}$

    Yes

    Range = $\{-3, -1, 5\}$ many-to-one
17. 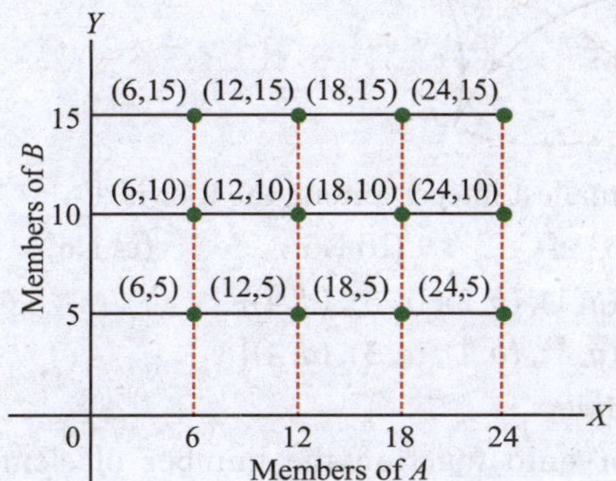

18. (a) 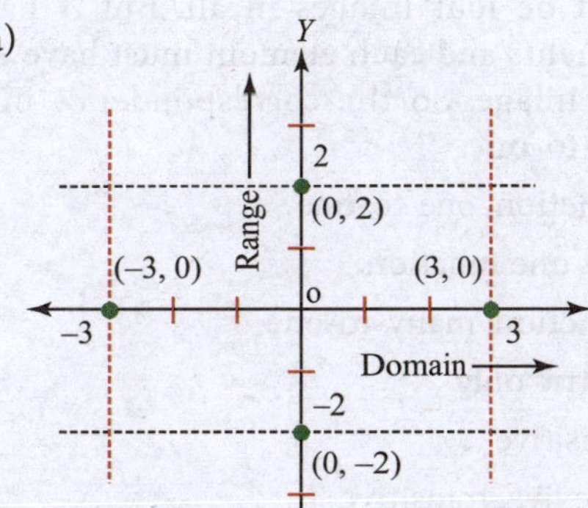

    (b) 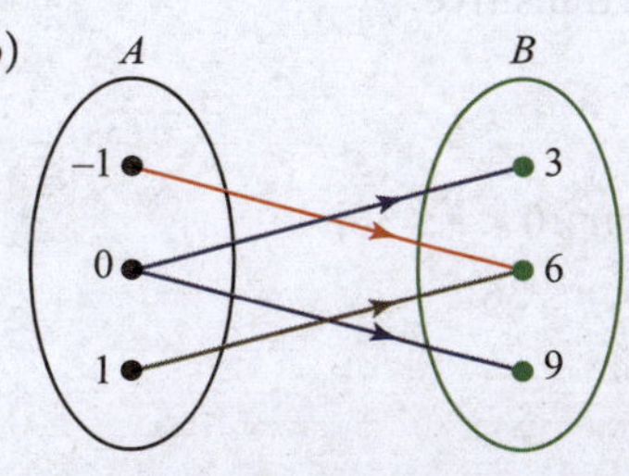

19.

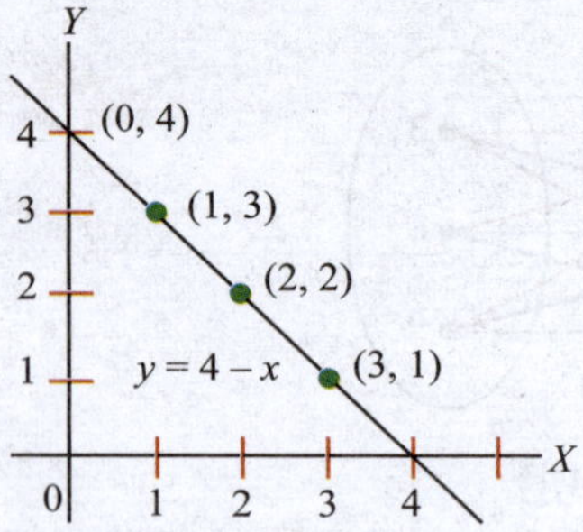

20. (a) {2, 4, 5}

(b) {7, 8, 9}

(c) {(2, 7), (2, 8), (2, 9), (4, 7), (4, 8), (4, 9), (5, 7), (5, 8), (5, 9)}

21.

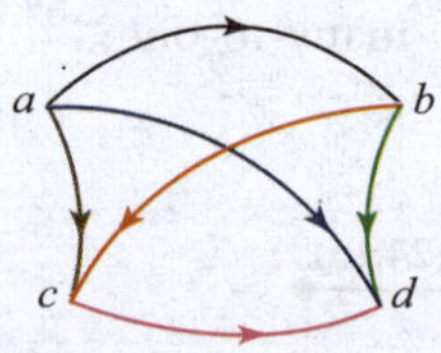

The smallest integer among the four is d.

22. (a) No (b) No (c) No

23. (a) {(*a*, 1), (*b*, 2), (*c*, 3), (*d*, 4)}

(b) {(*a*, 1), (*b*, 1), (*c*, 3), (*d*, 4)}

(c) *Hint:*

For onto function the number of elements in the range should be four. Hence, there must be four images in all. But *A* has four elements and each element must have exactly one image. So the correspondence must be one-to-one.

24. (a) Function, one-to-one

(b) Not one function

(c) Function, many-to-one

25. Symmetric only

26. (a) Transitive

(b) Reflexive, transitive

(c) Reflexive, transitive

(d) Transitive

27. (A, D)

28. $\{(r, V): V = 4\pi r^2, 0 < r \leq 36\}$

Domain $= 0 < r \leq 36$

It is a function

29. (A) Many–one

(B) Many–one

(C) Not a function

(D) Not a function

30. 6

31. (B) $a^2 + ab + b^2 = 8$ and $b = 2$

$\Rightarrow \quad a^2 + 2a + 4 = 8$

$\Rightarrow \quad a^2 + 2a - 4 = 0$

$$\therefore \quad a = \frac{-2 \pm \sqrt{4 - 4(1)(-4)}}{2(1)} = \frac{-2 \pm 2\sqrt{5}}{2} = -1 \pm \sqrt{5}$$

32. (A)

33. (D) $A \times B$ is a set of those ordered pairs in which first element belongs to set *A* and second element belongs to set *B*.

34. (B) Elements of set $A = \{1, 2, 3\}$ and B={3, 5}

$A \times B = \{(1, 3), (1, 5), (2, 3), (2, 5), (3, 3), (3, 5)\}$

35. (C) Clearly, the elements of $A \times A$ should be taken from $A = \{-1, 0, 1\}$.

36. (D) $A = \{-1, 0, 1\}$

$A \times A = \{(-1, -1), (-1, 0), (-1, 1), (0, -1), (0, 0), (0, 1), (1, -1), (1, 0), (1, 1)\}$

37. (C) Domain of relation is the set of those elements of set *A* which have image in B.

38. (A) $R \subseteq A \times B$

Number of subsets of $(A \times B) = 2^{2\times 4} = 2^8$

39. (A) Clearly, R={(1, 8), (2, 6), (3, 4), (4, 2)}

40. (B) Prime number less than 8 are {2, 3, 5, 7}.

41. (D) $R = \{(a, b) : a, b \in Z \text{ and } a - b \in Z\}$

(i) Clearly, $(a, a) \in R$ for all $a \in Z$ such that $a - a \in Z$. So, *R* is reflexive.

(ii) For $(a, b) \in R$, we have $(b, a) \in R$ such that $b - a \in Z$. So, *R* is symmetric.

(iii) $(a, b) \in R \Rightarrow a - b \in Z$

$(b, c) \in R \Rightarrow b - c \in Z$

$(a, c) \in R \Rightarrow a - c = (a - b) + (b - c) \in Z$

So, R is transitive.

42. (A) If S_1 is perpendicular to S_2 then S_2 is also perpendicular to S_1.

So, $(S_1, S_2) \in R \Rightarrow (S_2, S_1) \in R$

43. (D)

(i) reflexive — any triangle is congruent to itself.

(ii) symmetric — If S_1 is congruent to S_2 then S_2 is also congruent to S_1.

(iii) transitive — If S_1 is congruent to S_2 and S_2 is congruent to S_3 then S_1 is also congruent to S_3.

44. (B) $f(x) = 2x + 3a$

$f(1) = 2 + 3a$

$f(2) = 4 + 3a$

$f(1) = 2f(2)$

$\Rightarrow \quad 2 + 3a = 8 + 6a$

$\Rightarrow \quad -6 = 3a$

$\Rightarrow \quad a = -2$

45. (C) $f(x) = ax + b$

$(-1, 3) \in f$

$\Rightarrow \quad f(-1) = -a + b = -3$

$\Rightarrow \quad b - a = -3$

And $(0, -1) \in f$.

$f(x) = ax + b$

$f(0) = b = -1$

Then, $a = b + 3$

$\Rightarrow \quad a = -1 + 3$

$\Rightarrow \quad a = 2$

46. (D)

$$f(2) = \frac{2}{2^2 - 1} = \frac{2}{3}$$

$$f\left(\frac{-1}{2}\right) = \frac{\frac{-1}{2}}{\left(\frac{-1}{2}\right)^2 - 1} = \frac{\frac{-1}{2}}{\frac{1}{4} - 1} = \frac{\frac{-1}{2}}{\frac{-3}{4}} = \frac{2}{3}$$

47. (C) Many-one $\rightarrow$ many to one corresponds

Into $\rightarrow$ range $\neq$ co-domain

48. (D) If $a \in R \Rightarrow (a, a) \in R$; So, R is reflexive

If $(a, b) \in R \Rightarrow (b, a) \in R$; So, R is symmetric

If $(a, b) \in R$ and $(b, c) \in R \Rightarrow (a, c)$; So, R is transitive

49. (B) $f(5) = 5^3 = 125$

$f(1) = 1^3 = 1$

$$\frac{f(5) - f(1)}{5 - 1} = \frac{125 - 1}{4} = \frac{124}{4} = 31$$

Hints and answers to Olympiad and NTSE level exercises

50. (C) $\dfrac{f(a)}{f(a+1)} = \dfrac{a/(a-1)}{(a+1)/a} = \dfrac{a^2}{a^2 - 1} = f(a^2)$.

51. (D) $y = \dfrac{ax + b}{cx - a}$

$\Rightarrow \quad x(cy - a) = b + ay$

$\Rightarrow \quad x = \dfrac{ay + b}{cy - a} = f(y)$.

52. (B) $x^2 - 6x + 7 = (x - 3)^2 - 2$

Obviously, minimum value is -2 and maximum is ∞. Hence, range of function is $[-2, \infty]$.

53. (C) $f(x)$ is to be defined when $x^2 - 1 > 0$

$\Rightarrow \quad x^2 > 1$

$\Rightarrow \quad x < -1$ or $x > 1$ and $3 + x > 0$

$\therefore \quad x > -3$ and $x \neq -2$

$\therefore \quad D_f = (-3, -2) \cup (-2, -1) \cup (1, \infty)$.

54. (D) $f(x) = \sqrt{x - x^2} + \sqrt{4 + x} + \sqrt{4 - x}$

Clearly, $f(x)$ is defined,

if $\quad 4 + x \geq 0 \quad \Rightarrow \quad x \geq -4$

$4 - x \geq 0 \quad \Rightarrow \quad x \leq 4$

$x(1 - x) \geq 0 \quad \Rightarrow \quad x \geq 0$ and $x \leq 1$

$\therefore \quad$ Domain of $f = (-\infty, 4) \cap [-4, \infty) \cap [0, 1]$

$= [0, 1]$.

55. (D) $1 - \dfrac{1}{x} > 0 \Rightarrow x > 1$. Also, $x \neq 0$.

$\therefore \quad$ Required interval $= (-\infty, 0) \cup [1, \infty)$.

56. (A) $F[f(x)] = F(\log_a x) = a^{\log_a x} = x$

$f[F(x)] = f(a^x) = \log_a a^x = x \log_a a = x$.

57. (D) Clearly $(f + g)(x) = f(x) + g(x) = 2$

$\forall x \in R =$ constant function

Hence, the function $(f + g)(x)$ defined on $R \rightarrow R$ is neither surjective nor injective.

58. (C) Let $y = \dfrac{x^2}{x^2+1}$

$\Rightarrow \quad (y-1)x^2 + 0x + y = 0, y \neq 1$

for real values of x.

We have $D \geq 0$

$\Rightarrow \quad -4y(y-1) \geq 0$

$\Rightarrow \quad y(y-1) \leq 0$

$\Rightarrow \quad y \in [0, 1)$

$0 \leq \dfrac{x^2}{x^2+1} < 1$

59. (C) $f(x) = \log(\sqrt{x-4} + \sqrt{6-x})$

$\Rightarrow \quad x - 4 \geq 0$

and $6 - x \geq 0$

$\Rightarrow \quad x \geq 4$

and $x \leq 6$

$\Rightarrow \quad$ Domain of $f(x) = [4, 6]$

Progressions

1. (a) $\dfrac{555}{2}$ (b) 153

(c) 0 (d) 90

2. $d = 3$

3. $S_n = (n+1)n$

4. $a = 5$

5. $S_{15} = 495$

6. $S_p = \left\{\dfrac{1}{a} + \dfrac{p}{a} + 2b\right\}\dfrac{p}{2}$

7. (a) $a + 42d$

(b) 115

(c) 101

8. (a) $n = 17 \Rightarrow t_{17} = 49$

(b) $n = 17 \Rightarrow t_{17} = 40$

(c) $n = 8 \Rightarrow t_8 = -19$

9. (a) $a = 45, d = -3, n = 30$

(b) $a = -15, d = \dfrac{3}{2}, n = 21$

(c) $a = n + m - 1, d = -1, n = m + n$

10. (a) $S_{40} = 2380$

(b) $S_{20} = 1440$

(c) $S_n = n[(a-b)^2 + (n-1)ab]$

11. 12

13. 3, 5, 7 or 7, 5, 3

14. (a) $S_n = \dfrac{-1}{x^2}\left(\dfrac{1-(1+x^2)^n}{(1+x^2)^n}\right)$

(b) $S_n = \dfrac{\left(\dfrac{x}{1+x^2}\right)^n - 1}{\left(\dfrac{x}{1+x^2}\right) - 1}$

15. 8th term $\dfrac{3^6}{2^7}$

16. 8th term = 384

17. 10th term $= \dfrac{1}{2}$

16th term $= \dfrac{1}{128}$

18. (a) HP since 1, 4, 7, 10 … forms an AP.

(b) Not an HP since 3, 6, 12, 24 … does not form an AP.

(c) Not an HP since 1, 2, 7, 5 does not form an AP.

19. (a) $T_n = \dfrac{1}{a+(n-1)d}$, $T_{15} = \dfrac{1}{a+41d}$

(b) $T_n = \dfrac{12}{21+4(n-1)}$, $T_{31} = \dfrac{12}{141}$

(c) $T_n = \dfrac{14}{7+(n-1)}$, $T_{10} = \dfrac{7}{8}$

20. (a) nth term $= \dfrac{\sqrt{2}}{2^{n-1}}$

$t_{10} = \dfrac{\sqrt{2}}{512}$

(b) nth term = $(0.3)(0.2)^{n-1}$

$t_8 = (0.3)(0.2)^7$

(c) nth term $= (2^{3-n})$

$t_{10} = 2^{-7}$

(d) nth term $= \left(\dfrac{1}{10}\right)^{n-1}$

$t_{20} = \left(\dfrac{1}{10}\right)^{19}$

21. (a) The first term is $a = \sqrt{3}$

Number of terms n is 12

(b) The first term is $\frac{5}{4}$

Number of terms is 12

(c) $n = 9$

22. (a) $S_{10} = 121(\sqrt{3}+1)$

(b) $S_{12} = \frac{4^{12}-1}{-5} = -3355443$

(c) $S_{16} = \frac{2}{3}\left[1-\left(\frac{1}{10}\right)^{16}\right]$

(d) $S_{11} = \frac{3}{2}(1-3^{-11})$

23. $S_n = \frac{n}{2}\left[\frac{2(2a^2-1)+2(n-1)(a^2-1)}{a}\right]$

26. $r = 2$

27. Numbers are 8, 12, 18 or 18, 12, 8

29. $S_n = \frac{1-a^n}{(1-a)^2} - \frac{na^{n-1}}{1-a}$

30. $S_x = \frac{1}{1-x} + \frac{2x(1-x^n)}{(1-x)^2}$

33. HM between 2 and 10 is $\frac{10}{3}$

34. $\Sigma n^4 = \frac{n^5}{5} + \frac{n^2(n+1)^2}{2} - \frac{n(n+1)(2n+1)}{3} + \frac{n(n+1)}{2} - \frac{n}{5}$

36. $S_n = n(n+2)(2n^2+4n-1)$

37. $n = 0$

38. $n = -\frac{1}{2}$

39. 4284

40. (b) (i) 6

(ii) ab

(iii) 8

41. (a) $\frac{1}{a}$ (b) $\frac{9}{2}$ (c) 3

42. (C) $a_3 = a_2 + a_1 = 1 + 1 = 2$

$a_4 = a_3 + a_2 = 2 + 1 = 3$

$a_5 = a_4 + a_3 = 3 + 2 = 5$

43. (C) $S_n = 2n^2 + 9n$

$S_{20} = 2(20)^2 + 9 \times 20 = 800 + 180 = 980$

$S_{19} = 2(19)^2 + 9 \times 19 = 2(361) + 171 = 893$

$a_{20} = S_{20} - S_{19} = 980 - 893 = 87$

44. (D) $a_1 = 3$

$a_2 = 3$

$a_3 = 3 + 3 = 6$

$a_4 = 3 + 6 = 9$

$a_5 = 6 + 9 = 15$

$a_6 = 9 + 15 = 24$

$a_7 = 24 + 15 = 39$

$a_8 = 24 + 39 = 63$

45. (C) Common difference,

$d = t_{n+1} - t_n$

$= 5(n+1) + 4 - (5n+4)$

$= 5n + 5 + 4 - 5n - 4$

$= 5$

46. (B) Common difference depends upon n.

47. (C) Let the first term of the AP be 'a' and the common difference be 'd'.

$$\frac{1}{n} = a + (m-1)d$$

$$\frac{1}{m} = a + (n-1)d$$

$$\Rightarrow \frac{1}{n} - \frac{1}{m} = \{(m-1)-(n-1)\}d$$

$$\Rightarrow \frac{m-n}{mn} = (m-1-n+1)d$$

$$\Rightarrow d = \frac{1}{mn}$$

First term will be

$$a = \frac{1}{n} - (m-1)\left(\frac{1}{mn}\right)$$

$$= \frac{1}{n} - \frac{m-1}{mn}$$

$$= \frac{1}{n} - \frac{1}{n} + \frac{1}{mn}$$

$$= \frac{1}{mn}$$

$$t_{mn} = \frac{1}{mn} + (mn-1)\frac{1}{mn}$$

$$= \frac{1}{mn} + 1 - \frac{1}{mn} = 1$$

48. (A)

$$\frac{s_n}{S_n}=\frac{3n-13}{5n+21}$$

$$\frac{t_n}{T_n}=\frac{a+(n-1)d}{A+(n-1)D}$$

$$=\frac{2a+2(n-1)d}{2A+2(n-1)D}$$

$$=\frac{2a+[(2n-1)-1]d}{2A+[(2n-1)-1]D}$$

$$=\frac{s_{2n-1}}{S_{2n-1}}=\frac{3(2n-1)-13}{5(2n-1)+21}$$

$$\frac{t_{24}}{T_{24}}=\frac{3(2\times 24-1)-13}{5(2\times 24-1)+21}=\frac{1}{2}$$

49. (C) First term, $a = 20$ and

Common difference, $d=\frac{-2}{3}$

$$s_n=\frac{n}{2}[2a+(n-1)d]$$

$$\Rightarrow\quad 300=\frac{n}{3}(60-n+1)$$

$$\therefore\quad n=25, 36$$

50. (A) The last term of 49th group = 1 + 2 + 3 + ... + 49

First term of 50th group

= 1 + (1 + 2 + 3 + ... + 49)

= 1226

$$s_{50}=\frac{50}{2}[2\times 1226+(50-1)\times 1]=62525$$

51. (D) As we know that $1 + 3 + 5 + ... + (2n - 1) = n^2$

The given equation reduces to

$$\left(\frac{p+1}{2}\right)^2+\left(\frac{q+1}{2}\right)^2=\left(\frac{r+1}{2}\right)^2$$

$$(p+1)^2 + (q+1)^2 = (r+1)^2$$

By using Pythagorean theorem, we get

$$p + 1 = 8;\ q + 1 = 6;\ r + 1 = 10$$

$$\therefore\quad p+q+r=21$$

52. (B) Since α, β, γ and $\delta\ (\alpha<\beta<\gamma<\delta)$ are in AP,

$\beta = \alpha + d$

$\gamma = \alpha + 2d$

$\delta = \alpha + 3d$

For the equation $x^2 - 2x + A = 0$, we have

$\alpha + \beta = 2,\ \alpha\beta = A$

$\Rightarrow\quad 2\alpha + d = 2,\ \alpha\beta = A$

For the equation $x^2 - 18x + B = 0$, we have

$\gamma + \delta = 18,\ \gamma\delta = B$

$\Rightarrow\quad 2\alpha + 5d = 18,\ \gamma\delta = B$

By solving these equations, $A = -3$, $B = 77$

53. (B) a, b, c and d form an increasing AP.

So, $b = a + D$, $c = a + 2D$, $d = a + 3D$;

where D is common difference of AP

$d = a^2 + b^2 + c^2$

$\Rightarrow\quad 5D^2 + (6a - 3)D + 3a^2 - a = 0$

$$\therefore\quad D=\frac{-3(2a-1)\pm\sqrt{-24a^2-16a+9}}{10}$$

Now, $-24a^2-16a+9\geq 0$

$\Rightarrow\quad 24a^2+16a-9\leq 0$

$$\therefore\quad \frac{-4-\sqrt{70}}{12}\leq a\leq\frac{-4+\sqrt{70}}{12}$$

$\therefore\quad a=-1, 0$

$\therefore\quad a+b+c+d=2$

54. (C) Let the first term of AP be A and common difference be D.

$T_p = a$

$A + (p - 1)D = a$ ...(i)

$T_q = b$

$A + (q - 1)D = b$ ...(ii)

$T_r = c$

$A + (r - 1)D = c$ ...(iii)

Solving (i), (ii) and (iii), we get

$D(p - q) = a - b$

$D(q - r) = b - c$

$D(r - p) = c - a$

Now,

$a(q - r) + b(r - p) + c(p - q)$

$$=\frac{1}{D}[a(b-c)+b(c-a)+c(a-b)]=0$$

55. (B) $s_{2n} = 3s_n$

$$\Rightarrow\quad \frac{2n}{2}(a+(2n-1)d)=3\frac{n}{2}(a+(n-1)d)$$

$\Rightarrow \quad 2(a+(2n-1)d)=3(a+(n-1)d)$

$\Rightarrow \quad 2a+(4n-2)d=3a+(3n-3)d$

$\Rightarrow \quad (4n-2-3n+3)d=a$

$\Rightarrow \quad (n+1)d=a$

$$\Rightarrow \quad \frac{s_{3n}}{s_n}=\frac{\frac{3n}{2}[a+(3n-1)d]}{\frac{n}{2}[a+(n-1)d]}$$

$$=\frac{3[(n+1)+(3n-1)]d}{[(n+1)+(n-1)]d}$$

$$=\frac{3(4n)}{2n}=6$$

56. (D)

$$\frac{1}{\sqrt{a_1}+\sqrt{a_2}}+\frac{1}{\sqrt{a_2}+\sqrt{a_3}}+\frac{1}{\sqrt{a_3}+\sqrt{a_4}}+\ldots+\frac{1}{\sqrt{a_{n-1}}+\sqrt{a_n}}$$

$$=\frac{\sqrt{a_1}-\sqrt{a_2}}{-d}+\frac{\sqrt{a_2}-\sqrt{a_3}}{-d}+\frac{\sqrt{a_3}-\sqrt{a_4}}{-d}+\ldots+\frac{\sqrt{a_{n-1}}-\sqrt{a_n}}{-d}$$

$$=\frac{1}{d}\Big[\sqrt{a_2}-\sqrt{a_1}+\sqrt{a_3}-\sqrt{a_2}+ +\sqrt{a_4}-\sqrt{a_3}+\ldots+\sqrt{a_n}-\sqrt{a_{n-1}}\Big]$$

$$=\frac{\sqrt{a_n}-\sqrt{a_1}}{d}=\frac{a_n-a_1}{d\left[\sqrt{a_n}+\sqrt{a_1}\right]}$$

$$=\frac{(n-1)d}{d\left[\sqrt{a_n}+\sqrt{a_1}\right]}$$

$$=\frac{n-1}{\sqrt{a_n}+\sqrt{a_1}}$$

57. (B) Let a be the first term and r be the common ratio of GP. Then

$t_{m+n}=p \Rightarrow ar^{m+n-1}=p$ ...(i)

$t_{m-n}=q \Rightarrow ar^{m-n-1}=q$ ...(ii)

From (i) and (ii), we get

$$ar^{m+n-1}\times ar^{m-n-1}=p\times q$$

$\Rightarrow \quad a^2r^{2(m-1)}=pq$

$\Rightarrow \quad ar^{m-1}=\sqrt{pq}$

58. (A) $s=a+aa+aaa+aaaa+\ldots$ upto n terms

$$=a[1+11+111+1111+\ldots \text{ upto } n \text{ terms}]$$

$$=\frac{a}{9}[9+99+999+9999+\ldots \text{ upto } n \text{ terms}]$$

$$=\frac{a}{9}\left[(10^1-1)+(10^2-1)+(10^3-1)+\ldots+(10^n-1)\right]$$

$$=\frac{a}{9}\left[\frac{10}{9}(10^n-1)-n\right]$$

59. (B) $s=\{0.b+0.bb+0.bbb+\ldots$ upto n terms$\}$

$$=\frac{b}{9}\left\{\begin{matrix}1+1+1+\ldots \text{ upto } n \text{ terms}\\ -\left(\frac{1}{10}+\frac{1}{10^2}+\ldots+\frac{1}{10^n}\right)\end{matrix}\right\}$$

$$=\frac{b}{9}\left[n-\frac{1}{9}\left\{1-\left(\frac{1}{10}\right)^n\right\}\right]$$

60. (A) x_1, x_2, x_3, x_4 are in GP

$x_2=x_1r$

$x_3=x_1r^2$

$x_4=x_1r^3$

Here, r is common ratio of GP.

$x_1+x_2=3, x_1x_2=A$

$\Rightarrow \quad x_1(1+r)=3, x_1^2r=A$

$x_3+x_4=12, x_3x_4=B$

$\Rightarrow \quad x_1r^2(1+r)=12, x_1^2r^5=B$

$r=2$

then A = 2

$B=1^2\cdot 2^5=32$

61. (A) a, b, c are in AP and sum of a, b, c is given.

$A=b-D, c=b+D$

Given

$$a+b+c=\frac{3}{2}$$

$$b=\frac{1}{2}$$

$$a=\frac{1}{2}-D, c=\frac{1}{2}+D$$

Now, given that a^2, b^2, c^2 are in GP.

$(b^2)^2=a^2c^2$

$\Rightarrow \quad b^2=\pm ac$

$$D = \pm\frac{1}{\sqrt{2}}$$

$$\Rightarrow D = \frac{-1}{\sqrt{2}}$$

$$\therefore D < 0$$

62. (B) Since a, b and c in AP,

$$b = \frac{a+c}{2}$$

Since b, c and d are in GP,

$$c^2 = bd$$

Since c, d and e are in HP,

$$d = \frac{2ce}{c+e}$$

By solving these equations $c = 6$, $b = 4$, $d = 9$.

63. (C) $\dfrac{a^{n+1}+b^{n+1}}{a^n+b^n} = \dfrac{a+b}{2}$

$$\Rightarrow 2\left[\left(\frac{a}{b}\right)^{n+1}+1\right] = \left[\left(\frac{a}{b}\right)^{n}+1\right]\left(\frac{a}{b}+1\right)$$

Let $\dfrac{a}{b} = d$. Then

$$2d^{n+1}+2 = (d^n+1)(d+1)$$

$$\Rightarrow (d^n-1)(d-1) = 0$$

Since $d \neq 1$,

$$d^n - 1 = 0$$

$$\therefore n = 0$$

64. (C) Let $A_1, A_2, A_3, A_4, \ldots, A_n$ be n AM's between 3 and 54.

If d is the common difference, then

$$d = \frac{54-3}{n+1} = \frac{51}{n+1}$$

Now

$$\frac{A_8}{A_{n-2}} = \frac{3}{5}$$

$$n = 16$$

65. (B) $\dfrac{a^{n+1}+b^{n+1}}{a^n+b^n} = \sqrt{ab}$

$$\Rightarrow \frac{\left(\frac{a}{b}\right)^{n+1}+1}{\left(\frac{a}{b}\right)^{n}+1} = \sqrt{\frac{a}{b}}$$

Putting $d = \dfrac{a}{b}$ into the equation, we get

$$\left(d^{1/2}-1\right)\left(d^{n+\frac{1}{2}}-1\right) = 0$$

$$d^{1/2} - 1 \neq 0$$

$$n + \frac{1}{2} = 0$$

$$\Rightarrow n = -1/2$$

66. (D)

$$\frac{a^{n+1}+b^{n+1}}{a^n+b^n} = \frac{2ab}{a+b}$$

take $\left(\dfrac{a}{b}\right) = d$

$$n = -1$$

67. (C) Since a, b, c are in AP,

$$b = \frac{a+c}{2}$$

$$\Rightarrow a + c = 2b$$

$$a^2(b + c) + b^2(c + a) + c^2(a + b)$$
$$= a^2b + a^2c + 2b^3 + c^2a + c^2b$$
$$= b(a^2 + c^2) + ac\,(a + c) + 2b^3$$
$$= b(a^2 + c^2) + ac\,(2b) + 2b^3$$
$$= b(a + c)^2 + 2b^3$$
$$= 4b^3 + 2b^3 = 6b^3$$

68. (A) $\dfrac{1}{2}+\dfrac{3}{4}+\dfrac{7}{8}+\dfrac{15}{16}+\ldots$ upto n terms

$$= \left(1-\frac{1}{2}\right)+\left(1-\frac{1}{4}\right)+\left(1-\frac{1}{8}\right)+\left(1-\frac{1}{16}\right) + \ldots \text{upto } n \text{ terms}$$

$$= \left(1-\frac{1}{2}\right)+\left(1-\frac{1}{2^2}\right)+\left(1-\frac{1}{2^3}\right)+\left(1-\frac{1}{2^4}\right) + \ldots + \left(1-\frac{1}{2^n}\right)$$

$$= n - \left(\frac{1}{2} + \frac{1}{2^2} + \frac{1}{2^3} + ... + \frac{1}{2^n}\right)$$

$$= n - \frac{\frac{1}{2}\left[1 - \left(\frac{1}{2}\right)^n\right]}{\left(1 - \frac{1}{2}\right)}$$

$$= n + \left(\frac{1}{2}\right)^n - 1$$

Hints and answers to Olympiad and NTSE level exercises

69. (B) Given that, $T_p = a + (p - 1)\,d = q$ ...(i)

and $T_q = a + (q - 1)\,d = p$ ... (ii)

From (i) and (ii), we get

$$d = -\frac{(p-q)}{(p-q)} = -1$$

Putting value of d in equation (i), we get

$$a = p + q - 1$$

Now, r^{th} term is given by A.P.

$$T_r = a + (r - 1)d$$
$$= (p + q - 1) + (r - 1)(-1)$$
$$= p + q - r$$

Note: Students should remember this question as a formula.

70. (B) We have $\sqrt{2} + \sqrt{8} + \sqrt{18} + \sqrt{32} + ...$

$$= 1\sqrt{2} + 2\sqrt{2} + 3\sqrt{2} + 4\sqrt{2} + ...$$
$$= \sqrt{2}\,[1 + 2 + 3 + 4 + ... \text{ upto } 24 \text{ terms}]$$
$$= \sqrt{2} \times \frac{24 \times 25}{2} = 300\sqrt{2}.$$

71. (A) Given that $S_n = 2n^2 + 5n$

Putting $n = 1, 2, 3, ...$,

$$S_1 = 2 \times 1 + 5 \times 1 = 7,$$
$$S_2 = 2 \times 4 + 10 = 8 + 10 = 18$$
$$S_3 = 18 + 15 = 33.$$

So, $T_1 = S_1 = a = 7$,

$$T_2 = S_2 - S_1 = 18 - 7 = 11,$$
$$T_3 = S_3 - S_2 = 33 - 18 = 15$$

Therefore, series is 7, 11, 15, ...

Now, n^{th} term

$$= a + (n - 1)\,d$$
$$= 7 + (n - 1)\,4 = 4n + 3.$$

Aliter: As we know

$$T_n = S_n - S_{n-1}$$
$$= (2n^2 + 5n) - \{2(n - 1)^2 + 5(n - 1)\}$$
$$= 2n^2 + 5n - 2n^2 + 4n - 2 - 5n + 5$$
$$= 4n + 3$$

72. (A) Given series is

$$3 + 4\frac{1}{2} + 6\frac{3}{4} + ... = 3 + \frac{9}{2} + \frac{27}{4} + ...$$

$$= 3 + \frac{3^2}{2} + \frac{3^3}{4} + \frac{3^4}{8} + \frac{3^5}{16} + ... \text{ (in G.P.)}$$

Here, $a = 3$, $r = \frac{3}{2}$; then sum of the five terms,

$$S_5 = \frac{a(r^n - 1)}{r - 1} = \frac{3\left[\left(\frac{3}{2}\right)^5 - 1\right]}{\frac{3}{2} - 1} = \frac{1\left[\frac{3^5}{32} - 1\right]}{\frac{1}{2}}$$

$$= 6\left[\frac{243 - 32}{32}\right] = \frac{211 \times 3}{16} = \frac{633}{16} = 39\frac{9}{16}$$

73. (C) $S_p = 0$

$$\therefore \quad \frac{p}{2}[2a + (p-1)d] = 0$$

$$\Rightarrow \quad d = \frac{-2a}{p-1} \qquad ...(1)$$

Sum of next q terms = sum of an A.P. whose first term will be $T_{p+1} = a + pd$

$$\Rightarrow \quad S = \frac{q}{2}[2(a + pd) + (q-1)d]$$

$$= \frac{q}{2}[2a + (p-1)d + (p+q)d]$$

$$= \frac{q}{2}\left[0 - (p+q)\frac{2a}{p-1}\right]$$

$$= -a\,\frac{(p+q)q}{p-1}$$

74. (C) x, y, z are in GP

$$\Rightarrow \quad y^2 = xz \qquad ...(i)$$

We have $a^x = b^y = c^z = \lambda$ (say)

$$\Rightarrow \quad x \log a = y \log b = z \log c = \log \lambda$$

$$\Rightarrow \quad x = \frac{\log \lambda}{\log a},\ y = \frac{\log \lambda}{\log b},\ z = \frac{\log \lambda}{\log c}$$

Putting the values of x, y and z in (i), we get

$$\left(\frac{\log\lambda}{\log b}\right)^2 = \frac{\log\lambda}{\log a}\frac{\log\lambda}{\log c}$$

$\Rightarrow$ $(\log b)^2 = \log a \log c$

$\Rightarrow$ $\log_b a = \log_c b$

75. (A) We have: $\frac{1}{4}+\frac{1}{8}+\frac{1}{16}+\ldots = \frac{1/4}{1-1/2} = \frac{1}{2}$

Hence, $0.2^{\log_{\sqrt{5}}\left(\frac{1}{4}+\frac{1}{8}+\frac{1}{16}+\ldots\right)}$

$= 0.2^{\log_{\sqrt{5}}\frac{1}{2}} = \left(\frac{1}{5}\right)^{\log_{\sqrt{5}}\frac{1}{2}} = (5^{-1})^{2\log_5\frac{1}{2}}$

$= (5)^{-2\log_5\frac{1}{2}} = (5)^{\log_5 4} = 4$

76. (a) $\to$ (q); (b) $\to$ (r); (c) $\to$ (p); (d) $\to$ (r)

(a) a, b, c are in G.P.

$\Rightarrow$ $b^2 = ac$

$\Rightarrow$ $2\log_{10} b = \log_{10} a + \log_{10} c$

$\Rightarrow$ $\frac{2}{\log_a 10} = \frac{1}{\log_a 10} + \frac{1}{\log_c 10}$

$\Rightarrow$ $\frac{2}{y} = \frac{1}{x} + \frac{1}{z}$

$\Rightarrow$ x, y, z are in H.P.

(b) $\frac{a+be^x}{a-be^x} = \frac{b+ce^x}{b-ce^x} = \frac{c+de^x}{c-de^x}$

$\Rightarrow$ $\frac{2a}{a-be^x} - 1 = \frac{2b}{b-ce^x} - 1 = \frac{2c}{c-de^x} - 1$

$\Rightarrow$ $\frac{a-be^x}{a} = \frac{b-ce^x}{b} = \frac{c-de^x}{c}$

$\Rightarrow$ $1 - \frac{b}{a}e^x = 1 - \frac{c}{b}e^x = 1 - \frac{d}{e}e^x$

$\Rightarrow$ $\frac{b}{a} = \frac{c}{b} = \frac{d}{c} = \text{r say}$

$\Rightarrow$ a, b, c, d are in G.P.

(c) Given, $2b = a + c$, $x^2 = ab$, $y^2 = bc$

Now, $x^2 + y^2 = b(a + c) = b \cdot 2b = 2b^2$

$\Rightarrow$ $x^2 + y^2 = 2b^2$

$\Rightarrow$ x^2, b^2, y^2 are in A.P.

(d) $x \log a = y \log b = z \log c = k$ (say)

Also, $y^2 = xz$

$\Rightarrow$ $\frac{k^2}{(\log b)^2} = \frac{k^2}{\log a \cdot \log c}$

$\Rightarrow$ $\log a$, $\log b$, $\log c$ are in G.P.

77. (C)

(p) $A_p = a + (p - 1)d$ ...(1)

$A_q = a + (q - 1)d$ ...(2)

$A_r = a + (r - 1)d$ ...(3)

$A_s = a + (s - 1)d$ ...(4)

$A_q = kA_p$

$A_r = k^2 A_p$

$A_s = k^3 A_p$ ($\because A_p, A_q, A_r, A_s$ in G.P.)

$(p-q) = \frac{A_p - A_q}{d} = A_p\frac{(1-k)}{d}$ [from (1) and (2)]

$(q-r) = A_p K\frac{(1-k)}{d}$ [from (2) and (3)]

$(r-s) = A_p k^2\frac{(1-k)}{d}$

$\Rightarrow$ $p - q$, $q - r$, $r - s$ are in G.P

(q) $\ln x$, $\ln y$, $\ln z$ are in G.P.

$\Rightarrow$ $\ln(\ln x)$, $\ln(\ln y)$, $\ln(\ln z)$ are in A.P.

$\Rightarrow$ $2x + \ln(\ln x)$, $3x + \ln(\ln y)$, $4x + \ln(\ln z)$ are in A.P.

(r) $n!$ $3 \times n!$ and $(n + 1)!$ are in G.P.

$\Rightarrow$ $9(n!)^2 = n!(n + 1)!$

$\Rightarrow$ $(n + 1) = 9$

$\Rightarrow$ $n = 8$

$\therefore$ $n! = 8!$

$5 \times n! = 5 \times 8!$

$(n + 1)! = 9!$

$9! + 8! = 5 \times 9!$

$\Rightarrow$ $n!$, $5 \times n!$ and $(n + 1)!$ are in A.P.

(s) $\frac{(b-c)^2 + (a-b)^2 + (c-a)^2}{3}$

$= \frac{(b-c-2a)^2 + (c+a-2b)^2 + (a+b-2c)^2}{3}$

$\Rightarrow$ $(b + c - 2a)^2 - (b - c)^2 + (c + a - 2b)^2 - (c - a^2) + (a + b - 2c)^2 - (a - b)^2 = 0$

$\Rightarrow$ $a = b = c$

78. (A) a, b, c, d are in H.P.

$\Rightarrow$ b is H.M. of a and c

c is H.M. of b and d

Using A.M. $\geq$ H.M, we get

$$\frac{a+c}{2} > b \quad \Rightarrow \quad a+c > 2b$$

and $$\frac{b+d}{2} > c \quad \Rightarrow \quad b+d > 2c$$

$$\Rightarrow \quad a+c+b+d > 2b+2c$$

$$\Rightarrow \quad a+d > b+c$$

Matrices

1. Refer to one newspaper.

2. (a) 2×3 (b) 2×1

(c) 1×4 (d) 4×2

(e) 3×3 (f) 4×4

3. (b) Row matrix

(c) Column matrix

4. (a) and (b), plot a graph in the x-y plane

(c) (1, 1, 1)

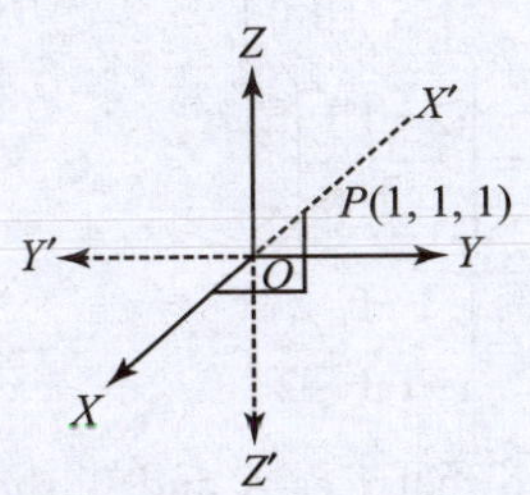

5. (d) and (e) same as above

6. (a) $\begin{bmatrix} 1 & -4 \\ 2 & 6 \\ 3 & 7 \end{bmatrix}$ (b) $[7 \quad 8]$

(c) $\begin{bmatrix} 3 \\ 7 \\ 8 \\ 9 \end{bmatrix}$ (d) $\begin{bmatrix} 1 & 3 & 5 & 7 \\ 2 & 4 & 6 & 9 \end{bmatrix}$

(e) $\begin{bmatrix} 1 & 4 & 7 \\ 2 & 5 & 8 \\ 3 & 6 & 9 \end{bmatrix}$ (f) $\begin{bmatrix} 0 & 1 & 1 & 0 \\ 1 & 0 & 0 & 0 \\ 0 & 1 & 0 & 0 \\ 1 & 1 & 1 & 1 \end{bmatrix}$

7. $A = \begin{bmatrix} a_{11} & a_{12} \\ a_{21} & a_{22} \\ a_{31} & a_{32} \end{bmatrix}$, $B = \begin{bmatrix} b_{11} & b_{12} & b_{13} \\ b_{21} & b_{22} & b_{23} \\ b_{31} & b_{32} & b_{33} \end{bmatrix}$

$C = \begin{bmatrix} c_{11} \\ c_{21} \\ c_{31} \\ c_{41} \\ c_{51} \end{bmatrix}$; $D = \begin{bmatrix} d_{11} & d_{12} & d_{13} & d_{14} & d_{15} \end{bmatrix}$

$E = \begin{bmatrix} f_{11} & f_{12} & f_{13} & f_{14} \\ f_{21} & f_{22} & f_{23} & f_{24} \\ f_{31} & f_{32} & f_{33} & f_{34} \\ f_{41} & f_{42} & f_{43} & f_{44} \end{bmatrix}$

8. (a) 10×10

(b) [0 2 4 6 8 10 12 14 16 18]

(c) Yes

(d) $a_{16} = 5, 8, 8, 5, 16$

9. (a) $\begin{bmatrix} 2 & 3 \\ 4 & 8 \end{bmatrix} \begin{bmatrix} x \\ y \end{bmatrix} = \begin{bmatrix} 5 \\ 9 \end{bmatrix}$

(b) $\begin{bmatrix} 3 & -4 & 3 \\ 7 & 2 & 6 \\ 2 & -7 & 8 \end{bmatrix} \begin{bmatrix} x \\ y \\ z \end{bmatrix} = \begin{bmatrix} 25 \\ 11 \\ 25 \end{bmatrix}$

(c) $\begin{bmatrix} 1 & 1 & 1 \\ 0 & -1 & 1 \\ -1 & 0 & 1 \end{bmatrix} \begin{bmatrix} x \\ y \\ z \end{bmatrix} = \begin{bmatrix} 7 \\ 3 \\ 5 \end{bmatrix}$

10. $A+B = \begin{bmatrix} 6 & 8 \\ 7 & 7 \end{bmatrix}$ $(A+B)' = \begin{bmatrix} 6 & 7 \\ 8 & 7 \end{bmatrix}$

$A' + B' = \begin{bmatrix} 6 & 7 \\ 8 & 7 \end{bmatrix}$

16. (a) $A^2 = \begin{bmatrix} 1 & 0 \\ -15 & 16 \end{bmatrix}$

(b) $A^3 = \begin{bmatrix} 1 & 0 \\ -63 & 64 \end{bmatrix}$

(c) $A + A^2 = \begin{bmatrix} 1 & 0 \\ -18 & 20 \end{bmatrix}$

(d) $A + A^2 + A^3 = \begin{bmatrix} 1 & 0 \\ -81 & 84 \end{bmatrix}$

17. $\begin{bmatrix} 12 & 3 \\ 1 & 0 \end{bmatrix}$

18. $\begin{bmatrix} 8 & 7 \\ 7 & 3 \end{bmatrix}$ and $\begin{bmatrix} 4 & 1 \\ 1 & 1 \end{bmatrix}$

20. 49, 68, 40.75 and 2090 cc

21. (A)

22. (C)

23. (A) All the elements below principal diagonal are zero.

24. (C) All non-diagonal elements of the matrix are zero.

25. (D) Sum of elements of leading diagonal of a square matrix is trace of matrix.

26. (B)

27. (B) $\text{diag}(3,5,7) = \begin{bmatrix} 3 & 0 & 0 \\ 0 & 5 & 0 \\ 0 & 0 & 7 \end{bmatrix}$ represents diagonal matrix.

28. (B) The matrix obtained by interchanging the rows and columns of a matrix is its transpose.

29. (B)

30. (C) Addition is only possible when orders of both the matrices are equal.

31. (C)

$$A = \begin{bmatrix} 0 & 2 \\ 3 & -4 \end{bmatrix} \Rightarrow kA = \begin{bmatrix} 0 & 2k \\ 3k & -4k \end{bmatrix}$$

$$kA = \begin{bmatrix} 0 & 3a \\ 2b & 24 \end{bmatrix} = \begin{bmatrix} 0 & 2k \\ 3k & -4k \end{bmatrix}$$

$k = -6,\ a = -4,\ b = -9$

32. (D)

$$c = \frac{B-A}{2} = \frac{1}{2}\left\{\begin{bmatrix} 3 & -1 & 2 \\ 4 & 2 & 5 \\ 2 & 0 & 3 \end{bmatrix} - \begin{bmatrix} 1 & 2 & -3 \\ 5 & 0 & 2 \\ 1 & -1 & 1 \end{bmatrix}\right\}$$

$$= \begin{bmatrix} 1 & -\frac{3}{2} & \frac{5}{2} \\ -\frac{1}{2} & 1 & \frac{3}{2} \\ \frac{1}{2} & \frac{1}{2} & 1 \end{bmatrix}$$

33. (A) $2x - y = \begin{bmatrix} 3 & -3 & 0 \\ 3 & 3 & 2 \end{bmatrix}$

On multiplying both sides by 2, we get

$$4x - 2y = \begin{bmatrix} 6 & -6 & 0 \\ 6 & 6 & 4 \end{bmatrix} \quad \text{...}(i)$$

$$x + 2y = \begin{bmatrix} 4 & 1 & 5 \\ -1 & 4 & -4 \end{bmatrix} \quad \text{..}(ii)$$

By adding (i) and (ii), we get

$$5x = \begin{bmatrix} 10 & -5 & 5 \\ 5 & 10 & 0 \end{bmatrix}$$

$$x = \begin{bmatrix} 2 & -1 & 1 \\ 1 & 2 & 0 \end{bmatrix}$$

$$y = \begin{bmatrix} 1 & 1 & 2 \\ -1 & 1 & -2 \end{bmatrix}$$

34. (A) For two matrices A and B, the product AB is defined if number of columns in A is equal to number of rows in B.

35. (A) $AB = \begin{bmatrix} 0 & 1 & 2 \\ 1 & 2 & 3 \\ 2 & 3 & 4 \end{bmatrix}\begin{bmatrix} 1 & -2 \\ -1 & 0 \\ 2 & -1 \end{bmatrix}$

$$= \begin{bmatrix} 3 & -2 \\ 5 & -5 \\ 7 & -8 \end{bmatrix}$$

36. (B)

37. (A) $A = \begin{bmatrix} \alpha & 2 \\ 2 & \alpha \end{bmatrix} = 0 \Rightarrow \alpha^2 - 4 = 0 \Rightarrow \alpha = 2, -2$

38. (B)

39. (A) $I_2 = \begin{bmatrix} 1 & 0 \\ 0 & 1 \end{bmatrix}$ and $A^2 = \begin{bmatrix} 5 & 8 \\ 8 & 13 \end{bmatrix}$

40. (A) $A^2 = \begin{bmatrix} \cos\theta & \sin\theta \\ -\sin\theta & \cos\theta \end{bmatrix} \times \begin{bmatrix} \cos\theta & \sin\theta \\ -\sin\theta & \cos\theta \end{bmatrix}$

$= \begin{bmatrix} \cos^2\theta - \sin^2\theta & 2\sin\theta\cos\theta \\ -2\sin\theta\cos\theta & \cos^2\theta - \sin^2\theta \end{bmatrix}$

41. (C) $x^2 - 4x + 3 = 0$

$\Rightarrow \ (x-3)(x-1) = 0 \ \Rightarrow \ x = 1, 3$

$x^2 - 8x + 15 = 0$

$\Rightarrow \ (x-3)(x-5) = 0 \ \Rightarrow \ x = 3, 5$

$x^2 - 6x + 5 = 0$

$\Rightarrow \ (x-1)(x-5) = 0 \ \Rightarrow \ x = 1, 5$

Clearly, $a = 1, b = 3, c = 5$

$\begin{bmatrix} a^2+c^2 & a^2+b^2 \\ b^2+c^2 & a^2+c^2 \end{bmatrix} = \begin{bmatrix} 1+25 & 1+9 \\ 9+25 & 1+25 \end{bmatrix}$

$= \begin{bmatrix} 26 & 10 \\ 34 & 26 \end{bmatrix}$

42. (B) $A = \begin{bmatrix} \cos\theta & -\sin\theta \\ \sin\theta & \cos\theta \end{bmatrix}$

$A^T = \begin{bmatrix} \cos\theta & \sin\theta \\ \sin\theta & \cos\theta \end{bmatrix}$

$A + A^T = \begin{bmatrix} 2\cos\theta & 0 \\ 0 & 2\cos\theta \end{bmatrix}$

$A + A^T = I_2$

$\Rightarrow \begin{bmatrix} 2\cos\theta & 0 \\ 0 & 2\cos\theta \end{bmatrix} = \begin{bmatrix} 1 & 0 \\ 0 & 1 \end{bmatrix}$

$\Rightarrow \ 2\cos\theta = 1$

$\Rightarrow \ \cos\theta = \frac{1}{2}$

$\Rightarrow \ \theta = \frac{\pi}{3}$

43. (C) Matrix, whose determinant is not equal to 0, is non-singular matrix.

44. (D) If matrix A is symmetric then, $A^T = A$

$\Rightarrow \begin{bmatrix} 4 & x+2 \\ 2x-3 & x+1 \end{bmatrix} = \begin{bmatrix} 4 & 2x-3 \\ x+2 & x+1 \end{bmatrix}$

45. (C) For a matrix A to be singular, determinant of matrix A should be equal to zero.

$\det \begin{bmatrix} 3-x & 2 & 2 \\ 2 & 4-x & 1 \\ -2 & -4 & -1-x \end{bmatrix} = 0$

$\Rightarrow \ (3-x)[(4-x)(-1-x)+4]$

$-2[2(-1-x)+2]+2[-8+2(4-x)] = 0$

$\Rightarrow \ (3-x)(x^2-3x) - 2(-2x) + 2(-2x) = 0$

$\Rightarrow \ (3-x)(x^2-3x) = 0$

46. (B) $3\begin{bmatrix} 2 & 3 \\ 4 & 5 \end{bmatrix} - 2\begin{bmatrix} 0 & 2 \\ 3 & 4 \end{bmatrix} = \begin{bmatrix} 6 & 9 \\ 12 & 15 \end{bmatrix} - \begin{bmatrix} 0 & 4 \\ 6 & 8 \end{bmatrix}$

$= \begin{bmatrix} 6 & 5 \\ 6 & 7 \end{bmatrix}$

47. (B) $\frac{1}{2}(A+B) = \frac{1}{2}\left\{\begin{bmatrix} 1 & 2 & 3 \\ 4 & 5 & 6 \\ 7 & 8 & 9 \end{bmatrix} + \begin{bmatrix} 9 & 8 & 7 \\ 6 & 5 & 4 \\ 3 & 2 & 1 \end{bmatrix}\right\}$

$= \frac{1}{2}\begin{bmatrix} 10 & 10 & 10 \\ 10 & 10 & 10 \\ 10 & 10 & 10 \end{bmatrix} = \begin{bmatrix} 5 & 5 & 5 \\ 5 & 5 & 5 \\ 5 & 5 & 5 \end{bmatrix}$

48. (B)

49. (A)

50. (D) $A^2 = \begin{bmatrix} 7 & 10 \\ 15 & 22 \end{bmatrix}, B^2 = \begin{bmatrix} 2 & 2 \\ 4 & 6 \end{bmatrix},$

$AB = \begin{bmatrix} 4 & 5 \\ 8 & 11 \end{bmatrix}, BA = \begin{bmatrix} 3 & 4 \\ 8 & 12 \end{bmatrix}$

$(A+B)^2 = \begin{bmatrix} 16 & 21 \\ 35 & 51 \end{bmatrix}$

$(A + B)^2 = A^2 + B^2 + AB + BA$

Hints and answers to Olympiad and NTSE level exercises

51. (A) $A^2 = \begin{bmatrix} 1 & 1 \\ 0 & 1 \end{bmatrix}\begin{bmatrix} 1 & 1 \\ 0 & 1 \end{bmatrix} = \begin{bmatrix} 1 & 2 \\ 0 & 1 \end{bmatrix},$

and $A^3 = A^2.A = \begin{bmatrix} 1 & 2 \\ 0 & 1 \end{bmatrix}\begin{bmatrix} 1 & 1 \\ 0 & 1 \end{bmatrix} = \begin{bmatrix} 1 & 3 \\ 0 & 1 \end{bmatrix}$

$$\Rightarrow \quad A^n = A^{n-1}.A = \begin{bmatrix} 1 & n-1 \\ 0 & 1 \end{bmatrix}\begin{bmatrix} 1 & 1 \\ 0 & 1 \end{bmatrix}.$$

$$= \begin{bmatrix} 1 & n \\ 0 & 1 \end{bmatrix}$$

52. (A) $A^2 = A.\, A = \begin{bmatrix} ab & b^2 \\ -a^2 & -ab \end{bmatrix}\begin{bmatrix} ab & b^2 \\ -a^2 & -ab \end{bmatrix}$

$$= \begin{bmatrix} a^2b^2 - a^2b^2 & ab^3 - ab^3 \\ -a^3b + a^3b & -a^2b^2 + a^2b^2 \end{bmatrix} = O$$

$\Rightarrow \quad A^3 = A \cdot A^2 = 0$ and $A^n = 0$, for all $n \geq 2$.

53. (C) $A = \begin{bmatrix} 2 & 0 & 0 \\ 0 & 2 & 0 \\ 0 & 0 & 2 \end{bmatrix}$

$$\Rightarrow \quad A^5 = \begin{bmatrix} 2^5 & 0 & 0 \\ 0 & 2^5 & 0 \\ 0 & 0 & 2^5 \end{bmatrix} = 2^4 \begin{bmatrix} 2 & 0 & 0 \\ 0 & 2 & 0 \\ 0 & 0 & 2 \end{bmatrix}$$

$= 16\, A.$

54. (B) $f(A) = I + A + A^2 + \ldots + A^{16}$

$$A = \begin{bmatrix} 0 & 5 \\ 0 & 0 \end{bmatrix}$$

$$\Rightarrow \quad A^2 = \begin{bmatrix} 0 & 5 \\ 0 & 0 \end{bmatrix}\begin{bmatrix} 0 & 5 \\ 0 & 0 \end{bmatrix} = \begin{bmatrix} 0 & 0 \\ 0 & 0 \end{bmatrix}$$

$$A^3 = A^2 \cdot A = \begin{bmatrix} 0 & 0 \\ 0 & 0 \end{bmatrix}$$

Similarly, $A^4 = A^5 = \ldots = A^{16} = \begin{bmatrix} 0 & 0 \\ 0 & 0 \end{bmatrix}$

$$f(A) = \begin{bmatrix} 1 & 0 \\ 0 & 1 \end{bmatrix} + \begin{bmatrix} 0 & 5 \\ 0 & 0 \end{bmatrix} + \begin{bmatrix} 0 & 0 \\ 0 & 0 \end{bmatrix} + \ldots + \begin{bmatrix} 0 & 0 \\ 0 & 0 \end{bmatrix}$$

$$= \begin{bmatrix} 1 & 5 \\ 0 & 1 \end{bmatrix}$$

55. (A) $I = A'A = \begin{bmatrix} 0 & x & x \\ 2y & y & -y \\ z & -z & z \end{bmatrix}\begin{bmatrix} 0 & 2y & z \\ x & y & -z \\ x & -y & z \end{bmatrix}$

$$= \begin{bmatrix} 2x^2 & 0 & 0 \\ 0 & 6y^2 & 0 \\ 0 & 0 & 3z^2 \end{bmatrix}$$

$\therefore \quad 2x^2 = 6y^2 = 3z^2 = 1$

$\therefore \quad x^2 + y^2 + z^2 = \frac{1}{2} + \frac{1}{6} + \frac{1}{3} = \frac{3+1+2}{6} = 1$

56. (B) Given $2A + 3B = \begin{bmatrix} 2 & -1 & 4 \\ 3 & 2 & 5 \end{bmatrix}$ ...(i)

And $A + 2B = \begin{bmatrix} 5 & 0 & 3 \\ 1 & 6 & 2 \end{bmatrix}$

$\Rightarrow \quad A2 + 4B = \begin{bmatrix} 10 & 0 & 6 \\ 2 & 12 & 4 \end{bmatrix}$ ...(ii)

Solving (i) and (ii), we get

$$-B = \begin{bmatrix} -8 & -1 & -2 \\ 1 & -10 & 1 \end{bmatrix}$$

$$\Rightarrow \quad B = \begin{bmatrix} 8 & 1 & 2 \\ -1 & 10 & -1 \end{bmatrix}.$$

57. (B) We have $A_\alpha A_\beta$

$$= \begin{bmatrix} \cos\alpha & \sin\alpha \\ -\sin\alpha & \cos\alpha \end{bmatrix}\begin{bmatrix} \cos\beta & \sin\beta \\ -\sin\beta & \cos\beta \end{bmatrix}$$

$$= \begin{bmatrix} \cos\alpha\cos\beta - \sin\alpha\sin\beta & \cos\alpha\sin\beta + \sin\alpha\cos\beta \\ -\sin\alpha\cos\beta - \cos\alpha\sin\beta & -\sin\alpha\sin\beta + \cos\alpha\cos\beta \end{bmatrix}$$

$$= \begin{bmatrix} \cos(\alpha+\beta) & \sin(\alpha+\beta) \\ -\sin(\alpha+\beta) & \cos(\alpha+\beta) \end{bmatrix} = A_{\alpha+\beta}$$

58. (C) $[1 \times y] \begin{bmatrix} 1 & 3 & 1 \\ 0 & 2 & -1 \\ 0 & 0 & 1 \end{bmatrix}\begin{bmatrix} 1 \\ x \\ y \end{bmatrix} = [0]$

$$\Rightarrow \quad [1 \;\; 3 + 2x \;\; 1 - x + y] \begin{bmatrix} 1 \\ x \\ y \end{bmatrix} = [0]$$

$\Rightarrow \quad [1 + 3x + 2x^2 + y - xy + y^2] = [0]$

$\Rightarrow \quad 2x^2 + y^2 + y + 3x - xy + 1 = 0$

If $y = 0, 2x^2 + 3x + 1 = 0$

$\Rightarrow (2x - 1)(x + 1) = 0$

$\Rightarrow x = -1/2, -1$ (Rational roots)

If $y = -1, 2x^2 + 4x + 1 = 0$

$\Rightarrow x = \frac{-4 \pm \sqrt{12}}{4} = \frac{-2 \pm \sqrt{3}}{2}$ (Irrational roots)

59. (C) Let $A = [A_{ij}]_{3 \times 3}$, where $a_{ij} = (k)^{i+j}$

So, $A = \begin{bmatrix} k^2 & k^3 & k^4 \\ k^3 & k^4 & k^5 \\ k^4 & k^5 & k^6 \end{bmatrix}$

(p) If A is singular, that $|A| = 0$

$$\Rightarrow k^2 \cdot k^3 \cdot k^4 \begin{vmatrix} 1 & 1 & 1 \\ k & k & k \\ k^2 & k^2 & k^2 \end{vmatrix} = 0, \forall k \in I$$

(q) If A is null matrix, then $k \in \{0\}$

(r) If A is skew-symmetric matrix which is not null-matrix, then it is not possible.

$\Rightarrow k \in \phi$

(s) If $A^2 = 3A$, then

$$\begin{bmatrix} k^2 & k^3 & k^4 \\ k^3 & k^4 & k^5 \\ k^4 & k^5 & k^6 \end{bmatrix}\begin{bmatrix} k^2 & k^3 & k^4 \\ k^3 & k^4 & k^5 \\ k^4 & k^5 & k^6 \end{bmatrix} = \begin{bmatrix} 3k^2 & 3k^3 & 3k^4 \\ 3k^3 & 3k^4 & 3k^5 \\ 3k^4 & 3k^5 & 3k^6 \end{bmatrix}$$

$$\Rightarrow \begin{bmatrix} k^4+k^6+k^8 & k^5+k^7+k^9 & k^6+k^8+k^{10} \\ k^5+k^7+k^9 & k^6+k^8+k^{10} & k^7+k^9+k^{11} \\ k^6+k^8+k^{10} & k^7+k^9+k^{11} & k^8+k^{10}+k^{12} \end{bmatrix} = \begin{bmatrix} 3k^2 & 3k^3 & 3k^4 \\ 3k^3 & 3k^4 & 3k^5 \\ 3k^4 & 3k^5 & 3k^6 \end{bmatrix}$$

$\Rightarrow k \in \{-1, 0, 1\}$

60. (A) Minimum number of zeros

$= \frac{n^2 - n}{2} = \frac{n(n-1)}{2}$

Determinants

1. 1

2. 1

3. –7

4. $a^2 - b^2$

5. $1 - \cos^2 A = \sin^2 A$

6. $a^2 + b^2 + c^2 - ab - ac - bc$

7. $\left(\frac{93}{155}, \frac{-62}{155}\right)$

8. $\left(\frac{45}{7}, \frac{10}{7}\right)$

9. $\left(\frac{-161}{17}, \frac{785}{17}\right)$

10. $\left(\frac{3}{2}, \frac{3}{2}\right)$

11. $\left(\frac{-3}{11}, \frac{-7}{22}\right)$

12. $ek - hf$; $dk - fg$; $dh - ge$; $bk - ch$; $ak - cg$; $ah - bg$; $bf - ce$; $af - cd$; $ae - bd$

13. 1

14. 40

15. 0

16. 51

17. 0

18. 0

19. (A) $\begin{vmatrix} \sec\theta & \tan\theta \\ \tan\theta & \sec\theta \end{vmatrix} = \sec^2\theta - \tan^2\theta = 1$

20. (A) $\begin{vmatrix} 5 & 10 & 15 \\ 10 & 20 & 30 \\ 25 & 30 & 35 \end{vmatrix}$

$= 2\begin{vmatrix} 5 & 10 & 15 \\ 5 & 10 & 15 \\ 25 & 30 & 35 \end{vmatrix}$ (Taking 2 common from R_2)

$= 0$ (R_1 and R_2 are identical)

21. (D) $D = \begin{vmatrix} 5 & 20 & 25 \\ 8 & 72 & 56 \\ 1 & 6 & 6 \end{vmatrix} = 5\times 8 \begin{vmatrix} 1 & 4 & 5 \\ 1 & 9 & 7 \\ 1 & 6 & 6 \end{vmatrix}$

Expanding the determinant along first row, we get

$D = 40\ \{1(54 - 42) + 4(7 - 6) + 5(6 - 9)\}$
$= 40\ \{12 + 4 - 15\}$
$= 40\ \{16 - 15\}$
$= 40$

22. (D) $\begin{vmatrix} 1 & a & b+c \\ 1 & b & c+a \\ 1 & c & a+b \end{vmatrix} = \begin{vmatrix} 1 & a & b \\ 1 & b & c \\ 1 & c & a \end{vmatrix} + \begin{vmatrix} 1 & a & c \\ 1 & b & a \\ 1 & c & b \end{vmatrix}$

Now,

$\begin{vmatrix} 1 & a & b \\ 1 & b & c \\ 1 & c & a \end{vmatrix} = 1(ab - c^2) + a(c - a) + b(c - b)$

$= ab - c^2 + ac - a^2 + bc - b^2$

$\begin{vmatrix} 1 & a & c \\ 1 & b & a \\ 1 & c & b \end{vmatrix} = 1(b^2 - ac) + a(a - b) + c(c - b)$

$= b^2 - ac + a^2 - ab + c^2 - bc$

After adding, we get 0.

23. (A) $\begin{vmatrix} 1 & a & a^2 \\ 1 & b & b^2 \\ 1 & c & c^2 \end{vmatrix} + \begin{vmatrix} 1 & a & -bc \\ 1 & b & -ca \\ 1 & c & -ab \end{vmatrix}$

$= \begin{vmatrix} 1 & a & a^2 \\ 1 & b & b^2 \\ 1 & c & c^2 \end{vmatrix} - \begin{vmatrix} 1 & a & bc \\ 1 & b & ca \\ 1 & c & ab \end{vmatrix}$

$= [1\{bc^2 - b^2c\} + a\{b^2 - c^2\} + a^2\{c - b\}]$
$\quad -[1\{ab^2 - ac^2\} + a\{ca - ab\} + bc\{c - b\}]$
$= bc^2 - b^2c + ab^2 - ac^2 + a^2c - a^2b$
$\quad - ab^2 + ac^2 - ca^2 - a^2b - bc^2 + b^2c$
$= 0$

24. (C) $\begin{vmatrix} -1 & 2 & 1 \\ 3+2\sqrt{2} & 2+2\sqrt{2} & 1 \\ 3-2\sqrt{2} & 2-2\sqrt{2} & 1 \end{vmatrix}$

Take $2\sqrt{2} = d$

$\begin{vmatrix} -1 & 2 & 1 \\ 3+d & 2+d & 1 \\ 3-d & 2-d & 1 \end{vmatrix} = -8d = -16\sqrt{2}$

25. (D) $\begin{vmatrix} 1 & 4 & 20 \\ 1 & -2 & 5 \\ 1 & 2x & 5x^2 \end{vmatrix} = 0$

$\Rightarrow\ 1(-10x^2 - 10x) + 4(5 - 5x^2) + 20(2x + 2) = 0$
$\Rightarrow\ -10x^2 - 10x + 20 - 20x^2) + 40x + 40 = 0$
$\Rightarrow\ -30x^2 + 30x + 60 = 0$
$\Rightarrow\ x^2 - x - 2 = 0$

26. (A) $\begin{vmatrix} a & a^2 & 1+a^3 \\ b & b^2 & 1+b^3 \\ c & c^2 & 1+c^3 \end{vmatrix}$

$= \begin{vmatrix} a & a^2 & 1 \\ b & b^2 & 1 \\ c & c^2 & 1 \end{vmatrix} + \begin{vmatrix} a & a^2 & a^3 \\ b & b^2 & b^3 \\ c & c^2 & c^3 \end{vmatrix}$

$= \begin{vmatrix} a & a^2 & 1 \\ b & b^2 & 1 \\ c & c^2 & 1 \end{vmatrix} + (abc)\begin{vmatrix} 1 & a & a^2 \\ 1 & b & b^2 \\ 1 & c & c^2 \end{vmatrix}$

$= \begin{vmatrix} 1 & a & a^2 \\ 1 & b & b^2 \\ 1 & c & c^2 \end{vmatrix} + (abc)\begin{vmatrix} 1 & a & a^2 \\ 1 & b & b^2 \\ 1 & c & c^2 \end{vmatrix}$

$= (1 + abc)\begin{vmatrix} 1 & a & a^2 \\ 1 & b & b^2 \\ 1 & c & c^2 \end{vmatrix}$

$= (1 + abc)(a - b)(b - c)(c - a)$

27. (A)

$\frac{1}{2}\left\|\begin{matrix} x_1 & y_1 & 1 \\ x_2 & y_2 & 1 \\ x_3 & y_3 & 1 \end{matrix}\right\| = \frac{\sqrt{3}}{4}\times(4)^2 = \sqrt{3}\times 4$

$\Rightarrow \begin{vmatrix} x_1 & y_1 & 1 \\ x_2 & y_2 & 1 \\ x_3 & y_3 & 1 \end{vmatrix}^2 = (2\times 4\times\sqrt{3})^2 = 64\times 3 = 192$

28. (B) $\begin{vmatrix} x & 3 & 6 \\ 3 & 6 & x \\ 6 & x & 3 \end{vmatrix} = 0$

$\Rightarrow x(18 - x^2) + 3(6x - 9) + 6(3x - 36) = 0$

$\Rightarrow 18x - x^3 + 18x - 27 + 18x - 216 = 0$

$\Rightarrow -x^3 + 54x - 243 = 0$

$\Rightarrow x^3 - 54x + 243 = 0$

$\begin{vmatrix} 2 & x & 7 \\ x & 7 & 2 \\ 7 & 2 & x \end{vmatrix} = 0$

$\Rightarrow 2(7x - 4) + x(14 - x^2) + 7(2x - 49) = 0$

$\Rightarrow 14x - 8 + 14x - x^3 + 14x - 343 = 0$

$\Rightarrow -x^3 + 42x - 351 = 0$

$\Rightarrow x^3 - 42x + 351 = 0$

$\begin{vmatrix} 4 & 5 & x \\ 5 & x & 4 \\ x & 4 & 5 \end{vmatrix} = 0$

$\Rightarrow 4(5x - 16) + 5(4x - 25) + x(20 - x^2) = 0$

$\Rightarrow 20x - 64 + 20x - 125 + 20x - x^3 = 0$

$\Rightarrow 60x - x^3 - 189 = 0$

$\Rightarrow x^3 - 60x + 189 = 0$

$x = -9$ satisfies all the three equations.

29. (D) $\Delta = \begin{vmatrix} x & \sin\theta & \cos\theta \\ -\sin\theta & -x & 1 \\ \cos\theta & 1 & x \end{vmatrix}$

$= x(-x^2 - 1) - \sin\theta(x\sin\theta - \cos\theta) + \cos\theta(-\sin\theta + x\cos\theta)$

$= -x^3 - x + x(\sin^2\theta + \cos^2\theta)$

$= -x^3$

30. (A)

$\Delta = \begin{vmatrix} abc & b^2c & c^2b \\ abc & c^2a & ca^2 \\ abc & a^2b & b^2a \end{vmatrix} = 0$

$-a^2b^2c^2(a^3 + b^3 + c^3 - 3abc) = 0$

$-a^2b^2c^2(a+b+c)(a^2+b^2+c^2-ab-bc-ca) = 0$

$-a^2b^2c^2(a+b+c)\left\{\frac{1}{2}\left\{(a-b)^2 + (b-c)^2 + (c-a)^2\right\}\right\} = 0$

$\therefore a+b+c = 0$

31. (B) $\begin{vmatrix} x & -6 & -1 \\ 2 & -3x & x-3 \\ -3 & 2x & x+2 \end{vmatrix} = 0$

$\Rightarrow x(-3x^2 - 6x - 2x^2 + 6x) - 6(-3x + 9 - 2x - 4) - 1(4x - 9x) = 0$

$\Rightarrow x(-5x^2) - 6(-5x + 5) + 5x = 0$

$\Rightarrow -5x^3 + 30x - 30 + 5x = 0$

$\Rightarrow -5x^3 + 35x - 30 = 0$

$\Rightarrow x^3 - 7x + 6 = 0$

32. (B) $\begin{vmatrix} 1 & 1 & 1 \\ x & y & z \\ x^3 & y^3 & z^3 \end{vmatrix}$

$= 1(yz^3 - y^3z) + 1(x^3z - xz^3) + 1(xy^3 - x^3y)$

$= (x - y)(y - z)(z - x)(x + y + z)$

33. (A) $\begin{vmatrix} pa & qb & rc \\ qc & ra & pb \\ rb & pc & qa \end{vmatrix}$

$= pa(rqa^2 - p^2bc) + qb(pb^2r - q^2ac) + rc(pqc^2 - abr^2)$

$= pqra^3 - p^3abc + pqrb^3 - q^3abc + pqrc^3 - abcr^3$

$= -abc(p^3 + q^3 + r^3) + pqr(a^3 + b^3 + c^3)$

$= -abc(3pqr) + pqr(3abc)$

$= 0$

Since $x + y + z = 0$

$\Rightarrow x^3 + y^3 + z^3 = 3xyz$

$a + b + c = 0$

$\Rightarrow a^3 + b^3 + c^3 = 3abc$

34. (A) $x + y = 4$

$3x - 2y = 9$

$\Delta = \begin{vmatrix} 1 & 1 \\ 3 & -2 \end{vmatrix} = -5$

$\Delta_x = \begin{vmatrix} 4 & 1 \\ 9 & -2 \end{vmatrix} = -17$

$\Delta_y = \begin{vmatrix} 1 & 4 \\ 3 & 9 \end{vmatrix} = -3$

$x = \frac{\Delta_x}{\Delta} = \frac{-17}{-5} = \frac{17}{5}$

$y = \frac{\Delta_y}{\Delta} = \frac{-3}{-5} = \frac{3}{5}$

35. (D) $\begin{vmatrix} b+c & c & b \\ c & c+a & a \\ b & a & a+b \end{vmatrix}$

$= (b+c)(ac+a^2+bc+ab-a^2)$

$\quad + c(ab-ac-bc)+b(ac-bc-ab)$

$= 4abc$

Hints and answers to Olympiad and NTSE level exercises

36. (C)

37. (A) Applying $C_1 \to C_1 + C_2$, we get

$$\begin{vmatrix} 1 & \cos^2 x & 1 \\ 1 & \sin^2 x & 1 \\ 2 & 12 & 2 \end{vmatrix} = 0.$$

38. (A) Use $C_1 \to C_1 - C_2 - C_3$

$$\Delta = \begin{vmatrix} 1 & \log_a b & \log_a c \\ \log_b a & 1 & \log_b c \\ \log_c a & \log_c b & 1 \end{vmatrix}$$

Now use change of base.

39. (B) $\begin{vmatrix} 1 & 4 & 20 \\ 1 & -2 & 5 \\ 1 & 2x & 5x^2 \end{vmatrix} = 0$

$$\Rightarrow \begin{vmatrix} 0 & 6 & 15 \\ 0 & -2-2x & 5(1-x^2) \\ 1 & 2x & 5x^2 \end{vmatrix} = 0$$

$$\begin{pmatrix} R_1 \to R_1 - R_2 \\ R_2 \to R_2 - R_3 \end{pmatrix}$$

$$\Rightarrow 3 \times 2 \times 5 \times \begin{vmatrix} 0 & 1 & 1 \\ 0 & -(1+x) & 1-x^2 \\ 1 & x & x^2 \end{vmatrix} = 0$$

$$\Rightarrow (1+x)\begin{vmatrix} 0 & 1 & 1 \\ 0 & -1 & 1-x \\ 1 & x & x^2 \end{vmatrix} = 0$$

$\Rightarrow x + 1 = 0$ or $x - 2 = 0$

$\Rightarrow x = -1, 2.$

40. (C) $\Delta = 0$

$$\Rightarrow bc.ca.ab \begin{vmatrix} a & b & c \\ b & c & a \\ c & a & b \end{vmatrix} = 0$$

$= -a^2b^2c^2\,(a^3 + b^3 + c^3 - 3abc) = 0$

$\Rightarrow a^2b^2c^2(a+b+c)\dfrac{1}{2}$

$\quad \times [(a - b)^2 + (b - c)^2 + (c - a^2)] = 0$

$\Rightarrow a + b + c = 0$

41. (B) $f(x) = 2(x - 3)(x - 5)$

$$\times \begin{vmatrix} 1 & x+3 & 3(x^2+3x+9) \\ 1 & x+5 & 4(x^2+5x+25) \\ 1 & 1 & 3 \end{vmatrix}$$

(Taking out $(x - 3)$, $(x - 5)$ and 2 from I row, II row and III column respectively)

$f(x) = 2(x - 3)(x - 5)$

$$\times \begin{vmatrix} 0 & (x+2) & 3(x^2+3x+8) \\ 0 & 2 & x^2+11x+73 \\ 1 & 1 & 3 \end{vmatrix},$$

$(R_1 \to R_1 - R_3$ and $R_2 \to R_2 - R_1)$

$= 2(x - 3)(x - 5)\,[1(x + 2)(x^2 + 11x + 73)$
$\quad - 6(x^2 + 3x + 8)]$

$= 2(x^2 - 8x + 15)(x^3 + 13x^2 + 95x + 146$
$\quad - 6x^2 - 18x - 48)$

$= 2(x^2 - 8x + 15)(x^3 + 7x^2 + 77x + 98)$

$= 2(x^5 - x^4 + 36x^3 - 413x^2 - 371x + 1470)$

$\therefore\ f(1)\cdot f(3) + f(3)\cdot f(5)\cdot f(1)$

$= 0 + 0 + 0 = 0 = f(3)$

42. (B) Putting $A = B$, we get that C_1 and C_2 are same.

Hence, $\Delta = 0$

Similarly, for $B = C$ or $C = A$

So, triangle is definitely isosceles.

43. (B) We have, $\begin{vmatrix} 1 & 1 & 1 \\ a & b & c \\ a^3 & b^3 & c^3 \end{vmatrix}$

$= (a - b)(b - c)(c - a)(a + b + c)$ ...(i)

Then

$$\Delta = \begin{vmatrix} 1 & 1 & 1 \\ (x-a)^2 & (x-b)^2 & (x-c)^2 \\ (x-b)(x-c) & (x-c)(x-a) & (x-a)(x-b) \end{vmatrix}$$

$= (a-b)(b-c)(c-a)(3x-a-b-c)$

Now given that a, b, c are all different then

$$D = 0 \Rightarrow x = \frac{1}{3}(a+b+c)$$

44. (C) Applying $C_1 \to C_1 + C_2 + C_3$, we get

$$\Delta = \begin{vmatrix} a+b+c-x & c & b \\ a+b+c-x & b-x & a \\ a+b+c-x & a & c-x \end{vmatrix}$$

$$= (a+b+c-x)\begin{vmatrix} 1 & c & b \\ 1 & b-x & a \\ 1 & a & c-x \end{vmatrix}$$

Applying $R_1 \to R_1 - R_2$, $R_2 \to R_2 - R_3$, we get

$$\Delta = (a+b+c-x)\begin{vmatrix} 0 & c-b+x & b-a \\ 0 & b-a-x & a-c+x \\ 1 & a & c-x \end{vmatrix}$$

$= (a+b+c-x)\,((x+c-b)(x+a-c)$
$\quad + (x+a-b)(b-a))$

$= (a+b+c-x)\,(x^2 + x(a-b)$
$\quad + (c-b)(a-c) + x(b-a) - (a-b)^2)$

$= (a+b+c-x) \times (x^2 + ac - c^2 - ab$
$\quad + bc - a^2 - b^2 - 2ab)$

$= (a+b+c-x)\,(x^2 - a^2 - b^2 - c^2 + ab + bc + ca)$

$= (a+b+c-x) \times$

$$\times\left(x^2 - \frac{1}{2}\{(a-b)^2 + (b-c)^2 + (c-a)^2\}\right)$$

Since $\Delta = 0$,

$\Rightarrow \quad x = a + b + c,$

$$\pm\sqrt{\frac{1}{2}((a-b)^2 + (b-c)^2 + (c-a)^2}$$

45. (B) $R_1 \to R_1 - R_2$, gives

$$\Delta = \begin{vmatrix} x-2 & 3(x-2) & -(x-2) \\ 2 & -3x & x-3 \\ -3 & 2x & x+2 \end{vmatrix}$$

$$= (x-2)\begin{vmatrix} 1 & 3 & -1 \\ 2 & -3x & x-3 \\ -3 & 2x & x+2 \end{vmatrix}$$

$$= (x-2)\begin{vmatrix} 1 & 3 & -1 \\ 0 & -3(x+2) & x-1 \\ 0 & 2x+9 & x-1 \end{vmatrix}$$

$\{R_2 \to R_2 - 2R_1, R_3 \to R_3 + 3R_1\}$

$= (x-2)\,\{-(3x+6)(x-1) - (x-1)(2x+9)\}$

$= -(x-2)(x-1)(5x+15)$

$\therefore \quad \Delta = 0$ gives $x = 2, 1, -3$

Rational Expressions

1. $\dfrac{x^2}{xyz}, \dfrac{y^2}{xyz}, \dfrac{z^2}{xyz}$

2. $\dfrac{(x+y)^2}{x^2-y^2}, \dfrac{x^3-y^3}{(x^2-y^2)}, \dfrac{x^2+y^2}{(x^2-y^2)}$

3. $\dfrac{x(x^2-x+1)}{x(x+1)(x-4)(x^2-x+1)},$

$\dfrac{x^2(x+1)}{x(x+1)(x^2-x+1)(x-4)},$

$\dfrac{(x-4)(x+1)(x-4)}{x(x+1)(x-4)(x^2-x+1)}$

4. (a) $\dfrac{2}{5} \quad x \neq -5$

(b) $\dfrac{5xz^3}{y^2} \quad y \neq 0$

(c) $\dfrac{x^2(x^2-y^2)}{x^2+3xy+y^2} \quad x^2+3xy+y^2 \neq 0$

(d) $\dfrac{x+2}{x-6} \quad x \neq 5, \; x \neq 6$

5. (a) $\dfrac{12x^2+11x-12}{12x^3} \quad x \neq 0$

(b) $\dfrac{x-2}{x(x+5)} \quad x \neq 0$

(c) $\dfrac{-14x+8}{(3x-2)(x-1)} \quad x \neq \dfrac{2}{3}$ and $x \neq 1$

(d) $\dfrac{-4}{2(2x-1)(x-1)(x+1)}; \quad x \neq 1$ and $\dfrac{1}{2}$

6. (a) $\dfrac{3(x+2)}{(x-1)(x-3)}$ $\quad x \neq 1$ and $x \neq 3$

(b) $\dfrac{2(a^3+x^3)}{a^4+a^2x^2+x^4} \Rightarrow \dfrac{2(a+b)}{a^2+ab+x^2}$

(c) 1

7. (a) $\dfrac{1}{x^3y^3}$

(b) $\dfrac{x+2}{x-1}$; $\quad x \neq -1$ and $x \neq -4$

(c) $\dfrac{x-2}{x^2-2}$

8. $\dfrac{8x^7}{1-x^8}$

9. $a^2-2ab+b^2$

10. $\dfrac{2(x+1)}{3x(x^2+x+1)}$

11. $\dfrac{x^2+2xy+y^2}{x^2-2xy+2y^2}$

12. $\dfrac{-b}{(a-b)(c-a)}$

13. $\dfrac{2(b+c-a)}{(a-b)(c-a)}$

14. 0

15. $\dfrac{8}{x^8-1}$

16. $\dfrac{-1}{x-a}$

17. $\dfrac{a+b+c}{(a+b-c)(a-b+c)(b+c-a)}$

18. (A) The expression is not defined for those values of x for which denominator is equal to zero.

$$(x-4)(x-1)^2 = 0$$

$$\Rightarrow \quad x = 4, 1$$

19. (C) The degree of numerator is greater than the degree of denominator.

20. (A) $\dfrac{y+10}{x-2} - \dfrac{8+y}{x-2} - \dfrac{2}{x-2}$

$$= \frac{y+10-8-y-2}{x-2} = \frac{0}{x-2} = 0$$

21. (D) $\dfrac{2x+1}{x^2-5} + \dfrac{x+11}{x^2-5} - \dfrac{x+1}{x^2-5}$

$$= \frac{2x+1+x+11-x-1}{x^2-5}$$

$$= \frac{2x+11}{x^2-5}$$

22. (B) $\dfrac{10}{(x+1)(x+2)} + \dfrac{11}{x+1} - \dfrac{8}{x+2}$

$$= \frac{10+11(x+2)-8(x+1)}{(x+1)(x+2)}$$

$$= \frac{10+11x+22-8x-8}{(x+1)(x+2)}$$

$$= \frac{3x+24}{(x+1)(x+2)}$$

$$= \frac{3(x+8)}{(x+1)(x+2)}$$

23. (B) $\dfrac{y+4}{y^2-1} + \dfrac{y-1}{y^2-3y-4} - \dfrac{y+1}{y^2-5y+4}$

$$= \frac{y+4}{(y-1)(y+1)} + \frac{y-1}{(y-4)(y+1)} - \frac{y+1}{(y-1)(y-4)}$$

$$= \frac{(y+4)(y-4)+(y-1)(y-1)-(y+1)(y+1)}{(y-1)(y+1)(y-4)}$$

$$= \frac{y^2-16+y^2-2y+1-y^2-2y-1}{(y-1)(y+1)(y-4)}$$

$$= \frac{y^2-4y-16}{(y-1)(y+1)(y-4)}$$

24. (A) $\dfrac{x^2-5x}{x^2-25} \div \dfrac{x^2-8x+15}{x-5}$

$$= \frac{x(x-5)}{(x-5)(x+5)} \times \frac{(x-5)}{(x-5)(x-3)}$$

$$= \frac{x}{(x+5)(x-3)};\ x \neq -5, 3, 5$$

25. (A) $\dfrac{x^2+x-6}{x^2-5x+6} \times \dfrac{x^3-27}{x^2-2xy+2y^2} \times \dfrac{x^4+4y^4}{x^2-9}$

$$=\frac{(x+3)(x-2)}{(x-2)(x-3)}\times\frac{(x-3)(x^2+9+3x)}{x^2-2xy+2y^2}\times\frac{x^4+4y^4+4x^2y^2-4x^2y^2}{x^2-9}$$

$$=\frac{(x^2+9+3x)}{(x^2-2xy+2y^2)}\times\frac{(x^2+2y^2)^2-(2xy)^2}{(x-3)}$$

$$(x^2+2y^2+2xy)$$

$$=\frac{(x^2+9+3x)}{(x^2-2xy+2y^2)}\times\frac{(x^2+2y^2-2xy)}{(x-3)}$$

$$=\frac{(x^2+9+3x)(x^2+2y^2+2xy)}{(x-3)}$$

26. (B)

$$\frac{2}{(x+1)(x-3)}-\frac{8(x+4)}{(x+1)(x+2)}+\frac{4x}{(x-3)(x+2)}=0$$

$$\Rightarrow \frac{2(x+2)-8(x-3)(x+4)+4x(x+1)}{(x+1)(x-3)(x+2)}=0$$

$$\Rightarrow 2x+4-8(x^2+x-12)+4x^2+4x=0$$

$$\Rightarrow 2x+4-8x^2-8x+96+4x^2+4x=0$$

$$\Rightarrow -4x^2-2x+100=0$$

$$\Rightarrow 2x^2+x-50=0$$

$$x=\frac{-1\pm\sqrt{1^2-4\times2(-50)}}{2\times2}$$

$$=\frac{-1\pm\sqrt{401}}{4}$$

27. (C) $\dfrac{x+3}{x^2-3x+2}+\dfrac{x+2}{x^2-4x+3}+\dfrac{x+1}{x^2-5x+6}=0$

$$\Rightarrow\frac{x+3}{(x-1)(x-2)}+\frac{x+2}{(x-1)(x-3)}+\frac{x+1}{(x-2)(x-3)}=0$$

$$\Rightarrow\frac{(x+3)(x-3)+(x+2)(x-2)+(x+1)(x-1)}{(x-1)(x-2)(x-3)}=0$$

$$\Rightarrow\frac{x^2-9+x^2-4+x^2-1}{(x-1)(x-2)(x-3)}=0$$

$$\Rightarrow 3x^2-14=0 \qquad (x\neq 1,2,3)$$

$$\Rightarrow 3x^2=14$$

$$\therefore x=\pm\sqrt{\frac{14}{3}}$$

28. (C) $\dfrac{1}{x}+\dfrac{2}{3x^2}+\dfrac{x-4}{4x^3}=0$

$$\Rightarrow \frac{12x^2+8x+3x-12}{12x^3}=0$$

$$\Rightarrow 12x^2+11x-12=0 \qquad (x\neq 0)$$

$$\Rightarrow x=\frac{-11\pm\sqrt{11^2-4\times12\times(-12)}}{2\times12}$$

$$\therefore x=\frac{-11\pm\sqrt{121+576}}{24}$$

$$=\frac{-11\pm\sqrt{697}}{24}$$

29. (B) $\dfrac{x-1}{2x^2+x-1}-\dfrac{x+1}{2x^2-3x+1}+\dfrac{4}{2(x^2-1)}=0$

$$\Rightarrow \frac{x-1}{(2x-1)(x+1)}-\frac{x+1}{(x-1)(2x-1)}+\frac{4}{2(x+1)(x-1)}=0$$

$$\Rightarrow \frac{2(x-1)^2-2(x+1)^2+4(2x-1)}{2(2x-1)(x+1)(x-1)}=0$$

$$\Rightarrow \frac{2x^2-4x+2-2x^2-4x-2+8x-4}{2(2x-1)(x+1)(x-1)}=0$$

$$\Rightarrow -4=0$$

which is not possible.

30. (D) $\dfrac{x+5}{(x-2)(x+3)}+\dfrac{x+8}{(x+3)(x-7)}=0$

$$\Rightarrow \frac{(x+5)(x-7)+(x+8)(x-2)}{(x-2)(x+3)(x-7)}=0$$

$$\Rightarrow \frac{x^2+5x-7x-35+x^2-2x+8x-16}{(x-2)(x+3)(x-7)}=0$$

$$\Rightarrow 2x^2+4x-51=0 \qquad (x\neq 2,-3,7)$$

$$\therefore x=\frac{-4\pm\sqrt{16-4\times2(-51)}}{2\times2}$$

$$=\frac{-4\pm2\sqrt{4+102}}{4}$$

$$=-1\pm\frac{\sqrt{106}}{2}$$

31. (C) $\frac{a}{(a-b)(a-c)}+\frac{b}{(b-c)(c-a)}+\frac{c}{(c-a)(c-b)}$

$=\frac{-a(b-c)+b(a-b)-c(a-b)}{(a-b)(b-c)(c-a)}$

$=\frac{-ab+ac+ba-b^2-ca+bc}{(a-b)(b-c)(c-a)}$

$=\frac{-b^2+bc}{(a-b)(b-c)(c-a)}$

$=\frac{-b(b-c)}{(a-b)(b-c)(c-a)}$

$=\frac{-b}{(a-b)(c-a)}$

32. (B) $\frac{x^2-2x-15}{x^2+7x+12}+\frac{x^2+3x-10}{x^2+9x+20}-\frac{x^2-4x-77}{x^2+11x+28}$

$=\frac{(x-5)(x+3)}{(x+4)(x+3)}+\frac{(x+5)(x-2)}{(x+4)(x+5)}-\frac{(x-11)(x+7)}{(x+4)(x+7)}$

$=\frac{(x-5)}{(x+4)}+\frac{(x-2)}{(x+4)}-\frac{(x-11)}{(x+4)}$

$=\frac{x-5+x-2-x+11}{(x+4)}$

$=\frac{(x+4)}{(x+4)}$

$=1$

33. (D) $\frac{1}{a^2-(b-c)^2}+\frac{1}{b^2-(c-a)^2}+\frac{1}{c^2-(a-b)^2}$

$=\frac{1}{(a-b+c)(a+b-c)}+\frac{1}{(b+c-a)(b-c+a)}+\frac{1}{(c+a-b)(c-a+b)}$

$=\frac{(c-a+b)+(a-b+c)+(a+b-c)}{(a-b+c)(a+b-c)(c-a+b)}$

$=\frac{(a+b+c)}{(a-b+c)(a+b-c)(c-a+b)}$

34. (A) $\frac{2x^4-4x^3-6x^2}{3x^4-6x^3-6x^2+12x}$

$=\frac{2x^2(x^2-2x-3)}{3x(x^3-2x^2-2x+4)}$

$=\frac{2x^2(x^2-3x+x-3)}{3x(x(x^2-2)-2(x^2-2))}$

$=\frac{2x^2\{x(x-3)+1(x-3)\}}{3x(x^2-2)(x-2)}$

$=\frac{2x^2(x-3)(x+1)}{3x(x^2-2)(x-2)}$

35. (C) $\frac{4-x}{x}+\frac{3}{2}=-\frac{4}{x}$

$\Rightarrow \frac{4-x}{x}+\frac{4}{x}=-\frac{3}{2}$

$\Rightarrow \frac{4-x+4}{x}=-\frac{3}{2}$

$\Rightarrow 2(8-x)=-3x$

$\Rightarrow 16-2x=-3x$

$\Rightarrow x=-16$

36. (B) $\frac{x}{x+3}+\frac{5}{x-7}=\frac{30}{x^2-4x-21}$

$\Rightarrow \frac{x}{x+3}+\frac{5}{x-7}-\frac{30}{x^2-4x-21}=0$

$\Rightarrow \frac{x}{x+3}+\frac{5}{x-7}-\frac{30}{x^2-7x+3x-21}=0$

$\Rightarrow \frac{x}{x+3}+\frac{5}{x-7}-\frac{30}{(x-7)(x+3)}=0$

$\Rightarrow \frac{x(x-7)+5(x+3)-30}{(x-7)(x+3)}=0$

$\Rightarrow \frac{x^2-7x+5x+15-30}{(x-7)(x+3)}=0$

$\Rightarrow x^2-2x-15=0 \quad (x\neq 7,-3)$

$\Rightarrow x^2-5x+3x-15=0$

$\Rightarrow x(x-5)+3(x-5)=0$

$\Rightarrow (x-5)(x+3)=0$

$\because x\neq -3$

$\therefore x=5$

37. (A) $\frac{2}{x-1}+\frac{3}{4}=\frac{5}{x-1}$

$\Rightarrow \frac{2}{x-1}-\frac{5}{x-1}=-\frac{3}{4}$

$\Rightarrow \frac{2-5}{x-1}=-\frac{3}{4}$

$\Rightarrow -\frac{3}{x-1}=-\frac{3}{4}$

$\Rightarrow \frac{1}{x-1}=\frac{1}{4}$

$\Rightarrow x-1=4$

$\therefore \quad x=5$

38. (D) $\frac{2}{a+1}+\frac{5}{a-1}=\frac{10}{a^2-1}$

$\Rightarrow \frac{2}{a+1}+\frac{5}{a-1}-\frac{10}{a^2-1}=0$

$\Rightarrow \frac{2(a-1)+5(a+1)-10}{(a+1)(a-1)}=0$

$\Rightarrow 2a-2+5a+5-10=0 \qquad (a\neq 1,-1)$

$\Rightarrow 7a=7$

$\therefore \quad a=1$

39. (C) $\frac{x}{x-2}-\frac{4}{x+2}=\frac{8}{x^2-4}$

$\Rightarrow \frac{x}{x-2}-\frac{4}{x+2}-\frac{8}{x^2-4}=0$

$\Rightarrow \frac{x(x+2)-4(x-2)-8}{(x+2)(x-2)}=0$

$\Rightarrow x^2+2x-4x+8-8=0 \qquad (x\neq 2,-2)$

$\Rightarrow x^2-2x=0$

$\Rightarrow x(x-2)=0$

$\because \quad x\neq 2$

$\therefore \quad x=0$

40. (C) $\frac{-3x}{x+1}+\frac{4x+1}{x}=\frac{-5}{x^2+x}$

$\Rightarrow \frac{-3x^2+(4x+1)(x+1)}{x(x+1)}=\frac{-5}{x(x+1)}$

$\Rightarrow \frac{-3x^2+4x^2+5x+1}{x(x+1)}=\frac{-5}{x(x+1)}$

$\Rightarrow x^2+5x+6=0 \qquad (x\neq 0,-1)$

$\therefore \quad x=-2,-3$

41. (A) $\frac{x^2-5x}{x^2-2x-15}\times\frac{x^2-9}{x^2-3x}$

$=\frac{x(x-5)}{(x+3)(x-5)}\cdot\frac{(x+3)(x-3)}{x(x-3)}$

$=1$

42. (B) $\frac{a^2-8a+16}{a}\times\frac{a^3}{4-a}$

$=\frac{(a-4)^2}{a}\times\frac{a^3}{(4-a)}$

$=-\frac{(a-4)^2}{a}\times\frac{a^3}{(a-4)}$

$=-a^2(a-4)$

$=a^2(4-a)$

43. (D) $\frac{x^2-4x-21}{x^2-49}\div\frac{x^2+8x+15}{x^2+2x-35}$

$=\frac{x^2-4x-21}{x^2-49}\times\frac{x^2+2x-35}{x^2+8x+15}$

$=\frac{(x-7)(x+3)}{(x-7)(x+7)}\times\frac{(x+7)(x-5)}{(x+3)(x+5)}$

$=\frac{x-5}{x+5}$

Hints and answers to Olympiad and NTSE level exercises

44. (D) $(a+b+c)^2=a^2+b^2+c^2+2(ab+bc+ca)$

or $\quad 0=a^2+b^2+c^2+2(ab+bc+ca)$

Now, $a+b+c=0$

or $\quad a=-(b+c)$

or $\quad a+b=-c$

Also, $\quad b+c=-a$ and $c+a=-b$

$\therefore \quad \frac{(a+b)^2}{ab}+\frac{(b+c)^2}{bc}+\frac{(c+a)^2}{ca}$

$=\frac{(-c)^2}{ab}+\frac{(-a)^2}{bc}+\frac{(-b)^2}{ca}$

$=\frac{a^3+b^3+c^3}{abc} \qquad (1)$

If $a+b+c=0$, then $a^3+b^3+c^3=3abc$

Substituting in (1), we get

$\frac{(a+b)^2}{ab}+\frac{(b+c)^2}{bc}+\frac{(c+a)^2}{ca}=\frac{3abc}{abc}=3$

45. (D) $\frac{a}{x+a}+\frac{b}{x-a}-\frac{c}{x^2+a^2}=\frac{(x-a)a+(x+a)b-c}{x^2-a^2}$

46. (D) $\frac{1}{(x+1)(x+2)}+\frac{1}{(x+2)(x+3)}+\frac{1}{(x+3)(x+1)}$

$=\frac{x+3+x+1+x+2}{(x+1)(x+2)(x+3)}$

$=\frac{3x+6}{(x+1)(x+2)(x+3)}$

$=\frac{3(x+2)}{(x+1)(x+2)(x+3)}$

$=\frac{3}{(x+1)(x+3)}$

47. (D) $\frac{x^3-3x^2}{9x^2-x^4}=\frac{x^2(x-3)}{x^2(9-x^2)}$

$=\frac{x-3}{(3-x)(3+x)}$

$=-\frac{1}{x+3}$

48. (D) $\frac{x^2-y^2}{(x+y)^2}=\frac{(x+y)(x-y)}{(x+y)^2}=\frac{(x-y)}{x+y}$

49. (B) $\frac{a^2-b^2}{a^2-3ab+2b^2}=\frac{(a+b)(a-b)}{a^2-2ab-ab+2b^2}$

$=\frac{(a+b)(a-b)}{a(a-2b)-b(a-2b)}$

$=\frac{(a+b)(a-b)}{(a-2b)(a-b)}=\frac{a+b}{a-2b}$

50. (A) $x^2+2x-8=x^2+4x-2x-8$

$=x(x+4)-2(x+4)(x-2)$

$x^2-x-20=x^2-5x+4x-20$

$=x(x-5)+4(x-5)=(x-5)(x+4)$

$\therefore\ \frac{x^2+2x-8}{x^2-x-20}=\frac{(x+4)(x-2)}{(x-5)(x+4)}=\frac{x-2}{x-5}$

51. (B) $\frac{x-5}{x+2}-\frac{3}{2-x}-\frac{14-x}{x^2-4}$

$=\frac{x-5}{x+2}+\frac{3}{x-2}+\frac{x-14}{x^2-4}$

$=\frac{(x-5)(x-2)+3(x+2)+x-14}{(x+2)(x-2)}$

$=\frac{x^2-7x+10+3x+6+x-14}{(x+2)(x-2)}$

$=\frac{x^2-3x+2}{(x+2)(x-2)}$

$=\frac{(x-2)(x-1)}{(x+2)(x-2)}=\frac{x-1}{x+2}$

52. (C) $\frac{2(m+2)}{4m+m^2-12}-\frac{1}{m-2}$

$=\frac{2(m+2)}{(m-2)(m+6)}-\frac{1}{m-2}$

$=\frac{2(m+2)-(m+6)}{(m-2)(m+6)}$

$=\frac{2m+4-m-6}{(m-2)(m+6)}$

$=\frac{(m-2)}{(m-2)(m+6)}$

$=\frac{1}{m+6}$

53. (B) $\frac{5}{x}=\frac{2}{a}+\frac{3}{b}$

$\Rightarrow\ \frac{5}{x}=\frac{2b+3a}{ab}$

$\Rightarrow\ x(2b+3a)=5ab$

$\therefore\ x=\frac{5ab}{2b+3a}$

Transformations

1. (a) $\begin{pmatrix}3\\-4\end{pmatrix}$ (b) $\begin{pmatrix}-3\\4\end{pmatrix}$ (c) $\begin{pmatrix}3\\-4\end{pmatrix}$

2. (a) $A(1, 3)$ and $B(2, 5)$ on the x-axis is $\begin{pmatrix}1\\-3\end{pmatrix}$ and $\begin{pmatrix}2\\-5\end{pmatrix}$

(b) $C(-3, 4)$ and $D(-4, -6)$ on the y-axis is $\begin{pmatrix}3\\-4\end{pmatrix}$ and $\begin{pmatrix}4\\-6\end{pmatrix}$

3. Reflection of $A(1, 1)$ on the y-axis is $\begin{pmatrix} -1 \\ 1 \end{pmatrix}$

Reflection of $B(3, 5)$ on the y-axis is $\begin{pmatrix} -3 \\ 5 \end{pmatrix}$

Reflection of $C(2, 7)$ on the y-axis is $\begin{pmatrix} -2 \\ 7 \end{pmatrix}$

4. (a)

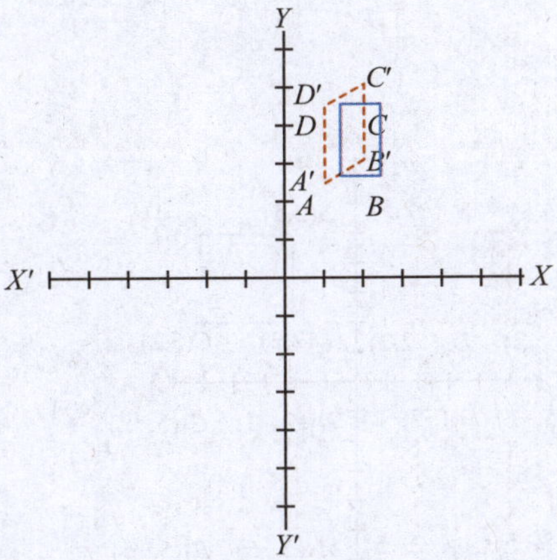

(b)

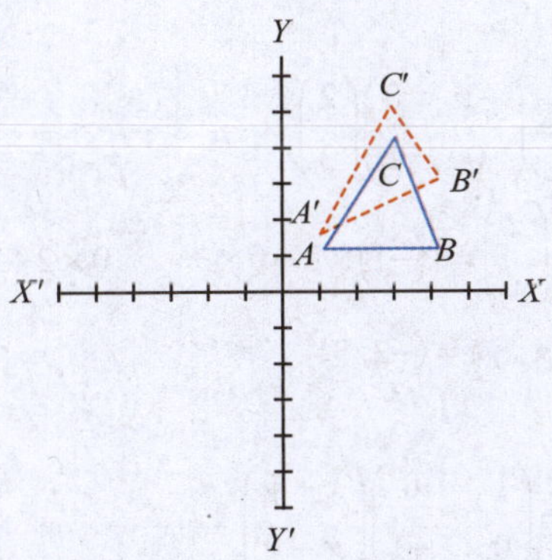

(c)

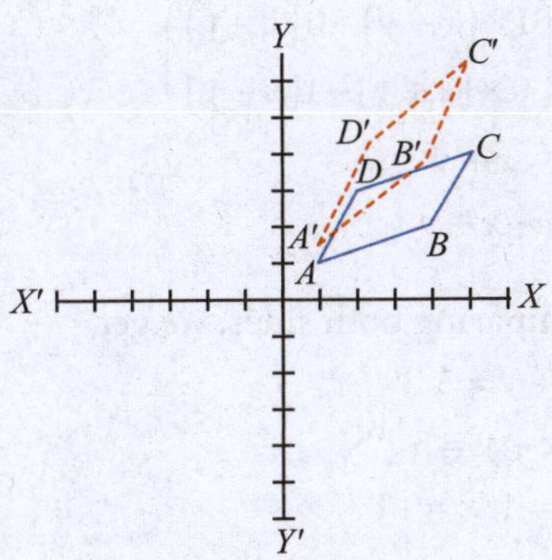

5. (a)

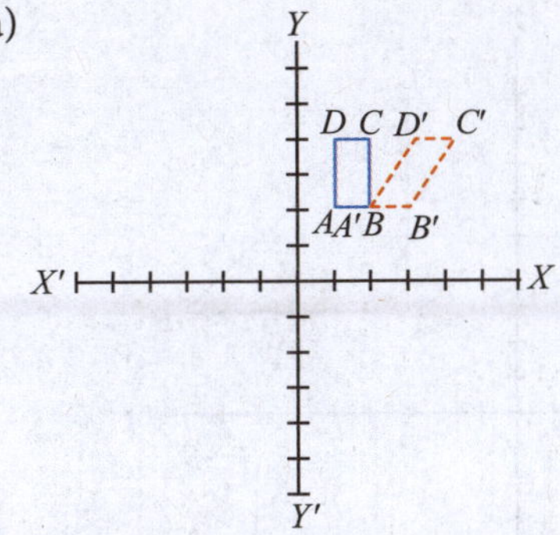

(b)

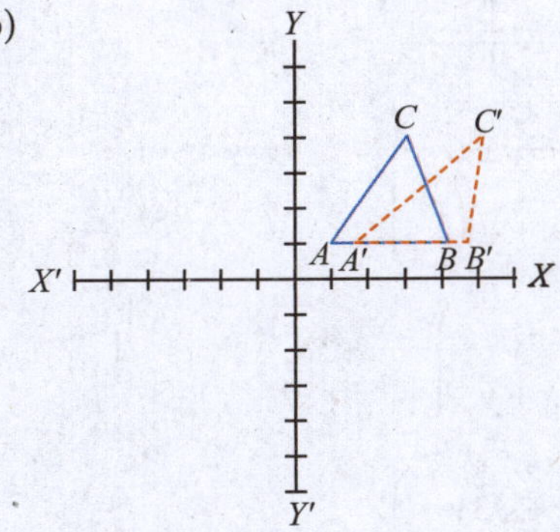

(c)

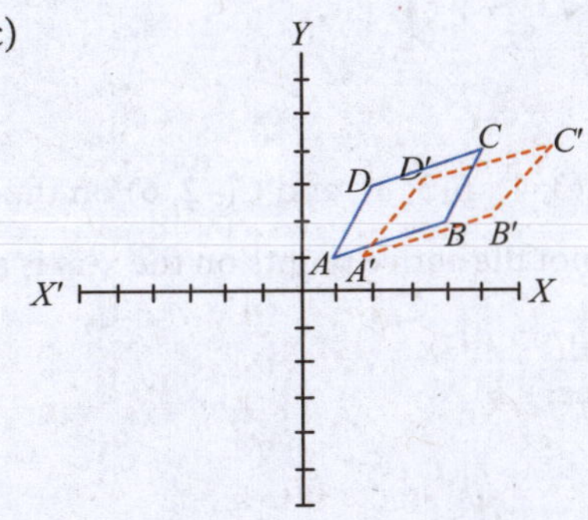

6.

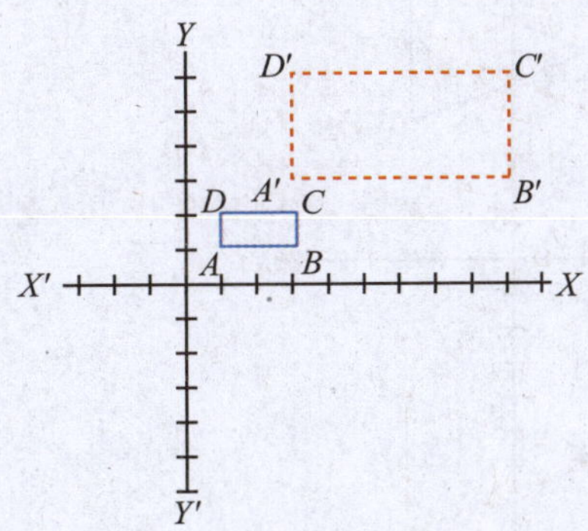

7.

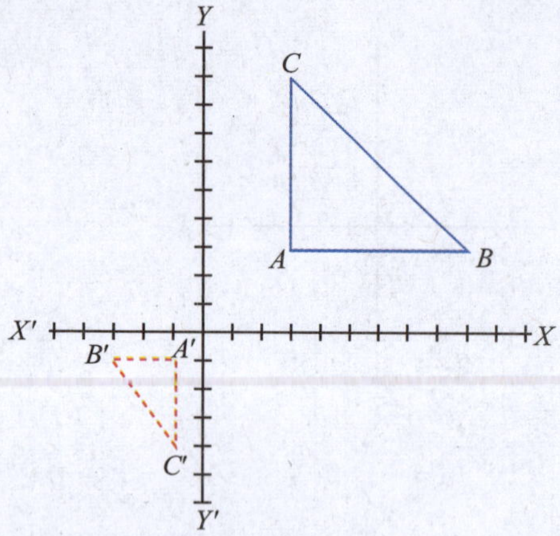

8. (a) $\frac{1}{2}\begin{pmatrix} \sqrt{3}x+y \\ -x+\sqrt{3}y \end{pmatrix}$

(b) $\frac{1}{2}\begin{pmatrix} x+y \\ -x+y \end{pmatrix}$

(c) $\frac{1}{2}\begin{pmatrix} x+\sqrt{3}y \\ -\sqrt{3}y+y \end{pmatrix}$

9. *Hint:*

(a) Plot $A(3, 4)$, $B(3, 8)$, and $C(-2, 6)$ on the graph.

(b) Also plot the above points on the x-axis and the y-axis.

(a) **Answer:**

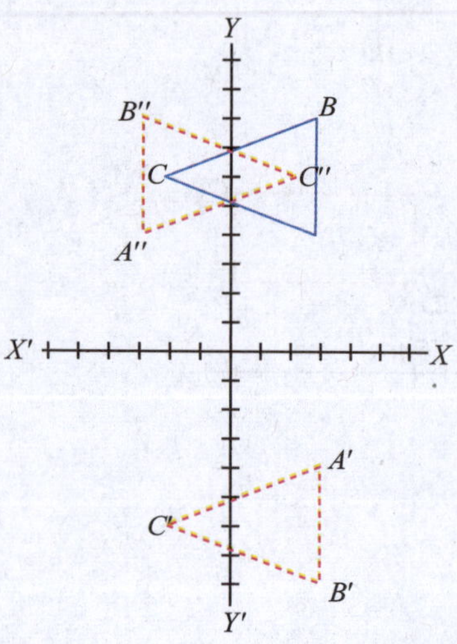

10.

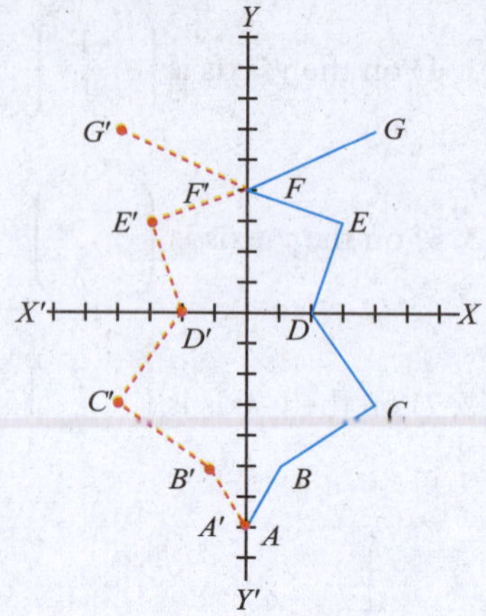

11.

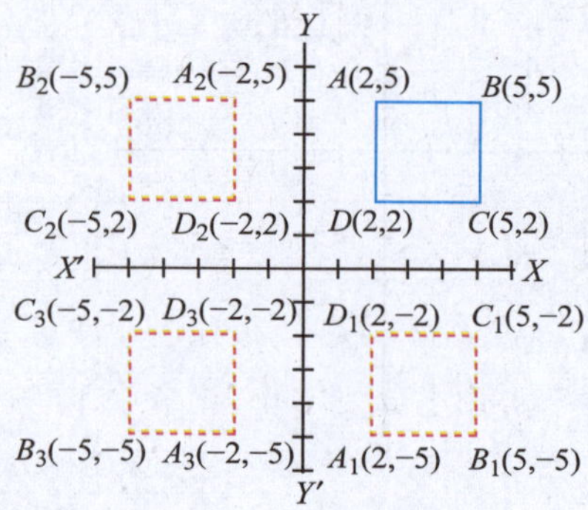

12. (D) $\begin{pmatrix} x \\ y \end{pmatrix} = \begin{pmatrix} -1 & 0 \\ 0 & 1 \end{pmatrix}\begin{pmatrix} 2 \\ 3 \end{pmatrix}$

$\Rightarrow \quad \begin{pmatrix} x \\ y \end{pmatrix} = \big((-1)\times 2+0\times 3 \qquad 0\times 2+1\times 3\big)$

$\therefore \quad (x, y) = (-2, 3)$

13. (A) $\begin{pmatrix} 3 \\ 5 \end{pmatrix} = \begin{pmatrix} 1 & 0 \\ 0 & -1 \end{pmatrix}\begin{pmatrix} x+y \\ x-y \end{pmatrix}$

$\begin{pmatrix} 3 \\ 5 \end{pmatrix} = \begin{pmatrix} 1\times(x+y)+0(x-y) \\ 0\times(x+y)-1(x-y) \end{pmatrix}$

$\begin{pmatrix} 3 \\ 5 \end{pmatrix} = \begin{pmatrix} x+y \\ -x+y \end{pmatrix}$

On comparing both sides, we get

$x + y = 3$

and $-x + y = 5$

$\therefore \quad y = 4, x = -1$

14. (B) $2\alpha + 3\beta = 3$ ...(i)

$3\alpha - 2\beta = -2$ ...(ii)

By $2 \times$ (i) + $3 \times$ (ii), we get

$4\alpha + 6\beta = 6$

$9\alpha + 6\beta = -6$

$13\alpha = 0$

$\therefore \quad \alpha = 0$ and $\beta = 1$

15. (C) $\begin{pmatrix} x \\ y \end{pmatrix} = \begin{pmatrix} -1 & 0 \\ 0 & -1 \end{pmatrix}\begin{pmatrix} 3 \\ 5 \end{pmatrix}$

$= \begin{pmatrix} -3\times1+0\times5 \\ 0\times3-1\times5 \end{pmatrix}$

$= \begin{pmatrix} -3 \\ -5 \end{pmatrix}$

16. (A)

17. (C) Reflection of (2, 2) through x-axis

$\begin{pmatrix} x \\ y \end{pmatrix} = \begin{pmatrix} 1 & 0 \\ 0 & -1 \end{pmatrix}\begin{pmatrix} 2 \\ 2 \end{pmatrix}$

$= \begin{pmatrix} 1\times2+0\times2 \\ 0\times2-1\times2 \end{pmatrix}$

$= \begin{pmatrix} 2 \\ -2 \end{pmatrix}$

Reflection of (4, 4) through x-axis

$\begin{pmatrix} x \\ y \end{pmatrix} = \begin{pmatrix} 1 & 0 \\ 0 & -1 \end{pmatrix}\begin{pmatrix} 4 \\ 4 \end{pmatrix}$

$= \begin{pmatrix} 1\times4+0\times4 \\ 0\times4-1\times4 \end{pmatrix}$

$= \begin{pmatrix} 4 \\ -4 \end{pmatrix}$

Now,

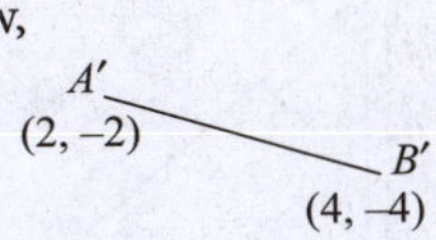

Slope $= \dfrac{-4-(-2)}{4-2} = \dfrac{-4+2}{2} = \dfrac{-2}{2} = -1$

Equation of reflected line segment $A'B'$ is

$y + 2 = -1(x - 2)$

$\Rightarrow \quad y + 2 = -x + 2$

$\Rightarrow \quad y = -x$

18. (A) Reflection of $A(2, 2)$ through origin.

$\begin{pmatrix} x \\ y \end{pmatrix} = \begin{pmatrix} -1 & 0 \\ 0 & -1 \end{pmatrix}\begin{pmatrix} 2 \\ 2 \end{pmatrix} = \begin{pmatrix} -2 \\ -2 \end{pmatrix}$

19. (D) $\begin{pmatrix} \alpha-\beta \\ \alpha+\beta \end{pmatrix} = \begin{pmatrix} 3 & 0 \\ 0 & 2 \end{pmatrix}\begin{pmatrix} 3\alpha+2 \\ 3\beta+5 \end{pmatrix}$

$\begin{pmatrix} \alpha-\beta \\ \alpha+\beta \end{pmatrix} = \begin{pmatrix} 9\alpha+6 \\ 6\beta+10 \end{pmatrix}$

$\alpha - \beta = 9\alpha + 6$

$\Rightarrow \quad 8\alpha + \beta = -6$

$\alpha + \beta = 6\beta + 10$

$\Rightarrow \quad \alpha - 5\beta = 10$

Now $\quad 4\alpha + 5\beta = -30$

$\alpha - 5\beta = 10$

$41\alpha = -20$

$\alpha = \dfrac{-20}{41}$

$5\beta = 10 + \left(\dfrac{-20}{41}\right)$

$\Rightarrow \quad \beta = \dfrac{410-20}{41\times5} = \dfrac{390}{41\times5} = \dfrac{78}{41}$

20. (C) $\begin{pmatrix} X \\ Y \end{pmatrix} = \begin{pmatrix} \cos 45^\circ & \sin 45^\circ \\ -\sin 45^\circ & \cos 45^\circ \end{pmatrix}\begin{pmatrix} 3 \\ 4 \end{pmatrix}$

$= \begin{pmatrix} \frac{1}{\sqrt{2}} & \frac{1}{\sqrt{2}} \\ \frac{-1}{\sqrt{2}} & \frac{1}{\sqrt{2}} \end{pmatrix}\begin{pmatrix} 3 \\ 4 \end{pmatrix} = \begin{pmatrix} \frac{3}{\sqrt{2}}+\frac{4}{\sqrt{2}} \\ \frac{-3}{\sqrt{2}}+\frac{4}{\sqrt{2}} \end{pmatrix} = \begin{pmatrix} \frac{7}{\sqrt{2}} \\ \frac{1}{\sqrt{2}} \end{pmatrix}$

21. (A) After reflection from x-axis = (2, –4)

Then, reflection through y-axis = (–2, –4)

And then through origin = (2, 4)

22. (A) Reflection of point A through the given transformation is

$\begin{pmatrix} X \\ Y \end{pmatrix} = \begin{pmatrix} \frac{-1}{3} & 1 \\ 1 & \frac{-1}{3} \end{pmatrix}\begin{pmatrix} 2 \\ 3 \end{pmatrix} = \begin{pmatrix} \frac{-2}{3}+3 \\ 2-\frac{3}{3} \end{pmatrix} = \begin{pmatrix} \frac{7}{3} \\ 1 \end{pmatrix}$

For point B,

$\begin{pmatrix} X \\ Y \end{pmatrix} = \begin{pmatrix} \frac{-1}{3} & 1 \\ 1 & \frac{-1}{3} \end{pmatrix}\begin{pmatrix} 6 \\ 4 \end{pmatrix} = \begin{pmatrix} \frac{-6}{3}+4 \\ 6-\frac{4}{3} \end{pmatrix} = \begin{pmatrix} 2 \\ \frac{14}{3} \end{pmatrix}$

For point C,

$$\begin{pmatrix} X \\ Y \end{pmatrix} = \begin{pmatrix} \frac{-1}{3} & 1 \\ 1 & \frac{-1}{3} \end{pmatrix} \begin{pmatrix} 7 \\ 8 \end{pmatrix} = \begin{pmatrix} \frac{-7}{3}+8 \\ 7-\frac{8}{3} \end{pmatrix} = \begin{pmatrix} \frac{17}{3} \\ \frac{13}{3} \end{pmatrix}$$

23. (B) $\begin{pmatrix} X \\ Y \end{pmatrix} = \begin{pmatrix} \cos\theta & \sin 2\theta \\ -\sin 2\theta & -\cos\theta \end{pmatrix} \begin{pmatrix} x \\ y \end{pmatrix}, \theta = 0°$

$$\begin{pmatrix} X \\ Y \end{pmatrix} = \begin{pmatrix} 1 & 0 \\ 0 & -1 \end{pmatrix} \begin{pmatrix} x \\ y \end{pmatrix}$$

Hints and answers to Olympiad and NTSE level exercises

24. (C) The new triangle will be a reflection of the old triangle.

25. (B) Reflection of (–3, 4) through origin is (3, 4).

Coordinates of new point,

$$\begin{pmatrix} X \\ Y \end{pmatrix} = \begin{pmatrix} \frac{1}{2} & 1 \\ 0 & 1 \end{pmatrix} \begin{pmatrix} 3 \\ 4 \end{pmatrix} = \begin{pmatrix} \frac{3}{2}+4 \\ 0+4 \end{pmatrix} = \begin{pmatrix} \frac{11}{2} \\ 4 \end{pmatrix}$$

26. (C) Reflection of (2, –5) through x-axis is (2, 5).

Coordinates of new point,

$$\begin{pmatrix} X \\ Y \end{pmatrix} = \begin{pmatrix} 0 & 1 \\ 0 & 1 \end{pmatrix} \begin{pmatrix} 2 \\ 5 \end{pmatrix} = \begin{pmatrix} 0+5 \\ 0+5 \end{pmatrix} = \begin{pmatrix} 5 \\ 5 \end{pmatrix}$$

27. (A) Reflection of (–1, –3) through y-axis is (1, –3).

Coordinates of new point,

$$\begin{pmatrix} X \\ Y \end{pmatrix} = \begin{pmatrix} 1 & -1 \\ 0 & -1 \end{pmatrix} \begin{pmatrix} 1 \\ -3 \end{pmatrix} = \begin{pmatrix} 1+3 \\ 0+3 \end{pmatrix} = \begin{pmatrix} 4 \\ 3 \end{pmatrix}$$

28. (C) Let O be the origin and the coordinates of a point P be (x, y).

Then $OP = \sqrt{x^2 + y^2}$

Coordinates of new point P' when P is reflected through x-axis,

$$\begin{pmatrix} X \\ Y \end{pmatrix} = \begin{pmatrix} 1 & 0 \\ 0 & -1 \end{pmatrix} \begin{pmatrix} x \\ y \end{pmatrix} = \begin{pmatrix} x+0 \\ 0-y \end{pmatrix} = \begin{pmatrix} x \\ -y \end{pmatrix}$$

Then $OP' = \sqrt{x^2 + y^2}$

Coordinates of new point P'' when P is reflected through y-axis,

$$\begin{pmatrix} X \\ Y \end{pmatrix} = \begin{pmatrix} -1 & 0 \\ 0 & 1 \end{pmatrix} \begin{pmatrix} x \\ y \end{pmatrix} = \begin{pmatrix} -x+0 \\ 0+y \end{pmatrix} = \begin{pmatrix} -x \\ y \end{pmatrix}$$

Then $OP'' = \sqrt{x^2 + y^2}$

29. (D) Let P the point located in second quadrant. Then its coordinates will be of the form $(-x, y)$, where $x > 0, y > 0$.

Coordinates of new point P' when P is reflected through origin,

$$\begin{pmatrix} X \\ Y \end{pmatrix} = \begin{pmatrix} -1 & 0 \\ 0 & -1 \end{pmatrix} \begin{pmatrix} -x \\ y \end{pmatrix} = \begin{pmatrix} x+0 \\ 0-y \end{pmatrix} = \begin{pmatrix} x \\ -y \end{pmatrix}$$

Clearly, point $(x, -y)$ will lie in the fourth quadrant.

30. (B) Coordinates of B will be (4, –6).

Coordinates of C will be

$$\begin{pmatrix} X \\ Y \end{pmatrix} = \begin{pmatrix} \frac{1}{2} & 1 \\ -1 & -\frac{1}{2} \end{pmatrix} \begin{pmatrix} 4 \\ -6 \end{pmatrix} = \begin{pmatrix} 2-6 \\ -4+3 \end{pmatrix} = \begin{pmatrix} -4 \\ -1 \end{pmatrix}$$

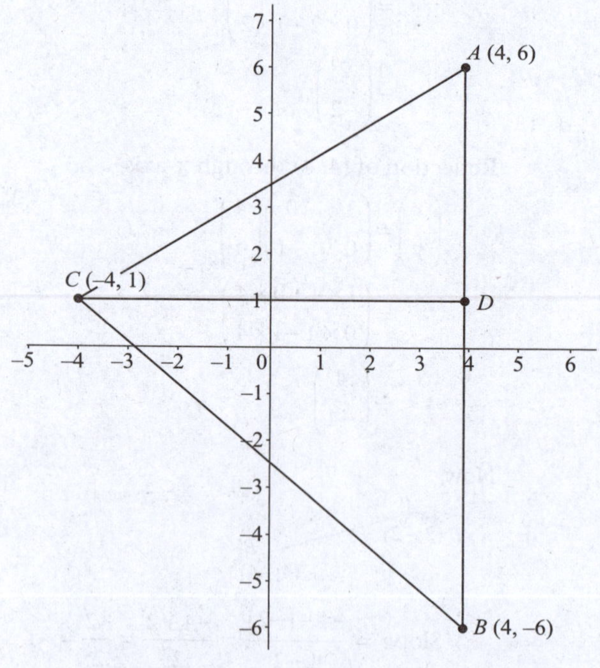

Area of $ABC = \frac{1}{2} \times 6 \times 8 + \frac{1}{2} \times 7 \times 8$

$= 24 + 28 = 52$ sq. unit

31. (D)

32. (D) Point obtained after first transformation,

$$\begin{pmatrix} X \\ Y \end{pmatrix} = \begin{pmatrix} \frac{2}{3} & -1 \\ 0 & -\frac{1}{4} \end{pmatrix} \begin{pmatrix} -6 \\ 4 \end{pmatrix} = \begin{pmatrix} -4-4 \\ 0-1 \end{pmatrix} = \begin{pmatrix} -8 \\ -1 \end{pmatrix}$$

Point obtained after first transformation,

$$\begin{pmatrix} X \\ Y \end{pmatrix} = \begin{pmatrix} 1 & \frac{-1}{2} \\ 0 & -1 \end{pmatrix} \begin{pmatrix} -8 \\ -1 \end{pmatrix} = \begin{pmatrix} -8+\frac{1}{2} \\ 0+1 \end{pmatrix} = \begin{pmatrix} -\frac{15}{2} \\ 1 \end{pmatrix}$$

33. (C) Area of triangle having vertices (x_1, y_1), (x_2, y_2), and (x_3, y_3) is given by

$$\Delta = \frac{1}{2}\left[x_1(y_2 - y_3) + x_2(y_3 - y_1) + x_3(y_1 - y_2)\right]$$

The coordinates (x_1, y_1) will transform to

$$\begin{pmatrix} X \\ Y \end{pmatrix} = \begin{pmatrix} \frac{1}{2} & \frac{-1}{2} \\ -1 & 1 \end{pmatrix} \begin{pmatrix} x_1 \\ y_1 \end{pmatrix}$$

$$= \begin{pmatrix} \frac{1}{2}x_1 - \frac{1}{2}y_1 \\ -x_1 + y_1 \end{pmatrix}$$

$$= \begin{pmatrix} \frac{x_1 - y_1}{2} \\ y_1 - x_1 \end{pmatrix}$$

i.e., $\left(\frac{x_1 - y_1}{2}, y_1 - x_1\right)$.

Similarly, other coordinates will be $\left(\frac{x_2 - y_2}{2}, y_2 - x_2\right)$ and $\left(\frac{x_3 - y_3}{2}, y_3 - x_3\right)$.

Now, area

$$\Delta = \frac{1}{2}\left[\left(\frac{x_1 - y_1}{2}\right)(y_2 - x_2 - y_3 + x_3) + \left(\frac{x_2 - y_2}{2}\right)(y_3 - x_3 - y_1 + x_1) + \left(\frac{x_3 - y_3}{2}\right)(y_1 - x_1 - y_2 + x_2)\right]$$

$$= 0$$

Circles and Cyclic Quadrilaterals

10. $\beta = 20°$

16. (A)

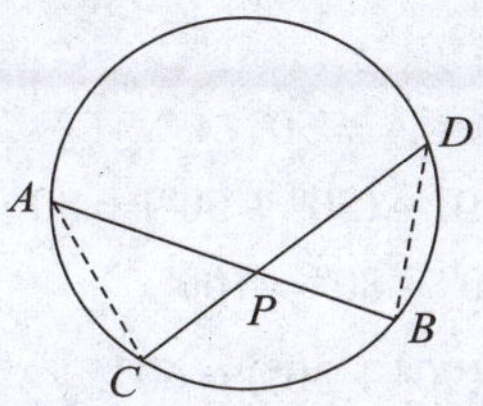

$\angle APC = \angle DPB$ [vertically opposite angles]

$\angle PAC = \angle PDB$ [angles in the same segment]

$\angle ACP = \angle DBP$ [angles in the same segment]

$\therefore \quad \Delta APC \sim \Delta DPB$

$\therefore \quad \frac{AP}{CP} = \frac{PD}{PB}$

$\Rightarrow \quad AP \times PB = CP \times PD$

17. (B) $\angle AOD = 2\angle ACD$ and $\angle AOD = 2\angle ABD$

$\therefore \quad \angle ACD = 2\angle ABD$

$\Rightarrow \quad 2(3b) = 2(2a)$

$\therefore \quad 3b = 2a$

$\angle ABC = (x - y)°$

18. (D) $x = 40° \times 2 = 80°$

$$y = \frac{360° - x}{2} = \frac{360° - 80°}{2} = \frac{280°}{2} = 140°$$

19. (C) From the figure, it is clear that AB and PQ are diameters.

So, using the theorem that angle in a semicircle is right angle, we get

$$\angle PAQ = \angle AQB = \angle QBP = \angle BPA = 9°$$

Also, $AB = PQ$.

Given that $AP = PB$.

Therefore, $APBQ$ is a square.

20. (A) $y = 180° - 120° = 60°$

$x = 2y = 120°$

21. (B) $\angle BAC = 62°$

$\angle APB = 180° - (46° + 62°)$

$= 180° - (108°)$

$= 72° = \beta$

22. (D) $\angle ADB = \angle ACB = 40°$ (angles in the same segment)

$\angle ABC = 180° - (40° + 20°)$
$= 180° - 60°$
$= 120°$

23. (D) $\angle ACO = 180° - 130° = 50°$
$OA = OC$ (radii)
$\Rightarrow \angle OAC = \angle OCA = 50°$
$\Rightarrow \angle AOC = 180° - (50° + 50°) = 80°$
$\angle COE = 180° - 80° = 100°$

24. (A) $\angle AOC = 180° - (90° + 30°) = 60°$
$\angle AOB = 2 \times 60° = 120°$
$\angle ADB = \frac{1}{2} \times 120° = 60°$

25. (B) $OA = OC$
$\Rightarrow \angle OAC = \angle OCA = 37°$
$\angle AOC = 180° - (37° + 37°)$
$= 180° - 74° = 106°$
$\angle ADC = \frac{360° - 106°}{2} = \frac{254°}{2} = 127°$

26. (C) $\angle OAD = \angle ODA = 30°$
$\angle AED = 180° - 60° = 120°$
$\angle AED = \frac{1}{2} \times 120° = 60°$

27. (B) $\angle BCP = 180° - (20° + 30°) = 180° - 50° = 130°$
$\angle ACB = 180° - 130° = 50°$
$\angle ACB = \angle ADB = 50°$ (angle in the same segment)

28. (B) Here $ABEF$ is a cyclic quadrilateral.
$\angle ABE = 180° - 80° = 100°$
$\angle CBE = 180° - 100° = 80°$
$\therefore \angle CDE = 180° - 80° = 100°$
$\therefore \angle COE = 2 \times 100° = 200°$

29. (A) $\angle CBD = 90°$ (angle in semicircle)
$\angle ABD = 180° - 90° = 90°$
$\angle AED = 180° - 90° = 90°$ (since ABCD is a cyclic quadrilateral)

30. (C) $\angle ADE = 60°$
$\therefore \angle ADC = 180° - 60° = 120°$
$\therefore \angle ABC = 180° - 120° = 60°$ (since ABCD is a cyclic quadrilateral)
$\angle AOC = 2 \times \angle ABC = 2 \times 60° = 120°$

31. (A) $\angle CEF = 180° - 55° = 125°$
$\angle CED = \angle AEF = 55°$
$\therefore \angle DCE = 180° - (55° + 30°) = 95°$
$\therefore \angle BCA = 180° - 95° = 85°$
$\angle AFE = 180° - (40° + 55°) = 85°$
$\angle DFB = 180° - 85° = 95°$
$\therefore \angle ABD = 180° - (95° + 30°)$
$= 180° - 125° = 55°$

Now, $\angle ABD + \angle CEF = 125° + 55° = 180°$
And $\angle DFB + \angle BCA = 95° + 85° = 180°$

32. (B) $\angle PQR = 180° - 50° = 130°$
$\angle PSR = 50°$
In ΔPSN, $\beta + 50° \angle SPN = 180°$
$\angle SPN = 130° - \beta$
In ΔRSM, $\alpha + 50° + \angle MRS = 180°$
$\angle MRS = 130° - \alpha$
Now,
$(130° - \beta) + (130° - \alpha) = 180°$
$260° - \beta - 3\beta = 180°$
$80° - 4\beta$
$\therefore \beta = 20°$

Hints and answers to Olympiad and NTSE level exercises

33. (C) $\angle QPT = \angle TSR = \angle q$

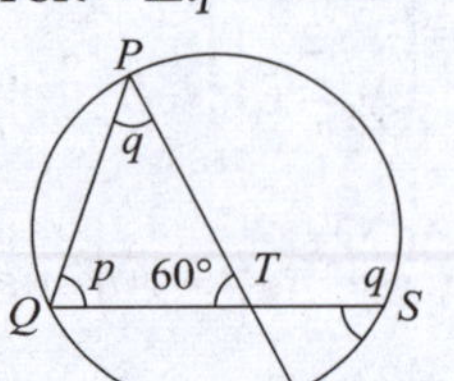

Sum of interior angles of a triangle = 180°
$\therefore$ In ΔPQT,
$60° + \angle p + \angle q = 180°$
or $\angle p + \angle q = 120°$

34. (A)

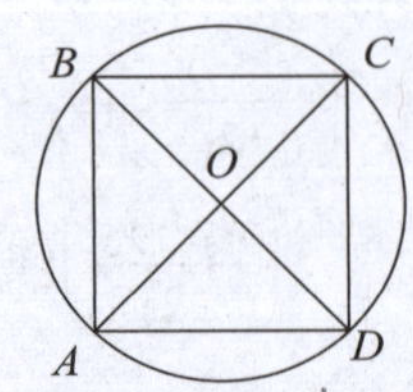

The two chords must be diameter.

Angle in a semicircle measures 90°. Therefore,

$\angle A = \angle B = \angle C = \angle D = 90°$

Therefore, *ABCD* is a rectangle.

35. (D) Given $\angle RQT = 110°$

$\angle PQR = 180° - 110° = 70°$

$\angle PRQ = \angle PQR = 70°$

[Since $PR = PQ$ and angles opposite to equal sides are equal]

$\angle POQ = 2\angle PRQ$

$\angle POQ = 140°$

Since $OP = OQ$ [Radius of the circle]

We have $\angle OPQ = \angle OQP$.

$\therefore$ $180° = \angle OPQ + \angle OQP + \angle POQ$

or $180° - \angle POQ = \angle OPQ + \angle OQP$

or $180° - 140° = 2\angle OPQ$

or $40° = 2\angle OPQ$

or $20° = \angle OPQ = \angle OQP$

36. (C) $\angle AOB = \dfrac{360}{3} = 120°$

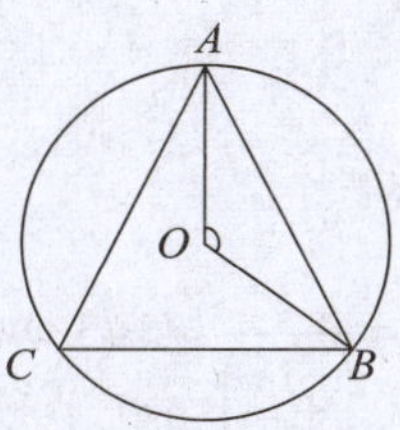

37. (C) $x + 2x = 180°$

or $3x = 180°$

or $x = 60°$

38. (C) $\angle PQR + \dfrac{1}{2}(114°) = 180°$

or $\angle PQR = 180° - 57° = 123°$

39. (A) $\angle AOD = 180° - 2 \times 20 = 140°$

40. (A) Area of kite = 2 × Area of ΔABC

$= 2 \cdot \dfrac{1}{2} \cdot x \cdot y = xy \text{ cm}^2$

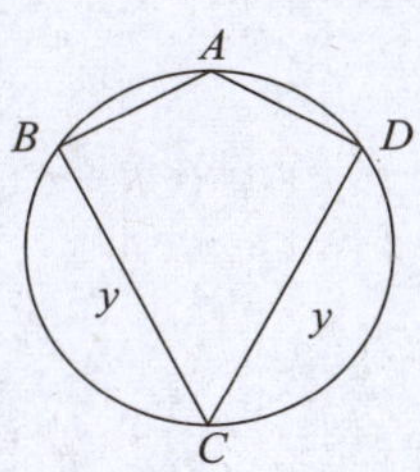

41. (A) In quadrilateral *FEBA*,

$\angle FEB = ½ \times \text{reflex } BAF = 104°$

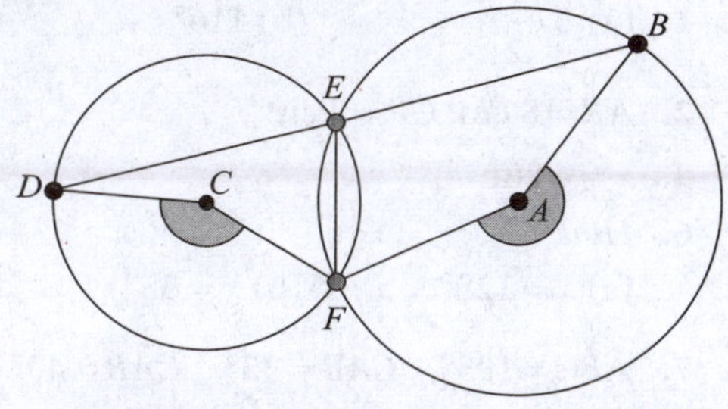

Since *DEB* is a straight line,

$\angle DEF + \angle FEB = 180°$

$\Rightarrow$ $\angle DEF = 76°$

In quadrilateral *DCFE*,

$\angle DCF = 2 \times \angle DEF = 152°$

42. (D) Let *A* be the centre of the circles. *QC* is tangent to smaller circle at *R*.

So, $\angle PCQ = 90°$

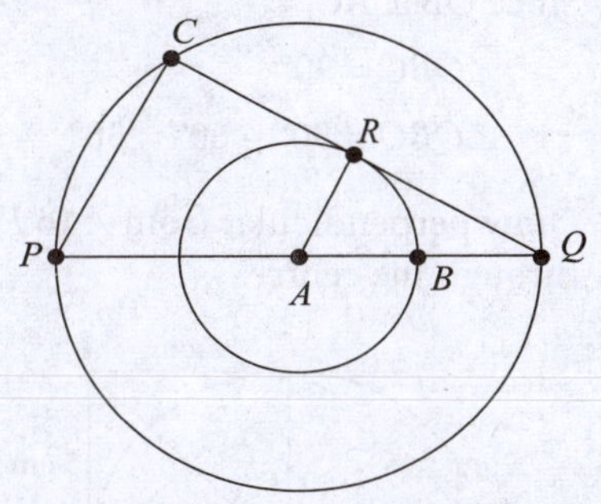

Also, *QC* is a chord of larger circle, so $AR \perp QC$ and $QR = CR$.

Since $PC \parallel AR$, ΔPQC and ΔAQR are similar.

$\therefore$ $PC = 2\,AR = 2b$

In ΔAQR, we have

$AR^2 + QR^2 = AQ^2$

$\Rightarrow$ $b^2 + QR^2 = a^2$

$\Rightarrow$ $QR^2 = a^2 - b^2$

$\therefore$ $QR = \sqrt{a^2 - b^2} = CR$

In ΔPCR, we have

$PC^2 + CR^2 = PR^2$

$\Rightarrow$ $4b^2 + a^2 - b^2 = PR^2$

$\Rightarrow$ $PR^2 = 3b^2 + a^2$

$\therefore$ $PR = \sqrt{3b^2 + a^2}$

Tangents to Circles

1. (a) $57\frac{1}{2}°$ (b) 116° (c) 70°

2. $AR = 8$ cm, $CP = 2$ cm

4. $x = 4$ cm

6. *Hint:*

(a) $x = 125°$ (b) $x = 55°$

7. $A\hat{B}C = 105°$, $C\hat{A}B = 35°$, $C\hat{A}B = 40°$

8. $R\hat{S}Q = 70°$, $R\hat{T}Q = 110°$, $R\hat{P}Q = 40°$

10. $x = 40$ cm

13. $A\hat{D}C = 70°$, $D\hat{C}B = 97°$

$D\hat{A}B = 83°$, $A\hat{B}C = 110°$

16. $10\sqrt{2}$ cm

17. (B) Since $OB \perp AC$,

$\therefore \quad \angle OBC = 90°$

$\angle QBO = 90° - 60° = 30°$

18. (C) Draw perpendicular from A to PQ. It will pass through the centre.

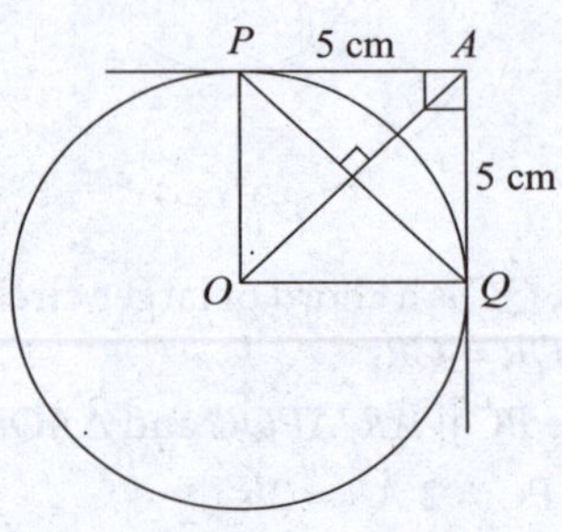

$$\angle OAQ = \frac{90°}{2} = 45°$$

In ΔAOQ,

$$\tan 45° = \frac{OQ}{AQ}$$

$\Rightarrow \quad OQ = AQ = 5$ cm

19. (D) $\angle AOB = 360° - 90° - 90° - 50° = 130°$

$$\angle OAB = \frac{180° - (130°)}{2} = \frac{50}{2} = 25°$$

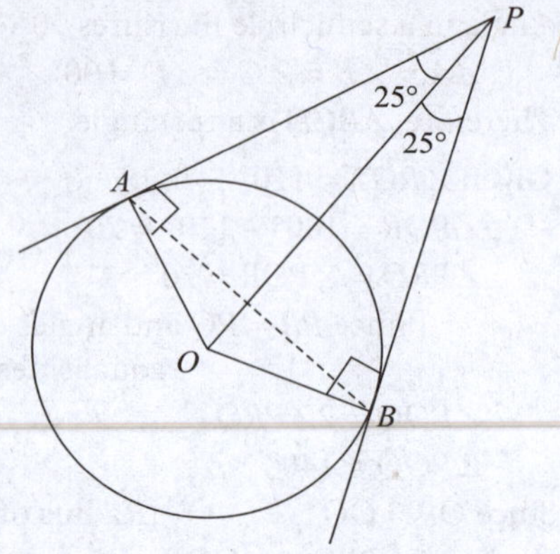

20. (B)

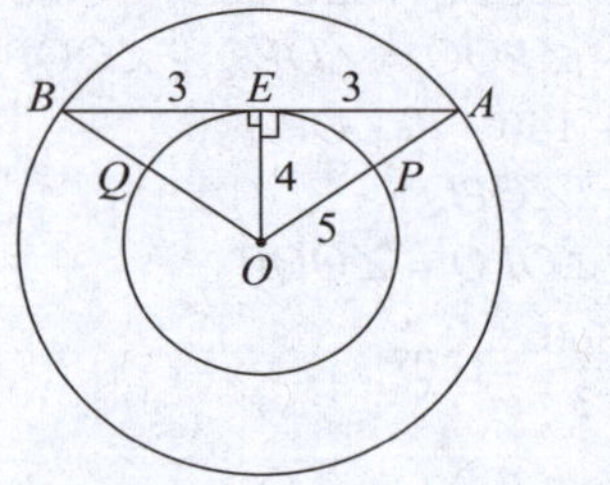

21. (D) $\tan 30° = \dfrac{OQ}{AQ}$

$$\therefore \quad AQ = \frac{OQ}{\tan 30°} = \frac{3}{\left(\frac{1}{\sqrt{3}}\right)} = 3\sqrt{3} \text{ cm}$$

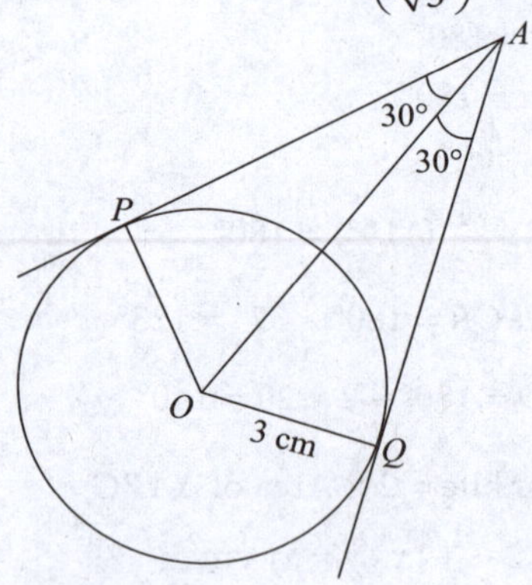

22. (C) $BC = 7$

$\Rightarrow \quad BQ + CQ = 7$

$\Rightarrow \quad BQ + 3 = 7 \qquad (CR = CQ = 3 \text{ cm})$

$\therefore \quad BQ = 4$

Now, $AB = x$

$\Rightarrow \quad AP + BP = x$

$\Rightarrow \quad AS + BQ = x \qquad (BQ = BP, AS = AP)$

$\therefore \quad x = 5 + 4 = 9$ cm

23. (B) $PR = PT = PQ$

$\therefore \quad RQ = PR + PQ = 2PT = 2 \times 3.8 = 7.6$ cm

24. (D) $\angle QOR = 180° - 46° = 134°$

25. (C)

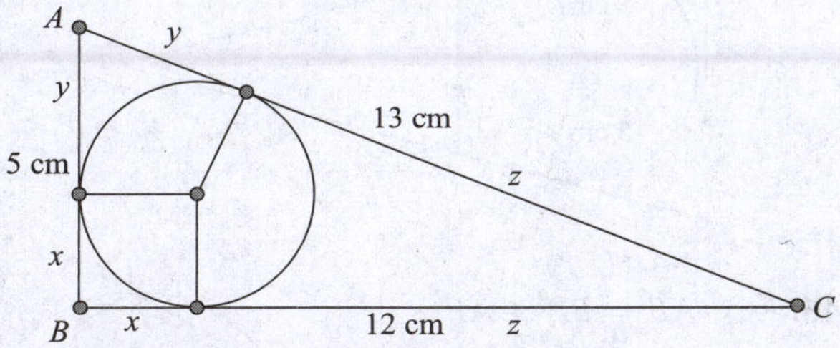

$x + y = 5$

$x + z = 12$

$y + z = 13$

$x + y + z = 15$

$z = 10, y = 3, x = 2$

26. (B) In ΔADE

$\Rightarrow \quad \tan\angle ADE = \frac{AE}{DE} = \frac{r}{5}$

$\Rightarrow \quad r = 5 \tan 45°$

$\Rightarrow \quad r = 5$ cm

27. (D) $AC = 11$ cm (Given)

$\Rightarrow \quad AR + RC = 11$

$\Rightarrow \quad AP + RC = 11$

$\Rightarrow \quad RC = 11 - 4 = 7$ cm

$BC = BQ + QC$

$\Rightarrow \quad BP + RC = 3 + 7 = 10$ cm

28. (C) $AE = AC - CE$

$= AB - PE = 8 - 3 = 5$ cm

29. (B) $a = 35°$ $(\because \angle OPQ = \angle OPR)$

$b = 180° - (90° + 35°) = 180° - 125° = 55°$

30. (D) $\angle OQP = 90° - 60° = 30°$

$\angle OQP = \angle OPQ$ $(\because OP = OQ)$

$= 30°$

31. (C) Draw a perpendicular from O to PS it will meet at Q.

In ΔQOP and QOS,

$OP = OS =$ radius of 2nd circle

$OQ = OQ$ (Common side)

$\angle OQP = \angle OQS = 90°$

$\therefore \quad \Delta QOP \cong QOS$ (RHS congruence criterion)

32. (D) $RO' = \sqrt{(O'S)^2 + (SR)^2} = 13$ cm

$OP = \sqrt{(PQ)^2 + (OQ)^2} = 5$ cm

$PR = PO + OO' + O'R$

$= 5 + (3 + 5) + 13$

$= 26$ cm

33. (C) $\angle OPQ = \angle OPR - \angle QPR = 90° - 60° = 30°$

$\angle POQ = 180° - (2 \times 30°) = 120°$

34. (B) $\angle OAP = 90°$

$OP = \sqrt{(OA)^2 + (AP)^2} = \sqrt{(6)^2 + (10)^2}$

$= \sqrt{36+100} = \sqrt{136}$ cm

$PB = \sqrt{(OP)^2 - (OB)^2} = \sqrt{136 - (3)^2}$

$= \sqrt{136-9} = \sqrt{127}$ cm

35. (D) $AB = AD + DB = 12$

$AD + BE = 12$ ...(i)

$AC = AF + FC = 10$

$AD + CE = 10$ ...(ii)

Adding (i) and (ii), we get

$2AD + CE + BE = 22$

$\Rightarrow \quad 2AD + BC = 22$

$\Rightarrow \quad 2AD = 22 - 8 = 14$

$\therefore \quad AD = 7$ cm

36. (C) $\angle ABP = \angle BAP = \frac{180° - 30°}{2} = 75°$

Since $PB \parallel AC$,

$\angle ACB + \angle CBP = 180°$

$\Rightarrow \quad 75° + \angle ABC + \angle ABP = 180°$

$\Rightarrow \quad 75° + \angle ABC + 75° = 180°$

$\Rightarrow \quad \angle ABC = 30$

37. (A) Since $OP \perp AB$

$\angle OPB = 90°$

$\Rightarrow \quad \angle OPQ + \angle QPB = 90°$

$\Rightarrow \quad \angle OPQ = 90° - \angle QPB = 90° - 50° = 40°$

$\therefore \quad \angle POQ = 180° - (2 \times 40°) = 100°$

38. (C) $\frac{OA}{O'A} = \frac{r+r+r}{r} = \frac{3r}{r} = \frac{3}{1} = 3$

39. (A) Since $OB = OP$ (radii of same circle)

$\angle OBP = \angle OPB = 30°$

$\therefore \quad \angle BOP = 120°$

Also, $\angle POA = 60°$

$OP = OA$ (radii of same circle)

$\angle OAP = \angle OPA = 60°$

$\angle TPA = 90 - 60° = 30°$

$\therefore \quad \angle PAT = 120°$

$\therefore \quad \angle PTA = 180° - (120° + 30°) = 30°$

40. (B) $OQ = OS$

$\Rightarrow \quad \angle OQS = \angle OSQ = 90° - 50° = 40°$

$OR = OS$

$\Rightarrow \quad \angle ORS = \angle OSR = 90° - 60° = 30°$

$\therefore \quad \angle QSR = \angle QSO + \angle RSO = 40° + 30° = 70°$

41. (C) $OQ = OP$

$\Rightarrow \quad \angle OPQ = \angle OQP = 90° - 60° = 30°$

$\angle POQ = 180° - 60° = 120°$

Major $\angle POQ = 240°$

$\Rightarrow \quad \angle PRQ = \dfrac{240°}{2} = 120°$

42. (B) $AR = BR = 4$ cm

$AR = AP = 4$ cm

$BC = QC - BQ = QC - BR = 11 - 4 = 7$ cm

43. (B) $CP = CQ = 11$ cm

$BQ = 11 - 7 = 4$ cm

$\therefore \quad BR = BQ = 4$ cm

44. (A) $OC \perp BCD$

$\angle OCD = 90°$

$\Rightarrow \quad \angle OCA + \angle ACD = 90°$

$\Rightarrow \quad \angle OAC + \angle ACD = 90°$

45. (B) $PT = \sqrt{13^2 - 5^2} = 12$ cm

$AP = AE = x$ (assume)

$AT = PT - AP = (12 - x)$cm

Since $OE \perp AB$

$\angle OEA = 90°$

$\angle AET = 90°$

$AT^2 = AE^2 + ET^2$

$\Rightarrow \quad (12 - x)^2 = x^2 + (13 - 5)^2$

$\Rightarrow \quad 24x = 80$

$\Rightarrow \quad x = \dfrac{10}{3}$

$\Rightarrow \quad AB = AE + BE = 2 \times \dfrac{10}{3} = \dfrac{20}{3}$ cm

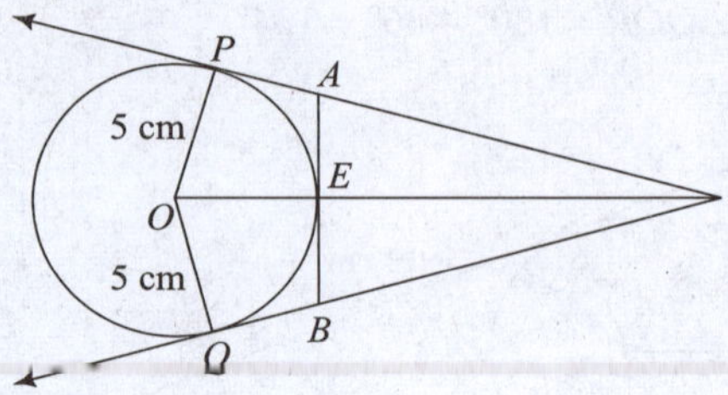

46. (A) $AB^2 = BP^2 + AP^2$

$36 = y^2 + x^2$ and

$OB^2 = OP^2 + BP^2$

$81 = (9 - x)^2 + y^2$

then $x = 2$ and $y = 4\sqrt{2}$

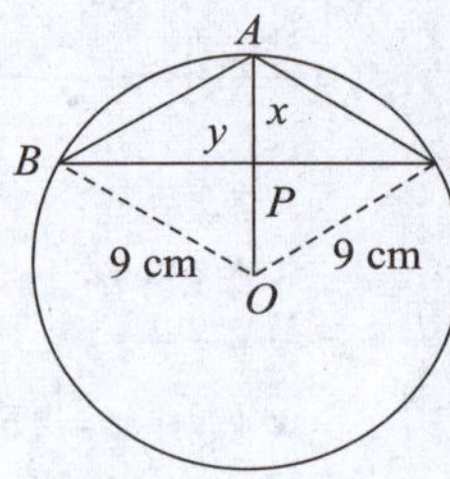

Area of triangle $= \dfrac{1}{2}(BC \times AP)$

$= \dfrac{1}{2} \times 8\sqrt{2} \times 2$ sq. cm

$= 8\sqrt{2}$ sq. cm

Hints and answers to Olympiad and NTSE level exercises

47. (C) $C = 2\pi r = 2 \times \dfrac{22}{7} \times \dfrac{21}{2} = 66$ cm

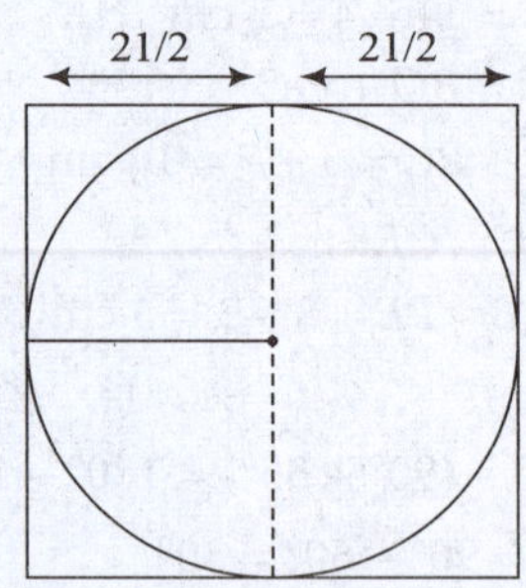

48. (D)

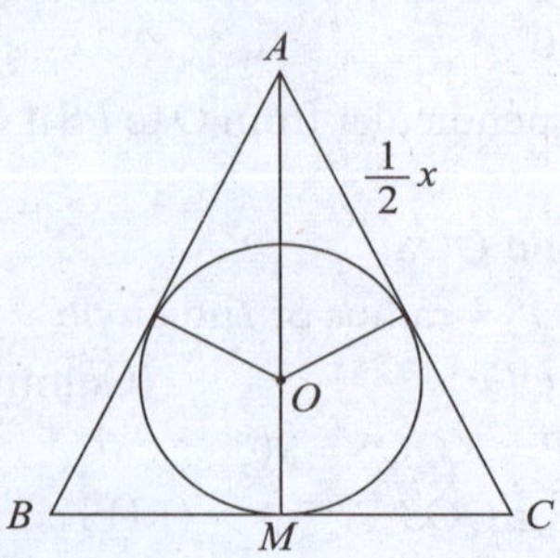

$$AM = \frac{\sqrt{3}}{2}x$$

or $$OM = \frac{1}{3} \cdot AM = \frac{1}{3} \cdot \frac{\sqrt{3}}{2} x = \frac{x}{2\sqrt{3}} \text{ cm}$$

49. (A) $AC = AQ$, $BC = BT$

$\Rightarrow$ Perimeter of $\Delta PAB = PA + PB + AB$

$= PA + PB + AC + CB$

$= PA + PB + AQ + BT$

$= PA + AQ + PB + BT$

$= 35 + 35 = 70$ cm

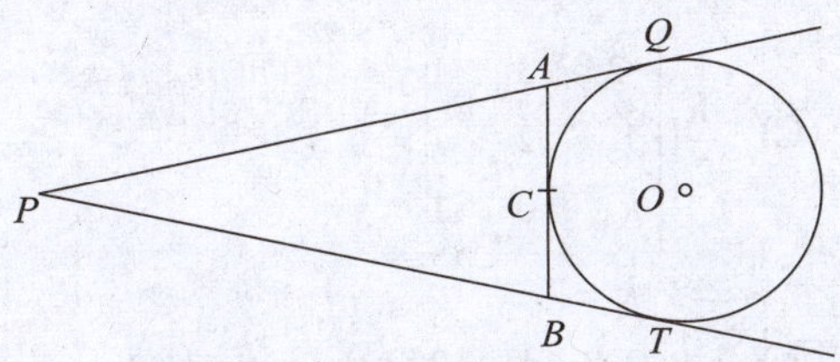

50. (A) $r = \sqrt{25^2 - 24^2}$

$= \sqrt{(25+24)(25-24)}$

$= \sqrt{49 \times 1} = 7$

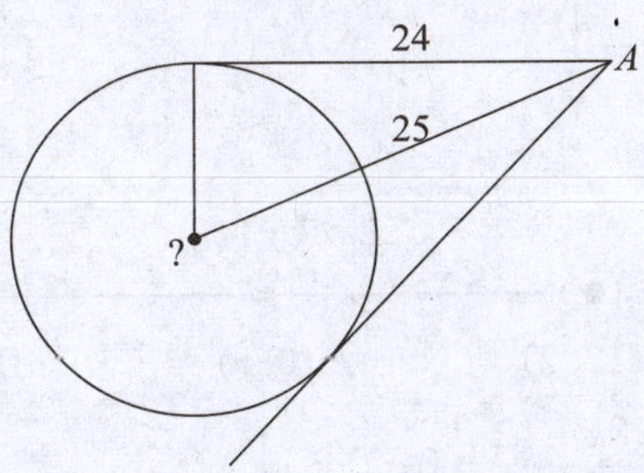

51. (D) $AB = \sqrt{12^2 - 5^2} = \sqrt{144 - 25} = \sqrt{119}$ cm

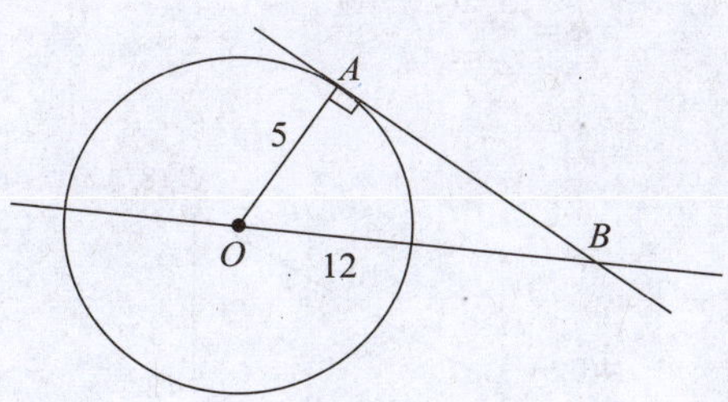

52. (B)

53. (B) $OB^2 = OA^2 + BA^2$

$\Rightarrow$ $OA^2 = OB^2 - BA^2$

$\Rightarrow$ $OA^2 = 13^2 - 12^2 = 25$

$\therefore$ $OA = 5$ cm

Area of circle $= \pi(5)^2 = 25\pi$ sq. cm

54. (D) $\angle OPA = 90° = \angle OQA$

So, $\angle POQ = 360° - 90° - 90° - 50° = 130°$

$\angle PRQ = ½ \times 130° = 65°$

55. (C) $\angle PBQ = 90°$

$\angle OPB = \angle OBP = 30°$

$\angle PQB = 180 - 30° - 90° = 60°$

$\therefore$ $\angle OBQ = 60°$

Therefore, triangle OBQ is equilateral.

$\therefore$ $OQ = OB = BQ = r$ (radius)

So, $\angle BQA = 180° - 60° = 120°$

Since radius OB is perpendicular to tangent AB,

$\angle OBA = 90°$

$\therefore$ $\angle QBA = 30°$

In ΔQAB,

$\angle QAB = 180° - 120° - 30° = 30°$

$\therefore$ $AQ = BQ = r$

So, $PQ : AQ = 2r : r = 2 : 1$

56. (A) Since $OP \perp AC$

$\angle OPC = 90°$

$\therefore$ $\angle OPR = 90° - 55° = 35°$

Since $OP = OR$ = radius

$\angle ORP = 35°$

Similarly, in ΔORQ

$\angle ORQ = 30°$

$\therefore$ $\angle PRQ = \angle ORP + \angle ORQ$

$= 35° + 30° = 65°$

Now, $\angle POQ = 2 \times \angle PRQ = 2 \times 65° = 130°$

In quadrilateral $POQA$,

$\angle OQA = \angle OPA = 90°$

$\therefore$ $\angle QAP = 360 - 90° - 90° - 130° = 50°$

Coordinate Geometry

1. (a) 5 (b) $\sqrt{13}$ (c) $\sqrt{(c-a)^2 + (d-b)^2}$

4. $CD = \frac{3\sqrt{10}}{2}$, $AE = \frac{3\sqrt{2}}{2}$, $BF = 3$

5. $5x + 3y - 18 = 0$

7. (3, 0)

8. 1 : 2

9. –6

10. $a = 2$

12. $(1, 1), \left(\frac{-5}{2}, \frac{9}{2}\right), \left(\frac{9}{2}, -\frac{5}{2}\right)$

13. $(\pm 3, 2)$

14. (5, 2)

15. (5, –2); (1, 2); (3, 4)

16. (D)

17. (A)

17. (A)

18. (B)

19. (B)

20. (C)

21. (B)

22. (C)

23. (D)

24. (C)

25. (A)

$$PQ = \sqrt{(\sqrt{5}+1-\sqrt{5}+2)^2 + \left(\sqrt{3}-1-\sqrt{3}-2\right)^2}$$

$$= \sqrt{9+9} = \sqrt{18} = 3\sqrt{2} \text{ units}$$

26. (B) $52 = (-3 - x)^2 + (4 - 7)^2$

$\Rightarrow \quad 25 = 9 + x^2 + 6x + 9$

$\Rightarrow \quad x^2 + 6x - 7 = 0$

$\Rightarrow \quad (x + 7)(x - 1) = 0$

$\therefore \quad x = -7, 1$

27. (C) $PA = PB$

$$\Rightarrow \sqrt{(x-6)^2 + (y+1)^2} = \sqrt{(x-2)^2 + (y-3)^2}$$

$$x^2 + 36 - 12x + y^2 + 1 + 2y$$

$$= x^2 + 4 - 4x + y^2 + 9 - 6y$$

$$8x - 8y - 24 = 0$$

$$x - y - 3 = 0$$

28. (C)

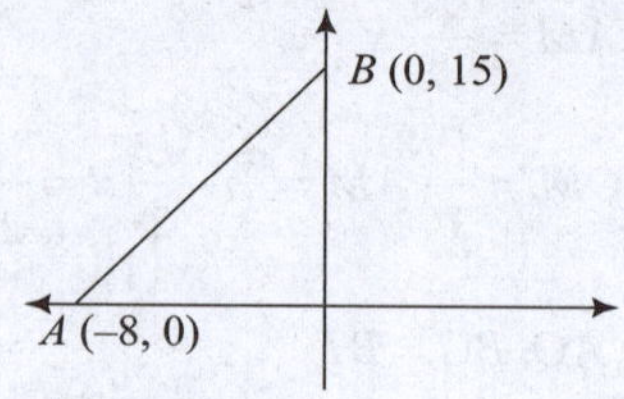

$$AB = \sqrt{(8)^2 + (15)^2}$$

$$= \sqrt{64 + 225}$$

$$= \sqrt{289}$$

$$= 17$$

29. (C) $\frac{1}{2}\begin{vmatrix} -2 & 3 & 1 \\ 1 & 2 & 1 \\ k & 0 & 1 \end{vmatrix} = 0$

$\Rightarrow \quad k = 7$

30. (A) Let P divide AB in the ratio $k : 1$.

Then,

$$\frac{7k + 1(-2)}{k+1} = 1 \Rightarrow \frac{7k - 2}{k+1} = 1$$

$$\Rightarrow 7k - 2 = k + 1 \Rightarrow 6k = 3$$

$$\Rightarrow k = \frac{1}{2}$$

31. (D)

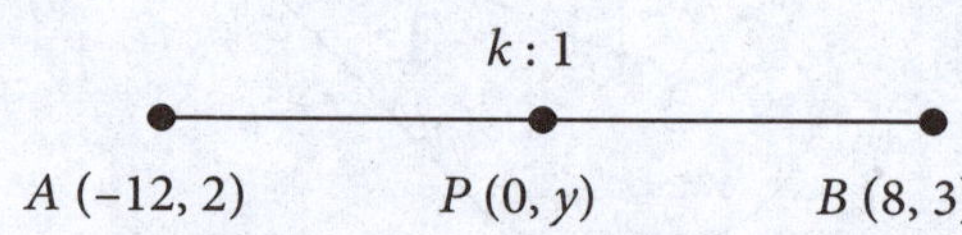

$$\frac{8k - 12}{k+1} = 0$$

$$8k = 12$$

$$k = \frac{3}{2}$$

32. (B)

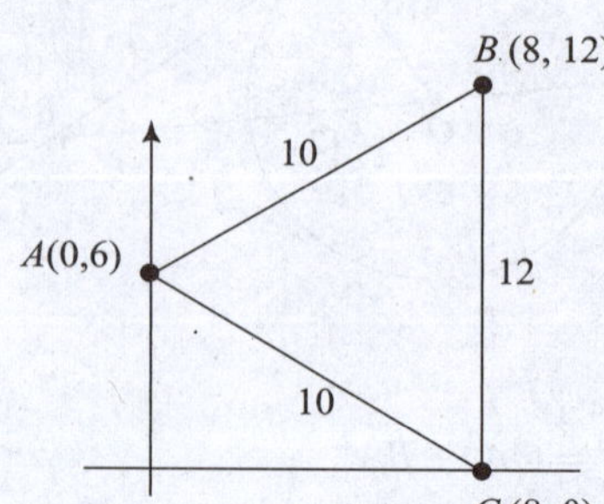

$$x\text{-coordinate} = \frac{10\times8+10\times8+12\times0}{10+10+12} = \frac{160}{32} = 5$$

$$y\text{-coordinate} = \frac{10\times12+12\times6+10\times0}{10+10+12} = \frac{192}{32} = 6$$

33. (A) Let $0(x, y)$ be the centre of the circle. Then

$OA^2 = OB^2$

$\Rightarrow (x-2)^2 + (y+9)^2 = (x-5)^2 + (y+8)^2$

$\Rightarrow 3x + y = 2$

$OA^2 = OC^2$

$\Rightarrow (x-2)^2 + (y+9)^2 = (x-2)^2 + (y-1)^2$

$\Rightarrow y = -4$

34. (D) Let (x, y) be the other end of the diameter.

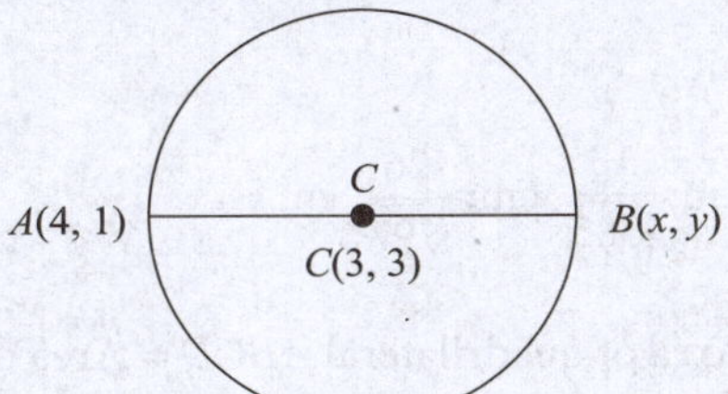

$$\left(\frac{x+4}{2}, \frac{y+1}{2}\right) \equiv (3,3)$$

$\therefore\ B \equiv (2, 5)$

Hints and answers to Olympiad and NTSE level exercises

35. (D) The given points are collinear because the point $\left(\frac{kc+la}{k+l}, \frac{kd+lb}{k+l}\right)$ divides the points (a, b) and (c, d) in the ratio of $k : l$.

36. (B) Let $A(0, 0)$, $B(a, 0)$ and $C\left(\frac{a}{2}, \frac{a\sqrt{3}}{2}\right)$

Hence,

$$AB = \sqrt{a^2 + 0} = a,$$

$$BC = \sqrt{\left(\frac{a}{2}\right)^2 + \left(\frac{a\sqrt{3}}{2}\right)^2} = a$$

and $AC = \sqrt{\left(\frac{a}{2}\right)^2 + \left(\frac{a\sqrt{3}}{2}\right)^2} = a$

Hence the triangle is equilateral.

37. (C) The point lies on axis of x, if $y = 0$.

Therefore, $\frac{1+3+c}{3} = 0 \Rightarrow c = -4$.

38. (A)

39. (C) $1 = \sqrt{4 + 12} = 4$

$\Rightarrow p^2 + q^2 = 16$ and $(p-2)^2 + (q - 2\sqrt{3})^2 = 16$

$\Rightarrow p + \sqrt{3}q = 4.$

Now from options only (4, 0) satisfies the equation

40. (D) In ΔABC, $A \equiv (-3, 0)$; $B \equiv (4, -1)$ and $C \equiv (5, 2)$

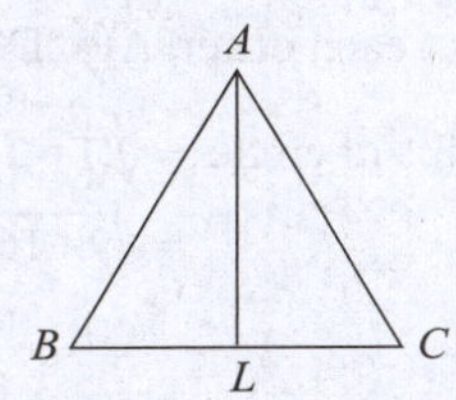

We know that $BC = \sqrt{(5-4)^2 + (2+1)^2} = \sqrt{1+9} = \sqrt{10}$

and area of ΔABC

$$= \frac{1}{2}[-3(-1-2)+4(2-0)+5(0+1)] = 11$$

$\therefore$ altitude $AL = \frac{2\Delta ABC}{BC} = \frac{2\times11}{\sqrt{10}} = \frac{22}{\sqrt{10}}$

41. (C) Mid point of AB is $(-1, 4)$, Slope of $AB = -1/2$

Perpendicular bisector of AB $y - 4 = 2(x + 1)$

$\Rightarrow 2x - y + 6 = 0$ ...(i)

Similarly perpendicular bisector of A (1, 3) and $C(5, -1)$ is $x - y - 2 = 0$ ...(ii)

Point of intersection of (i) and (ii) is $P = (-8, -10)$

Then $PA = \sqrt{(1+8)^2 + (3+10)^2} = \sqrt{81+169}$

42. (A) Given centroid $G\left(-\frac{2}{\sqrt{3}}, \frac{2}{\sqrt{3}}\right)$ and one vertex $A(2, 2)$

$\therefore$ Distance length of side of triangle

$= \sqrt{3}\ GA$

$$= \sqrt{3}\sqrt{\left(-\frac{2}{\sqrt{3}}-2\right)^2+\left(\frac{2}{\sqrt{3}}-2\right)^2}$$

$$= 2\sqrt{\left(1+\sqrt{3}\right)^2+\left(1-\sqrt{3}\right)^2}$$

$$= 2\sqrt{8}$$

$$= 4\sqrt{2}$$

43. (B)

(i) Points $(a, 0)$ and $(0, b)$ lie on perpendicular axes.

(ii) (0, 8) and (0, −8) lie on y-axis and their midpoint is (0, 0).

(iii) Let $A(-8, 8)$, $B(2, 8)$, $C(6, 2)$ and $D(-4, 2)$.
$AB = CD$, $AD = BC$
Also, midpoint of AC and BD is same i.e., (−1, 5).
Since opposite sides are equal and diagonals bisect each other, ABCD is a parallelogram.

(iv) Radius of circle $= \sqrt{(1+2)^2+(1-5)^2}$

$= \sqrt{9+16} = 5$

44. (C)

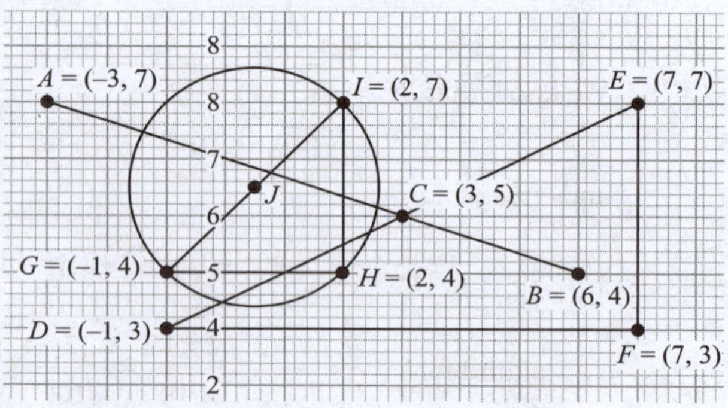

Area of a Triangle

1. (a) $12\frac{1}{2}$ unit^2

(b) 42 unit^2

2. (a) 96 unit^2

(b) [illegible]6 unit^2

3. (a) Slopes are equal $= \frac{-2}{3}$

Therefore, they lie on a straight line.

(b) Slopes are equal $= \frac{-1}{2}$

Therefore, they lie on a straight line.

4. (A)

5. (C)

6. (B)

7. (a) $2ac$ unit^2

(b) a^2 unit^2

9. 2

10. $7x + y - 11 = 0$

11. −37

12. (a) $a^2(m_1 - m_2)(m_2 - m_3)(m_3 - m_1)$

(b) $\frac{a^2}{2}(m_1 - m_2)(m_2 - m_3)(m_3 - m_1)$

(c) $\frac{a^2(m_1 - m_2)(m_2 - m_3)(m_3 - m_1)}{2m_1m_2m_3}$

13. −2

14. 2 or $\frac{2}{3}$

15. 97 unit^2

16. $\frac{13}{\sqrt{2}}$ cm; $\frac{13}{\sqrt{85}}$ cm; $\frac{26}{\sqrt{97}}$ cm

17. (C) Area of quadrilateral $ABCD$ = Area of (ΔABC) + Area of (ΔADC)

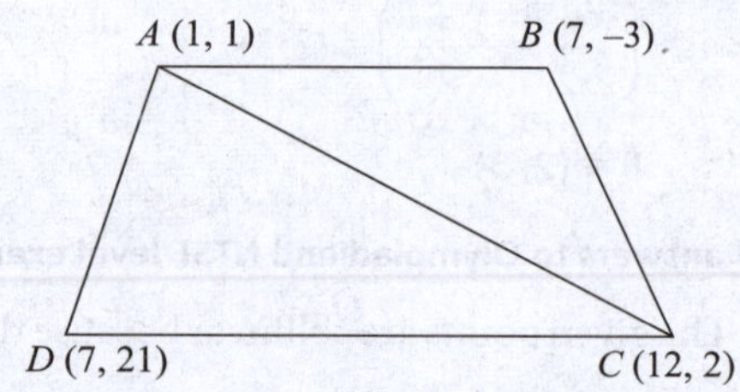

18. (C) If the points A, B and C are collinear then area of triangle ABC will be zero.

$$\frac{1}{2}\begin{vmatrix} -2 & 3 & 1 \\ 1 & 2 & 1 \\ K & 0 & 1 \end{vmatrix} = 0$$

$$\Rightarrow \begin{vmatrix} -2 & 3 & 1 \\ 1 & 2 & 1 \\ K & 0 & 1 \end{vmatrix} = 0$$

$$\Rightarrow \quad -2(2) + 3(K - 1) + 1(-2K) = 0$$

$$\Rightarrow \quad -4 + 3K - 3 - 2K = 0$$

$\Rightarrow\ -7+K=0$

$\Rightarrow\ K=7$

19. (A) $\frac{1}{2}\begin{vmatrix} x & -1 & 1 \\ 2 & 1 & 1 \\ 4 & 5 & 1 \end{vmatrix}=0$

$\Rightarrow\ x(1-5)-1(4-2)+1(10-4)=0$

$\Rightarrow\ -4x-2+6=0$

$\Rightarrow\ x=1$

20. (B) $\Delta=\frac{1}{2}\begin{vmatrix} a & b+c & 1 \\ a & b-c & 1 \\ -a & c & 1 \end{vmatrix}=2ac$

21. (B) $\Delta=\frac{1}{2}\begin{vmatrix} a & c+a & 1 \\ a & c & 1 \\ -a & c-a & 1 \end{vmatrix}$

$$\Rightarrow\ \frac{1}{2}\left\{\begin{matrix} a(c-c+a)+(c+a)(-a-a)+ \\ 1\left(ac-a^2+ac\right) \end{matrix}\right\}$$

$$\Rightarrow\ \frac{1}{2}\left|\left\{a^2+(c+a)(-2a)-a^2+2ac\right\}\right|$$

$$\Rightarrow\ \frac{1}{2}\left|\left\{a^2-2ac-2a^2-a^2+2ac\right\}\right|$$

$$\Rightarrow\ \frac{1}{2}\left|2a^2\right|=a^2$$

22. (A) Area of triangle

$$\Delta=\frac{1}{2}\begin{vmatrix} 0 & 4 & 1 \\ 3 & 6 & 1 \\ -8 & -2 & 1 \end{vmatrix}$$

$$\Rightarrow\ \frac{1}{2}\{0(6+2)+4(-8-3)+1(-6+48)\}$$

$$\Rightarrow\ \frac{1}{2}|-44+42|$$

$$\Rightarrow\ \frac{1}{2}\times 2$$

$$\Rightarrow\ 1$$

23. (A) Area of triangle is given by

$$\Delta=\frac{1}{2}\begin{vmatrix} a\cos\theta_1 & b\sin\theta_1 & 1 \\ a\cos\theta_2 & b\sin\theta_2 & 1 \\ a\cos\theta_3 & b\sin\theta_3 & 1 \end{vmatrix}$$

$$=\frac{1}{2}ab\begin{vmatrix} \cos\theta_1 & \sin\theta_1 & 1 \\ \cos\theta_2 & \sin\theta_2 & 1 \\ \cos\theta_3 & \sin\theta_3 & 1 \end{vmatrix}$$

$$=\frac{1}{2}ab\left|\begin{matrix} \cos\theta_1(\sin\theta_2-\sin\theta_3)+ \\ \sin\theta_1(\cos\theta_3-\cos\theta_2)+ \\ (\sin\theta_3\cos\theta_2-\sin\theta_2\cos\theta_3) \end{matrix}\right|$$

$$=\frac{1}{2}ab\left|\begin{matrix} \cos\theta_1\left\{2\cos\left(\frac{\theta_2+\theta_3}{2}\right)\sin\left(\frac{\theta_2-\theta_3}{2}\right)\right\} \\ +\sin\theta_1\left\{2\sin\left(\frac{\theta_3+\theta_2}{2}\right)\sin\left(\frac{\theta_2-\theta_3}{2}\right)\right\} \\ +\sin(\theta_3-\theta_2)\} \end{matrix}\right|$$

$$=\frac{1}{2}ab\left|\begin{matrix} \cos\theta_1\left\{2\cos\left(\frac{\theta_2+\theta_3}{2}\right)\sin\left(\frac{\theta_2-\theta_3}{2}\right)\right. \\ +\sin\theta_1\left\{2\sin\left(\frac{\theta_3+\theta_2}{2}\right)\sin\left(\frac{\theta_2-\theta_3}{2}\right)\right. \\ \left.+2\sin\left(\frac{\theta_3-\theta_2}{2}\right)\cos\left(\frac{\theta_3-\theta_2}{2}\right)\right\} \end{matrix}\right|$$

$$=\frac{1}{2}ab\times 2\sin\left(\frac{\theta_2-\theta_3}{2}\right)\left|\cos\theta_1\cos\left(\frac{\theta_2+\theta_3}{2}\right)+\sin\theta_1\sin\left(\frac{\theta_2+\theta_3}{2}\right)-\cos\left(\frac{\theta_3-\theta_2}{2}\right)\right|$$

$$=ab\sin\left(\frac{\theta_2-\theta_3}{2}\right)\left|\cos\left(\theta_1-\frac{\theta_2+\theta_3}{2}\right)-\cos\left(\frac{\theta_3-\theta_2}{2}\right)\right|$$

$$=ab\sin\left(\frac{\theta_2-\theta_3}{2}\right)\left|\cos\left(\frac{2\theta_1-\theta_2-\theta_3}{2}\right)-\cos\left(\frac{\theta_3-\theta_2}{2}\right)\right|$$

$$= ab\sin\left(\frac{\theta_2-\theta_3}{2}\right)\left|\begin{matrix} 2\sin\left(\dfrac{\dfrac{\theta_3-\theta_2}{2}-\dfrac{2\theta_1-\theta_2-\theta_3}{2}}{2}\right) \\ \sin\left(\dfrac{\dfrac{\theta_3-\theta_2}{2}+\dfrac{2\theta_1-\theta_2-\theta_3}{2}}{2}\right)\end{matrix}\right|$$

$$= ab\sin\left(\frac{\theta_2-\theta_3}{2}\right)\left|2\sin\left(\frac{\theta_3-\theta_1}{2}\right)\sin\left(\frac{\theta_1-\theta_2}{2}\right)\right|$$

$$= 2ab\sin\left(\frac{\theta_1-\theta_2}{2}\right)\sin\left(\frac{\theta_2-\theta_3}{2}\right)\sin\left(\frac{\theta_3-\theta_1}{2}\right)$$

24. (C) Area of triangle is given by

$$\Delta = \frac{1}{2}\begin{vmatrix} am_1m_2 & a(m_1+m_2) & 1 \\ am_2m_3 & a(m_2+m_3) & 1 \\ am_3m_1 & a(m_3+m_1) & 1 \end{vmatrix}$$

$$= \frac{a^2}{2}\begin{vmatrix} m_1m_2 & m_1+m_2 & 1 \\ m_2m_3 & m_2+m_3 & 1 \\ m_3m_1 & m_3+m_1 & 1 \end{vmatrix}$$

$$= \frac{a^2}{2}\left|\begin{matrix} m_1m_2(m_2+m_3-m_3-m_1)+ \\ (m_1+m_2)(m_3m_1-m_2m_3)+ \\ m_2m_3{}^2+m_1m_2m_3- \\ m_1m_2m_3-m_1m_3{}^2 \end{matrix}\right|$$

$$= \frac{a^2}{2}\left|\begin{matrix} m_1m_2(m_2-m_1)+(m_1+m_2) \\ m_3(m_1-m_2)+m_3{}^2(m_2-m_1) \end{matrix}\right|$$

$$= \frac{a^2}{2}\left|(m_2-m_1)\{m_1m_2-m_1m_3-m_2m_3+m_3{}^2)\}\right|$$

$$= \frac{a^2}{2}\left|(m_2-m_1)\{m_1(m_2-m_3)-m_3(m_2-m_3)\}\right|$$

$$= \frac{a^2}{2}\left|(m_2-m_1)(m_1-m_3)(m_2-m_3)\right|$$

$$= \frac{a^2}{2}\left|(m_1-m_2)(m_2-m_3)(m_3-m_1)\right|$$

25. (B)

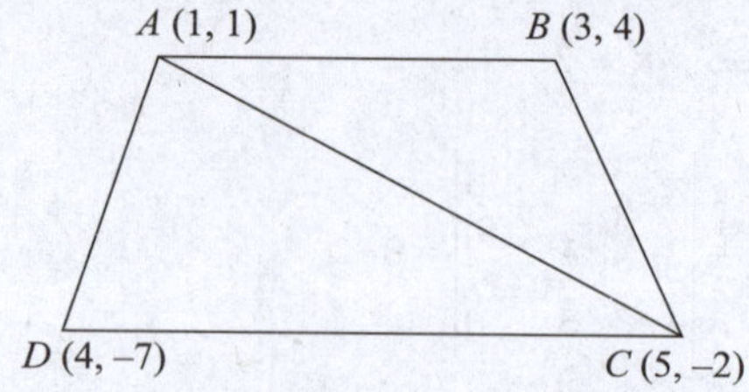

Area of quadrilateral $ABCD$ = Area of triangle ABC + Area of triangle ACD

26. (A)

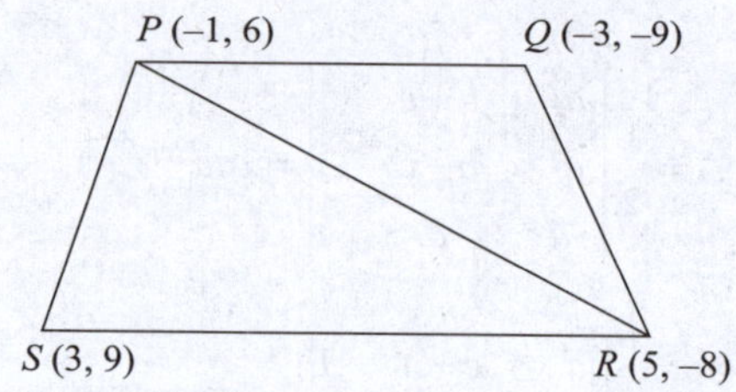

Area of quadrilateral $PQRS$ = Area of (ΔPQR) + Area of (ΔPRS)

27. (C) Area of ΔPQR

$$= \frac{1}{2}\begin{vmatrix} -\frac{1}{2} & 3 & 1 \\ -5 & 6 & 1 \\ -8 & 8 & 1 \end{vmatrix}$$

$$= \frac{1}{2}\left|\left\{-\frac{1}{2}(6-8)+3(-8+5)+1(-40+48)\right\}\right|$$

$$= \frac{1}{2}\left|\{1-9+8\}\right| = \frac{1}{2}(9-9) = 0$$

Hints and answers to Olympiad and NTSE level exercises

28. (D) Area of ΔABC

$$= \frac{1}{2}\begin{vmatrix} a & b+c & 1 \\ b & c+a & 1 \\ c & a+b & 1 \end{vmatrix}$$

$$= \frac{1}{2}\left\{\begin{matrix} a(c+a-a-b)+(b+c)(c-b)+ \\ 1(ab+b^2-c^2-ac) \end{matrix}\right\}$$

$$= \frac{1}{2}\{ac-ab+c^2-b^2+ab+b^2-c^2-ac\}$$

$$= 0$$

29. (A) $\dfrac{\Delta DBC}{\Delta ABC}=\dfrac{1}{2}\Rightarrow 2\Delta DBC=\Delta ABC$

$\Rightarrow \quad 2(14x-7)=\dfrac{49}{2}\Rightarrow x=\dfrac{77}{56}=\dfrac{11}{8}.$

30. (A)

31. (B) Let P be (x, y). $\Delta = \pm 12$

$\Rightarrow \quad x + 3y + 1 = 0$

and $x + 3y - 23 = 0$.

32. (D) $A=\dfrac{1}{2}\begin{vmatrix}1 & p^2 & 1\\ 0 & 1 & 1\\ p & 0 & 1\end{vmatrix}=\dfrac{1}{2}\,[1(1-0)+p(p^2-1)]$

$=\dfrac{1}{2}\,(p^3-p+1)$

Hence $A=\dfrac{1}{2}\,|\,p^3-p+1\,|$

Now, minimum value of modulus is zero. Since $A\,(p)$ is a cubic it must vanish for some p other than given as A, B, C.

33. (C) Taking the coordinates of vertices O, P, Q, R as $(0, 0)$, $(a, 0)$, (a, a), $(0, a)$ respectively we get the coordinates of M as $(a, a/2)$ and of N as $(a/2, a)$

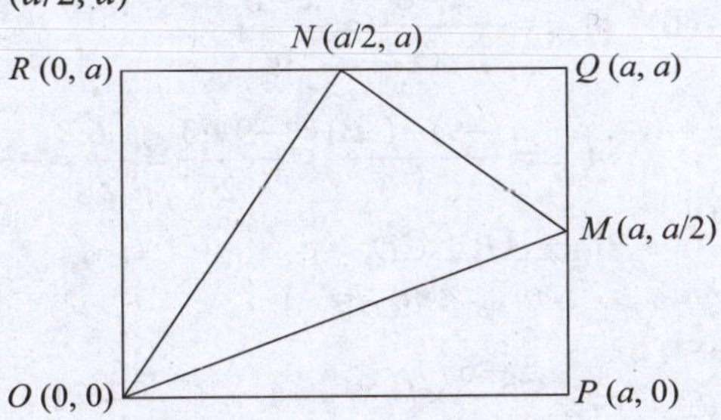

$\therefore$ Area of the ΔOMN

$=\dfrac{1}{2}\begin{vmatrix}0 & 0 & 1\\ a & a/2 & 1\\ a/2 & a & 1\end{vmatrix}=\dfrac{3a^2}{8}$

Area of the square $= a^2$

$\therefore \quad$ the required ratio is $8 : 3$

34. Let $ABCD$ be the given quadrilateral with vertices $A(-4, 5)$, $B(0, 7)$, $C(5, -5)$ and $D(-4, -2)$.

Then, by plotting A, B, C and D on the Cartesian plane and joining AB, BC, CD, and DA, the given quadrilateral can be drawn as shown in the given figure.

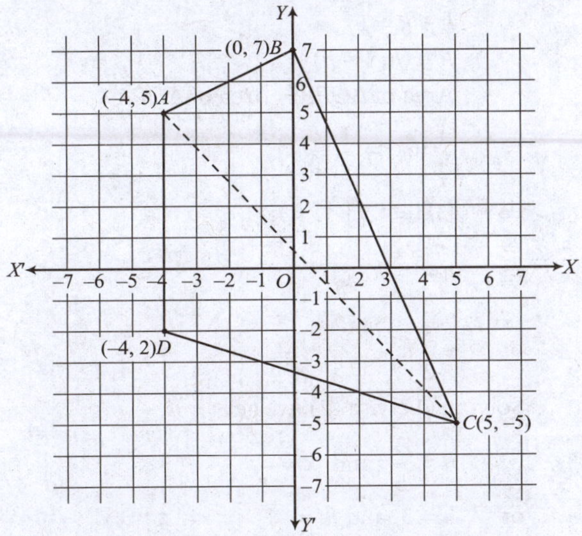

Area $(ABCD)$ = Area (ΔABC) + Area (ΔACD)

From the figure, we have

Area of $\Delta ABC=\dfrac{1}{2}\,|\,(-4)\,(7+5)+0\,(-5-5)+5\,(5-7)\,|$ unit^2 = 29 unit^2

Area of $\Delta ACD=\dfrac{1}{2}\,|\,(-4)\,(-5+2)+5\,(-2-5)+(-4)\,(5+5)\,|$ unit$^2=\dfrac{63}{2}$ unit^2

Thus, area $(ABCD)=\left(29+\dfrac{63}{2}\right)$ unit^2

$=\dfrac{121}{2}$ unit^2

35. (C) $\dfrac{\text{Area of }\Delta DBC}{\text{Area of }\Delta ABC}=\dfrac{1}{2}$

$\Rightarrow \quad \dfrac{\frac{1}{2}[x(5+2)-3(-2-2x)+4(2x-5)]}{\frac{1}{2}[6(5+2)-3(-2-3)+4(3-5)]}=\dfrac{1}{2}$

$\Rightarrow \quad \dfrac{7x+6+6x+8x-20}{42+15-8}=\dfrac{1}{2}$

$\Rightarrow \quad \dfrac{3x-2}{7}=\dfrac{1}{2}$

$\Rightarrow \quad x=\dfrac{11}{6}$

36. (C) Let $P \equiv (h, k)$.

P lies on the line $x + y = 1$.

$\therefore \quad h + k = 1 \qquad \text{...(1)}$

$\therefore \quad P \equiv (h, 1 - h)$

Area of ΔOPB = Area of ΔOPA

$\Rightarrow \quad \frac{1}{2}|6k| = \frac{1}{2}|4h|$

$\Rightarrow \quad |3k| = |2h|$

$\Rightarrow \quad h = \frac{3}{2}k$

or $\quad -\frac{3}{2}k \qquad \text{...(2)}$

Solving (1) with (2), we get

$h = \frac{3}{5}$ and $k = \frac{2}{5}$

or $\quad h = 3$ and $k = -2$

37. (C) Area of triangle formed with midpoints

$$= \frac{1}{2}[-1(1-5) + 6(5+2) + 3(-2-1)]$$

$$= \frac{37}{2}$$

$\therefore \quad$ Area of $\Delta ABC = 4 \times \frac{37}{2} = 74$ sq. units

Equation of a Straight Line – I

1. -4 and 3
2. $\frac{1}{2}$
5. $x + y - 4 = 0$
6. $\sqrt{3}y - x - 2\sqrt{3} = 0$
8. $2x + 5y - 20 = 0$
9. $2x + 3y - 12 = 0$
10. $x + y - 11 = 0$
11. 2
13. $x = 2$
14. $2x - 3y + 12 = 0$
15. $4x + 3y - 24 = 0$ or $x + y - 7 = 0$
16. $x - y - 5 = 0$
17. $8x - 5y + 60 = 1$
18. $3x + 4y + 12 = 0$ or $4x + 3y + 12 = 0$
20. $(6, -1)$

21. (D) Slope = tan 135° = –1

$$\Rightarrow \quad -1 = \frac{(-3)-5}{x-2} \quad \Rightarrow -1 = \frac{-8}{x-2}$$

$$\Rightarrow \quad x - 2 = 8 \quad \Rightarrow x = 10$$

22. (C) Slope of line segment

$$AB = \frac{(-2)-(-1)}{4-3} = \frac{-2+1}{1} = \frac{-1}{1} = -1$$

$\tan\theta = -1$

$\Rightarrow \quad \theta = 135°$

θ is the angle with x-axis.

23. (D) $m_{AB} = m_{BC}$

$$\Rightarrow \quad \frac{1-(-1)}{2-x} = \frac{5-1}{4-2}$$

$$\Rightarrow \quad \frac{2}{2-x} = \frac{4}{2}$$

$$\Rightarrow \quad 2 - x = 1 \quad \Rightarrow x = 1$$

24. (C) $m_{AB} = m_{CD}$

$$\Rightarrow \quad \frac{3-2}{x-7} = \frac{-1-0}{4-6} \Rightarrow \frac{1}{x-7} = \frac{-1}{-2}$$

$$\Rightarrow \quad \frac{1}{x-7} = \frac{1}{2} \Rightarrow x = 9$$

25. (B) $m_{AB} = \frac{x-6}{4-(-2)} = \frac{x-6}{6}$

$$m_{CD} = \frac{(-9)-(-3)}{5-3} = \frac{-9+3}{2} = \frac{-6}{2} = -3$$

Since $AB \perp CD$,

$$m_{AB} \times m_{CD} = -1$$

$$\Rightarrow \quad \frac{x-6}{6} \times (-3) = -1 \Rightarrow \frac{x-6}{2} = 1$$

$\Rightarrow \quad x - 6 = 2$

$\Rightarrow \quad x = 8$

26. (B) Slope of line AB,

$$m_{AB} = \frac{8-(-2)}{(-1)-(4)} = \frac{10}{-5} = -2$$

Since AB passes through the point $A(-1, 8)$, equation of line AB is

$$y - 8 = -2(x + 1)$$

$\Rightarrow \quad y - 8 = -2x - 2$

$\Rightarrow \quad y + 2x - 6 = 0$

27. (C) E is the midpoint of AC.

$$E=\left(\frac{2-2}{2},\frac{5-1}{2}\right)=(0,2)$$

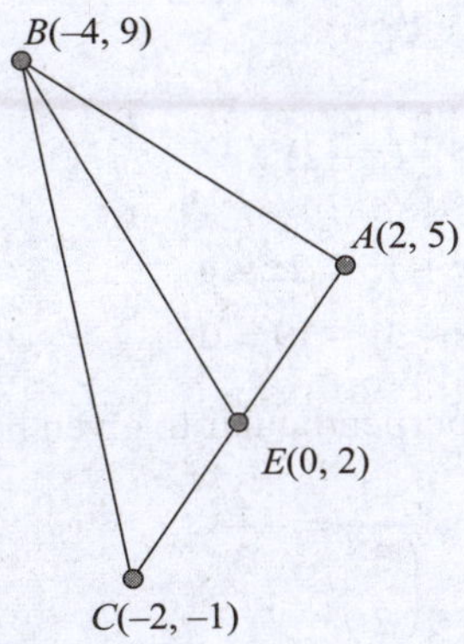

Slope of line BE

$$m_{BE}=\frac{9-2}{-4-0}=\frac{7}{-4}=\frac{-7}{4}$$

Equation of line BE,

$$y-2=\frac{-7}{4}(x-0)$$

$$4y-8=-7x$$

$$\Rightarrow\quad 7x+4y-8=0$$

28. (B) Let P be the midpoint of AB.

$$P\equiv\left(\frac{2+6}{2},\frac{3-5}{2}\right)\equiv(4,-1)$$

Slope of $AB=\frac{3-(-5)}{2-6}=\frac{8}{-4}=-2$

Slope of perpendicular through $P=\frac{-1}{(-2)}=\frac{1}{2}$

Equation of perpendicular bisector of AB is

$$y+1=\frac{1}{2}(x-4)$$

$$\Rightarrow\quad 2y+2=x-4$$

$$\Rightarrow\quad 2y-x+6=0$$

29. (B) Let us consider $BP\perp AC$.

Slope of $AC=\frac{1-5}{2-4}=\frac{-4}{-2}=2$

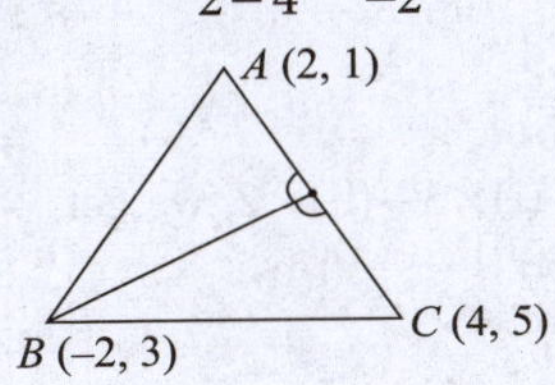

Slope of $BP=-\frac{1}{2}$

Equation of BP is

$$y-3=-\frac{1}{2}(x+2)$$

$$\Rightarrow\quad 2y-6=-x-2$$

$$\Rightarrow\quad 2y+x-4=0$$

30. (A) Let P be the midpoint of BC.

$$P\equiv\left(\frac{2-1}{2},\frac{-3-2}{2}\right)\equiv\left(\frac{1}{2},\frac{-5}{2}\right)$$

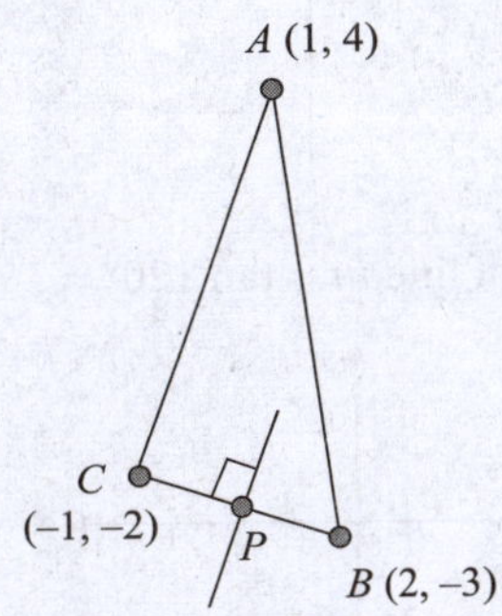

Slope of $BC=\frac{-3+2}{2+1}=\frac{-1}{3}$

Then slope of right bisector of BC through $P=3$

Equation of right bisector

$$y+\frac{5}{2}=3\left(x-\frac{1}{2}\right)$$

$$\Rightarrow\quad 2y+5=6x-3$$

$$\Rightarrow\quad 2y-6x+8=0$$

$$\Rightarrow\quad 3x-y-4=0$$

31. (B) Midpoint of $PQ=\left(\frac{-2+6}{2},0\right)=(2,0)$

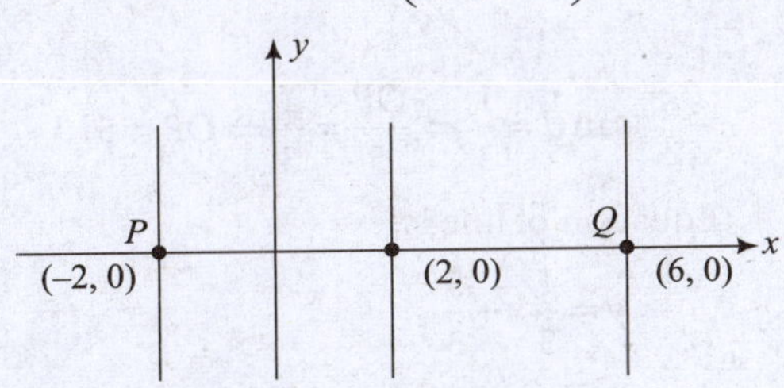

32. (D) $y-(-4)=0\quad\Rightarrow\quad y+4=0$

33. (B) Equation of the line

$$y-3=\frac{1}{\sqrt{3}}(x-0)$$

$$y\sqrt{3}-3\sqrt{3}=x$$

$$\Rightarrow \quad x-y\sqrt{3}+3\sqrt{3}=0$$

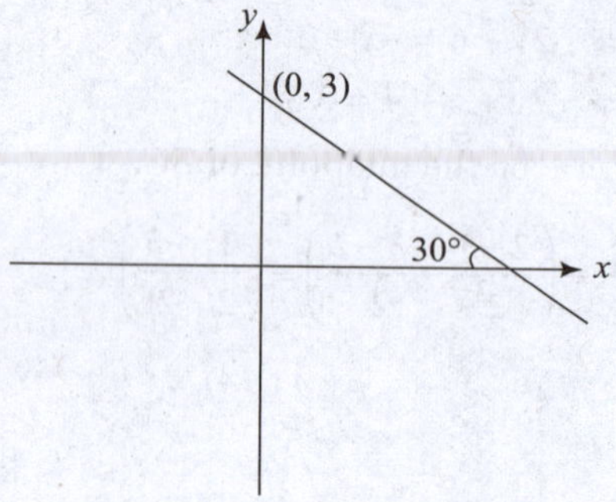

34. (C) Slope of line $m = \tan 120° = -\sqrt{3}$

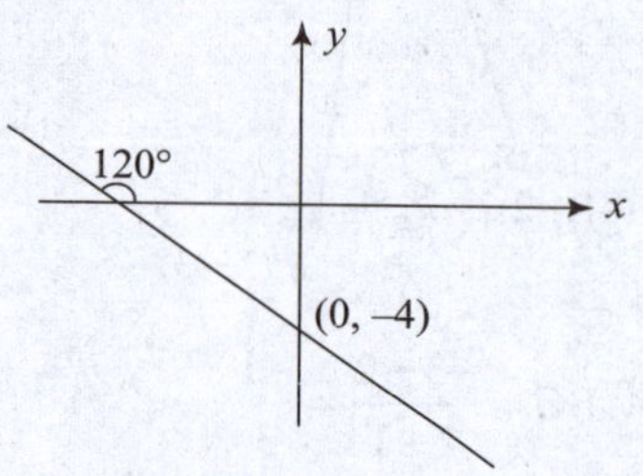

Equation of a line,

$$y + 4 = -\sqrt{3}\,(x - 0)$$

$$y + 4 = -\sqrt{3}x$$

$$\Rightarrow \quad y + \sqrt{3}x + 4 = 0$$

35. (A)

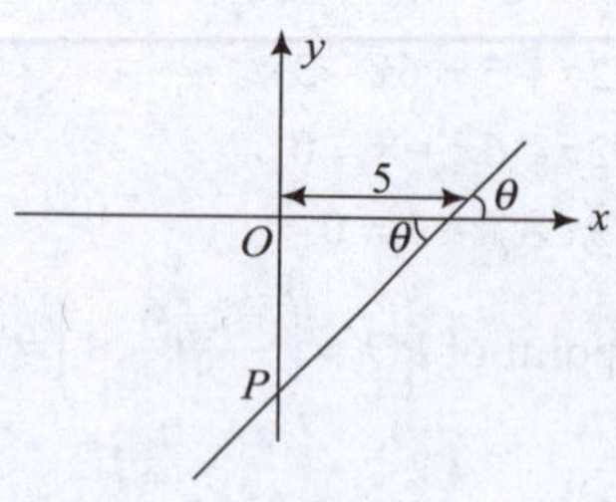

$$\tan\theta=\frac{1}{3}\Rightarrow\frac{OP}{5}=\frac{1}{3}\Rightarrow OP=5/3$$

Equation of line is,

$$y=\frac{1}{3}x-\frac{5}{3}$$

$$\Rightarrow \quad 3y = x - 5$$

$$\Rightarrow \quad x - 3y - 5 = 0$$

36. (C) Product of slopes of lines = $\left(-\frac{2}{3}\right)\times\frac{3}{2}=-1$

37. (D) $\tan\theta=\frac{-1}{\sqrt{3}}\Rightarrow\theta=150°$

38. (C) $y+5=\frac{2}{3}(x-2)$

$$\Rightarrow \quad 3y + 15 = 2x - 4$$

$$\Rightarrow \quad 2x - 3y - 19 = 0$$

39. (A) Slope perpendicular to given line

$$=\frac{-1}{\left(\frac{3}{2}\right)}=-\frac{2}{3}$$

$$y=-\frac{2}{3}x-3$$

$$\Rightarrow \quad 3y = -2x - 9$$

$$\Rightarrow \quad 2x + 3y + 9 = 0$$

40. (C) $\frac{x}{2}+\frac{y}{-3}=1$

$$\Rightarrow \quad -3x + 2y = -6$$

$$\Rightarrow \quad 3x - 2y - 6 = 0$$

41. (D) $\frac{x}{4}+\frac{y}{6}=1$

$$\Rightarrow \quad 6x + 4y = 24$$

$$\Rightarrow \quad 3x + 2y - 12 = 0$$

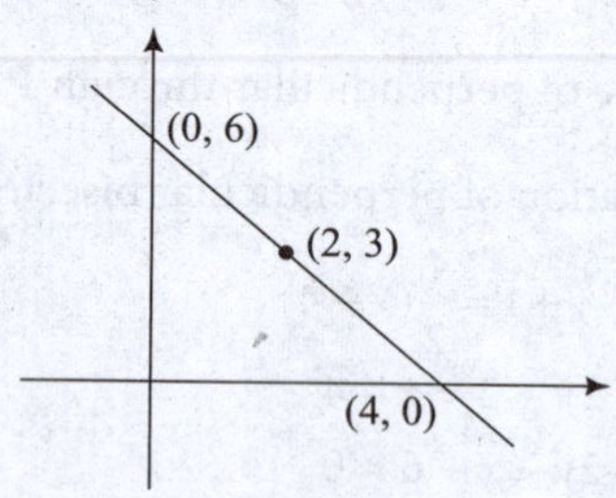

42. (B)

$$\frac{8}{a}-\frac{9}{b}=1 \qquad \text{...(i)}$$

$$\frac{12}{a}-\frac{15}{b}=1 \qquad \text{...(ii)}$$

From (i)×3−(ii)×2, we get

$$\frac{24}{a}-\frac{27}{b}=3$$

$$\frac{24}{a} - \frac{30}{b} = 2$$
$$- \quad + \quad -$$
$$\frac{3}{b} = 1$$

$\therefore \quad b = 3$

So, $\frac{8}{a} - \frac{9}{3} = 1$

$\Rightarrow \quad \frac{8}{a} = 4$

$\therefore \quad a = 2$

Hints and answers to Olympiad and NTSE level exercises

43. (C) Here $c = -1$ and $m = \tan\theta = \tan 45° = 1$
(Since the line is equally inclined to the axes, so $\theta = 45°$)
Hence equation of straight line is $y = \pm(1.x) - 1$
$\Rightarrow \quad x - y - 1 = 0$ and $x + y + 1 = 0$.

44. (B) Let the required equation is $y = -x + c$ which is perpendicular to $y = x$ and passes through (3, 2). So $2 = -3 + c \Rightarrow c = 5$. Hence required equation is $x + y = 5$.

45. (C) Here $D(1, 1)$ therefore equation of line AD is given by $2x + y - 3 = 0$. Thus the line perpendicular to AD is $x - 2y + k = 0$ and it passes through B, so $k = 0$. Hence required equation is $x - 2y = 0$.

46. (D) Slope of $DE = \frac{7-3}{5-1} = 1 \Rightarrow$ Slope of $AB = 1$

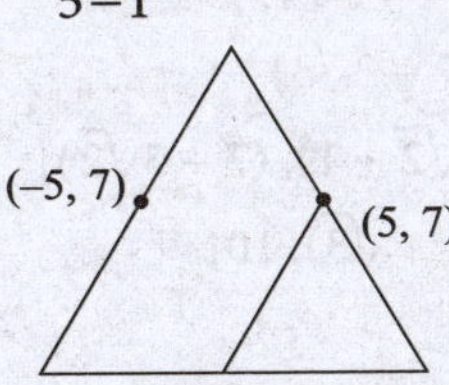

Hence equation of AB is
$y - 7 = (x + 5) \quad \Rightarrow x - y + 12 = 0$.

47. (A) Obviously, $x_1 = \frac{a}{2}$ and $y_1 = \frac{b}{2}$.

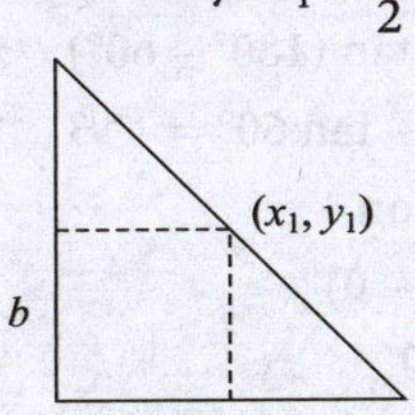

Therefore the equation of line AB is $\frac{x}{a} + \frac{y}{b} = 1$

$$\Rightarrow \quad \frac{x}{2x_1} + \frac{y}{2y_1} = 1 \Rightarrow xy_1 + yx_1 = 2x_1y_1.$$

48. (B) Midpoint $\equiv \left(\frac{1+1}{2}, \frac{3-7}{2}\right) = (1, -2)$

Therefore required line is
$2x - 3y = k \Rightarrow 2x - 3y = 8$.

49. (A) Point of intersection of the lines is (3, −2).
Hence the equation is
$2x - 7y = 2(3) - 7(-2) = 20$.

50. (B) Slope of given line is $\frac{1}{2}$.

Thus $\tan 45° = \pm \dfrac{m - \frac{1}{2}}{1 + m \cdot \frac{1}{2}}$

$\Rightarrow \quad m = \pm 3$.

Hence option (B) is correct.

51. a → q, r, s; b → p; c → q, s; d → q

(a) $h_1 = \left|\frac{10}{\sqrt{10}}\right| = \sqrt{10}$

$h_2 = \left|\frac{10}{\sqrt{10}}\right| = \sqrt{10}$

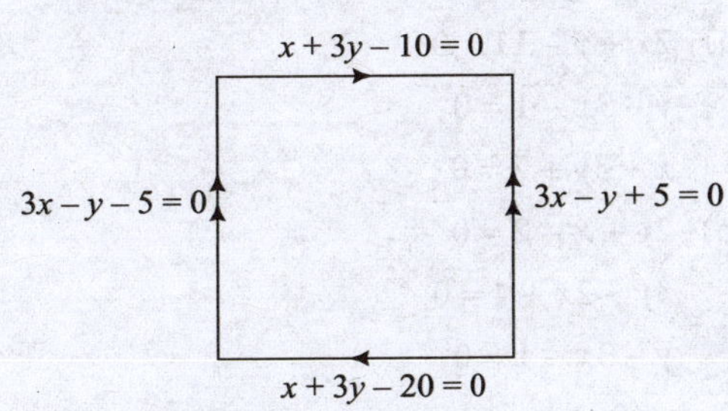

$\Rightarrow \quad$ a square of side $\sqrt{10}$

$\Rightarrow \quad$ area = 10 $\quad \Rightarrow \quad Q, R, S$

(b) $m_{AB} = \frac{5}{-1} = -5$

$m_{DC} = \frac{9}{-1} = -9$

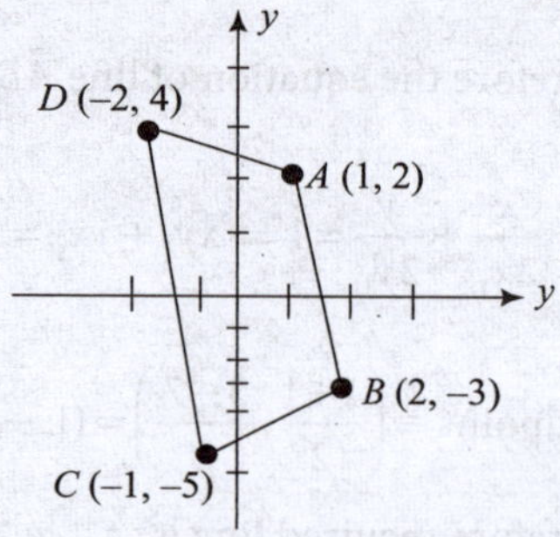

not a parallelogram $\Rightarrow P$

Similarly (C) and (D) can be checked

52. (B) Applying the formula, the required foot is

$$\left(\frac{1^2\times2-1\times1\times4+1}{1^2+1^2},\frac{1^2\times4-1\times1\times2+1}{1^2+1^2}\right)$$

$$=\left(-\frac{1}{2},\frac{3}{2}\right).$$

Trick : Here, in options only point $\left(-\frac{1}{2},\frac{3}{2}\right)$ is satisfying the given equation of line.

Equation of a Straight Line – II

1. (a) $x+y=0$

(b) $x-y+2=0$

(c) $10x+7y-22=0$

(d) $(b+d)x+(c-a)y+(bc+ad)=0$

(e) $ax-by+ab=0$

2. (a) $7x+y-11=0$

$y-3x-1=0$

$x+3y+7=0$

(b) $2y+x-2=0$

$3y-2x+4=0$

$y-3x-1=0$

3. $y-x=0$

$x+y-10=0$

4. I. (a) $\sqrt{3}x+y=10$

(b) $x+y=7\sqrt{2}$

(c) $x+\sqrt{3}y=16$

II. (a) $x-\sqrt{3}y+2a=0$

(b) $x-y+9\sqrt{2}=0$

(c) $\sqrt{3}x-y+11=0$

III. (a) $\sqrt{3}x+y+2m=0$

(b) $\sqrt{2}x+\sqrt{2}y+7=0$

(c) $x+\sqrt{3}y+12=0$

IV. (a) $x-\sqrt{3}y=2q$

(b) $\sqrt{2}x-\sqrt{2}y-15=0$

(c) $\sqrt{3}x-y=18$

5. (A) **6.** (B) **7.** (A) **8.** (C) **9.** (C) **10.** (B)

11. $x-6y+22=0$; $x=6$; $x+6y-34=0$

12. $x-7y+26=0$;

$3x-7+2=0$;

$x+3y+14=0$

13. $\left(\frac{1}{5},\frac{-3}{5}\right)$

14. (a) $(3(1+\sqrt{2}), (5+3\sqrt{2}))$

(b) $(3(1+\sqrt{3}), 8)$

(c) $(6, 5+3\sqrt{3})$

15. $x_1=\frac{-8}{5}$, $y_1=\frac{-6}{5}$

16. $x=2$, $y=2$; $(2, 2)$

17. (a) $(-6, 7+3\sqrt{3})$

(b) $[-3(\sqrt{2}+1), (7+3\sqrt{2})]$

(c) $[-3(1+\sqrt{3}), 10]$

18. (B)

19. (C)

20. (A)

21. (B) Angle with +ve direction of x-axis

$= 180° - 60° = 120°$

Slope $= \tan 120° = \tan(180° - 60°)$

$= -\tan 60° = -\sqrt{3}$

So, required equation

$y-3=-\sqrt{3}(x-0)$

$\Rightarrow y+\sqrt{3}x-3=0$

22. (A)

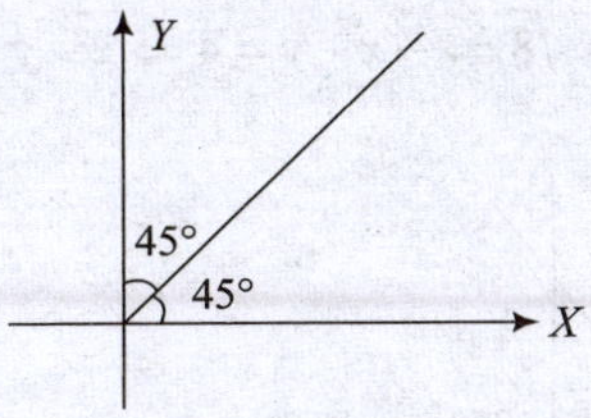

$y - 0 = \tan 45° (x - 0)$

$y = 1(x)$

$y = x$

23. (C) $(y + 4) = \tan 30° (x - 2)$

$$y+4=\frac{1}{\sqrt{3}}(x-2)$$

$$\sqrt{3}y+4\sqrt{3}=x-2$$

$$\sqrt{3}y+4\sqrt{3}-x+2=0$$

24. (C)

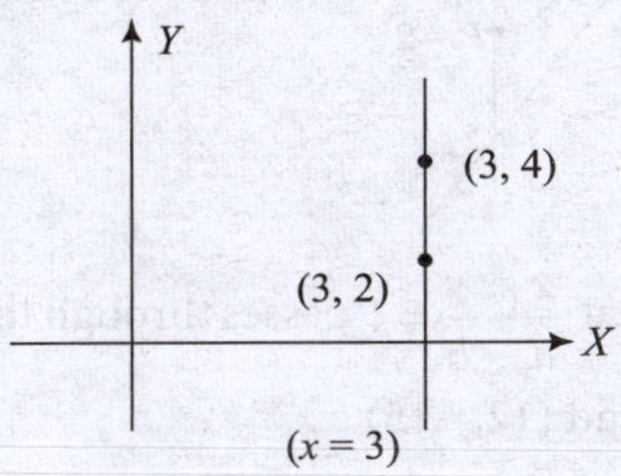

25. (C)

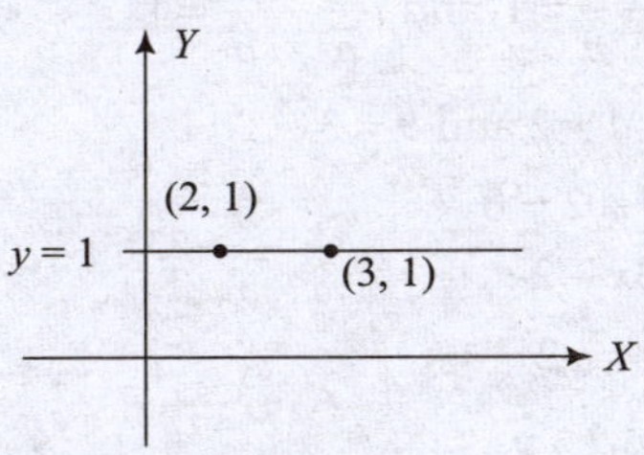

26. (D) $\frac{x}{2}+\frac{y}{-3}=1$

$\Rightarrow \quad -3x + 2y = -6$

$\Rightarrow \quad 3x - 2y + 6 = 0$

27. (A) Let the intercepts made by the line on the x-axis and y-axis be a and $14 - a$, respectively.

$$\frac{x}{a}+\frac{y}{14-a}=1$$

This line passes through point (3, 4).

So, $\frac{3}{a}+\frac{4}{14-a}=1$

$\Rightarrow \quad a^2 - 13a + 42 = 0$

$\therefore \quad a = 6, 7$

Equation of the line is

$$\frac{x}{6}+\frac{y}{14-6}=1 \quad \Rightarrow \quad \frac{x}{6}+\frac{y}{8}=1$$

$\Rightarrow \quad 8x + 6y = 48 \quad \Rightarrow 4x + 3y = 24$

or $\frac{x}{7}+\frac{y}{14-7}=1 \quad \Rightarrow \quad \frac{x}{7}+\frac{y}{7}=1$

$\Rightarrow \quad x + y - 7 = 0$

28. (C) Let the line segment meet the axes at $(A, 0)$ and $(0, B)$.

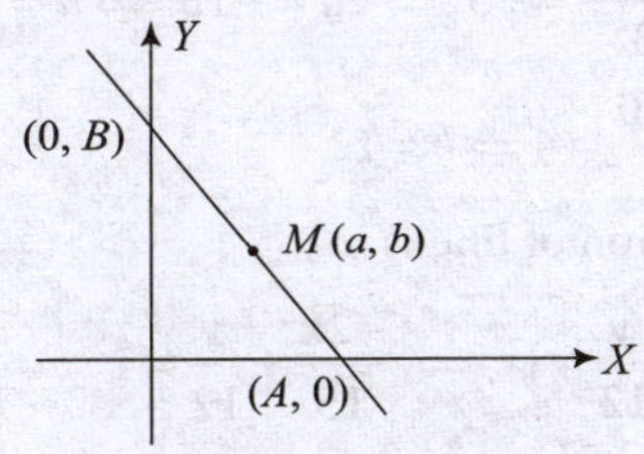

Then equation of line will be $\frac{x}{A}+\frac{y}{B}=1$.

$M(a, b)$ is the mid-point of a line segment.

So, $\frac{0+A}{2}=a \Rightarrow A=2a$

$\frac{0+B}{2}=b \Rightarrow B=2b$

So, the required line is

$$\frac{x}{2a}+\frac{y}{2b}=1 \Rightarrow bx+ay=2ab$$

29. (B) Let a and $-a$ be the intercepts.

$$\frac{x}{a}+\frac{y}{-a}=1$$

$x - y = a$

This passes through the point (–3, 7).

So, $(-3) - 7 = a \Rightarrow a - 10$

Therefore, $x - y = -10 \Rightarrow y + 10 = 0$

30. (A) Let a and b be the x-intercept and y-intercept, respectively.

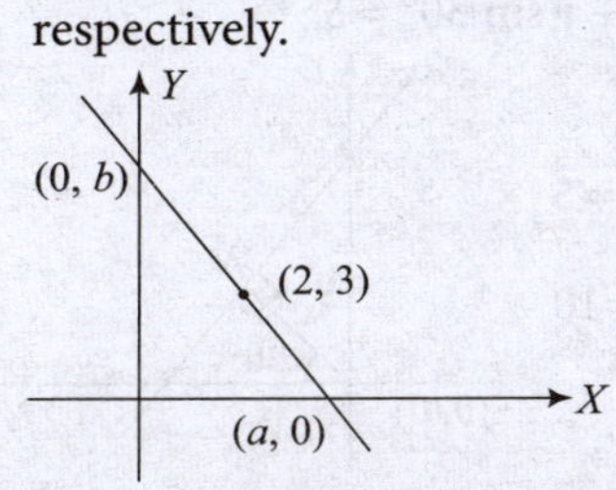

Then $\frac{a+0}{2}=2 \Rightarrow a=4$

$\frac{0+b}{2}=3 \Rightarrow b=6$

Equation of line is

$\frac{x}{4}+\frac{y}{6}=1$

$6x+4y=24 \Rightarrow 3x+2y=12$

31. (B) Let a and b be the x-intercept and y-intercept, respectively.

Then $\frac{2a+0}{3}=-5 \Rightarrow 2a=-15 \Rightarrow a=-\frac{15}{2}$

And $\frac{b+0}{3}=4 \Rightarrow b=12$

So, equation of line is

$\frac{x}{-15/2}+\frac{y}{12}=1 \Rightarrow \frac{-2x}{15}+\frac{y}{12}=1$

$\Rightarrow \frac{-8x+5y}{60}=1 \Rightarrow -8x+5y=60$

$\Rightarrow 8x-5y+60=0$

32. (B) Point of intersection of the lines $4x+3y-3=0$ and $2x-3y+1=0$ is $x=\frac{1}{3}, y=\frac{5}{9}$

Let the equation of required line be $\frac{x}{a}+\frac{y}{a}=1$.

Since it passes through $\left(\frac{1}{3},\frac{5}{9}\right), a=\frac{8}{9}$.

33. (C) Area of a triangle $=\frac{1}{2}\times 2a\times 2b = 2ab$ sq. units

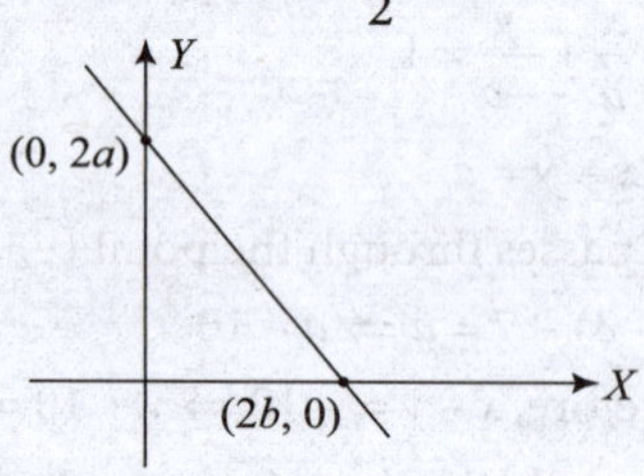

34. (B) $x\cos 30° + y\sin 30° = 5$

$\frac{x\sqrt{3}}{2}+\frac{y}{2}=5$

$\sqrt{3}x+y=10$

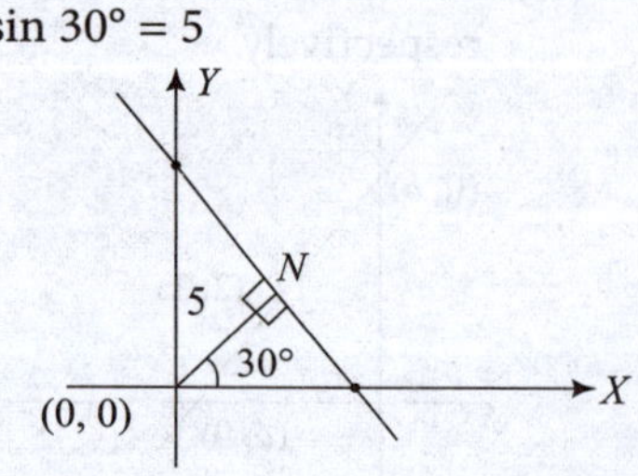

35. (A) $x\cos 135° + y\sin 135° = \sqrt{8}$

$\frac{-x}{\sqrt{2}}+\frac{y}{\sqrt{2}}=\sqrt{8} \Rightarrow -x+y=4 \Rightarrow x-y+4=0$

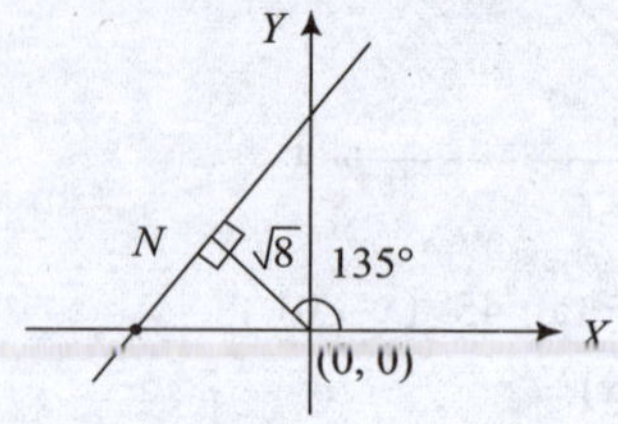

36. (D) $x\cos 240° + y\sin 240° = 2$

$\frac{-x}{2}-\frac{\sqrt{3}y}{2}=2$

$x+\sqrt{3}y+4=0$

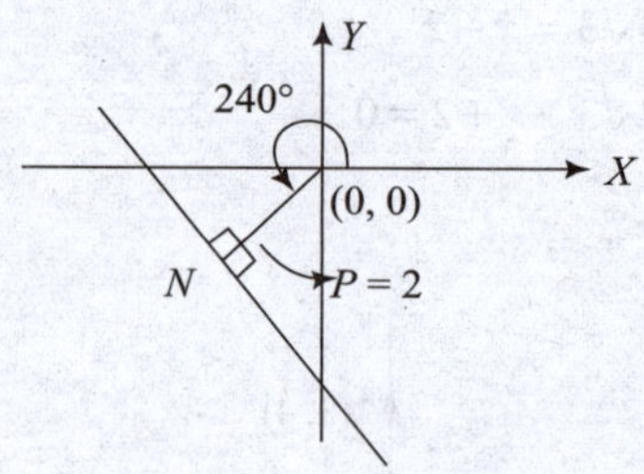

37. (A) Given that $\frac{x}{a}+\frac{y}{b}=1$ passes through the points (8, –9) and (12, –15).

So, $\frac{8}{a}-\frac{9}{b}=1$ and $\frac{12}{a}-\frac{15}{b}=1$

We get, $a = 2$ and $b = 3$.

38. (A) $\sqrt{3}x+y+2=0$

$y=-\sqrt{3}x-2$

39. (A) $\sqrt{3}x+y=-2$

$\frac{x}{\left(-\frac{2}{\sqrt{3}}\right)}+\frac{y}{(-2)}=1$

$\Rightarrow -\frac{\sqrt{3}}{2}x-\frac{1}{2}y=1$

$\Rightarrow x\cos 210^\circ + y\sin 210^\circ = 1$

$\therefore \alpha = 210°$ and $P = 1$

40. (D) $-x-\sqrt{3}y=5$

Dividing both sides by $\sqrt{1^2+\left(\sqrt{3}\right)^2}=2$, we get

$$\left(-\frac{1}{2}\right)x+\left(-\frac{\sqrt{3}}{2}\right)y=\frac{5}{2}$$

$$x\cos 240^\circ + y\sin 240^\circ = \frac{5}{2}$$

41. (A) Let 'a' and 'b' be the x-intercept and y-intercept, respectively.

$$\frac{1}{2}ab=6 \Rightarrow ab=12$$

And $\quad a^2+b^2=25$

By solving, we get $a = 4$ or $a = 3$

So, the equation of line is

$$\frac{x}{4}+\frac{y}{3}=1$$

or $\quad \frac{x}{3}+\frac{y}{4}=1$

$\Rightarrow \quad 3x+4y-12=0$

$\Rightarrow \quad 4x+3y-12=0$

42. (C) $y+4=0$

$-y = 4$

$x\cos 270^\circ + y\sin 270^\circ = 4$

43. (B) The point of intersection of $x - 7y + 15 = 0$ and $2x + y = 0$ is

$x=-1, y=2$

Equation of the line is

$x = -1 \Rightarrow x + 1 = 0$

44. (B) The point of intersection of the lines $3x + y = 2$ and $2x - y = 3$ is $(1, -1)$.

Since all three lines pass through the same point,

$k(1) + 2(-1) = 3 \Rightarrow k = 5$

45. (C) Slope of $y = x$ is 1 and angle of inclination $\theta = 45^\circ$.

$$\cos\theta=\frac{1}{\sqrt{2}}, \sin\theta=\frac{1}{\sqrt{2}}$$

The points $\left(2\pm 2\left(\frac{1}{\sqrt{2}}\right)\right), \left(2\pm 2\left(\frac{1}{\sqrt{2}}\right)\right)$

$$\left(2\pm\sqrt{2}\right), \left(2\pm\sqrt{2}\right)\left(2+\sqrt{2}\right), \left(2+\sqrt{2}\right)$$

or $\quad \left(2-\sqrt{2}\right), \left(2-\sqrt{2}\right)$

46. (B) Area $=\frac{1}{2}\times(4-1)\times(3-1)$

$$=\frac{1}{2}\times 3\times 2 = 3 \text{ sq. unit}$$

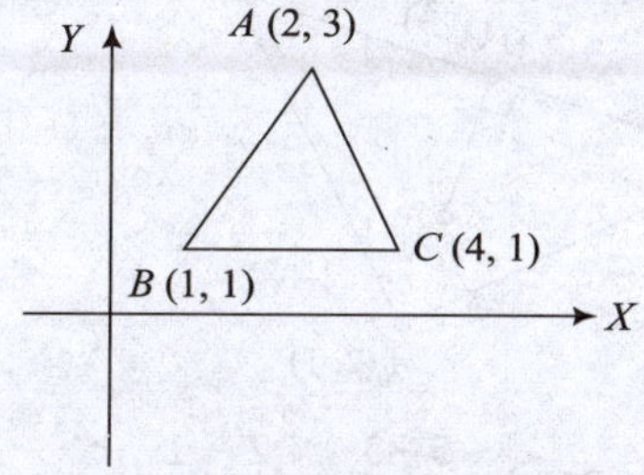

47. (C) D is the mid-point of BC.

$$D=\left(\frac{-4-2}{2},\frac{9-1}{2}\right)=(-3,4)$$

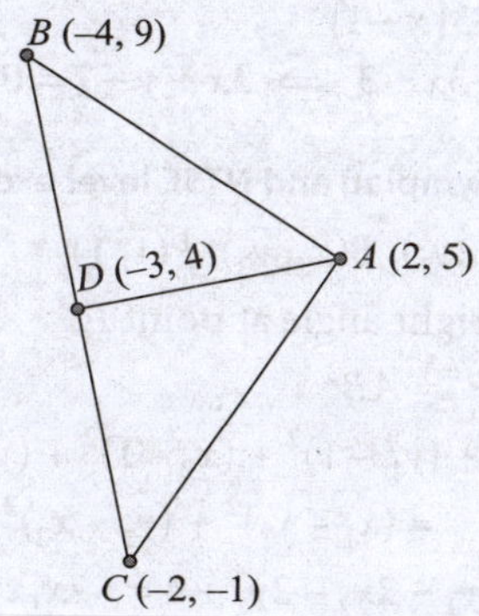

Slope of $AD = \frac{5-4}{2-(-3)} = \frac{1}{5}$

Equation of line

$$y-5=\frac{1}{5}(x-2)$$

$\Rightarrow \quad 5y-25=x-2$

$\Rightarrow \quad x-5y+23=0$

48. (B) Slope of $AC = \frac{1-5}{2-4} = \frac{-4}{-2} = 2$

Slope of altitude to $AC = -\frac{1}{2}$

Equation of altitude through B,

$$y-3=-\frac{1}{2}(x+2)$$

$\Rightarrow \quad 2y-6=-x-2$

$\Rightarrow \quad 2y+x-4=0$

49. (B) D is the mid-point of BC.

$$D \equiv \left(\frac{-2+4}{2}, \frac{3+5}{2}\right) \Rightarrow D \equiv (1,4)$$

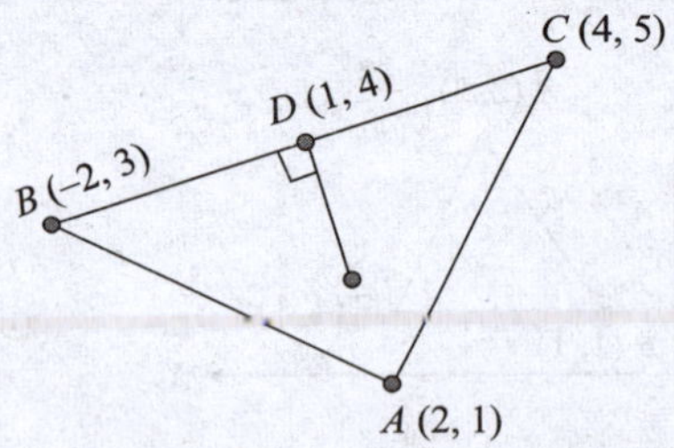

Slope of $BC = \dfrac{5-3}{4-(-2)} = \dfrac{2}{6} = \dfrac{1}{3}$

Slope of bisector through $D = -3$

Equation of line bisector through D

$$y-4=-3(x-1)$$

$$\Rightarrow \quad y-4=-3x+3 \Rightarrow 3x+y-7=0$$

Hints and Answers to Olympiad and NTSE level exercises

50. (B) We have $A(x_1, y_1)$, $B(x_2, y_2)$, $P(1, 1)$.

AB subtends right angle at point P.

$\therefore \quad AP^2 + BP^2 = AB^2$

$\therefore \quad (x_1 - 1)^2 + (y_1 - 1)^2 + (x_2 - 1)^2 + (y_2 - 1)^2 = (x_1 - x_2)^2 + (y_1 - x_2)^2$

$\therefore \quad -2x_1 - 2y_1 - 2x_2 - 2y_2 + 4 = -2x_1x_2 - 2y_1y_2$

$\therefore \quad x_1x_2 + y_1y_2 = x_1 + x_2 + y_1 + y_2 - 2$

51. (B) The line $\dfrac{x}{p} + \dfrac{y}{q} = 1$ where 'p' is a prime and 'q' is a positive integer passes through (4, 3).

$\therefore \quad \dfrac{4}{p} + \dfrac{3}{q} = 1$

$\therefore \quad q = \dfrac{3p}{p-4} = 3 + \dfrac{12}{p-4}$

Since, 'q' is a positive integer, $p > 4$ and $(p - 4)$ divides 12, there are only two such prime number $p = 5$ and $p = 7$.

52. (B) Let the point be (x, y).

Then $(x - a)^2 + y^2 - (x + a)^2 - y^2 = 2k^2$

$\Rightarrow \quad -4ax - 2k^2 = 0$

$\Rightarrow \quad 2ax + k^2 = 0.$

53. (C) From the given equations $3(my + 2) + 2y = 10$ or $y(3m + 2) = 4$

$$y = \frac{4}{3m+2} \quad \text{since } y \in I$$

$\Rightarrow \quad 3m + 2 = \pm 1, \pm 2, \pm 4$

$\Rightarrow \quad m = -1, -\dfrac{1}{3}, -\dfrac{4}{3}, 0, \dfrac{2}{3}, -2$

54. (D) The given line is $bx - ay = ab$

Obviously it cuts x-axis at $(a, 0)$. The equation of line perpendicular to (i) is $ax + by = k$, but it passes through $(a, 0) \Rightarrow k = a^2$.

Hence required equation of line is $ax + by = a^2$

i.e., $\dfrac{x}{b} + \dfrac{y}{a} = \dfrac{a}{b}$

55. (B) Angle between both the lines is

$$\tan^{-1} m \pm \tan^{-1} m = \tan^{-1} \frac{2m}{1-m^2} \text{ m or } \tan^{-1} 0$$

Therefore equation of lines are

$$y = 0, y = \frac{2mx}{1-m^2}.$$

56. (D)

57. (D) Line $x \cos\alpha + y \sin\alpha = P$ meets axis at $A(P \sec \alpha, 0)$ and $B(0, \operatorname{cosec} \alpha)$

Let mid-point of AB is (h, k).

$\therefore \quad h = P/(2\cos \alpha)$ and $k = P(2\sin \alpha)$

Eliminating A, we get locus as $\dfrac{1}{x^2} + \dfrac{1}{y^2} = \dfrac{4}{P^2}$

58. (B) The set of lines is $4ax + 3by + c = 0$, where $a + b + c = 0$

Eliminating c, we get $4ax + 3by - (a + b) = 0$

$\Rightarrow \quad a(4x - 1) + b(3y - 1) = 0$

This passes through the intersection of the lines $4x - 1 = 0$

and $3y - 1 = 0$, i.e., $x = \dfrac{1}{4}, y \dfrac{1}{3}$ i.e., $\left(\dfrac{1}{4}, \dfrac{1}{3}\right)$.

59. (A) $\rightarrow$ (p); (B) $\rightarrow$ (p, q, r), (C) $\rightarrow$ (p, r), (D) $\rightarrow$ (q, s)

(A) point $O(0, 0)$ and $B(2, 0)$

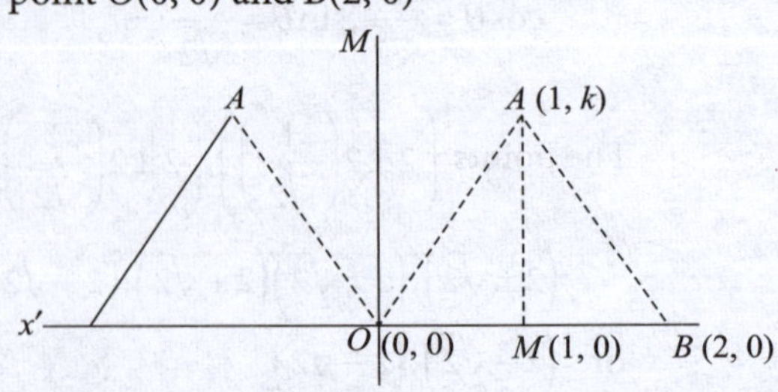

From figure here, ΔABC is equilateral, then $\tan 60° = k$

or $k = \sqrt{3}$ (for 1st quad.)

or $k = -\sqrt{3}$ (for 4th quad.)

Then possible coordinates are $(1, \pm\sqrt{3})$

Similarly for 2nd quad. point is $(-1, \sqrt{3})$

For 3rd quad. $(-1, -\sqrt{3})$

(B) **Case (i)**

If $OA = AB$, then $\angle A = 30°$

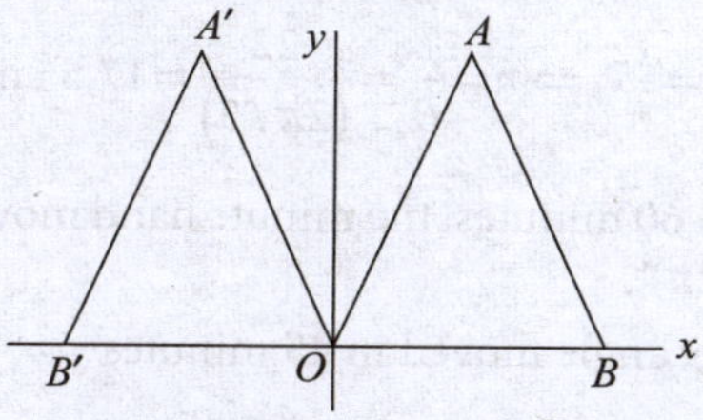

$\therefore \quad \angle AOB = 75°$

$\therefore \quad \dfrac{AM}{OM} = \tan 75°$

$AM = OM \tan 75°$

$k = 1 \times (2 + \sqrt{3})$

$\therefore \quad k = 2 + \sqrt{3}$

point A $(1, 2 + \sqrt{3})$

By symmetric all possible points are $(\pm 1, \pm(2 - \sqrt{3}))$

Case (ii) :

$AO = OB \qquad \therefore \quad \angle AOB = 120°$

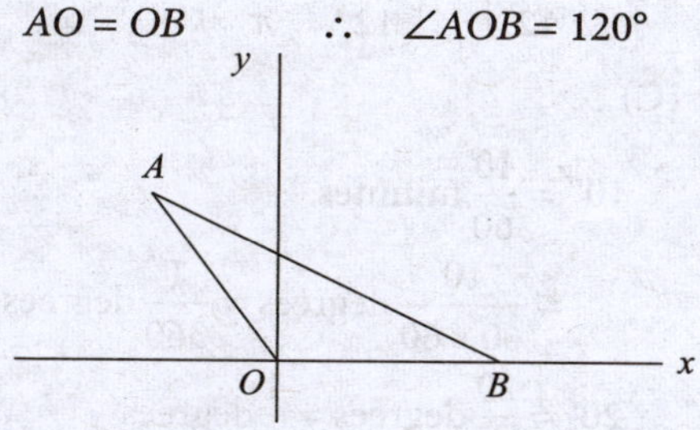

$|AM| = |OM \tan 120°| = -\sqrt{3}$

Point A is $(1, -\sqrt{3})$

By symmetry all possible points are $(\pm 1, \pm\sqrt{3})$.

(C)

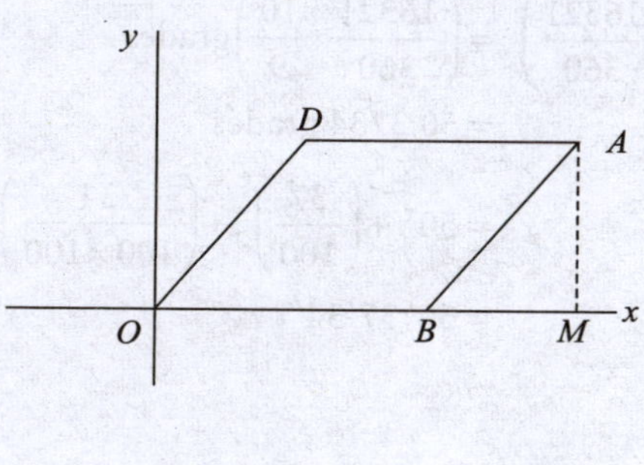

Let $\angle DOB = \angle ABM = \theta$

Area of $\Delta OAB = \dfrac{1}{2} OB \times AM = \dfrac{1}{2}\sqrt{3}$

$\Rightarrow \quad 2 \times 2 \sin\theta = \sqrt{3}$

$\Rightarrow \quad \sin\theta = \dfrac{\sqrt{3}}{4}$

$\Rightarrow \quad AM = \sqrt{3}$ and $BM = 1$

Hence A has coordinates $(3, \sqrt{3})$

By symmetry all possible coordinates are $(\pm 3, \pm\sqrt{3})$

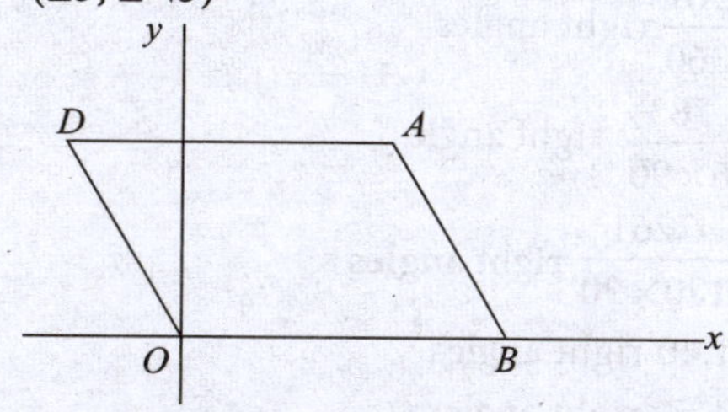

From the above figure A has coordinates $(1, \sqrt{3})$

By symmetry all possible coordinates are $(\pm 1, \pm\sqrt{3})$

(D)

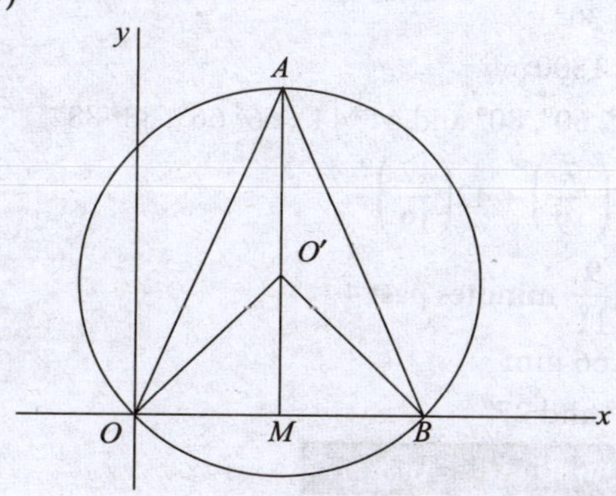

$OB = 2$ units $= OO' =$ radius

$\Rightarrow \quad OM = \dfrac{2}{2} = 1$ unit

In $\Delta OO'M$

$O'M = \sqrt{4 - 1} = \sqrt{3}$

$\because \quad \Delta OAB$ is isosceles hence point A lie on $\perp$ bisector of OB.

$\therefore \quad AM = \sqrt{3} + 2 = O'M + OA$

$\therefore \quad$ Coordinate A will be $(1, 2 + \sqrt{3})$ in 1st quadrant

By symmetry all possible coordinates of A are $(\pm 1, \pm(2 + \sqrt{3}))$

Measurement of Angles

1. (a) $33^g 33' 33\frac{1}{33}''$

(b) 90^g

(c) $153^g 88' 88\frac{8}{9}''$

2. $40'$

3. (a) $\frac{11}{15}$ right angle

(b) $\frac{301}{360}$ right angles

(c) $\frac{783}{6\times 90}$ right angle

(d) $\frac{7261}{120\times 90}$ right angles

4. (a) 1.40 right angles

(b) 1.50 right angles

(c) 0.505005 right angles

(d) 2.201515 right angles

5. $133°20'$

6. (a) $60°$

(b) $30°$

(c) $1800°$

7. $40°, 60°, 80°$ and $44^g 44'', 66^g 66'', 88^g 88''$

8. $47\left(\frac{7}{19}\right)^{\circ}$; $42\left(\frac{12}{19}\right)^{\circ}$

9. $21\frac{9}{11}$ minutes past 4

10. 14.66 mm

11. $45°$ and $27°$

12.

Question	Answer
(a)	(iv)
(b)	(i)
(c)	(v)
(d)	(iii)
(e)	(ii)

13. (a)

14. $\frac{1}{14\pi}$ second

15. $135°$

16. $33°20'$ and $10°48'$

17. $1^c = 57°17'44.8''$ nearly

18. The places are 6704.76 km apart

19. 2.9845^c

20. 40229 km approximately rounded off to km

21. 23 km approximately

22. 110.25 m

24. (A) $\ell = 22$ cm

$$\theta = 72^{\circ} = \left(72\times\frac{\pi}{180}\right)^c = \left(\frac{2\pi}{5}\right)^c$$

$$\theta = \frac{\ell}{r} \Rightarrow r = \frac{\ell}{\theta} = \frac{22}{(2\pi/5)} = 17.5 \text{ cm}$$

25. (D) In 60 minutes, the minute hand moves through $(2\pi)^c$.

So, angle moved in 45 minutes

$$= \left(\frac{2\pi}{60}\times 45\right)^c = \left(\frac{3\pi}{2}\right)^c$$

Distance moved by the tip of the minutes hand in 45 minutes.

$$\ell = r\theta = \left(1.4\times\frac{3\pi}{2}\right) = 6.6 \text{ cm}$$

26. (B) $36° = \left(36\times\frac{\pi}{180}\right)^c = \left(\frac{\pi}{5}\right)^c$

27. (A) $\left(\frac{5\pi}{12}\right)^c = \left(\frac{5\pi}{12}\times\frac{180°}{\pi}\right) = 75°$

28. (C)

$$10'' = \frac{10}{60} \text{ minutes}$$

$$= \frac{10}{60\times 60} \text{ degrees} = \frac{1}{360} \text{ degrees}$$

$$20' = \frac{20}{60} \text{ degrees} = \frac{1}{3} \text{ degrees}$$

$$\therefore\ 45°20'10'' = \left(45 + \frac{1}{360} + \frac{1}{3}\right) \text{degrees} = \left(\frac{16321}{360}\right)^{\circ}$$

$$\left(\frac{16321}{360}\right)^{\circ} = \left(\frac{16321}{360}\times\frac{10}{9}\right) \text{grades}$$

$$= 50.3734 \text{ grades}$$

$$= 50^g + \left(\frac{37}{100}\right)^g + \left(\frac{34}{100\times 100}\right)^g$$

$$= 50^g 37' 34''$$

29. (C) Let the angles be $3x$, $4x$ and $5x$.

$\therefore \quad 3x + 4x + 5x = 180°$

$\Rightarrow \quad 12x = 180°$

$\therefore \quad x = 15°$

Smallest angle $= 15° \times 3 = 45°$

30. (D) $r = 5$ cm

$$\theta = 33°5' = 33 + \frac{15}{60} = \frac{133}{4} \text{ degrees}$$

$$\theta(\text{in radian}) = \frac{133}{4} \times \frac{\pi}{180} = \frac{209}{360} \text{ radians}$$

$$\theta = \frac{\ell}{r}$$

$$\Rightarrow \quad \ell = r\theta = \frac{209}{360} \times 5 = 2\frac{65}{72} \text{ cm}$$

31. (C) Diameter of wheel = 4 feet

Radius of wheel = 2 feet

Circumference of wheel = $2\pi r$

$= 2\pi \times 2 = 4\pi$ feet

Number of revolutions made in 1 second = 6

Distance covered in 1 second

$= 4\pi \times 6 = 24\pi$ feet

Speed of train = 24π feet/sec

32. (B) Measure of each interior angle of a regular polygon of n sides $= \dfrac{2n-4}{n} \times 90°$

Measure of each interior angle of regular pentagon

$$= \frac{2(5)-4}{5} \times 90° = \frac{6}{5} \times 90°$$

$$= \frac{6}{5} \times 90 \times \frac{\pi}{180} \text{ radians}$$

$$= \frac{3}{5}\pi \text{ radians}$$

33. (C)

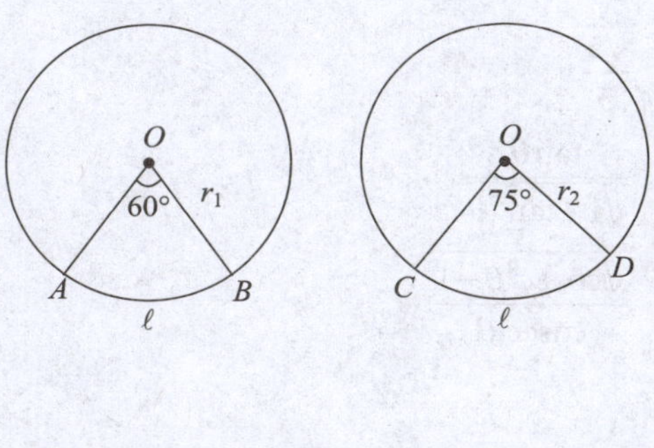

$$r_1 \times \frac{60}{180} \times \pi = r_2 \times \frac{75}{180} \times \pi \quad \therefore \quad \frac{r_1}{r_2} = \frac{5}{4}$$

34. (B) Difference between acute angles $= \left(\dfrac{\pi}{15}\right)^c = 12°$

Let the two acute angles be x and y. Then

$x + y = 90$

$x - y = 12$

Solving these, we get

$x = 51°$ and $y = 39°$

Hence, the angles are 51°, 39°, and 90°.

35. (B) $\ell = 66$ m;

$$\theta = 72° = \left(\frac{72}{180} \times \pi\right) \text{radian} = \frac{2\pi}{5} \text{ radian}$$

$$r = \frac{\ell}{\theta} = 66 \times \frac{5}{2\pi} = 66 \times \frac{5 \times 7}{2 \times 22} = 52.5 \text{ m}$$

36. (B) $r = 14$ cm

$$\theta = 36° = \frac{\cancel{36} \times \pi}{\cancel{180}_5} = \frac{\pi}{5} \text{radian}$$

$$\ell = r\theta = 14 \times \frac{\pi}{5} = \frac{\cancel{14}^2}{5} \times \frac{22}{\cancel{7}} = 8.8 \text{ cm}$$

37. (D) Since four angles are in AP, we can take them as

$a - 3d, a - d, a + d, a + 3d$

Then,

$(a - 3d) + (a - d) + (a + d) + (a + 3d) = 360°$

$\Rightarrow \quad 4a = 360°$

$\therefore \quad a = 90°$

And $a + 3d = 2(a - 3d)$

$\Rightarrow \quad a + 3d = 2a - 6d$

$\Rightarrow \quad 9d = a$

$\Rightarrow \quad 9d = 90°$

$\therefore \quad d = 10°$

Therefore, the angles are 60°, 80°, 100°, 120°

38. (B) $\ell = 121$ cm;

$r = 180$ cm;

$$\theta = \frac{\ell}{r} = \frac{121}{180} \text{ radians}$$

$$= \frac{121}{180} \times \frac{180}{\pi} \text{ degrees}$$

$$= \frac{121}{22} \times 7 \text{ degrees}$$

39. (C) $r = 1500$ m

ℓ = distance covered in 10 seconds

$= \dfrac{66\times1000}{60\times60}\times10$ m

$\theta = \dfrac{\ell}{r} = \dfrac{66\times1000\times10}{60\times60\times1500}$ radians

$= \dfrac{66}{6\times6\times15}\times\dfrac{180}{\pi}$ degrees = 7 degrees

40. (B) Number of revolutions in one second $= \dfrac{180}{60} = 3$

Angle covered in 1 revolution = 2π

Total angle = $2\pi\times3 = 6\pi$ radian

41. (C) Let the number of sides of polygon be n.

Sum of total angle $= \dfrac{n}{2}[2\times120^\circ+(n-1)5^\circ]$

$= \dfrac{n}{2}(5n+235^\circ)$

Total interior angle = $(2n - 4)\times90^\circ$

Then $\dfrac{n}{2}(5n+235^\circ) = 2n-4\times90^\circ$

$\therefore \quad n = 9, 16$

Hints and answers to Olympiad and NTSE level exercises

42. (B) Sum of the angles of a polygon = $(2n - 4)$ 90°

Sum of the angles of a rectangle = $(2\times5 - 4)$ 90°

$= 6\times90^\circ = 540^\circ$

43. (A) $x + y = 70$

$x - y = 10$

$+ \quad + \quad +$

$2x = 80$

$x = 40$

Since, $x - y = 10$, we get $y = 30$.

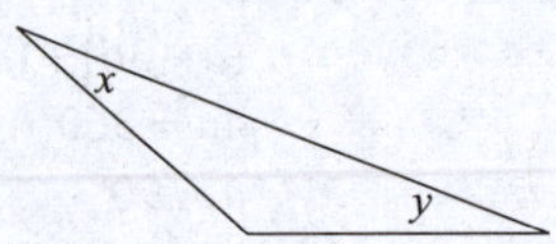

44. (D) Given $x = y + z$

We know

$x + y + z = 180$

$\therefore \quad 2(y + z) = 180$

or $\quad y + z = 90$

or $\quad x = 90$

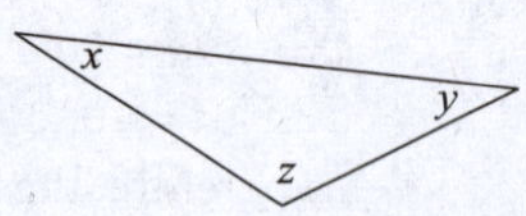

45. (B) $2\angle A = 3\angle A$

and $5\angle B = 2\angle C$

$\angle A + \angle B + \angle C = 180^\circ$

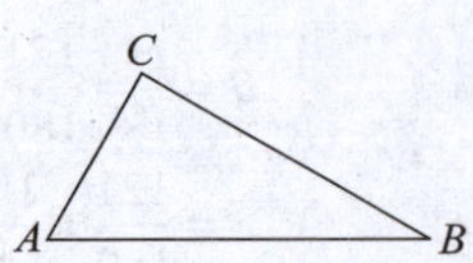

$\therefore \quad \dfrac{2}{3}\angle B + \angle B + \dfrac{5}{2}\angle B = 180^\circ$

$5\angle B = 180^\circ$ or $\angle B = 36^\circ$

$\therefore \quad \angle A = \dfrac{3}{2}\times30 = 54^\circ$

and $\angle C = \dfrac{5}{2}\times36 = 90^\circ$

Therefore, the triangle is right-angled.

46. (C) If one angle is 90° in a triangle, the other two must be complementary.

47. (D) $\angle A = \dfrac{1}{2}(\angle B + \angle C)$

$\angle A + \angle B + \angle C = \dfrac{1}{2}(\angle B + \angle A)$

or $\quad \angle B + \angle C = 90^\circ$

or $\quad \angle A = 90^\circ$

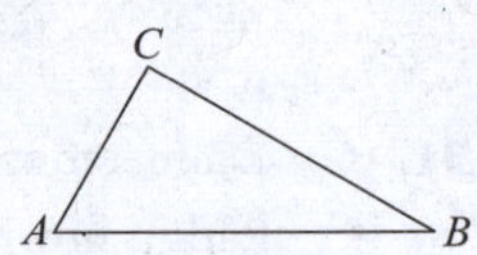

48. (D) 12 – 1 = 11

49. (C) Supplement of 150° is 30° and complement angle of 30° is 60°.

50. (B) If one angle is 90°, the other two are 45° and 45°.

51. (C) $\angle A + \angle A + 25 + 4\angle A + 5 = 180$

or $\quad 6\angle A = 180 - 30 = 150$

or $\quad \angle A = \dfrac{150}{6} = 25^\circ$

$\therefore \quad 4\angle A + 5^\circ = 100^\circ + 5^\circ = 105^\circ$

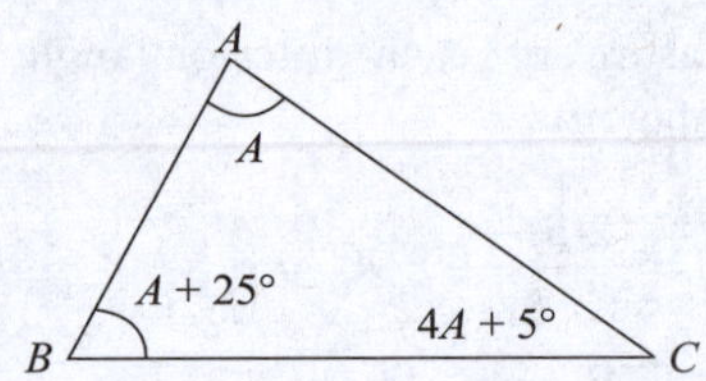

Trigonometry

11. $\dfrac{3}{4}$

12. $\dfrac{\sqrt{5}}{3}$

13. $\dfrac{\tan\theta}{\sqrt{1+\tan^2\theta}}$

14. $\dfrac{\sqrt{\operatorname{cosec}^2\theta-1}}{\operatorname{cosec}\theta}$

22. (B) $\sec\theta - \tan\theta$

$$= (\sec\theta - \tan\theta) \times \frac{(\sec\theta + \tan\theta)}{(\sec\theta + \tan\theta)}$$

$$= \frac{(\sec^2\theta - \tan^2\theta)}{(\sec\theta + \tan\theta)} = \frac{1}{(\sec\theta + \tan\theta)} = \frac{1}{a}$$

23. (C) $(1 + \tan A + \sec A)(1 + \cot A - \operatorname{cosec} A)$

$$= \left(1 + \frac{\sin A}{\cos A} + \frac{1}{\cos A}\right)\left(1 + \frac{\cos A}{\sin A} - \frac{1}{\sin A}\right)$$

$$= \left(\frac{\cos A + \sin A + 1}{\cos A}\right)\left(\frac{\sin A + \cos A - 1}{\sin A}\right)$$

$$= \frac{(\cos A + \sin A)^2 - 1}{\sin A \cos A}$$

$$= \frac{\cos^2 A + \sin^2 A + 2\sin A\cos A - 1}{\sin A \cos A}$$

$$= \frac{1 + 2\sin A\cos A - 1}{\sin A\cos A}$$

$$= 2$$

24. (C) $\sin^2\theta + \cos^2\theta = 1$

$$\cos^2\theta = 1 - \sin^2\theta$$

$$= 1 - \left(\frac{5}{7}\right)^2 = \frac{49 - 25}{49} = \frac{24}{49}$$

$\therefore \quad \cos\theta = \dfrac{2\sqrt{6}}{7}$ (Since θ lies in 1st quadrant)

$$\therefore \quad \tan\theta = \frac{\sin\theta}{\cos\theta} = \frac{5/7}{2\sqrt{6}/7} = \frac{5}{2\sqrt{6}}$$

25. (D) $(\cos\theta + \sin\theta)^2 + (\cos\theta - \sin\theta)^2$

$$= \cos^2\theta + \sin^2\theta + 2\sin\theta\cos\theta + \cos^2\theta + \sin^2\theta - 2\sin\theta\cos\theta$$

$$= 2\cos^2\theta + 2\sin^2\theta$$

$$= 2(\cos^2\theta + \sin^2\theta)$$

$$= 2$$

26. (D) $\dfrac{1}{1-\sin\alpha} + \dfrac{1}{1+\sin\alpha} = \dfrac{1+\sin\alpha + 1 - \sin\alpha}{1-\sin^2\alpha}$

$$= \frac{2}{\cos^2\alpha} = 2\sec^2\alpha$$

27. (B) $\dfrac{\sin\theta}{1+\cos\theta} + \dfrac{1+\cos\theta}{\sin\theta}$

$$= \frac{\sin^2\theta + (1+\cos\theta)^2}{\sin\theta(1+\cos\theta)}$$

$$= \frac{\sin^2\theta + 1 + \cos^2\theta + 2\cos\theta}{\sin\theta(1+\cos\theta)}$$

$$= \frac{2(1+\cos\theta)}{\sin\theta(1+\cos\theta)} = 2\operatorname{cosec}\theta$$

28. (B) $\sqrt{\dfrac{1+\cos\theta}{1-\cos\theta}} = \sqrt{\dfrac{1+\cos\theta}{1-\cos\theta} \times \dfrac{1+\cos\theta}{1+\cos\theta}}$

$$= \sqrt{\frac{(1+\cos\theta)^2}{\sin^2\theta}}$$

$$= \frac{1+\cos\theta}{\sin\theta}$$

29. (A) $\dfrac{\sec\theta - 1}{\sec\theta + 1}$

$$= \frac{\frac{1}{\cos\theta} - 1}{\frac{1}{\cos\theta} + 1} = \frac{1-\cos\theta}{1+\cos\theta} \times \frac{1-\cos\theta}{1-\cos\theta}$$

$$= \frac{(1-\cos\theta)^2}{1-\cos^2\theta} = \frac{(1-\cos\theta)^2}{\sin^2\theta} = \left(\frac{1-\cos\theta}{\sin\theta}\right)^2$$

$$= (\operatorname{cosec}\theta - \cot\theta)^2 = (\cot\theta - \operatorname{cosec}\theta)^2$$

30. (C) $1 + \dfrac{\cot^2\theta}{1+\operatorname{cosec}\theta} = 1 + \dfrac{\operatorname{cosec}^2\theta - 1}{\operatorname{cosec}\theta + 1}$

$$= 1 + \operatorname{cosec}\theta - 1$$

$$= \operatorname{cosec}\theta$$

31. (C) $\dfrac{\cos\theta}{\operatorname{cosec}\theta - 1} + \dfrac{\cos\theta}{\operatorname{cosec}\theta + 1}$

$$= \cos\theta\left[\frac{\operatorname{cosec}\theta + 1 + \operatorname{cosec}\theta - 1}{(\operatorname{cosec}\theta - 1)(\operatorname{cosec}\theta + 1)}\right]$$

$$= \cos\theta\left[\frac{2\operatorname{cosec}\theta}{\operatorname{cosec}^2\theta - 1}\right]$$

$$= \cos\theta\left[\frac{2}{\sin\theta}\frac{1}{\cot^2\theta}\right]$$

$$= 2\frac{\cos\theta}{\sin\theta}.\tan^2\theta$$

$$= 2\tan\theta$$

32. (D) $x \cos A = 1$

$x = \sec A$ and $y = \tan A$

$x^2 - y^2 = (\sec A)^2 - (\tan A)^2 = 1$

33. (A) $\dfrac{(1+\sin\theta)(1-\sin\theta)}{(1+\cos\theta)(1-\cos\theta)}$

$$= \frac{(1-\sin^2\theta)}{(1-\cos^2\theta)} = \frac{\cos^2\theta}{\sin^2\theta} = (\cot\theta)^2$$

$$= \left(\frac{7}{8}\right)^2 = \frac{49}{64}$$

34. (D) $\cos^4\theta - \sin^4\theta = \dfrac{1}{2}$

$$\Rightarrow (\cos^2\theta + \sin^2\theta)(\cos^2\theta - \sin^2\theta) = \frac{1}{2}$$

$$\Rightarrow (\cos^2\theta - \sin^2\theta) = \frac{1}{2}$$

$$\Rightarrow 2\cos^2\theta - 1 = \frac{1}{2}$$

$$\Rightarrow \cos^2\theta = \frac{3}{4}$$

$$\Rightarrow \cos\theta = \frac{\sqrt{3}}{2}$$

35. (C) $(a \sin\theta - b\cos\theta)^2 + (a\cos\theta + b\sin\theta)^2$

$$= a^2\sin^2\theta + b^2\cos^2\theta - 2ab\sin\theta\cos\theta +$$

$$a^2\cos^2\theta + b^2\sin^2\theta + 2ab\sin\theta\cos\theta$$

$$= a^2(\sin^2\theta + \cos^2\theta) + b^2(\sin^2\theta + \cos^2\theta)$$

$$= a^2 + b^2$$

36. (D) $\dfrac{\sin\theta}{1+\cos\theta} + \dfrac{\sin\theta}{1-\cos\theta} = 2$

$$\Rightarrow \frac{\sin\theta(1-\cos\theta)+\sin\theta(1+\cos\theta)}{(1+\cos\theta)(1-\cos\theta)} = 2$$

$$\Rightarrow \frac{2\sin\theta}{1-\cos^2\theta} = 2$$

$$\Rightarrow \frac{2\sin\theta}{\sin^2\theta} = 2$$

$$\Rightarrow \frac{2}{\sin\theta} = 2$$

$$\Rightarrow \sin\theta = 1$$

$$\therefore \quad \theta = \frac{\pi}{2}$$

37. (A) $\dfrac{\cos\theta}{1+\sin\theta} + \dfrac{1+\sin\theta}{\cos\theta} = 4$

$$\Rightarrow \frac{\cos^2\theta + (1+\sin\theta)^2}{\cos\theta(1+\sin\theta)} = 4$$

$$\Rightarrow \frac{\cos^2\theta + 1 + \sin^2\theta + 2\sin\theta}{\cos\theta(1+\sin\theta)} = 4$$

$$\Rightarrow \frac{2(1+\sin\theta)}{\cos\theta(1+\sin\theta)} = 4$$

$$\Rightarrow 2\sec\theta = 4$$

$$\Rightarrow \sec\theta = 2$$

$$\therefore \quad \theta = \frac{\pi}{3}$$

38. (B) $(1 + \cos A)(1 - \cos A)(1 + \cot^2 A)$

$$= (1 - \cos^2 A)(1 + \cot^2 A)$$

$$= \sin^2 A \operatorname{cosec}^2 A$$

$$= 1$$

39. (C) $\dfrac{\operatorname{cosec} A}{1-\operatorname{cosec} A} + \dfrac{\operatorname{cosec} A}{\operatorname{cosec} A + 1}$

$$= \operatorname{cosec} A\left[\frac{1+\operatorname{cosec} A + 1 - \operatorname{cosec} A}{1-\operatorname{cosec}^2 A}\right]$$

$$= \operatorname{cosec} A\left(\frac{2}{-\cot^2 A}\right)$$

$$= -\operatorname{cosec} A\left(\frac{2\sin^2 A}{\cos^2 A}\right)$$

$$= -\left(\frac{2\sin A}{\cos^2 A}\right) = -2\tan A\sec A$$

40. (B) $\sqrt{\dfrac{1+\sin\theta}{1-\sin\theta}} + \sqrt{\dfrac{1-\sin\theta}{1+\sin\theta}}$

$$= \sqrt{\frac{1+\sin\theta}{1-\sin\theta}\times\frac{1+\sin\theta}{1+\sin\theta}} + \sqrt{\frac{1-\sin\theta}{1+\sin\theta}\times\frac{1-\sin\theta}{1-\sin\theta}}$$

$$= \frac{1+\sin\theta}{\cos\theta} + \frac{1-\sin\theta}{\cos\theta}$$

$$= \frac{2}{\cos\theta} = 2\sec\theta$$

41. (A) $(1+\cot A-\operatorname{cosec} A)(1+\tan A+\sec A)=2\sec A$

$$\Rightarrow \left(1+\frac{\cos A}{\sin A}-\frac{1}{\sin A}\right)\left(1+\frac{\sin A}{\cos A}+\frac{1}{\cos A}\right)=2\sec A$$

$$\Rightarrow \left(\frac{\sin A+\cos A-1}{\sin A}\right)\left(\frac{\cos A+\sin A+1}{\cos A}\right)=2\sec A$$

$$\Rightarrow \frac{(\sin A+\cos A)^2-1}{\sin A\cos A}=2\sec A$$

$$\Rightarrow \frac{\sin^2 A+\cos^2 A+2\sin A\cos A-1}{\sin A\cos A}=2\sec A$$

$$\Rightarrow 2=2\sec A$$

$$\Rightarrow \sec A=1$$

$$\therefore\ A=90^\circ$$

42. (B) $(\sec\theta\cot\theta)^2-(\cos\theta\operatorname{cosec}\theta)^2$

$$=\frac{1}{\cos^2\theta}\cdot\frac{\cos^2\theta}{\sin^2\theta}-\frac{\cos^2\theta}{\sin^2\theta}$$

$$=\frac{1}{\sin^2\theta}-\frac{\cos^2\theta}{\sin^2\theta}$$

$$=\frac{1-\cos^2\theta}{\sin^2\theta}$$

$$=\frac{\sin^2\theta}{\sin^2\theta}$$

$$=1$$

43. (C) $x=a\cos\theta$ and $y=a\sin\theta$

$x^2=a^2\cos^2\theta$ and $y^2=a^2\sin^2\theta$

$x^2+y^2=a^2(\sin^2\theta+\cos^2\theta)$

$\therefore\ x^2+y^2=a^2$

44. (C) $x=A\sec\alpha$ and $y=B\tan\alpha$

$\frac{x}{A}=\sec\alpha$ and $\frac{y}{B}=\tan\alpha$

$\frac{x^2}{A^2}=\sec^2\alpha$ and $\frac{y^2}{B^2}=\tan^2\alpha$

Now,

$$\sec^2\alpha-\tan^2\alpha=\frac{x^2}{A^2}-\frac{y^2}{B^2}=1$$

45. (D) $x=a\cos\theta$ and $y=b\sin\theta$

$x^2=a^2\cos^2\theta$ and $y^2=b^2\sin^2\theta$

$\frac{x^2}{a^2}=\cos^2\theta$ and $\frac{y^2}{b^2}=\sin^2\theta$

$$\frac{x^2}{a^2}+\frac{y^2}{b^2}=(\sin^2\theta+\cos^2\theta)$$

$$\therefore\ \frac{x^2}{a^2}+\frac{y^2}{b^2}=1$$

46. (B) $(\sin\theta+\cos\theta)^2+(\sin\theta-\cos\theta)^2$

$=\sin^2\theta+\cos^2\theta+2\sin\theta+\cos\theta+\sin^2\theta+\cos^2\theta-2\sin\theta\cos\theta$

$=1+1$

$=2$

47. (C) $\sec^2\theta(1+\sin\theta)(1-\sin\theta)$

$=\sec^2\theta(1-\sin^2\theta)$

$=\sec^2\theta\cdot\cos^2\theta$

$=1$

48. (D) $x^2+y^2+z^2$

$$=(r\sin\theta\cos\phi)^2+(r\sin\theta.\sin\phi)^2+(r\cos\theta)^2$$

$$=r^2\sin^2\theta\cos^2\phi+r^2\sin^2\theta\sin^2\phi+r^2\cos^2\theta$$

$$=r^2\sin^2\theta(\cos^2\phi+\sin^2\phi)+r^2\cos^2\theta$$

$$=r^2\sin^2\theta+r^2\cos^2\theta$$

$$=r^2$$

49. (A) $\frac{\cos A}{1-\tan A}+\frac{\sin A}{1-\cot A}$

$$=\frac{\cos A}{1-\frac{\sin A}{\cos A}}+\frac{\sin A}{1-\frac{\cos A}{\sin A}}$$

$$=\frac{\cos^2 A}{\cos A-\sin A}+\frac{\sin^2 A}{\sin A-\cos A}$$

$$=\frac{\cos^2 A-\sin^2 A}{\cos A-\sin A}$$

$$=\cos A+\sin A$$

50. (C) $a\sec\alpha-c\tan\alpha=d$ and $b\sec\alpha+d\tan\alpha=c$

$$\frac{a}{\cos\alpha}-\frac{c\sin\alpha}{\cos\alpha}=d$$

$\Rightarrow\ d\cos\alpha+c\sin\alpha=a$ ...(i)

and $\frac{b}{\cos\alpha}+\frac{d\sin\alpha}{\cos\alpha}=c$

$\Rightarrow\ c\cos\alpha-d\sin\alpha=b$ ...(ii)

Squaring and adding (i) and (ii), we get

$$d^2\cos^2\alpha + c^2\sin^2\alpha + 2dc\sin\alpha\cos\alpha$$
$$+c^2\cos^2\alpha + d^2\sin^2\alpha - 2dc\sin\alpha\cos\alpha$$
$$= a^2 + b^2$$
$$\Rightarrow \quad d^2\left(\cos^2\alpha + \sin^2\alpha\right) + c^2\left(\cos^2\alpha + \sin^2\alpha\right)$$
$$= a^2 + b^2$$
$$\Rightarrow \quad c^2 + d^2 = a^2 + b^2$$

51. (B) Let us assume that

$$4\sin\theta - 3\cos\theta = x \qquad \ldots\text{(i)}$$

Given equation is

$$3\sin\theta + 4\cos\theta = 5 \qquad \ldots\text{(ii)}$$

Squaring and adding (i) and (ii), we get

$$16\sin^2\theta + 9\cos^2\theta - 24\sin\theta\cos\theta$$
$$+ 9\sin^2\theta + 16\cos^2\theta + 24\sin\theta\cos\theta$$
$$= x^2 + 25$$
$$\Rightarrow \quad 16\left(\cos^2\theta + \sin^2\theta\right) + 9\left(\cos^2\theta + \sin^2\theta\right)$$
$$= x^2 + 25$$
$$\Rightarrow \quad 16 + 9 = x^2 + 25$$
$$\Rightarrow \quad x^2 = 0$$
$$\therefore \quad x = 0$$

Hints and answers to Olympiad and NTSE level exercises

52. (B) Squaring the given relation and putting $\tan\theta = t$,

$$(m+2)^2 t^2 + 2(m+2)(2m-1)t + (2m+1)^2$$
$$= (2m+1)^2 (1 + t^2)$$
$$\Rightarrow \quad 3(1 - m^2)t^2 + (4m^2 + 6m - 4)t - 8m = 0$$
$$\Rightarrow \quad (3t - 4)[(1 - m^2)t + 2m] = 0$$

which is true if $t = \tan\theta = \dfrac{4}{3}$

or $\quad \tan\theta = \dfrac{2m}{m^2 - 1}$.

53. (D) The expression can be written as

$$\frac{1 + \cos y - \sin^2 y}{1 + \cos y} + \frac{(1 - \cos^2 y) - \sin^2 y}{\sin y(1 - \cos y)}$$
$$= \frac{\cos y(1 + \cos y)}{1 + \cos y} + 0 = \cos y.$$

54. (D) $p = \dfrac{2\sin\theta}{1 + \cos\theta + \sin\theta},\ q = \dfrac{\cos\theta}{1 + \sin\theta}$

$$\Rightarrow \quad p + q = \frac{\cos\theta}{1 + \sin\theta} + \frac{2\sin\theta}{1 + \sin\theta + \cos\theta}$$
$$\Rightarrow \quad p + q = 1.$$

55. (D) Given that $\tan\theta - \cot\theta = a$ ...(i)

and $\sin\theta + \cos\theta = b$ ...(ii)

Now $(b^2 - 1)^2 (a^2 + 4)$

$$= \left\{(\sin\theta + \cos\theta)^2 - 1\right\}^2 \left\{(\tan\theta - \cot\theta)^2 + 4\right\}$$
$$= [1 + \sin 2\theta - 1]^2 [\tan^2\theta + \cot^2\theta - 2 + 4]$$
$$= \sin^2 2\theta(\operatorname{cosec}^2\theta + \sec^2\theta)$$
$$= 4\sin^2\theta\cos^2\theta\left[\frac{1}{\sin^2\theta} + \frac{1}{\cos^2\theta}\right] = 4$$

Trick : Obviously the value of expression $(b^2 - 1)^2 (a^2 + 4)$ is independent of θ, therefore put any suitable value of θ. Let $\theta = 45°$, we get $a = 0$, $b = \sqrt{2}$ so that $[(\sqrt{2})^2 - 1]^2 (0^2 + 4) = 4$.

56. (B) $A + B = 45°$, therefore $2A = 90° - 2B$

$\therefore \quad \cos 2A = \sin 2B$.

57. (D) As given $\dfrac{\sin A + \sin B}{\cos A + \cos B} = \dfrac{C}{D}$

$$\Rightarrow \quad \frac{2\sin\frac{A+B}{2}\cdot\cos\frac{A-B}{2}}{2\cos\frac{A+B}{2}\cdot\cos\frac{A-B}{2}} = \frac{C}{D}$$
$$\Rightarrow \quad \tan\frac{A+B}{2} = \frac{C}{D}$$

Thus, $\sin(A+B) = \dfrac{2\tan\frac{A+B}{2}}{1 + \tan^2\frac{A+B}{2}}$

$$= \frac{2\frac{C}{D}}{1 + \frac{C^2}{D^2}} = \frac{2CD}{(C^2 + D^2)}$$

58. (B) $\sqrt{\left(\dfrac{1 - \sin\theta}{1 + \sin\theta}\right)} + \sqrt{\left(\dfrac{1 + \sin\theta}{1 - \sin\theta}\right)}$ is the sum of two positive quantities and hence, the result must be positive. But for $\dfrac{\pi}{2} < \theta < \pi$, we have the sum equal to $\frac{1-\sin\theta+1+\sin\theta}{\sqrt{1-\sin^2\theta}}=\frac{2}{\cos\theta}$; which is negative.

($\because$ $\cos\theta$ is negative for θ lying in 2nd quadrant).

So, required value

$=\frac{-2}{\cos\theta}=-2\sec\theta, \left(\frac{\pi}{2}<\theta<\pi\right)$

59. (A) Given that $2y\cos\theta = x\sin\theta$ ...(i)

and $2x\sec\theta - y\,\text{cosec}\,\theta = 3$ ...(ii)

$\Rightarrow \frac{2x}{\cos\theta}-\frac{y}{\sin\theta}=3$

$\Rightarrow 2x\sin\theta - y\cos\theta - 3\sin\theta\cos\theta = 0$...(iii)

Solving (i) and (iii), we get $y = \sin\theta$ and $x = 2\cos\theta$

Now, $x^2 + 4y^2 = \cos^2\theta + 4\sin^2\theta$

$= 4(\cos^2\theta + \sin^2\theta) = 4.$

60. (A) $\to$ (q), (B) $\to$ (s), (C) $\to$ (p), (D) $\to$ (r)

(A) $A = \sin^2\theta + \cos^4\theta$

$=\frac{1-\cos 2\theta}{2}+\left(\frac{1+\cos 2\theta}{2}\right)^2$

$=\frac{1}{2}-\frac{1}{2}\cos 2\theta+\frac{1}{4}+\frac{1}{2}\cos 2\theta+\frac{1}{4}\cos^2 2\theta$

$=\frac{3}{4}+\frac{1}{4}\left(\frac{\cos 4\theta+1}{2}\right)=\frac{3}{4}+\frac{1}{8}+\frac{1}{8}\cos 4\theta$

Now, $-1\le\cos 4\theta\le 1$

$\Rightarrow \frac{-1}{8}\le\frac{\cos 4\theta}{8}\le\frac{1}{8}$

$\Rightarrow \frac{3}{4}+\frac{1}{8}-\frac{1}{8}\le\frac{3}{4}+\frac{1}{4}\left(\frac{1+\cos 4\theta}{2}\right)\le\frac{3}{4}+\frac{1}{8}+\frac{1}{8}$

$\Rightarrow \frac{3}{4}\le A\le 1$

(B) $A = 3\cos^2\theta + \sin^4\theta$

$=3\frac{1+\cos 2\theta}{2}+\left(\frac{1-\cos 2\theta}{2}\right)^2$

$=\frac{3+3\cos 2\theta}{2}+\frac{1-2\cos 2\theta+\cos^2 2\theta}{4}$

$=\frac{7+4\cos 2\theta+\cos^2 2\theta}{4}$

$=\frac{(\cos 2\theta+2)^2+3}{4}$

Now, $1\le\cos 2\theta+2\le 3$

$\Rightarrow 1\le\frac{(\cos 2\theta+2)^2+3}{4}\le 3$

(C) $A = \sin^2\theta - \cos^4\theta$

$=\frac{1-\cos 2\theta}{2}-\left(\frac{1+\cos 2\theta}{2}\right)^2$

$=\frac{1}{2}-\frac{1}{2}\cos 2\theta-\frac{1}{4}-\frac{1}{2}\cos 2\theta-\frac{1}{4}\cos^2 2\theta$

$=\frac{1}{4}-\cos 2\theta-\frac{1}{4}\cos^2 2\theta$

$=-\left(\frac{1}{4}\cos^2 2\theta+\cos 2\theta-\frac{1}{4}\right)$

$=\frac{5}{4}-\left(\frac{1}{2}\cos 2\theta+1\right)^2$

Now, $-\frac{1}{2}\le\frac{1}{2}\cos 2\theta\le\frac{1}{2}$

$\Rightarrow \frac{1}{2}\le\frac{1}{2}\cos 2\theta+1\le\frac{3}{2}$

$\Rightarrow \frac{1}{4}\le\left(\frac{1}{2}\cos 2\theta+1\right)^2\le\frac{9}{4}$

$\Rightarrow -\frac{9}{4}\le-\left(\frac{1}{2}\cos 2\theta+1\right)^2\le-\frac{1}{4}$

$\Rightarrow -1\le\frac{5}{4}-\left(\frac{1}{2}\cos 2\theta+1\right)^2\le 1$

(D) $A = \tan^2\theta + 2\cot^2\theta$

$= (\tan\theta - \sqrt{2}\cot\theta)^2 + 2\sqrt{2} \ge 2\sqrt{2}$

61. (A) $\left(\frac{\sin 2A}{1+\cos 2A}\right)\left(\frac{\cos A}{1+\cos A}\right)$

$=\frac{2\sin A\cos A}{2\cos^2 A}\frac{\cos A}{1+\cos A}=\frac{\sin A}{1+\cos A}=\tan\frac{A}{2}.$

Standard and Allied Angles

1. $\frac{5}{2}$

2. 1

3. $\frac{\sqrt{6}-\sqrt{2}}{4}$

11. $\frac{\sqrt{2}(\sqrt{3}+1)}{4}$

12. $\frac{\sqrt{2}(\sqrt{3}+1)}{4}$

13. $\sin A$

14. $\sin A$

15. $\tan A$

16. 9

17. $\frac{11}{12}$

18. $\frac{\sqrt{3}}{2}$

19. 6

20. $\frac{1}{\sqrt{3}}$

21. $\frac{1}{\sqrt{2}}$

22. $-\frac{1}{\sqrt{2}}$

23. -1

24. (a) 9° (b) 22½° (c) 10°

26. (a) 0 (b) -1 (c) $2\tan A$

27. (a) $\sin A\cos A + 2$ (b) -1
(c) $\sin\theta$

28. (a) sin 7° (b) $-\tan 40$
(c) $-\cos 25°$ (d) $-\cot 24°$
(e) cosec 24°

29. (a) Second quadrant
(b) Third quadrant
(c) Second quadrant
(d) Fourth quadrant
(e) Fourth quadrant

31. (a) 1 (b) 5

35. $BC = 8\sqrt{3}$ cm, $AC = 16\sqrt{3}$ cm

36. $AD = \frac{10\sqrt{3}}{\sqrt{3}+1}$

37. $BC = 9$ cm, $AB = 9\sqrt{3}$ cm

38. (A)
$$\cot(-135^\circ) = -\cot(135^\circ) = -\cot(90^\circ + 45^\circ) = -(-\tan 45^\circ) = \tan 45^\circ = 1$$

39. (B)
$$\text{cosec}(-1410^\circ) = -\text{cosec}(1410^\circ) = -\text{cosec}(4\times 360^\circ - 30^\circ) = \text{cosec}\,30^\circ = 2$$

40. (B)
$$\cos\theta + \sin(270^\circ+\theta) - \sin(270^\circ-\theta) + \cos(180^\circ+\theta)$$
$$= \cos\theta - \cos\theta + \cos\theta - \cos\theta = 0$$

41. (A)
$$\frac{\cos(90^\circ+\theta)\sec(270^\circ+\theta)\sin(180^\circ+\theta)}{\text{cosec}(-\theta)\cos(270^\circ-\theta)\tan(180^\circ+\theta)}$$
$$= \frac{(-\sin\theta)(\text{cosec}\,\theta)(-\sin\theta)}{(-\text{cosec}\,\theta)(-\sin\theta)(\tan\theta)} = \cos\theta$$

42. (C)
$$\cos\left(\frac{3\pi}{2}+\theta\right)\cos(2\pi+\theta)\left[\cot\left(\frac{3\pi}{2}-\theta\right)+\cot(2\pi+\theta)\right]$$
$$= (\sin\theta.\cos\theta)[\tan\theta + \cot\theta]$$
$$= \sin^2\theta + \cos^2\theta = 1$$

43. (B) $\sin^2 60° + \tan 45° - \cos 60° + \text{cosec}^2 45°$
$$= \left(\frac{\sqrt{3}}{2}\right)^2 + 1 - \left(\frac{1}{2}\right) + (\sqrt{2})^2$$
$$= \frac{3}{4} + 1 - \frac{1}{2} + 2 = \frac{5}{2} + \frac{3}{4} = \frac{10+3}{4} = \frac{13}{4}$$

44. (C) $\sin^2 30° + \sin^2 45° + \sin^2 60° + \tan 45°$
$$= \left(\frac{1}{2}\right)^2 + \left(\frac{1}{\sqrt{2}}\right)^2 + \left(\frac{\sqrt{3}}{2}\right)^2 + 1$$
$$= \frac{1}{4} + \frac{1}{2} + \frac{3}{4} + 1 = \frac{5}{2}$$

45. (C) $A + B + C = 180°$
$$A + B = 180^\circ - C$$
$$\frac{A+B}{2} = \frac{180^\circ - C}{2}$$
$$\tan\left(\frac{A+B}{2}\right) = \tan\left(90^\circ - \frac{C}{2}\right) = \cot\frac{C}{2}$$

46. (D)
$$\frac{\cos(\pi+\theta)\cos(-\theta)}{\cos(\pi-\theta)\cos\left(\frac{\pi}{2}+\theta\right)}$$
$$= \frac{-\cos\theta\cos\theta}{(-\cos\theta)(-\sin\theta)} = -\cot\theta$$

47. (A) $$\frac{\dfrac{\sin(180^\circ+\theta)\cos(90^\circ+\theta)\tan(270^\circ-\theta)}{\cot(360^\circ-\theta)}}{\dfrac{\sin(360^\circ-\theta)\cos(360^\circ+\theta)\text{cosec}(-\theta)}{\sin(270^\circ+\theta)}}$$

$$=\frac{(-\sin\theta)(-\sin\theta)(\cot\theta)(-\cot\theta)}{(-\sin\theta)(\cos\theta)(-\text{cosec}\,\theta)(-\cos\theta)}=1$$

48. (B) $$\frac{\cos 135^\circ-\cos 120^\circ}{\cos 135^\circ+\cos 120^\circ}$$

$$=\frac{\cos(90^\circ+45^\circ)-\cos(180^\circ-60^\circ)}{\cos(90^\circ+45^\circ)+\cos(180^\circ-60^\circ)}$$

$$=\frac{-\sin 45^\circ+\cos 60^\circ}{-\sin 45^\circ-\cos 60^\circ}$$

$$=\frac{-\frac{1}{\sqrt{2}}+\frac{1}{2}}{-\frac{1}{\sqrt{2}}-\frac{1}{2}}=\frac{\frac{-2+\sqrt{2}}{2\sqrt{2}}}{\frac{-2-\sqrt{2}}{2\sqrt{2}}}$$

$$=\frac{-2+\sqrt{2}}{-2-\sqrt{2}}=\frac{2-\sqrt{2}}{2+\sqrt{2}}=\frac{2-\sqrt{2}}{2+\sqrt{2}}\times\frac{2-\sqrt{2}}{2-\sqrt{2}}$$

$$=\frac{\left(2-\sqrt{2}\right)^2}{4-2}=\frac{4+2-4\sqrt{2}}{2}=3-2\sqrt{2}$$

49. (A) $3\tan^2 45^\circ-\sin^2 60^\circ-\frac{1}{2}\cot^2 30^\circ+\frac{1}{8}\sec^2 45^\circ$

$$=3(1)^2-\left(\frac{\sqrt{3}}{2}\right)^2-\frac{1}{2}\left(\sqrt{3}\right)^2+\frac{1}{8}\left(\sqrt{2}\right)^2=1$$

50. (C) $\tan 1^\circ \tan 2^\circ \tan 3^\circ \ldots \tan 87^\circ \tan 88^\circ \tan 89^\circ$

$=(\tan 1^\circ \tan 89^\circ)(\tan 2^\circ \tan 88^\circ)\ldots\tan 45^\circ$

$=\{\tan 1^\circ \tan(90^\circ-1^\circ)\}.\{\tan 2^\circ \tan(90^\circ-2^\circ)\}\ldots\tan 45^\circ$

$=(\tan 1^\circ \cot 1^\circ)(\tan 2^\circ \cot 2^\circ)\ldots\tan 45^\circ$

$=1\times1\times1\times\ldots\times1=1$

51. (B) $\sin^2 5^\circ+\sin^2 10^\circ+\sin^2 15^\circ+\ldots+\sin^2 90^\circ$

$$=\left(\sin^2 5^\circ+\sin^2 85^\circ\right)+\left(\sin^2 10^\circ+\sin^2 80^\circ\right)+\ldots+\left(\sin^2 40^\circ+\sin^2 50^\circ\right)+\left(\sin^2 45^\circ+\sin^2 90^\circ\right)$$

$$=\left(\sin^2 5^\circ+\cos^2 5^\circ\right)+\left(\sin^2 10^\circ+\cos^2 10^\circ\right)+\ldots+\left(\sin^2 40^\circ+\cos^2 40^\circ\right)+\left(\left(\frac{1}{\sqrt{2}}\right)^2+(1)^2\right)$$

$$=(1+1+1+1+1+1+1+1)+\frac{1}{2}+1=9\frac{1}{2}$$

52. (D) $\cos^2\frac{\pi}{16}+\cos^2\frac{3\pi}{16}+\cos^2\frac{5\pi}{16}+\cos^2\frac{7\pi}{16}$

$$=\cos^2\frac{\pi}{16}+\cos^2\frac{3\pi}{16}+\cos^2\left(\frac{\pi}{2}-\frac{3\pi}{16}\right)+\cos^2\left(\frac{\pi}{2}-\frac{\pi}{16}\right)$$

$$=\cos^2\frac{\pi}{16}+\cos^2\frac{3\pi}{16}+\sin^2\frac{3\pi}{16}+\sin^2\frac{\pi}{16}$$

$$=1+1=2$$

53. (D) $\cot^2 30^\circ-2\cos^2 60^\circ-\frac{3}{4}\sin^2 45^\circ-4\sin^2 30^\circ$

$$=\left(\sqrt{3}\right)^2-2\left(\frac{1}{2}\right)^2-\frac{3}{4}\left(\frac{1}{\sqrt{2}}\right)^2-4\left(\frac{1}{2}\right)^2$$

$$=3-2\left(\frac{1}{4}\right)-\frac{3}{8}-1$$

$$=3-\frac{1}{2}-\frac{3}{8}-1=2-\frac{1}{2}-\frac{3}{8}$$

$$=\frac{16-4-3}{8}=\frac{9}{8}$$

54. (B) $$\frac{\sec 480^\circ\cdot\text{cosec}\,570^\circ\cdot\tan 330^\circ}{\sin 600^\circ\cdot\cos 660^\circ\cdot\cot 405^\circ}$$

$$=\frac{\sec(3\times180^\circ-60^\circ)\cdot\text{cosec}(3\times180^\circ+30^\circ)\cdot\tan(2\times180^\circ-30^\circ)}{\sin(3\times180^\circ+60^\circ)\cdot\cos(4\times180^\circ-60^\circ)\cdot\cot(2\times180^\circ+45^\circ)}$$

$$=\frac{(-\sec 60^\circ)(-\text{cosec}\,30^\circ)(-\tan 30^\circ)}{(-\sin 60^\circ)(\cos 60^\circ)(\cot 45^\circ)}$$

$$=\frac{-(2)(-2)\left(-\frac{1}{\sqrt{3}}\right)}{-\left(\frac{\sqrt{3}}{2}\right)\left(\frac{1}{2}\right)(1)}=\frac{16}{3}$$

55. (C) $\sin^2 6° + \sin^2 12° + \sin^2 18° + ... + \sin^2 84° + \sin^2 90°$

$= (\sin^2 6^\circ + \sin^2 84^\circ) + (\sin^2 12^\circ + \sin^2 78^\circ) + ... + \sin^2 90^\circ$

$= (\sin^2 6^\circ + \cos^2 6^\circ) + (\sin^2 12^\circ + \cos^2 12^\circ) + ... + \sin^2 90^\circ$

$= 1+1+1+1+1+1+1+(1)^2 = 8$

56. (A) $\tan 9° \cdot \tan 27° \cdot \tan 45° \cdot \tan 63° \cdot \tan 81°$

$= \left(\tan 9^\circ \cdot \tan 81^\circ\right)\left(\tan 27^\circ \cdot \tan 63^\circ\right) \cdot \tan 45^\circ$

$= \left(\tan 9^\circ \cdot \tan\left(90^\circ - 9^\circ\right)\right) \left(\tan 27^\circ \cdot \tan\left(90^\circ - 27^\circ\right)\right) \cdot \tan 45^\circ$

$= \left(\tan 9^\circ \cdot \cot 9^\circ\right)\left(\tan 27^\circ \cdot \cot 27^\circ\right) \cdot \tan 45^\circ$

$= 1 \times 1 \times 1 = 1$

57. (B) $\tan\alpha \cdot \tan 2\alpha \cdot \tan 3\alpha \cdot ... \cdot \tan(2n-2)\alpha \cdot \tan(2n-1)\alpha$

$= \tan\alpha \cdot \tan 2\alpha \cdot \tan 3\alpha \cdot ... \cdot \tan n\alpha \cdot ... \cdot \tan\left(2n\alpha - 2\alpha\right) \cdot \tan\left(2n\alpha - \alpha\right)$

$= \tan\alpha \cdot \tan 2\alpha \cdot \tan 3\alpha \cdot ... \cdot \tan\frac{\pi}{4} \cdot ... \cdot \tan\left(\frac{\pi}{2} - 2\alpha\right) \cdot \tan\left(\frac{\pi}{2} - \alpha\right)$

$= \tan\alpha \cdot \tan 2\alpha \cdot \tan 3\alpha \cdot ... \cdot \cot 2\alpha \cdot \cot\alpha$

$= \left(\tan\alpha \cdot \cot\alpha\right)\left(\tan 2\alpha \cdot \cot 2\alpha\right) ...$

$= 1 \times 1 \times 1 \times 1 \times 1 ... = 1$

58. (A) $2(\sin^6\theta + \cos^6\theta) - 3\ 2(\sin^4\theta + \cos^4\theta) + 1$

$= 2\left\{\left(\sin^2\theta + \cos^2\theta\right)^3 - 3\sin^2\theta\cos^2\theta \left(\sin^2\theta + \cos^2\theta\right)\right\} - 3\left\{\left(\sin^2\theta + \cos^2\theta\right)^2 - 2\sin^2\theta\cos^2\theta\right\} + 1$

$= 2\left\{1 - 3\sin^2\theta\cos^2\theta\right\} - 3\left\{1 - 2\sin^2\theta\cos^2\theta\right\} + 1$

$= 2 - 6\sin^2\theta\cos^2\theta - 3 + 6\sin^2\theta\cos^2\theta + 1$

0.

59. (D) $\tan^2\frac{\pi}{3} + 2\cos^2\frac{\pi}{4} + 3\sec^2\frac{\pi}{6} + 4\cos^2\frac{\pi}{2}$

$= \left(\sqrt{3}\right)^2 + 2\left(\frac{1}{\sqrt{2}}\right)^2 + 3\left(\frac{2}{\sqrt{3}}\right)^2 + 4(0)^2$

$= 3 + 1 + 4 + 0 = 8$

60. (B) $\sin\frac{\pi}{6}\cos 0 + \sin\frac{\pi}{4}\cos\frac{\pi}{4} + \sin\frac{\pi}{3}\cos\frac{\pi}{6}$

$= \frac{1}{2} \times 1 + \frac{1}{\sqrt{2}} \times \frac{1}{\sqrt{2}} + \frac{\sqrt{3}}{2} \times \frac{\sqrt{3}}{2}$

$= \frac{1}{2} + \frac{1}{2} + \frac{3}{4} = 1 + \frac{3}{4} = \frac{7}{4}$

61. (B) $\frac{\sin(-A)}{\sec\left(180^\circ + A\right)} - \frac{\tan\left(90^\circ + A\right)}{\cot A} + \frac{\cos\left(360^\circ - A\right)}{\sin A} + \frac{\cos A}{\sin\left(90^\circ + A\right)}$

$= \frac{-\sin A}{-\sec A} - \frac{\left(-\cot A\right)}{\cot A} + \frac{\cos A}{\sin A} + \frac{\cos A}{\cos A}$

$= \sin A\cos A + 1 + \cot A + 1$

$= \sin A\cos A + \cot A + 2$

62. (A) $\frac{\cos\left(90^\circ + A\right)\sec\left(-A\right)\tan\left(180^\circ - A\right)}{\sec\left(180^\circ + A\right)\sin\left(180^\circ + A\right)\cot\left(90^\circ - A\right)}$

$= \frac{-\sin A\ \sec A\ \left(-\tan A\right)}{-\sec A\ \left(-\sin A\right)\tan A}$

$= 1$

63. (B) $\sin 10° \cos 80° + \cos 10° \sin 80°$

$= \sin 10^\circ \cos(90^\circ - 10^\circ) + \cos 10^\circ \sin(90^\circ - 10^\circ)$

$= \sin 10^\circ \sin 10^\circ + \cos 10^\circ \cos 10^\circ$

$= \sin^2 10^\circ + \cos^2 10^\circ$

$= 1$

64. (A) $\tan^2 45° - \sin^2 30° = x \sin^2 45° \cdot \tan 60°$

$x = \frac{\tan^2 45^\circ - \sin^2 30^\circ}{\sin^2 45^\circ . \tan 60^\circ}$

$= \frac{1 - \frac{1}{4}}{\frac{1}{2}\sqrt{3}} = \frac{3/4}{\sqrt{3}/2} = \frac{\sqrt{3}}{2} = \sin 60^\circ$

65. (C) $x \sin 60^\circ \cos^2 45^\circ = \frac{\tan^2 45^\circ \operatorname{cosec} 30^\circ \cot 45^\circ}{\sec 45^\circ \sec^2 60^\circ}$

$\Rightarrow x\left(\frac{\sqrt{3}}{2}\right)\left(\frac{1}{\sqrt{2}}\right)^2 = \frac{(1)\times(2)\times(1)}{\sqrt{2}\times(2)^2}$

$\Rightarrow x\left(\frac{\sqrt{3}}{4}\right) = \frac{2}{4\sqrt{2}} \quad \therefore x = \sqrt{\frac{2}{3}}$

66. (B) $\tan 60^\circ = \dfrac{\text{AD}}{\text{BD}}$

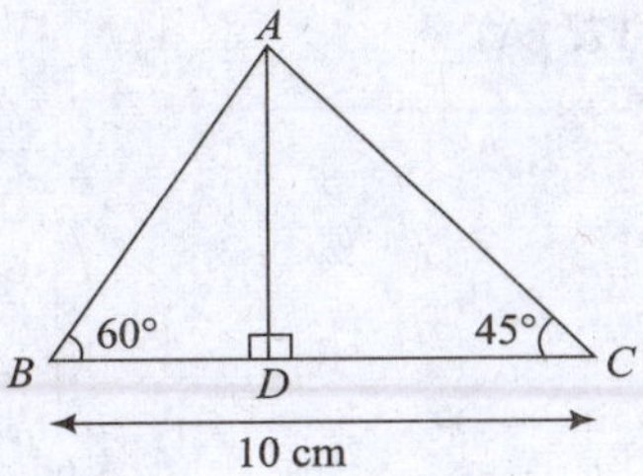

$\text{BD} = \text{AD} \cot 60^\circ$

$= \dfrac{\text{AD}}{\sqrt{3}}$

$\tan 45^\circ = \dfrac{AD}{DC}$

$DC = AD \cot 45^\circ$

$DC = AD$

Now, $BD + DC = BC$

$\dfrac{AD}{\sqrt{3}} + AD = 10$

$AD = \dfrac{10\sqrt{3}}{\sqrt{3}+1}$

Hints and answers to Olympiad and NTSE level exercises

67. (D) $\cos 105° + \sin 105° = \cos (90° + 15°) + \sin (90° + 15)$

$= \cos 15° - \sin 15° = \dfrac{\sqrt{3}+1}{2\sqrt{2}} - \dfrac{\sqrt{3}-1}{2\sqrt{2}}$.

$= \dfrac{2}{2\sqrt{2}} = \dfrac{1}{\sqrt{2}}$

68. (A) $\tan\left(\dfrac{\pi}{4} + \theta\right) - \tan\left(\dfrac{\pi}{4} - \theta\right)$

$= \dfrac{1+\tan\theta}{1-\tan\theta} - \dfrac{1-\tan\theta}{1+\tan\theta}$

$= \dfrac{4\tan\theta}{1-\tan^2\theta} = 2\left(\dfrac{2\tan\theta}{1-\tan^2\theta}\right) = 2\tan 2\theta$.

69. (B) $\sin 50° - \sin 70° + \sin 10°$

$= -2 \cos 60° \sin 10° + \sin 10°$

$= \sin 10° (1 - 2 \cos 60°) = 0$

70. (B) We know that

$\tan(20° + 40°) = \dfrac{\tan 20° + \tan 40°}{1 - \tan 20° \tan 40°}$

$\Rightarrow \sqrt{3} = \dfrac{\tan 20° + \tan 40°}{1 - \tan 20° \tan 40°}$

$\Rightarrow \sqrt{3} - \sqrt{3} \tan 20° \tan 40° = \tan 20° + \tan 40°$

$\Rightarrow \tan 20° + \tan 40° + \sqrt{3} \tan 20° \tan 40° = \sqrt{3}$.

71. (C) $\tan x + \tan\left(\dfrac{\pi}{3} + x\right) + \tan\left(\dfrac{2\pi}{3} + x\right)$

$= \tan x + \dfrac{\tan x + \sqrt{3}}{1 - \sqrt{3}\tan x} + \dfrac{\tan x - \sqrt{3}}{1 + \sqrt{3}\tan x}$

$= \tan x + \dfrac{8\tan x}{1 - 3\tan^2 x} = \dfrac{3(3\tan x - \tan^3 x)}{1 - 3\tan^2 x}$

$= 3\tan 3x$

Therefore, the given equation is $3 \tan 3x = 3$

$\Rightarrow \tan 3x = 1$.

72. (B) $2\cos\dfrac{\pi}{13}.\cos\dfrac{9\pi}{13} + \cos\dfrac{3\pi}{13} + \cos\dfrac{5\pi}{13}$

$= 2\cos\dfrac{\pi}{13}.\cos\dfrac{9\pi}{13} + 2\cos\dfrac{4\pi}{13}\cos\dfrac{\pi}{13}$

$= 2\cos\dfrac{\pi}{13}\left[\cos\dfrac{9\pi}{13} + \cos\dfrac{4\pi}{13}\right]$

$= 2\cos\dfrac{\pi}{13}\left[2\cos\dfrac{\pi}{2} \cdot \cos\dfrac{5\pi}{26}\right] = 0 \quad \left[\because \cos\dfrac{\pi}{2} = 0\right]$

73. (C) $2 \cos 10° + \sin 100° + \sin 1000° + \sin 10000°$

$= 2 \cos 10° + \sin(90° + 10°) + \sin(3 \times 360° - 80°) + \sin(28 \times 360° - 80°)$

$= 2 \cos 10° + \cos 10° - \sin 80° - \sin 80°$

$= 3 \cos 10° - 2\cos 10°$

$= \cos 10°$

74. (C) $\sec 11° \cdot \sec 19° - 2 \cot 71°$

$= \dfrac{1}{\cos 11° \cos 19°} - \dfrac{2\sin 19°}{\cos 19°}$

$= \dfrac{1 - 2\sin 19° \cos 11°}{\cos 11° \cos 19°} = \dfrac{1 - (\sin 30° + \sin 8°)}{\cos 11° \cos 19°}$

$= \dfrac{1 - (1/2) - \sin 8°}{\cos 11° \cos 19°} = \dfrac{\sin 30° - \sin 8°}{\cos 11° \cos 19°}$

$= \dfrac{2\sin 11° \cos 19°}{\cos 11° \cos 19°} = 2 \tan 11°$

75. (A) $\dfrac{\cot 84° \cot 48°}{\cot 66° \cot 78°}$

$= \dfrac{(2 \cos 84° \sin 66°)(2\sin 78° \cos 48°)}{(2\sin 84° \cos 66°)(2 \cos 78° \sin 48°)}$

$= \dfrac{(\sin 150° - \sin 18°)(\sin 126° + \sin 30°)}{(\sin 150° + \sin 18°)(\sin 126° - \sin 30°)}$

$= \dfrac{(\sin 30° - \sin 18°)(\sin 54° + \sin 30°)}{(\sin 30° + \sin 18°)(\sin 54° - \sin 30°)}$

$$= \frac{\left(\frac{1}{2}-\frac{\sqrt{5}-1}{4}\right)\left(\frac{\sqrt{5}+1}{4}+\frac{1}{2}\right)}{\left(\frac{1}{2}+\frac{\sqrt{5}-1}{4}\right)\left(\frac{\sqrt{5}+1}{4}-\frac{1}{2}\right)}$$

$$= \frac{(3-\sqrt{5})(3+\sqrt{5})}{(\sqrt{5}+1)(\sqrt{5}-1)} = \frac{9-5}{5-1} = \frac{4}{4} = 1$$

76. (C) $\sin\frac{9\pi}{14}\sin\frac{11\pi}{14}\sin\frac{13\pi}{14}$

$$= \sin\frac{5\pi}{14}\sin\frac{3\pi}{14}\sin\frac{\pi}{14}$$

$$= \cos\frac{\pi}{7}\cos\frac{2\pi}{7}\cos\frac{3\pi}{7}$$

$$= -\cos\frac{\pi}{7}\cos\frac{2\pi}{7}\cos\frac{4\pi}{7} = \frac{\sin\frac{8\pi}{7}}{8\sin\frac{\pi}{7}} = \frac{1}{8}$$

Heights and Distances

Apply appropriate *t*-ratios in the respective figures and get the required result for all questions.

1. 30°
2. 11.8 m (approximately)
3. 17.7 m (approximately)
4. 136.6 m (approximately)
5. 13 m; 7.5 m from the second pillar
6. 12 m and 7 m
7. 74.6 m (approximately)
8. 1366 m (approximately)
9. 42.3 m (approximately)
10. 20.8 m and 32.8 m (approximately)
11. 6.4 min (approximately)
12. 315.5 m (approximately)
13. $\frac{a\tan\alpha\times\tan\beta}{\tan\alpha+\tan\beta}$
14. 440 km/hr
15. (B)
16. (B) $20\cot 30° = d$

$\Rightarrow d = 20\sqrt{3}$

17. (A)

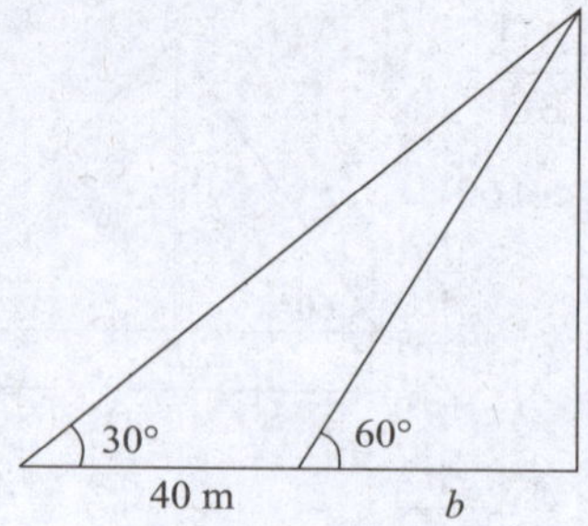

$b = h\cot 60^\circ$ $\qquad b+40 = h\cot 30^\circ$

$$\Rightarrow \frac{b}{b+40} = \frac{\cot 60^\circ}{\cot 30^\circ} \quad \therefore b = 20\,\text{m}$$

18. (D) Let the height of the house be h.

Height of the tower = 60 m

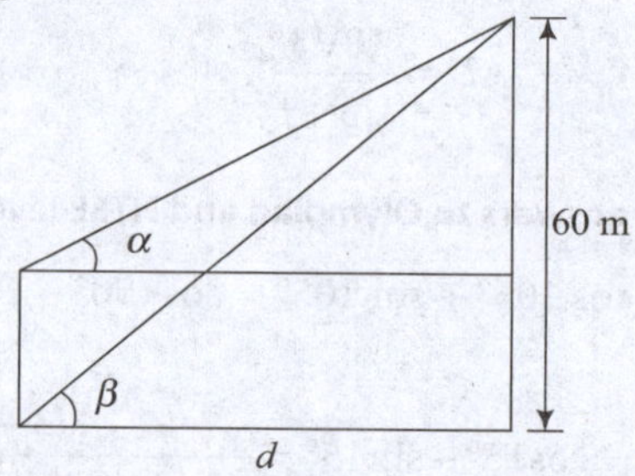

$60 = d\tan\beta$ and $60 - h = d\tan\alpha$

$$\frac{60}{60-h} = \frac{\tan\beta}{\tan\alpha}$$

$$\Rightarrow -h = \frac{60\tan\alpha - 60\tan\beta}{\tan\beta}$$

$$\Rightarrow h = \frac{60\sin(\beta-\alpha)}{\cos\alpha\cos\beta\frac{\sin\beta}{\cos\beta}} \quad \therefore h = \frac{60\sin(\beta-\alpha)}{\cos\alpha\sin\beta}$$

19. (A)

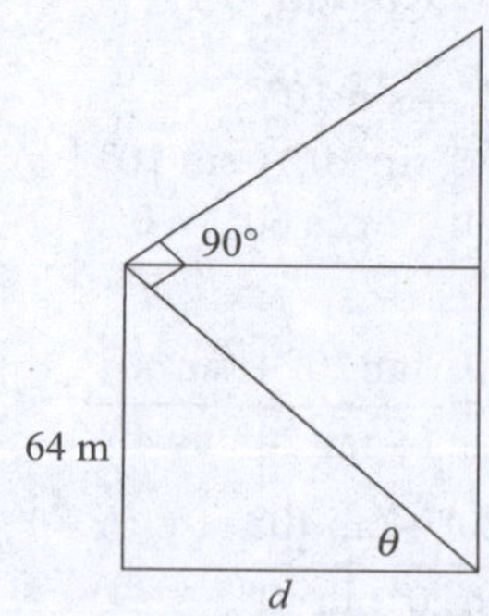

$64\cot\theta = d$ $\qquad (100-64)\tan\theta = d$

$(64)(36) = d^2$ $\qquad d = 8\times 6 = 48$ m

20. (B)

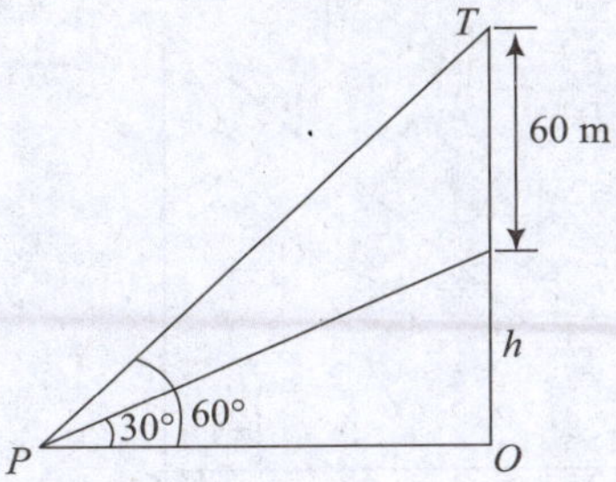

$$(60 + h)\cot 60^\circ = h \cot 30^\circ \quad \Rightarrow h = 30 \text{ m}$$

21. (D)

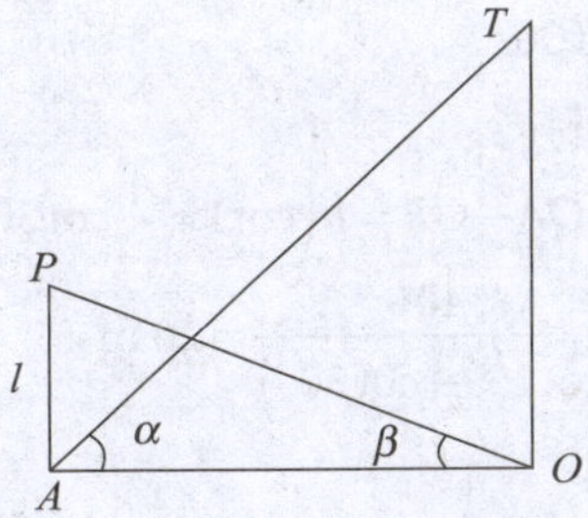

$$\tan\beta = \frac{l}{OA} \quad \Rightarrow OA = \frac{l}{\tan\beta}$$

$$\tan\alpha = \frac{OT}{OA} \quad \Rightarrow OT = OA$$

$$\tan\alpha = \frac{l}{\tan\beta}\cdot\tan\alpha = l\cot\beta\tan\alpha$$

22. (B)

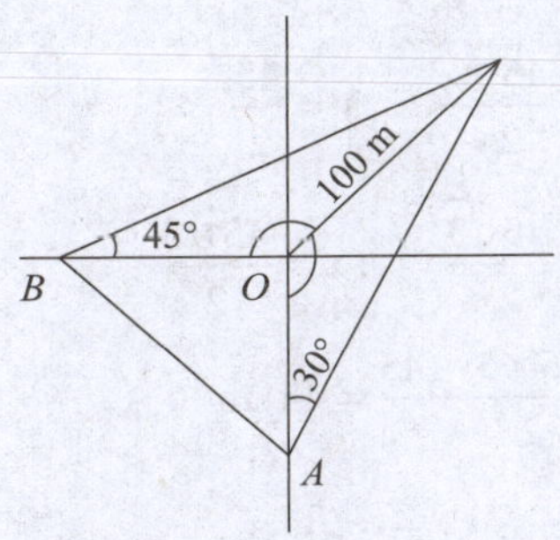

$$OB = 100\cot 45^\circ$$

$$OA = 100\cot 30^\circ$$

$$AB = \sqrt{OA^2 + OB^2} = 200 \text{ m}$$

23. (B) $\tan\alpha = \dfrac{h}{\sqrt{3}h} = \dfrac{1}{\sqrt{3}} \quad \therefore\ \alpha = 30^\circ$

24. (D) Length of the ladder $= \dfrac{6\sqrt{3}}{\sin 60^\circ} = \dfrac{6\sqrt{3}}{\frac{\sqrt{3}}{2}} = 12$ m

25. (C)

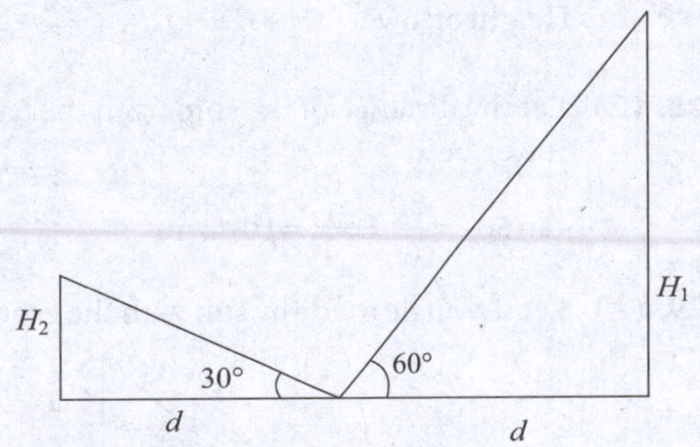

$$H_1 = d\tan 60^\circ \qquad H_2 = d\tan 30^\circ$$

$$\frac{H_1}{H_2} = \frac{\tan 60^\circ}{\tan 30^\circ} = \frac{3}{1}$$

26. (B)

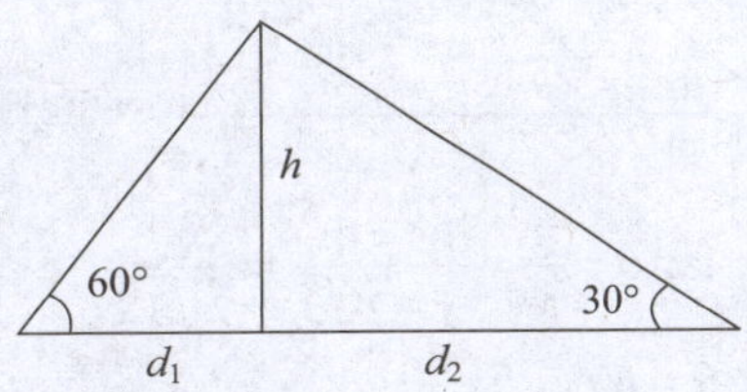

$$d_2 = h\cot 30^\circ = 500\sqrt{3}$$

$$d_1 = \frac{500}{\sqrt{3}}$$

$$\text{Diameter, } D = 500\sqrt{3} + \frac{500\sqrt{3}}{3}$$

$$\frac{2000\sqrt{3}}{3} = \frac{2000}{\sqrt{3}} \text{ m}$$

27. (D)

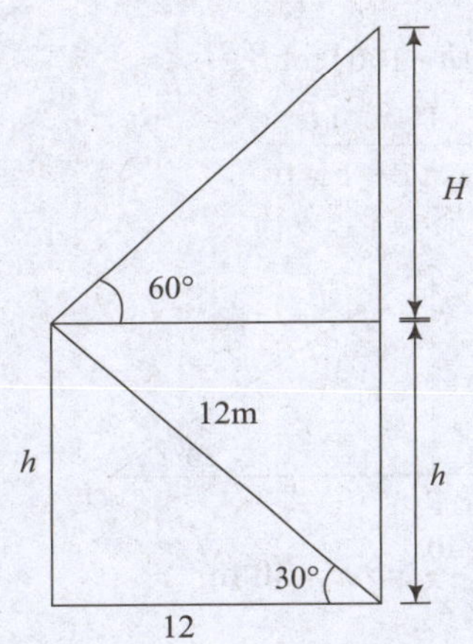

$$h = 12\tan 30^\circ = \frac{12}{\sqrt{3}}$$

$$H = 12\tan 60^\circ = 12\sqrt{3}$$

Height of tower $= H + h = 12\sqrt{3} + \frac{12}{\sqrt{3}} = 16\sqrt{3}$ m

28. (C) Let the distance of the ship from the base of the tower be x.

$$\tan 60^\circ = \frac{30}{x} \Rightarrow x = 10\sqrt{3}\text{ m}$$

29. (A) Let the angle made by sun with the ground be θ.

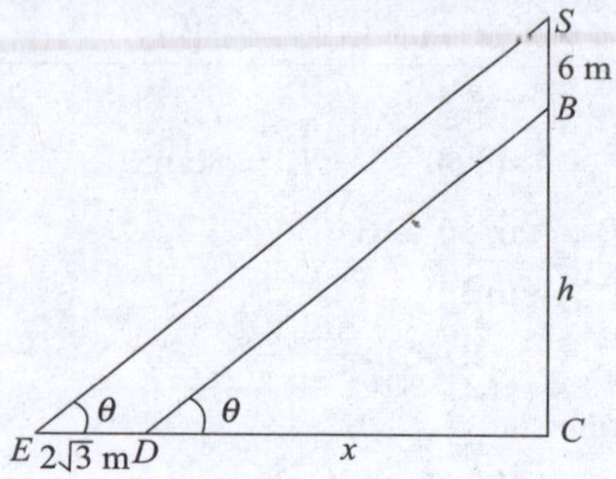

$$\tan\theta = \frac{h}{x} = \frac{h+6}{x+2\sqrt{3}} = \frac{6}{2\sqrt{3}}$$

$\therefore\ \theta = 60^\circ$

30. (C) Let the height of the cliff be h.

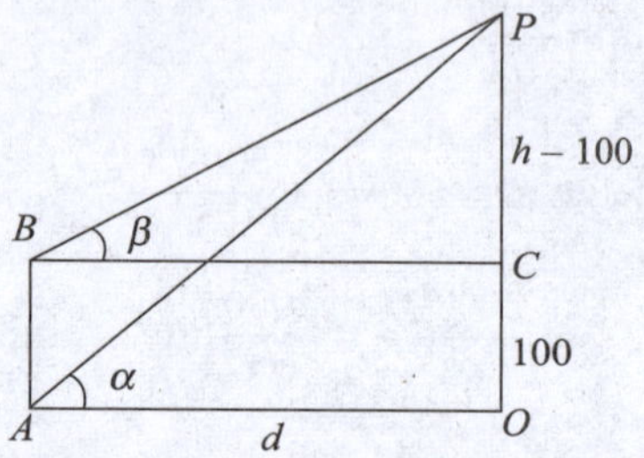

$$h\cot\alpha = (h-100)\cot\beta$$

$$h = \frac{100\cot\beta}{\cot\beta - \cot\alpha}$$

31. (D)

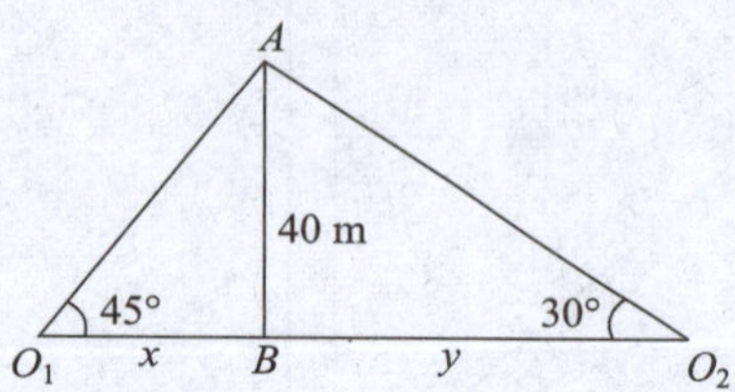

$$\tan 45^\circ = \frac{40}{x} \Rightarrow x = 40\text{ m}$$

$$\cot 30^\circ = \frac{y}{40} \Rightarrow y = 40\cot 30^\circ = 40\sqrt{3}\text{ m}$$

Distance between the men = $40 + 40\sqrt{3}$

= 109.28 m

32. (B)

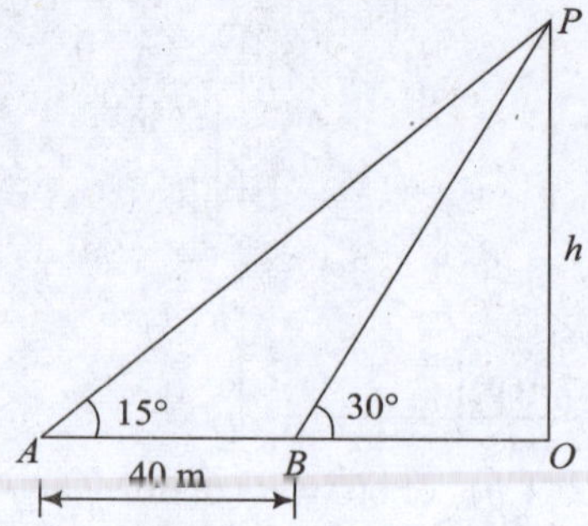

$OB = h\cot 30^\circ$

$OA = h\cot 15^\circ$

$$AB = OA - OB = h\left(\cot 15^\circ - \cot 30^\circ\right)$$

$$\therefore\ h = \frac{40}{\left(\cot 15^\circ - \cot 30^\circ\right)} = 20\text{ m}$$

33. (A)

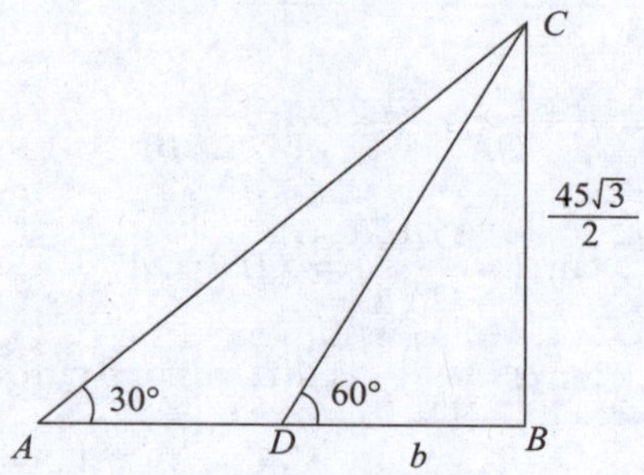

$$\tan 60^\circ = \frac{45\sqrt{3}}{2y} \quad \therefore y = \frac{45}{2}$$

$$\tan 30^\circ = \frac{45\sqrt{3}}{2x} \quad \therefore x = \frac{45\times 3}{2}$$

$$x - y = \frac{45\times 3}{2} - \frac{45}{2} = 45\text{ m}$$

34. (A)

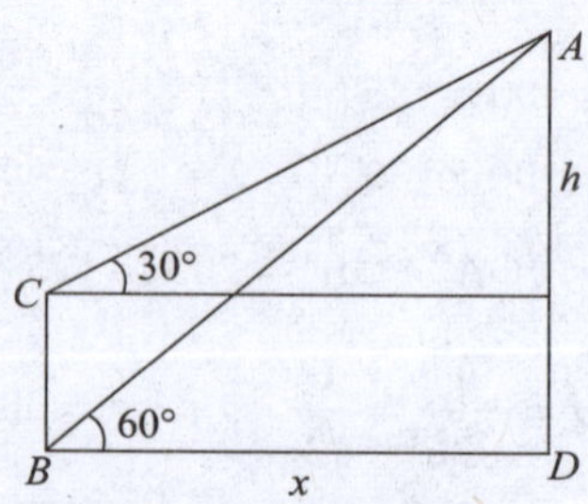

Total distance from temple

$$= \sqrt{x^2 + 240^2} = \sqrt{\frac{h^2}{3} + 240^2}$$

Now,

$$\tan 30^\circ = \frac{h}{\sqrt{\frac{h^2}{3}+240^2}}$$

$$\Rightarrow \frac{h}{\sqrt{\frac{h^2}{3}+240^2}} = \frac{1}{\sqrt{3}} \Rightarrow \frac{h^2}{\frac{h^2}{3}+240^2} = \frac{1}{3}$$

$$\therefore \quad h = 60\sqrt{6} \text{ m}$$

Hints and answers to Olympiad and NTSE level exercises

35. (C) $OA = h \cot 60^\circ$, $OB = h \cot 30^\circ$,

$OB = OA = 20 = h (\cot 30^\circ - \cot 60^\circ)$

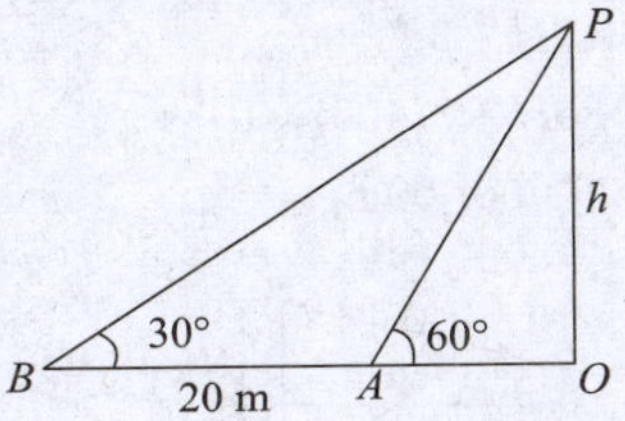

$$\Rightarrow h = \frac{20}{\left(\sqrt{3}-\frac{1}{\sqrt{3}}\right)} = \frac{20\sqrt{3}}{2} = 10\sqrt{3} \text{ m.}$$

36. (C) $d = h \cot 30^\circ - h \cot 60^\circ$ and time = 3 min.

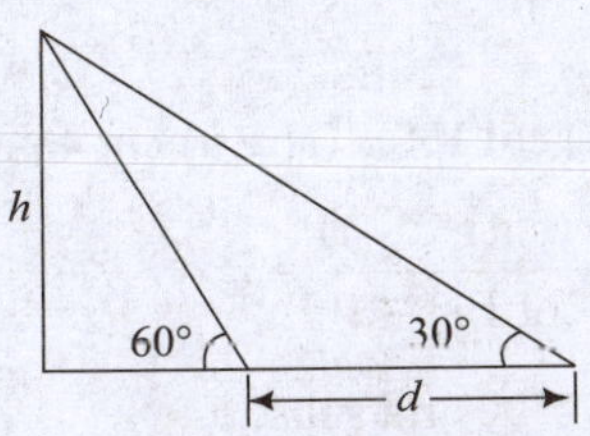

$$\therefore \quad \text{Speed} = \frac{h(\cot 30^\circ - \cot 60^\circ)}{3} \text{ per minute}$$

It will travel distance $h \cot 60^\circ$ in

$$\frac{h \cot 60^\circ \times 3}{h(\cot 30^\circ - \cot 60^\circ)} = 1\frac{1}{2} \text{ minutes.}$$

37. (A) $\dfrac{\sin 38^\circ}{l} = \dfrac{\sin(SPO)}{2.05}$

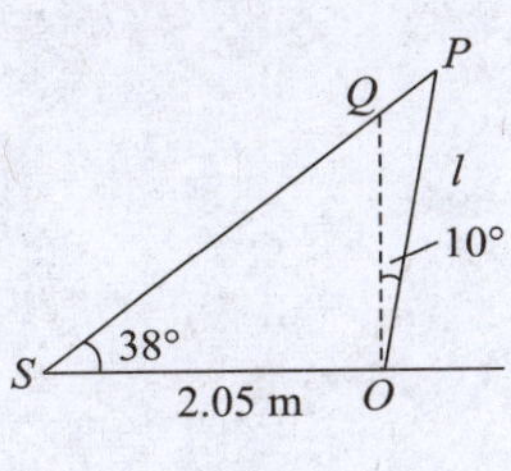

$$= \frac{\sin(180^\circ - 38^\circ - 90^\circ - 10^\circ)}{2.05}$$

$$\Rightarrow \quad l = \frac{2.05 \sin 38^\circ}{\sin 42^\circ}$$

38. (B) $d = H \cot 30^\circ - H \cot 60^\circ$;

Time taken = 10 second

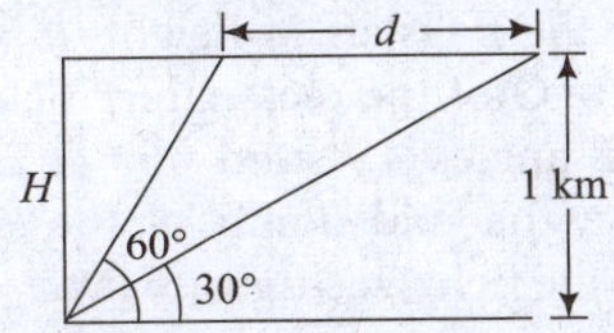

$$\text{Speed} = \frac{\cot 30^\circ - \cot 60^\circ}{10} \times 60 \times 60 = 240\sqrt{3}.$$

39. (B) $d_2 = h \cot 30^\circ = 500\sqrt{3}, d_1 = \dfrac{500}{\sqrt{3}}$

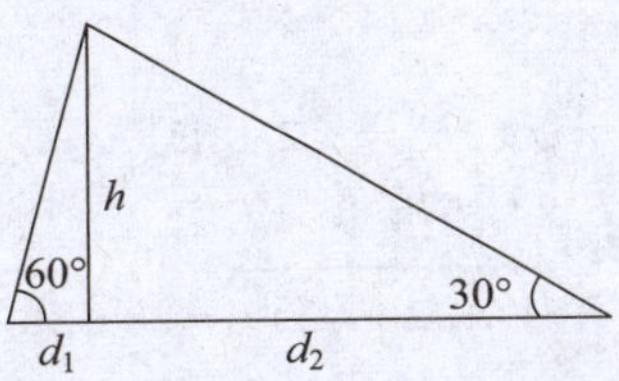

$$\text{Diameter } D = 500\sqrt{3} + \frac{500}{3}\sqrt{3} = \frac{2000}{\sqrt{3}} \text{ m.}$$

40. (A) Total distance from temple = $\sqrt{x^2 + (240)^2}$

where

$$x = \frac{h}{\tan 60^\circ} = \frac{h}{\sqrt{3}}$$

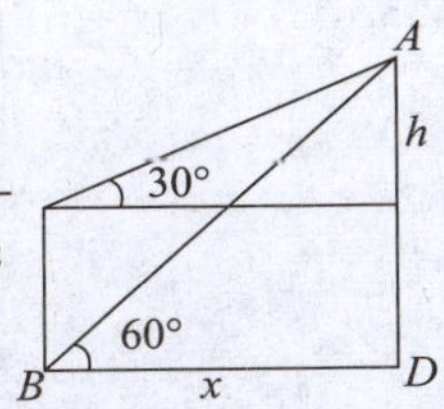

$$\text{So distance} = \sqrt{\frac{h^2}{3} + (240)^2}$$

$$\text{but } \frac{h}{\sqrt{\frac{h^2}{3}+(240)^2}} = \frac{1}{\sqrt{3}}$$

$$\Rightarrow \frac{h^2}{\frac{h^2}{3}+(240)^2} = \frac{1}{3}$$

After solving, $h = 60\sqrt{6}$ m.

41. (A) Let BC be the declivity and BA be the tower. Using sine rule in ΔABC, we have

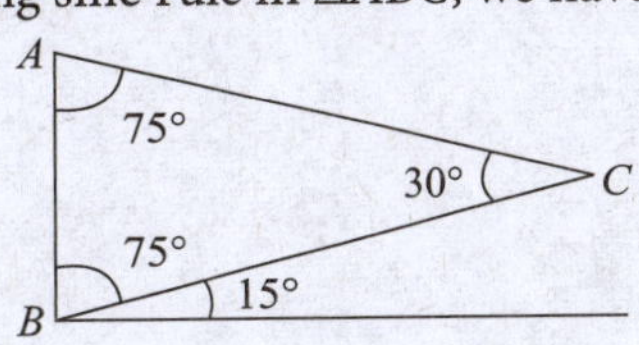

$$\frac{BC}{\sin 75°} = \frac{AB}{\sin 30°}$$

$$\Rightarrow \quad AB = \frac{80 \sin 30°}{\sin 75°}$$

$$\Rightarrow \quad AB = \frac{40 \times 2\sqrt{2}}{\sqrt{3}-1} = 40(\sqrt{6}-\sqrt{2})$$

42. (C) Let OP be the flagstaff of height 'h' standing at the centre O of the rectangular field $ABCD$ subtending angles 15° and 45° at E and F respectively, the mid points of the sides AD and DC of the field (see figure), then

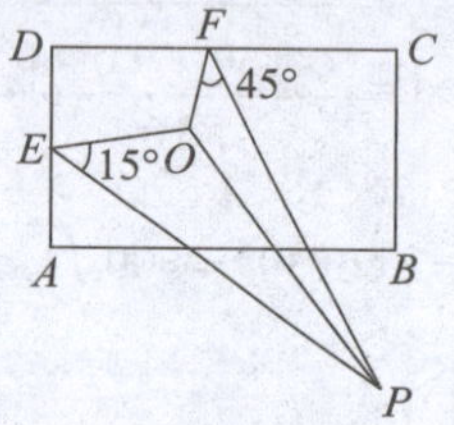

$$OE = h \cot 15° = h(2 + \sqrt{3})$$

and $OF = h \cot 45° = h$

$$\Rightarrow \quad EF = h\sqrt{1^2 + (2+\sqrt{3})^2} = 2h\sqrt{2+\sqrt{3}}$$

$$\Rightarrow \quad 1200 = AC = 2EF = 4h\sqrt{(2+\sqrt{3})}$$

$$\Rightarrow \quad h = \frac{300}{\sqrt{2+\sqrt{3}}} = 300\sqrt{2-\sqrt{3}} \text{ m}$$

43. (C)

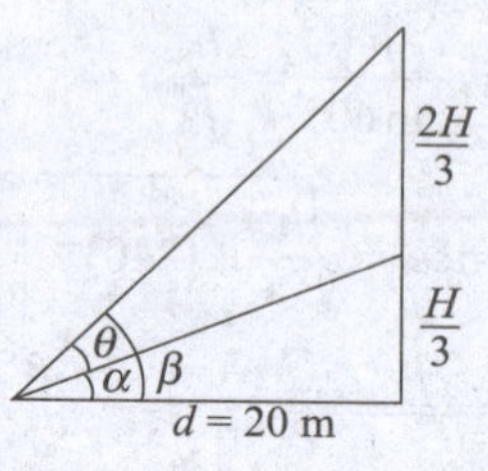

$$\frac{H}{3}\cot\alpha = d \text{ and } H \cot \beta = d$$

or $$\frac{H}{3d} = \tan\alpha \text{ and } \frac{H}{d} = \tan\beta$$

$$\tan(\beta - \alpha) = \frac{1}{2} = \frac{\frac{H}{d} - \frac{H}{3d}}{1 + \frac{H^2}{3d^2}}$$

$$\Rightarrow \quad 1 + \frac{H^2}{3d^2} = \frac{4H}{3d}$$

$$\Rightarrow \quad H^2 - 4dH + 3d^2 = 0$$

$$\Rightarrow \quad H^2 - 80H + 3(400) = 0$$

$$\Rightarrow \quad H = 20 \text{ m or } 60 \text{ m}$$

44. (C)

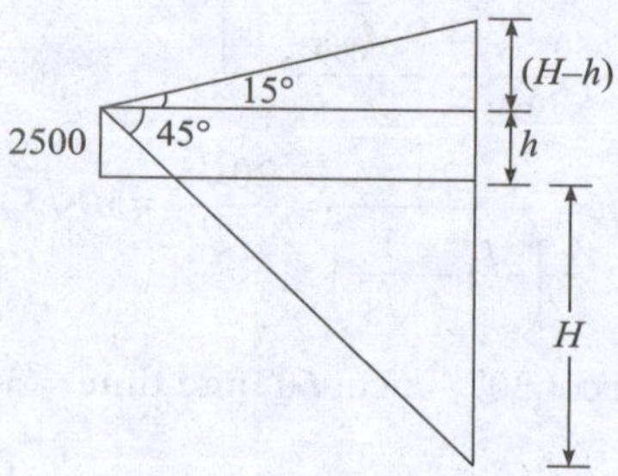

$$(H - h) \cot 15° = (H + h) \cot 45°$$

or $$H = \frac{h(\cot 15° + 1)}{(\cot 15° - 1)}$$

Since $h = 2500$ and substitute

$\cot 15° = 2 + \sqrt{3}$ we get, $H = 2500\sqrt{3}$ m